W9-BTI-453

DISCOVERIES IN THE HUMAN BRAIN

DISCOVERIES IN THE HUMAN BRAIN

Neuroscience Prehistory, Brain Structure, and Function

Louise H. Marshall, PhD

AND

Horace W. Magoun, PhD

Brain Research Institute, Los Angeles, CA

HUMANA PRESS TOTOWA, NEW JERSEY

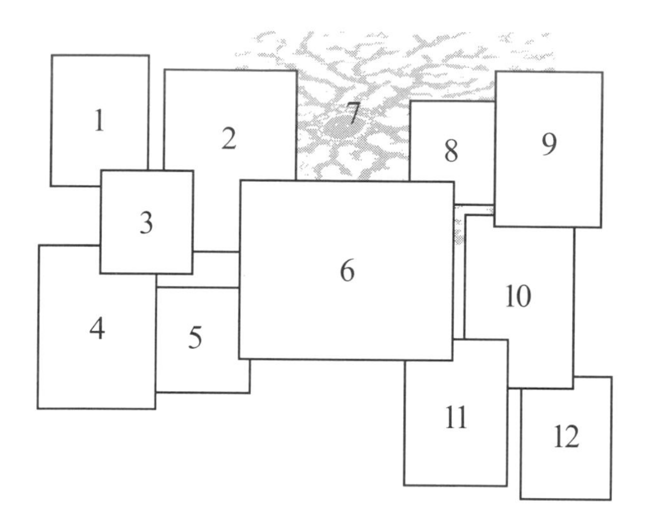

Jacket Picture Guide and Page Reference

1. Charles Darwin, p. 2
2. Andreas Vesalius, p. 32
3. Descarte's pineal gland, p. 37
4. Cushing and guest at the Harvey Cushing Society, p. 85
5. Taung Child fossil specimen, p. 16
6. Franz Gall's drawing of the human brain, p. 118
7. Pio del Rio-Hortega's protoplasmic neuroglia illustration, p. 153
8. Rita Levi Montalcini's photomicrograph of the "halo effect," p. 133
9. Baron Constantin von Economo, p. 109
10. Pio del Rio-Hortega's fibrous neuroglia illustration, p. 153
11. Elizabeth Crosby, p. 221
12. Pictograph of a left human hand on a cave wall in southwestern France, p. 22

999 Riverview Drive, Suite 208
Totowa, New Jersey 07512

This publication is printed on acid-free paper. ∞
ANSI Z39.48-1984 (American National Standards Institute)
Permanence of Paper for Printed Library Materials.

For additional copies, pricing for bulk purchases, and/or information about other Humana titles, contact Humana at the above address or at any of the following numbers: Tel.: 973-256-1699; Fax: 973-256-8341; E-mail: humana@mindspring.com or visit our Website: http://humanapress.com

Dust jacket, cloth cover, and type template design by Patricia F. Cleary.
Template finalization by Kathryn A. Bleckman.
Page layout and art production by Donna Niethe.

ISBN 0-89603-435-6

Printed in the United States of America. 10 9 8 7 6 5 4 3 2 1

PREFACE

You can climb back up a stream of radiance to the sky, and back through history up the stream of time.
—Robert Frost

From the last years of the second millennium, we can look back on antecedent events in neuroscience with amazement that so much of modern biomedical science was anticipated, or even said or done, in an earlier time. That surprise can be matched by appreciation for what the pioneer investigators, with no inkling that they were creating a discipline, contributed to its emergence as a productive force in human progress. In today's reductionist atmosphere, in which research at the molecular level is producing breathtaking new knowledge throughout biology, the student may lose sight of the grand sweep of the continuum and the powerful momentum that is expressed by the phylogenetic scale and its manifestation of form following function. It is conceivable that awareness of that evolutionary tension may mitigate some of the complications that now threaten the pursuit of science in general, such as extremes of bioethics, priority of discovery, patenting of fragments of research, and the realities of research funding.

A recurrent theme throughout this chronicle of significant discoveries about the brain is the relationship of form and function. From the ancient teleological beliefs to the pre-Darwinian climate of evolutionary thought–a single primeval form from which all animals descend–there have been "the formal and the functional standpoints" (Russell, 1916, p. 305). While the morphologists focused on "absolute form" and the phylogenetic construction of genealogical trees, Darwin's coming offered natural selection as a concept and favored the functional consequences of environmental pressure. Phylogeny, which "stands as a central theme for evolutionary biology" (Gould, 1977, p. 2), and its correlate–form follows function–wind through the history of neuroscience.

The co-author, Horace Winchell Magoun, died in 1991 before the present manuscript was completed. Under his plan, the volume embodies the topics that he judged to be important in brain history leading into the end of the century, and was undertaken in response to the enthusiasm generated by exhibition at several national and international meetings of a series of large posters for which Magoun wrote a 27-page brochure. The posters were viewed by a multitude of young neuroscientists who wanted more, as well as by mature investigators who were warmly pleased to see familiar names and faces from the past. The acclaim was accompanied by a veritable deluge of requests for an illustrated, expanded publication.

The success of a historical poster is initially a visual process, to focus the viewers' attention, to arrest their passing-by. Also important is the emotional effect–is the title provocative? And finally, does the poster intellectually satisfy the viewer's aroused curiosity by telling a concise story with relevant images and quotations that convey an idea's development? This book, so conceived, carries out the format/genre with what we believe are tempting visuals and crisp text. We intend to impart an urge to know more, and hence to reinforce the reader's enjoyment of the rich heritage of neuroscience. The first two brief chapters delineate sparingly those overarching topics that apply to all biology–evolution and phylogeny–or the relationship of environmental pressure and change, illustrated by reconstructions of brains from prehistoric crania, and early evidence of handedness, arts, and skills. The last chapter, one of the lengthiest, describes the discoveries related to three great integrating systems which, in their most developed form, encompass the richness of human existence. This implies a chronology in the arrangement of chapters, but the timekeeping is loose, and follows concepts rather than calendars. The intervening chapters deal with human brain structures in the order in which they were revealed. First, the ventricles perceived as the seat of brain-power, then a shift to the cerebral substance itself; next, the convolutions and lobes, with inquiry into their functions, and a chapter on the nonsymmetricality of the nervous system, either morphologically or functionally. This introduces fine structure and the neuron doctrine, as well as neurochemistry, and leads to the "little brain" or cerebellum. Next, dipping below the cerebral cortex, discoveries about the thalamus, hypothalamus, and pituitary gland are explored, each with its own beginning, yet interdigitating as new facts accrue. Each chapter concludes with an account of the knowledge prevailing during the third quarter of the twentieth century; to bring the story further would have invited an unmanageable mass of information in this broadest and most active biomedical discipline.

Discoveries in the Human Brain was planned to facilitate and encourage a wider knowledge of how neuroscience evolved from the three streams of neural, behavioral, and communicative sciences. We believe that becoming acquainted with some of the exquisite drawings and historical countenances, accompanied by readings from the expressive statements of those who were early engaged in brain and behavior inquiry, may generate an appreciation for the sense of wonder that accompanies, and indeed drives, scientific endeavor. *Discoveries in the Human Brain* should provide the tinder and show where to look for additional fuel.

Educational utility dictated precise citations of the original sources and their known translations. We have leaned heavily on that "bible" of brain history by Clarke and O'Malley (University of California Press, 1968), now reprinted and expanded by Clarke (Norman, 1996) and on two additional American classics, by Haymaker and Schiller (Thomas, 1970) and by McHenry (Thomas, 1969), publications that lead the way to more detailed inquiry. Completion of the manuscript has been the fruit of a long, but not dreary, process, impossible without the help of experts in many fields, to all of whom I am deeply grateful. Some translations and preliminary drafts of early chapters were contributed by Marion Anker. Charles H. Sawyer, Elizabeth R. Lomax, Paul D. MacLean, and Ynez V. O'Neill kindly read my first attempts in their areas of expertise and added greatly to the manuscript's strength. Additional helpful individuals were Laura S. Allen, the late William Oldendorf, Russell A. Johnson, Shirley Lavenberg, Virginia Hansen, Victoria Shabanzadeh, and Supinder Bedi, all at the University of California, Los Angeles; also Wendy Saywood, Oxford University; Christian Spenger, University of Bern; Michael R. Cuénod, University of Geneva; and Jane Roberts, Windsor Castle Royal Library. I am especially appreciative of helpful comments from those reviewers who took the time to read all the way through; nonetheless, all errors of any kind are to be laid solely at my door.

Pictorial materials were furnished by many individuals and repositories, and are acknowledged as appropriate. Nonattributed impressions are from the extensive holdings of the Division of History and Special Collections of the Louise M. Darling Biomedical Library, University of California, Los Angeles, the staff of which has our sincere gratitude, as well as the Office of Instructional Materials of the medical school. Publication has been made possible through the Frances Margaret O'Malley bequest to the Brain Research Institute, which not only provided the heavy financial outlay for illustrative materials, but also a generous subvention to assure superb quality at a non-astronomic price. The cooperation and help of the talented people at Humana Press are deeply appreciated. And not least, we thank Carmine D. Clemente and Arnold B. Scheibel, former directors of the Brain Research Institute and the Neuroscience History Archives, for their sustained encouragement.

Louise H. Marshall
October, 1997

CONTENTS

List of Color Plates

Color plates appear as an insert following page 84.

Plate 1 (Fig. 2.18 from Chapter 2). Right-handedness in the Amerindians is illustrated in the Mayan "Mural of Bonampak" (ca. AD 850) in southern Mexico, depicting blood sacrifices to a displeased deity. (Adapted from Davidson, 1962, pp. 408–409.)

Plate 2 (Fig. 3.2 from Chapter 3). A drawing from a Latin manuscript of the late twelfth to early thirteenth centuries. Above the two eyes and their converging nerves is the brain with its coverings and the attached "cells," only one of which is labeled. On the right, the diamond shape is inscribed "dwelling place of the brain or the place of reason." (M. R. James, 1907, vol. I, pp. 218, 219. From Goncille and Caisu MS 190/223 f.br.)

Plate 3 (Fig. 4.8 from Chapter 4). The frontispiece of Felix Vicq d'Azyr's *Traité d'anatomie et de physiologie* (1786) foretells the delicacy and detail of this beautiful publication (*see* Fig. 4.7), only the first volume of which appeared owing to the author's early death.

Plate 4 (Fig. 5.2 from Chapter 5). An early account of a speech disorder, the miracle of Zacharias as described by St. Luke, relates that when an angel appeared beside the altar to tell Zacharias that his wife, Elizabeth, would bear a child, Zacharias dropped his censer and exclaimed "That cannot be." Immediately, he was struck dumb for disbelieving Heaven's messenger, although his writing was unimpaired, as his companions indicate. (From the Gospel Book of Henry III, AD 1043–1045, Patrimonio Nacional, Madrid.)

1 Introduction

The Basic Postulates

CONTENTS

The nervous system has not developed during phylogeny with the brain of man as its fixed pattern or goal.

(Kappers, Huber, and Crosby, 1936, p. xiii)

The story of the discoveries of the anatomy and physiology of the human brain rests on a firm foundation of evolutionary processes. The transformations that have brought modern humankind to its current dominant intellectual status on Earth follow a pattern governed by certain fundamental rules of nature that thread their way through the fabric that constitutes the history of brain and behavior—terms that define modern neuroscience. An awareness of those rules or postulates facilitates our understanding of how the antecedent discoveries have shaped present-day knowledge of the brain and its awesome functional capacities. An overview of three broad postulates is introduced here as prelude to a more specific description of some of the events that entered into what we know about the brain at the end of the twentieth century.

NATURE OF THE CONTINUUM

The accumulated evidence from the fossil and comparative records rationalizes the acceptance of the first postulate: The evolutionary process is continuing beyond the present moment and humans and other animals occupy positions at points along a "bushy" continuum. In that progression, be it smooth or incremental as biological evolutionalists continue to argue (*see* Gould, 1995, Chap. 11), the highly complex human brain is regarded as the most significant evidence of enhanced adaptation to the physical and social pressures of the niche in which our species resides. Variations in climate and habitat have exercised a direct effect on this planet's living organisms, modifying the species that are responsive to change and eliminating those that are not. The adaptive processes have been accompanied by structural and physiological modifications culminating in the great diversity of life forms and behaviors known in the world today. The challenge of understanding evolution lies in reconciling its vast temporal scale with its omnipresence—there may be imperceptible change in what appears at the moment to be stable, a kind of "jerky" continuum of change interspersed with periods of relative stability (*see* Eldredge and Gould, 1972).

A rudimentary knowledge of the anatomy of both vertebrate and invertebrate forms can be traced in Western civilization from its Greek origins through an Arabic–Latin revival, fed by a deeply rooted interest in the morphology of animal bodies, especially the human (F. J. Cole, 1949, p. 126). By the seventeenth century and with the new optical magnification introduced by Leeuwenhoek and his fellow Dutch scientists, and no slackening in the burning curiosity to learn how animals (and plants) are structured, there was a great flowering of comparative anatomical studies. Centuries of description, however, yielded isolated facts and homologies whose overall significance was not discernible. As Cole stated (ibid., p. 471): "Descriptive anatomy had served its purpose and could do no more. Until an evolutionary *principle* was demonstrated, further random research could but swell the accumulation of data which awaited integration into a science."

CHARLES DARWIN

In the nineteenth century, a unifying purpose was brought to the accumulation of random observations by two theories that proposed to explain the *Scala naturae*, or arrangement of living creatures in an array of increasing complexity, familiar to students of nature since the preceding century. The French naturalist, Jean-Baptiste Lamarck (1744–1829), who first popularized biology in his country, challenged the widely held belief that species were fixed by divine creation and, instead, emphasized inheritance through several generations of characteristics acquired by the individual organism's experience. The more enduring theory of evolution through random natural selection was formulated by Charles Robert Darwin (1809–1882; Fig. 1.1) in his revolutionary publication, *On the Origin of Species by Means of Natural Selection or the Preservation of Favored Races in the Struggle for Life* (1859). He stressed the great variation of heritable characteristics as having evolved through the diversity of environmental demands for species survival:

> Although much remains obscure, and will long remain obscure, I can entertain no doubt, after the most deliberate study and dispassionate judgment of which I am capable, that the view which most naturalists entertain and which I formerly entertained—namely, that each species has been independently created—is erroneous. I am fully convinced that species are not immutable; but that those belonging to what are called the same genera are lineal descendants of some other and generally extinct species, in the same manner as the acknowledged varieties of any one species are the descendants of that species. Furthermore, I am convinced that Natural Selection has been the main but not the exclusive means of modification (ibid., p. 6).

Fig. 1.1. Although several of the ideas concerning evolution were also proposed independently by Alfred Russell Wallace, it was Charles Darwin (lithograph by Y. H. Maguire, 1849) who presented both theory and evidence for the evolution of species (1859) and later applied it to human beings (1872).

In his later writings, Darwin considered more specifically the phylogenetic aspects of brain, mind, and behavior without developing a detailed analysis. Inquiries into the history of brain organization were a post-Darwinian occurrence as writers after him carried the concepts further in evolutionary terms (Magoun, 1960).

The second postulate recognizes a hierarchy of levels of structure and function in the nervous system. This concept was essential background for the development of evolutionary thinking and was influenced by Charles Lyell's newly introduced theory (1830–1833) of geologic evolution as seen in levels of stratification of earth forms. The idea was expressed by the influential English philosopher, Herbert Spencer (1820–1903) in his views of the structured development of the nervous system. He envisioned (1855) a succession of neural strata, with each higher increment serving ever more complex functions and dominating those below. To explain adaptive changes from an amorphous, unorganized homogeneity in the lowest animals to

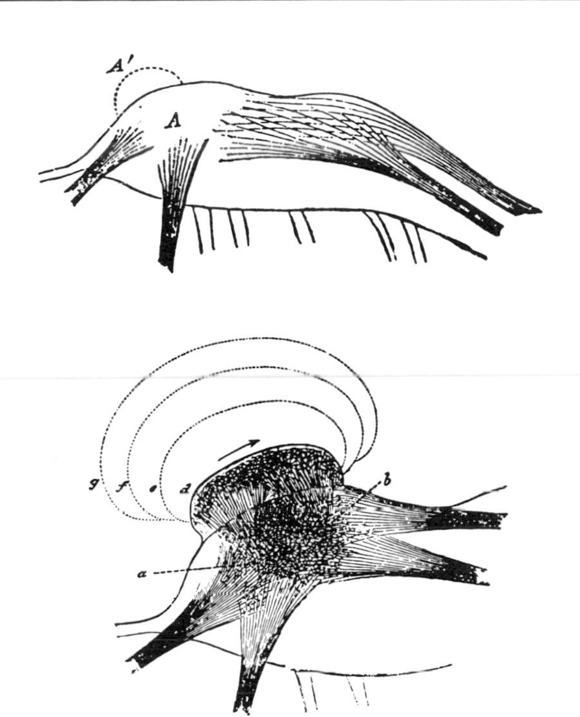

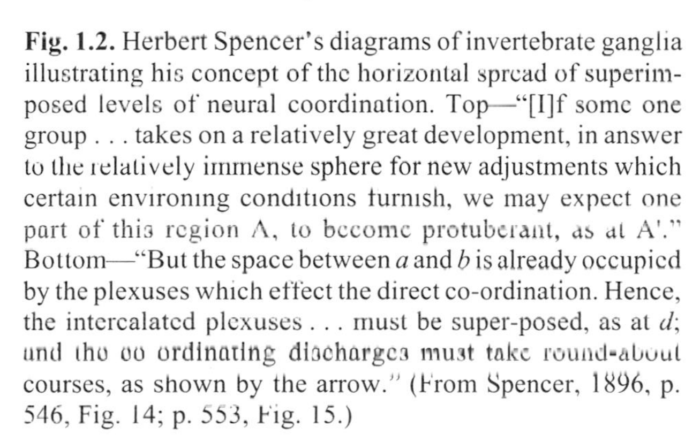

Fig. 1.2. Herbert Spencer's diagrams of invertebrate ganglia illustrating his concept of the horizontal spread of superimposed levels of neural coordination. Top—"[I]f some one group . . . takes on a relatively great development, in answer to the relatively immense sphere for new adjustments which certain environing conditions furnish, we may expect one part of this region A, to become protuberant, as at A'." Bottom—"But the space between *a* and *b* is already occupied by the plexuses which effect the direct co-ordination. Hence, the intercalated plexuses . . . must be super-posed, as at *d*; and the co-ordinating discharges must take round-about courses, as shown by the arrow." (From Spencer, 1896, p. 546, Fig. 14; p. 553, Fig. 15.)

Fig. 1.3. Ivan Sechenov returned from his medical studies abroad with a galvanometer and induction coil and initiated electrophysiology in Russia. His fame rested on scientific achievements as well as on intellectual qualities which kept him in trouble with the Czarist government.

a definite, coherent heterogeneity in the "intelligent" forms, Spencer sketched a neural plexus (Fig. 1.2) that adjusted to increased environmental demands by the successive addition of higher coordinating layers which, of necessity, were protuberant and superimposed because of the pre-emption of the original space.

Similar ideas, promulgated by Spencer previous to Darwin's publication, were also expressed by Thomas Laycock (1812–1876), an English physiologist writing from Edinburgh:

> [A]s we ascend [the scale] still higher in animal life, the instincts gradually lose their unknowing character, and the mental faculties emerge, with their appropriate organic basis in the encephalon. . . . Finally . . . we find Man in his highest development; evincing in art and science the results of the operation of mental powers, which in the lower animals are purely instinctive, in the lowest organisms simply vital processes (Laycock, 1860, vol. 2, p. 61).

Laycock was aware of the possibility of conflict between the successive levels: "[E]ven with the highest and strongest of human motives, . . . it is often difficult to curb them [the animal instincts] effectually. Those classed under the head of organic [primordial] instincts or corporeal appetites . . . are the furthest removed from the will and the consciousness" (ibid., p. 198).

The writings of Spencer and Laycock, authoritative additions to the ferment of evolutionary thought in the midnineteenth century, were highly influential in determining the directions of scientific inquiry in Britain and on the continent. It was an opportune time for the young Russian, Ivan Mikhailovich Sechenov (1829–1905; Fig. 1.3), to use a small legacy and study with Carl Ludwig (Vienna), Helmholtz and DuBois-Reymond (Berlin), and others. He returned to St. Petersburg armed with new instruments and ideas and completed his experiments, carried out with frogs for the most part, demonstrating that chemical stimulation of the thalamus depresses spinal reflexes and that the inhibition is reversible (1863). Additional important contributions were the observation of spontaneous fluctuations of current, which he attributed to the intrinsic activity of spinal centers and the summation of subthreshold stimuli (Brazier, 1961, p. 75).

Sechenov claimed that all actions, conscious and unconscious alike, originate as reflexes, and the mental or "psychic" reflexes are based on physiologic phenomena and provide evidence of a hierarchic organization of brain function. The final sentence of his most famous work, *Reflexes of the Brain*, states: "Now let anyone try to contend that psychical activity and its expression—muscular movement—are possible, even for a single moment, without external sensory stimulation!" (Sechenov, transl. 1960, p. 139). Such materialism offended the Czarist view that psychic life is spiritual, and Sechenov's paper (and subsequent book) were censored. He persisted in his chosen direction, however, and this self-confident and careful neuroscientist earned unusual praise: "All major landmarks in the history of world and Russian physiology are closely linked to [his] name. . . . Pavlov . . . called him the 'father of Russian physiology'" (Koshtoyants, 1960, p. 7).

A century after Lyell and Darwin, an active neuroscientist speculated about the pervasiveness of the stratification concept:

> One cannot but be curious concerning the derivation of the views of Hughlings Jackson in neurology, of Pavlov in physiology, of Freud in psychiatry, and of Edinger in anatomy, each of which accounted for the phylogenetic elaboration of the central nervous system in terms of a series of superimposed levels, added successively as the evolutionary scale was ascended (Magoun, 1960, p. 188).

Evidence of the primordial evolutionary process has been examined in many modern survivors, including the most primitive, the lower invertebrates living in an aquatic environment and lacking a backbone. In those forms, the nervous system is distributed diffusely throughout the body and motor responses to sensory stimuli are effected through simple, short reflex pathways. In some worms, regarded as the lowest and earliest animals to have what might be called a brain, neurons seem to cluster around the statocyst at one end of the animal, producing a bulge in the flat body surface, as in Fig. 1.4. The transition to a vertebrate brain is believed to have commenced about 100 million years before the Mammalia appeared. It was about 300 million additional years before an organized brain evolved, which in its highest development is deemed "the most complex of all machines yet known" (Hodgson, 1977, p. 23).

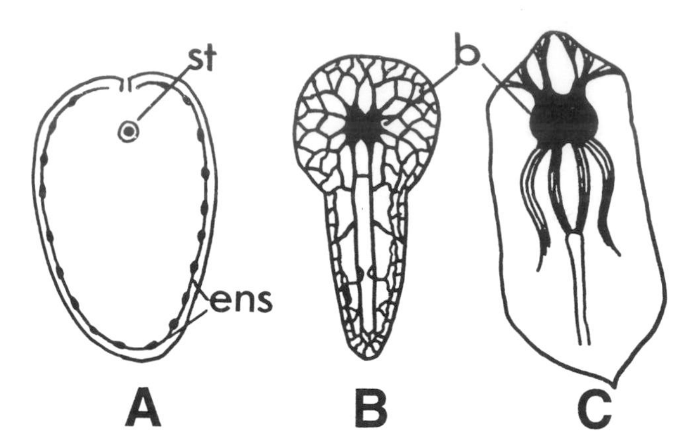

Fig. 1.4. States in evolution of the brain among flatworms. **A**—Epidermal nervous system (ens) and statocyst (st). **B**—Bilobe brain (b) surrounding statocyst. **C**—Increased cephalization and loss of nerve net. (From Hodgson, 1977, p. 24, Fig. 1.)

GALEN OF PERGAMUM

The evolutionary change to the brain encased in a bony cranium was predicated on a third postulate—function determines structure—a universally accepted concept that has been applied in many spheres and ages of human activity. For example, Chicago's great urban architect, Louis Sullivan, phrased his version of the postulate as "form follows function." Advocacy of this principle relative to human anatomy appeared among the ancient thinkers. Aristotle (384–322 BC), who "may fairly be said to be the founder of comparative anatomy . . . because he described the structure of many animals and classified them in a scientific way" (Russell, 1916, p. 2), recognized mankind as the most complex animal and initiated his studies on human structure. "In Aristotle's view the gradation of organic forms is the consequence, not the cause, of the gradation observable in their activities" (ibid., p. 15). In turn, Galen acknowledged the primacy of function over form to an exquisite degree in this passage about the human hand, written in AD 177:

> [L]et us investigate this very important part of man's body, examining it to determine not simply whether it is useful . . . or suitable for an intelligent animal, but whether it is in every respect so constituted that it would not have been better had it been made differently. One . . . characteristic of a prehensile instrument constructed in the best manner is the ability to grasp readily. . . . For this purpose, then, which was better—for the hand to be cleft into many divisions or to remain wholly undivided? (translated by May, 1968, p. 72).

There is no doubt about the historical impact of Galen of Pergamum on the direction of biomedical science: for more than 14 centuries, his writings were accepted without questioning by physician–anatomists. A prolific and enthusiastic writer, this Greek physician resided mostly in Rome at the beginning of the first millennium. In addition to his medical practice, which included gladiators with serious head wounds, he dissected many nonhuman animals, compared his observations with what his predecessors had written, and lectured and wrote down a vast accumulated knowledge for the edification of what turned out to be an overly loyal posterity.

Galen's viewpoint was strictly teleological, and "[t]he line between anatomy and physiology was more difficult to draw in Galen's day than in ours; yet even then one might focus one's attention upon the function rather than upon the organs. . . ." (Sarton, 1954, p. 45). More specifically, "the early part of his *De usu partium* might be described as a long hymn to the Divine Wisdom in fitting the hand for its functions" (Singer, in Galen, 1956, p. xix). That tribute was to one section of what is arguably the most important of Galen's writings, deemed "supremely valuable not only because it is a rich source [of beliefs before and including Galen] but because of its unconscious portrayal of a personality worth knowing in any age and the tremendous influence it has exerted down through the centuries. [As] an anatomist of the first rank . . .[i]n neurology Galen made some of his finest contributions" (May, in Galen, 1968, pp. 12, 43).

Fig. 1.5. Ludwig Edinger proposed that the mammalian brain consists of one part for "elementary" functions and another part that evolved phylogenetically to attain its highest development in *Homo sapiens*. Painting by Clovis Corinth, 1909; impression from *Gemälde des historischen Museums*, Frankfurt-am-Main: Waldemar Kramer, 1957.

"When [Galen] died, experimental science too fell dead" (Singer, in Galen, 1956, p. xxiii). Why further progress in anatomy and physiology was delayed for so long after Galen is perhaps best understood by considering him in the context of his times: the Greco-Roman empire had slid into its long decline, closely followed by the Dark Ages, and not until the Renaissance was the flame of experimental science rekindled.

Returning to the nineteenth century and the comparative anatomists who "discovered" phylogeny, which became the foundation stone of evolution, they accepted the principle of "function determines form" as a basic postulate of their science, as it remains today. In the modern era, the very active Edinger Institute at Frankfort-am-Main launched the European school of comparative anatomy. Its founder, Ludwig Edinger (1855–1918; Fig. 1.5), seems to have been acquainted with most of the prominent figures in late nineteenth century German neurology, having learned the myelin stain from Weigert himself and worked with Flechsig, among other neurologists. In his early research, Edinger traced neural fibers from the spinal cord to the thalamus, rather than to the cerebellum, thus challenging Meynert. Then he turned his attention to lower vertebrates, studying both their behaviors and their brain tissues and comparing them with more complex species, and realized that in its simpler functions the stem of the brain is not constructed differently in the lower and higher vertebrates, including man. The difference in the behaviors of higher vertebrates, compared with lower forms, had therefore to be found in the increasing elaboration of their cerebral cortex (Fig. 1.6). Edinger concluded that the vertebrate brain consists of two parts, the "old" (paleo) brain and the neopallium, more recently developed in response to more elaborate reactions to the environment.

An example of the relationship of function and structure occurs among modern reptiles in which poor sight and hearing are counterbalanced by highly developed senses of smell and taste served by a network of nerve fibers in the olfactory bulb. As illustrated in Fig. 1.7, this ancient part of the brain is disproportionately large in primitive animals, but as the cerebral hemispheres become more complex due to differentiation, specialization, and integration, the early olfactory system is diminished and supplanted by new structures.

The initial interest of comparative neurologists in the primitive nervous system of invertebrates soon was extended to the peripheral nervous system—particularly the cranial nerves—in lower vertebrates (fish and amphibians). Later and in the Mammalia, analyses were made of the central connections with brain stem and spinal cord. Two pioneering Americans took this first step in comprehending the organization and function of the elaborate brain of man, the short-lived Clarence Luther Herrick (1858–1904), and his younger brother, Charles Judson Herrick (1868–1960; *see* p. 186 of this volume and Fig. 9.11). Their key role in the development of an American school of comparative neurology, interdisciplinary in its range of interests but primarily neuroanatomical in its initial focus, was a part of the foundation on which the new knowledge of function and form was built,

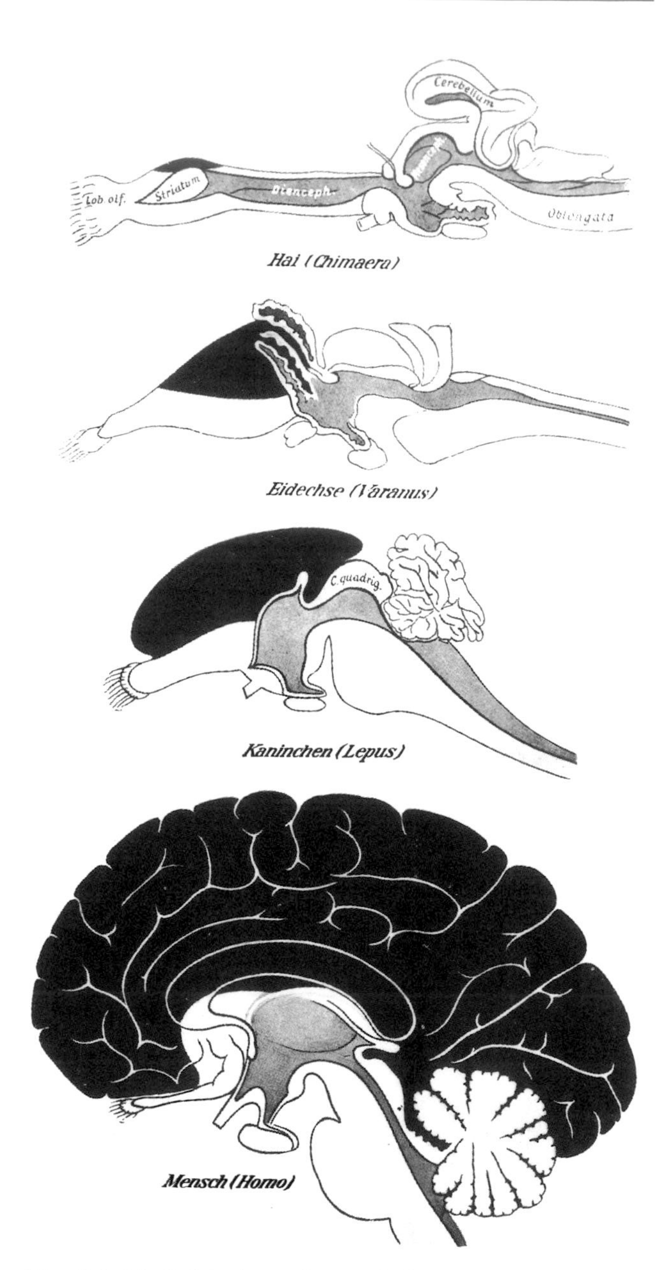

Fig. 1.6. Medial brain surfaces of (from top; not in scale) shark, lizard, rabbit, and human to illustrate the relative importance of the neopallium (solid) in the phylogenetic scale. (From Edinger, 1909, facing p. 190, Taf. I.)

culminating in the emergence of neuroscience. The younger Herrick was convinced that specialized use is one of the forces that determines structure and for his career-long study of a few species living today, he selected those that represented a specialized adaptation. Thus, to trace the nerves serving olfaction, he chose an animal highly skilled in finding its prey by smell, the shark. He also believed strongly that the evolution of adaptive behavior

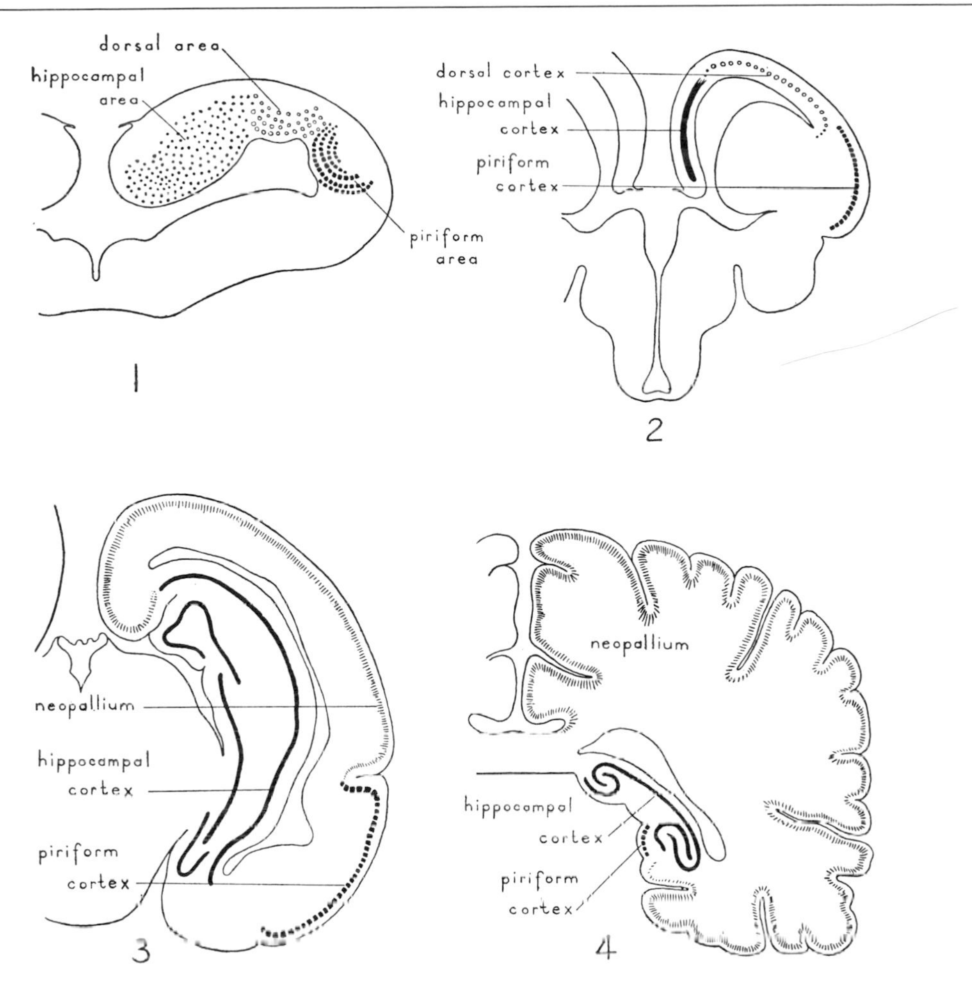

Fig. 1.7. Diagrams of the relative extent of the olfactory and nonolfactory pallial fields as seen in cross sections of brains of vertebrates of various phylogenetic stages. 1—A primitive amphibian with entire pallial field olfactory (×12). 2—A box-tortoise with differentiated cortex, all of which is under olfactory influence (×7). 3—The opossum with olfactory cortex (hippocampal and piriform) more extensive than nonolfactory cortex (neopallium) (×4). 4—Human, with enlargement of neopallium and reduction of olfactory cortex (×0.5). (From C. J. Herrick, 1933, p. 8.)

occurs parallel with that of structure: the more complex the behavior, the more elaborate the nervous system. Herrick, in his pursuit of "neurology as a biological subject" (Kingsland, 1993, p. 446), "envisioned a multidisciplinary attack on all aspects of human physiology, behavior, and psychology, with comparative anatomy contributing insight about the evolutionary development of the nervous system" (ibid., p. 449). Herrick's view was a refinement of more general gropings that had been recorded by earlier thinkers such as Spencer, Wallace, and Darwin, and defines him (and his elder brother) as neuroscientists.

OVERVIEW OF THE BASE

The weight of evidence identifies the neopallium as the great central "organ" of the higher species. It receives almost the entire sensory inflow; from the external and internal environment it stores impressions; and from it emanate those forces that produce modulated action, and, in Penfield's words (1975), "the mystery of the mind." The contributions of three great figures in comparative anatomy, Edinger in Germany and the Herrick brothers in America, are highlighted to illustrate the beginnings, in the late nineteenth century, of a serious effort to relate brain phylogeny with evolutionary behavior. The neopallium, however, constitutes only part of the nervous system contained within the cranium and any history of the myriad major and minor discoveries revealing how the brain functions must necessarily be selective. These accounts proceed with an inescapable overlay of evolutionary influences expressing the three postulates: evolution is a multiple-branched, stop-and-go continuum, it operates at a slow—phylogenetic—tempo, and function determines structure. Those threads may be followed through the fabric woven from the antecedent discoveries that have revealed some of what today is known about the brain. And although there are many causal factors to consider in evolutionary theory, the press of ecology, which is itself many-faceted, is generally viewed as predominant. The relative stasis or "punctuated equilibrium" at any one moment may not be clear, but the long-term effects of natural selection are the foundation of modern evolutionary analysis, and our three postulates are reassuring in that they offer humankind a niche in the grand scheme.

2 Evolution of the Mammalian Brain

Contents

The head, with its structures so obviously intended to serve as an avenue of expression for the mind.

(Herrick, 1891a, p. 101)

It was apparent from the earliest sightings of apes that a remarkable degree of resemblance exists between human and anthropoid forms (Fig. 2.1, right). Natives sharing the apes' natural habitat in Java and Africa had named them "orang-utang" and "chimpanzee," respectively, both meaning "man of the woods." In an early seventeenth century account, an English privateer, Andrew Battell, after being held captive in Africa by the Portuguese, reported that apes walked erect, built shelters, and buried their dead (Ravenstein, 1901). In a more focused, scientific contribution, an English physician and anatomist, Edward Tyson (1650–1708), published in 1699 his complete dissection of a young chimpanzee (Fig. 2.1, left), which he erroneously thought to be an orang-utang:

> I have made a *Comparative* Survey of this *Animal*, with a *Monkey*, an *Ape*, and a *Man*. By viewing the same Parts of all these together, we may the better observe *Nature's Gradation* in the formation of *Animal Bodies*; and the Transitions made from one to another; than which nothing can more conduce to the Attainment of the true Knowledge, both of the *Fabrick*, and the *Uses* of the Parts. By following *Nature's* Clew in this wonderful *Labyrinth* of the *Creation*, we may be more easily admitted into her *Secret Recesses*, which Thread if we miss, we must needs err and be bewilder'd (Tyson, 1699, Preface).

In his comparison, made "with the greatest exactness, observing each Part in both; it was very surprising to me to find so great a resemblance of the one to the other, that nothing could be more" (ibid., p. 54; also quoted in F. J. Cole, 1949, p. 221), Tyson noted the disproportionate "magnitude" of the brain to body bulk in his "Pygmie." Thus, looking beyond their gross physical appearances, Tyson saw the most significant of the differences among apes and humans—the brain.

The keen interest of the early naturalists to learn about the "ape-man" often took the form of watching individual primates in zoos and animal colonies and, more recently, observing them in the wild. One of the first of those intrepid investigators, Richard Lynch Garner (1848–1920) from Philadelphia, studied vocalization in nonhuman primates from the safety of a demountable cage (Fig. 2.2). In his book, *Gorillas and Chimpanzees*, describing his experiences and written in the attempt to dissuade the general public from the erroneous idea of human beings' direct descent from apes, Garner compared his subjects' brains with that of man:

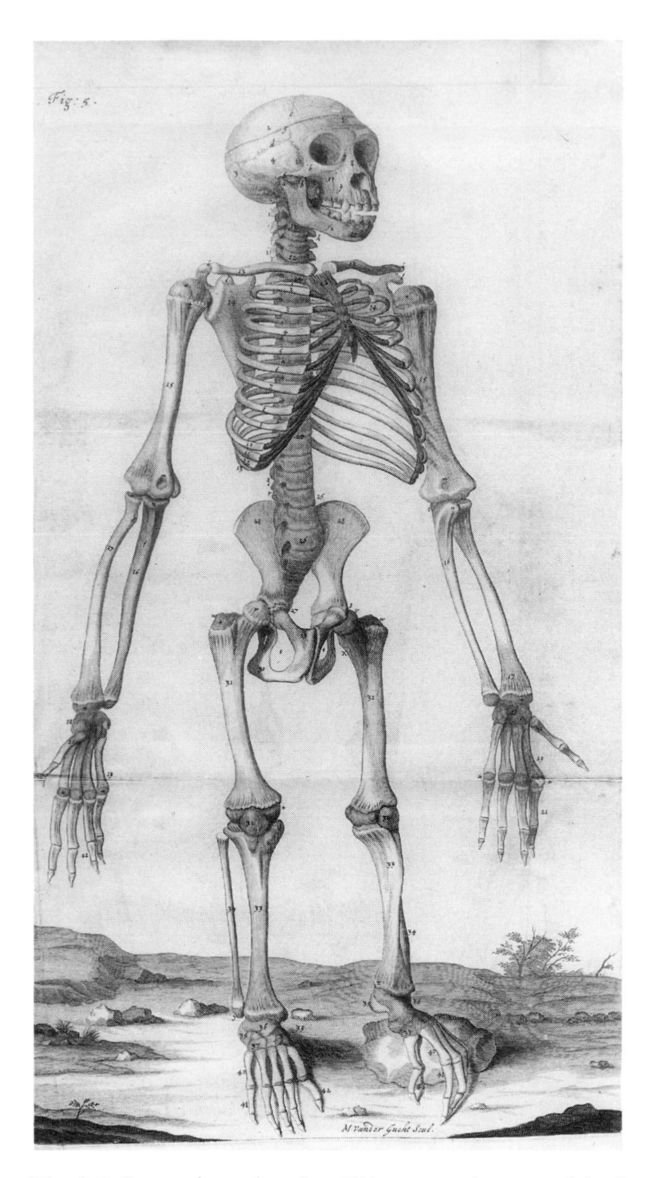

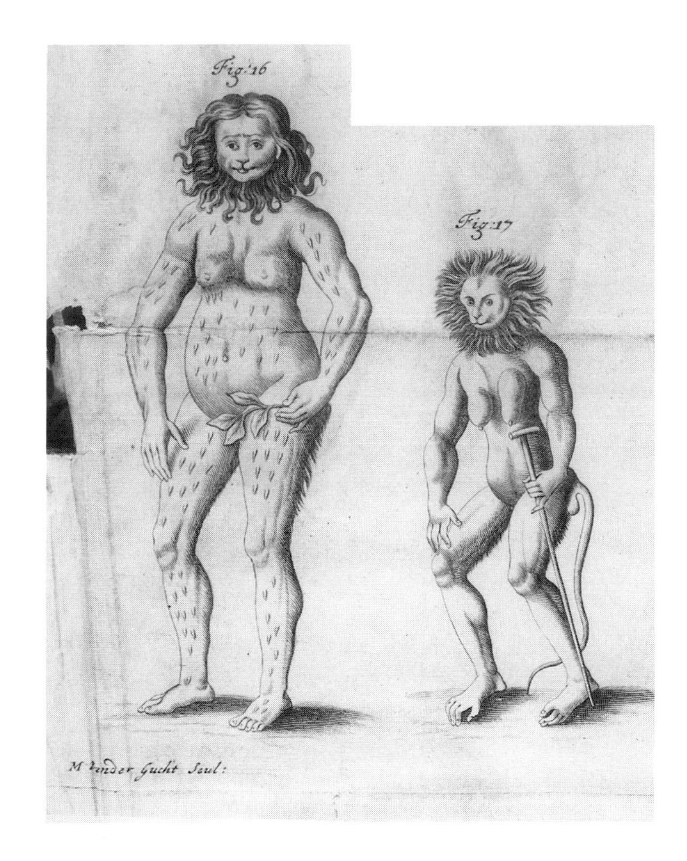

Fig. 2.1. Engravings showing 17th-century images of the "man of the woods." Left—"Represents the *Sceleton*, or the *Bones*." Center—"Represents the *Figure* that *Jacob Bontuis* gives of the *Orang-Outang* in *Piso*." Right—"Is taken out of Gesner, which he tells us, he met with in a *German* Book, wrote about the *Holy Land*." (From Tyson, 1699, Figs. 5 × 2/3; 16 × 1; 17 × 1.)

It has the same lobes, convolutions and centres. The texture is slightly coarser. The small details are less intricate and their lines somewhat less distinct. But these also differ to a certain extent in different men. In man and apes the same nerves are present and connect the same organs of sensation, volition and motion. In all essential points they are one. . . .To the casual observer the general resemblance is apparent, but to the student the unity becomes evident (Garner, 1896, pp. 10,11).

PHYLOGENY LEADS THE WAY

A gradation of animal forms as a consequence of a gradation of activities is an old idea, dating from Aristotle (E. S. Russell, 1916, p. 15; *see* p. 4, of this volume). In contrast with plants, "Animals, however, that not only live but feel, present a greater multiformity of parts, and this diversity is greater in some animals than in others, being most varied in those to whose share had fallen not mere life, but life of a high degree. Now such an animal is man" (ibid., p. 16).

Fig. 2.2. Photograph of R. L. Garner, an early ethologist, and his companion in front of the demountable cage from which he studied simian communication in the Congo. (From Garner, 1896, facing p. 22.)

The term "phylogeny" was coined by Ernest Haeckel in 1866 and was defined succinctly by Darwin in the fifth edition of *Origin of Species*: "Professor Häckel . . . has recently brought his great knowledge & abilities to bear on what he calls Phylogeny, or the line of descent of all organic beings" (Darwin, 1872, p. xiv). The history of lineages from a stem species was a means of arranging the descriptive data into a scale that Haeckel thought illuminated the evolutionary process. As a modern writer said: "Scale is everything in history and geology" (Gould, 1995, p. 142). The phylogenetic approach to evolution may be vertical or horizontal depending on the specimens being compared. A vertical study of phylogeny is illustrated in Fig. 2.3, a series of drawings of a fossil cranium representative of an early species (the discovery is discussed on p. 16) in the human lineage. The series demonstrates the attempt of Sir Arthur Keith (1866–1955), eminent British anatomist and anthropologist, to show that this specimen of *Australopithecus africanus* revealed more ape-like than hominid characters, "and yet there are suggestions, as Professor Dart maintains, that in its convolutionary organization of the brain of *Australopithecus* rose above that of either gorilla or chimpanzee" (Keith, 1931, p. 81). Keith's comparison reflects at least four million years of alternating stability and

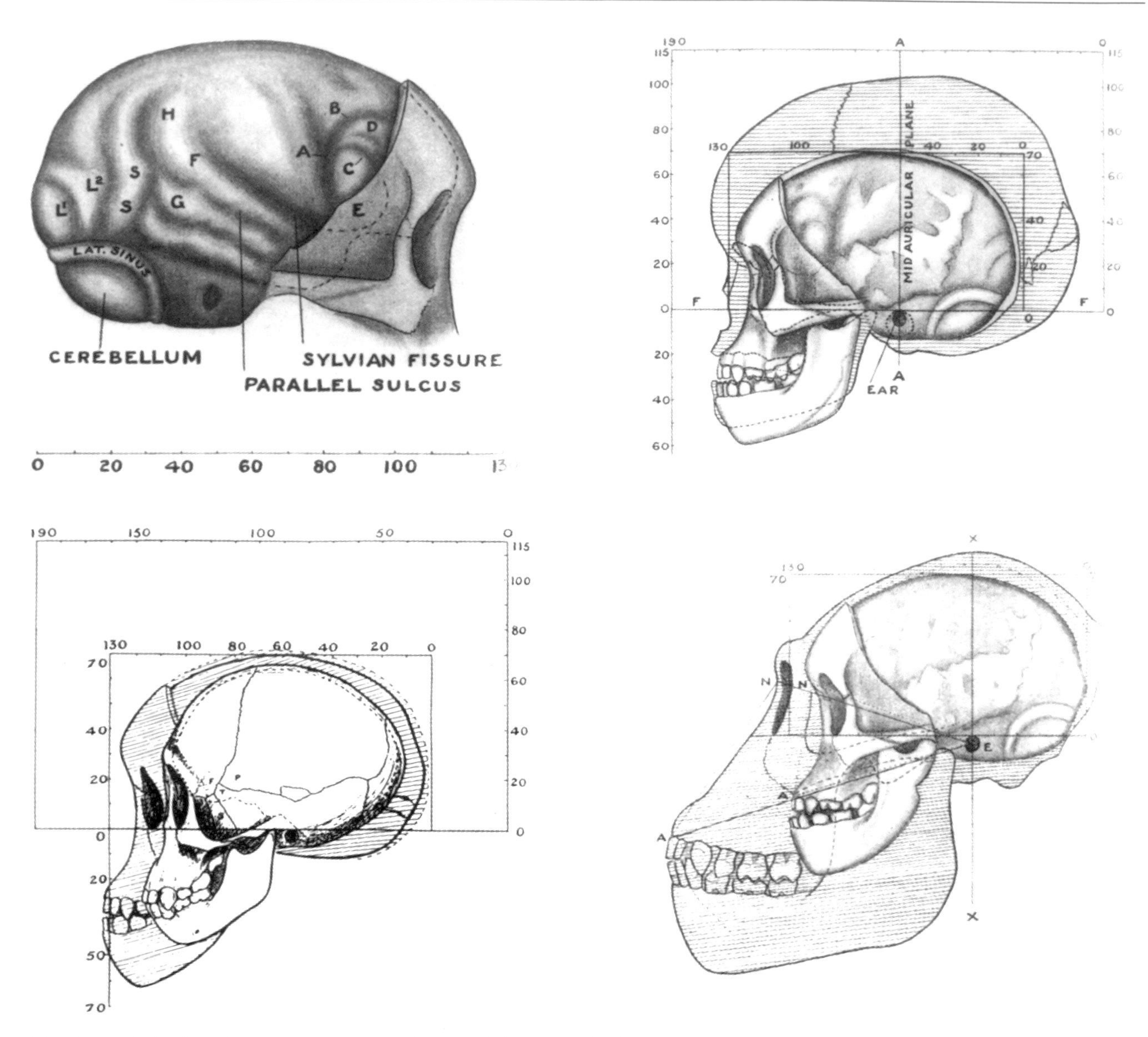

Fig. 2.3. Early drawings comparing the "Taungs" juvenile fossil specimen of *Australopithecus africanus*, identified by Raymond Dart in 1924, with human and anthropoid ape skulls. Above left—Natural endocranial cast of brain fitted into facial skull showing only distinct convolutions and fissures. Above right—Profile of the Taungs specimen superimposed on that of an Australian aborigine child at the same stage of tooth eruption. Below left—Chimpanzee skull superimposed on the Taungs specimen. Below right—Outline of Taungs child's skull and brain cast over that of a young gorilla. (From Keith, 1931, p. 80, Fig. 18; pg. 88, Fig. 21; p. 92, Fig. 22; p. 96, Fig. 24.) *See also* von Bonin's extensive investigation (1934) of skull capacities.

change in brain size, espoused as "punctated" evolution (Eldredge and Gould, 1977). Phylogeny also allows horizontal studies of homologies among forms living at a given time. Comparison of the brain size of modern apes and human beings reveals a fourfold difference—the ratio of brain to body weight is 1:50 in man and about 1:200 in apes.

A description of phylogeny as a process was expressed by an English anthropologist, Roger Lewin:

[T]he progression through more and more advanced animal groups—from amphibians through reptiles to mammals—is marked at each step by a substantial leap in the degree of encephalization displayed by each group as a whole. These stepwise mental increments between the major animal classes reflect gestalt jumps in the complexity of neural processing involved in the animal's daily lives [sic]. Each

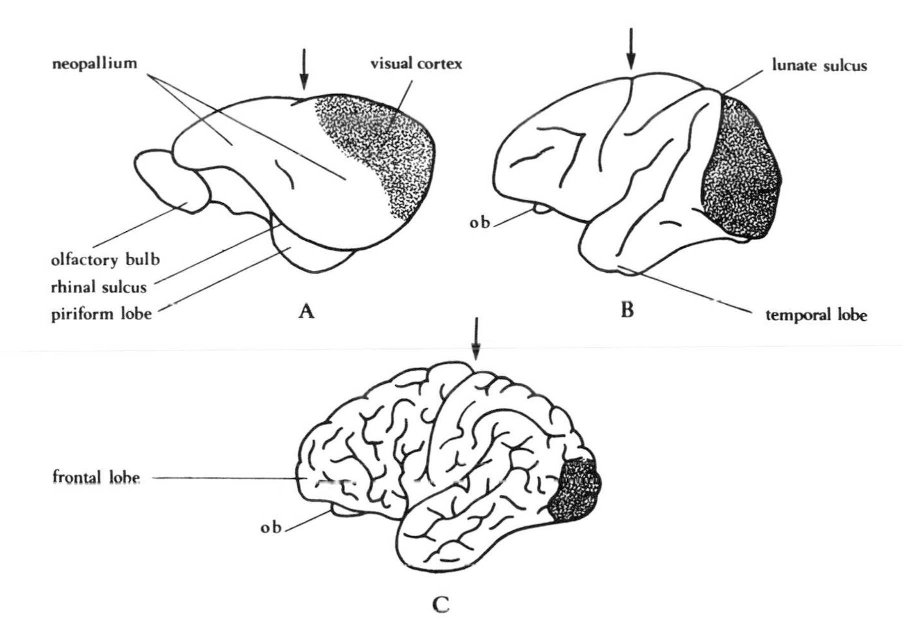

Fig. 2.4. Schematics of left cerebral hemispheres of A—tree shrew, B—monkey, and C—human brains, drawn the same size. Note again (*see* Fig. 1.7, p. 7, of this volume), the decrease in size of the olfactory bulb (ob) and primary visual cortex (shaded) with ascending phylogenetic scale and increased cortical convolutions and frontal lobes. Arrows indicate the boundary of the frontal lobe. (Modified from B. H. Campbell, 1974, p. 253.)

> increment has been accompanied by an ever greater learning capacity as opposed to genetically determined fixed action patterns (Lewin, 1984, p. 81).

Lewin was especially concerned with a second important point, the organization of the brain, specifically the relative proportions of the various lobes. In the human pattern, the frontal, temporal, and parietal lobes are predominant and the occipital lobe is relatively small, whereas in the ape this pattern is reversed. The differences are illustrated, for example, by comparisons of the primary visual cortical areas in phylogenetically different species (Fig. 2.4). In humans, the primary projection area for vision constitutes 2% to 3% of the total cortical area, in apes 8% to 10%, and further down the evolutionary tree, 15% in lemurs (Holloway, 1968, p. 148). Detailed documentation of the relative sizes of primary sensory and motor areas are to be found in Russia at the great primate brain collection of the Moscow Brain Institute. The measurements, accumulated during six decades by institute scientists, show that whereas the primary sensory areas decrease with higher phylogenetic position, the secondary and tertiary sensory regions increase parallel with the development of specific human cognitive processes such as speech and symbolic thinking (Blinkov and Glezer, 1968).

The changes in functional patterns and relative brain size are accompanied by an increase in the association areas of the cerebrum, interposed between the afferent (incoming) and efferent (outgoing) signals and mediating their redistribution. As C. Judson Herrick concluded in his monograph on the two most thoroughly studied brains, the rat's and man's: "The enormous increase in size of the human cortex is chiefly in associational fields. Here, then, is to be sought the structural organization upon which depend human culture and the progress of civilization" (1926, p. 265).

The belief that a complex social life correlates with a large brain (in proportion to body size) has been widely held to "explain" the chimpanzee's intellectual superiority over that of other apes. This advantage is not confined to chimpanzees, however. Observations in the wild of the gentle bottlenose dolphin, another "big-brain" mammal, reveal surprising similarities between their social habits and those of chimpanzees in their natural habitat (Booth, 1988). Although dolphins and chimpanzees evolved separately from a common ancestor about 60 million years ago, in very different environments, their behavioral similarities and relatively large brains with a highly convoluted cortex reinforce the concepts that social interaction and brain development are interdependent.

The influence of brain development on the morphology of the vertebrate skull, as shown in Fig. 2.3, was the topic of a comprehensive study of the development of vertebrate skulls by the English biologist, Gavin Rylands De Beer (1899–

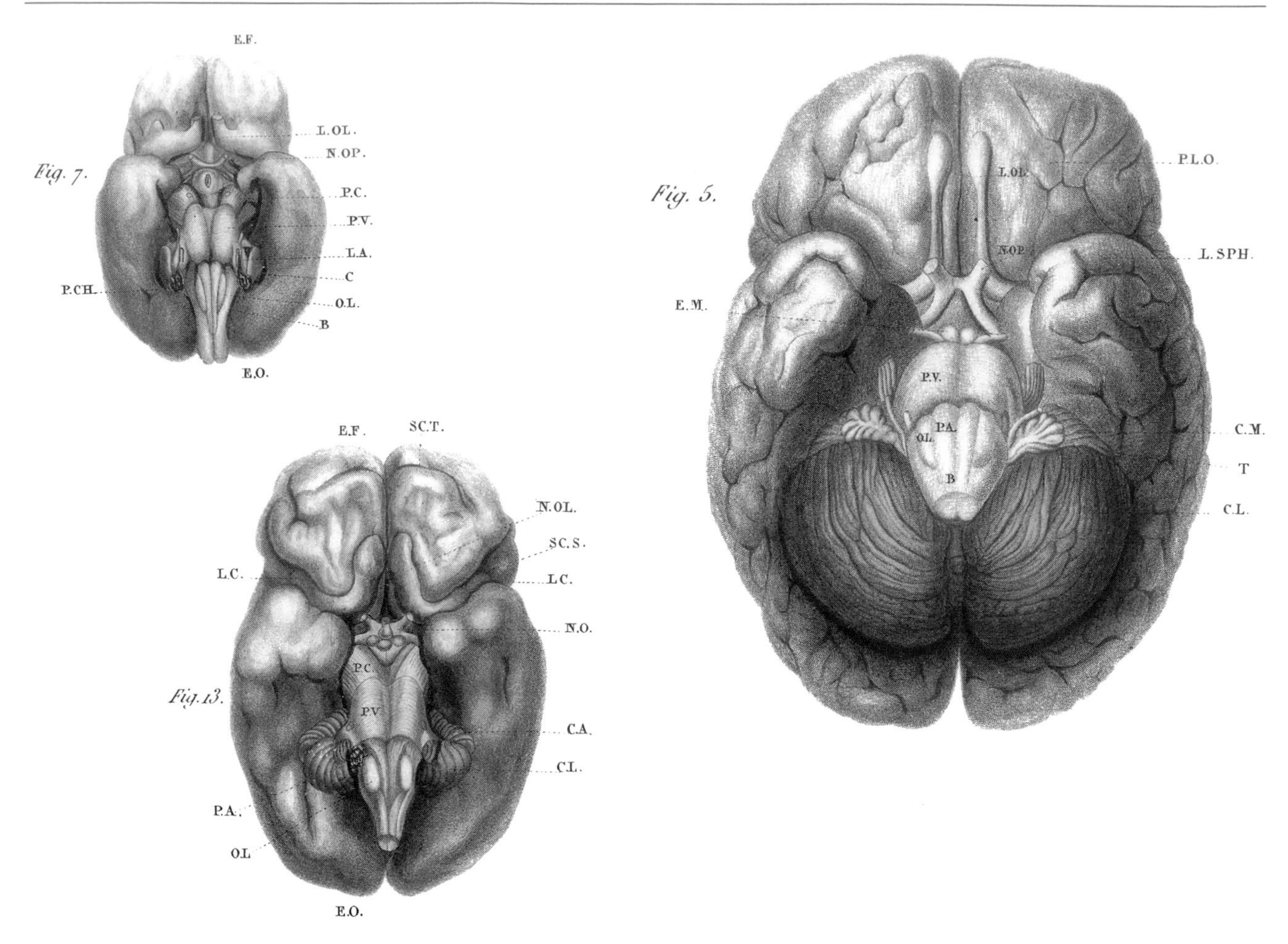

Fig. 2.5. Ventral views of brains of human fetuses shown in scale to illustrate the absolute and relative reduction in size of the olfactory apparatus as gestation progresses. From left: at three months gestation (L.OL), at six months (N.OL), at term (L.OL). (Adapted from Leuret et Gratiolet, *Atlas*, 1859–1857 [sic], Plate XXIX, Figs. 7, 13; Plate XXXI, Fig. 5.)

1972). He drew his conclusions after noting the correlation of bone formation and size of the brain in medical cases of microcephaly, anencephaly, and hydrocephaly. In addition to those "natural" examples, De Beer cited (1937, p. 476 passim) the 1902 report by the German zoologist and embryologist Hans Spemann (1869–1941) of the formation of two osteocrania in Triton after dual brains were experimentally produced.

ONTOGENY "RECAPITULATES"

While the comparative neuroanatomists were exploring the animal species in their search for examples of the direct relationship between brain size and complexity and higher intelligence, the embryologists were finding evidence of the individual organism's humble beginnings in the morphological changes taking place during prenatal life. Those findings were consistent with and reinforced the early nineteenth century popularity of "Naturwissenschaft," or the romantic natural philosophy which held that there is a unifying simplicity that underlies the complexity of the living world. A short-lived movement, *Naturphilosophie* was especially appealing to many German biologists (Clarke and Jacyna, 1987) who sought to identify one principle or form common to plants and animals. As pointed out by F. J. Cole (1949, p. 23), in pre-evolutionary biology there had been little mortar to bind together the miscellany of observations and the philosophies that sprang from them.

Although Nicholaus Steno (1638–1686) had written 200 years earlier that "In the Foetus of Animals we see how the brain is gradually formed" (transl. in Steno, 1950, p. 40), systematic studies of fetal brain structure were not undertaken until the

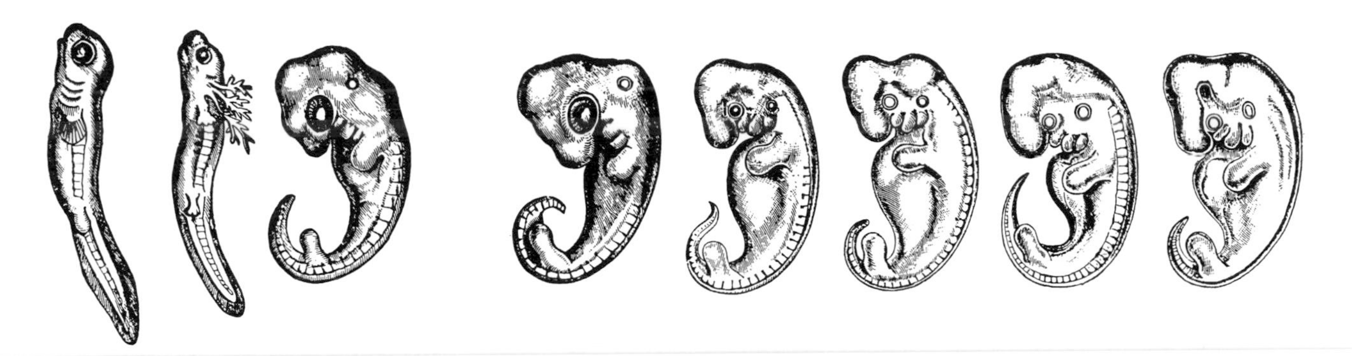

Fig. 2.6. A vigorous supporter of Darwin's ideas, Ernst Haeckel proposed that an organism passes through embryonic stages that repeat its evolutionary ancestry. He illustrated this theory by pointing out the similarities in embryos at various stages of development in fish (left), salamander, tortoise, bird, pig, ox, rabbit, and human. (From Haeckel, 1891, between pp. 252–253; also in Hutton, 1978, pp. 16–17.)

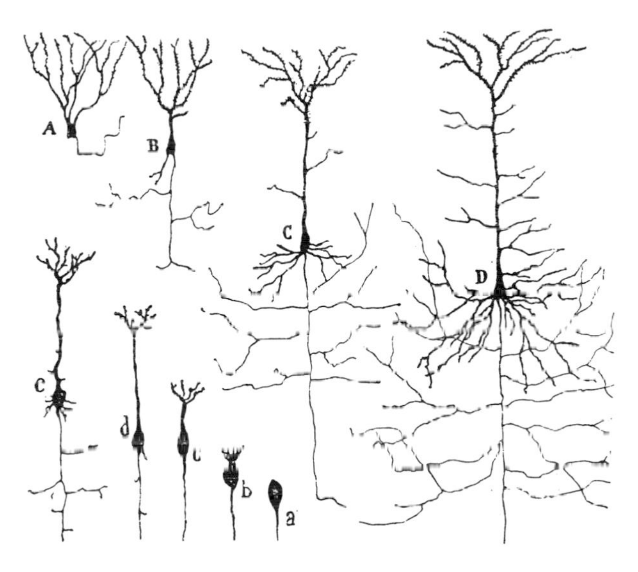

Fig. 2.7. Schema of phylogenetic (above) and ontogenetic (below) evolution of pyramidal cells. Above: A—frog, B—lizard, C—rat, D—man. Below: a—neuroblast without protoplasmic stalk, b—beginning stalk of terminal plume, c—further developed, d—appearance of axonal collaterals, e—formation of proplasmic processes and of the stalk. (From Ramón y Cajal, 1892, p. 466, Fig. 10; also 1911, vol. II, Figs. 545 and 861.)

nineteenth century. Perhaps the most important were carried out by the German anatomist Friedreich Tiedemann (1781–1861). In the introduction to his influential studies of mammalian fetal brains at various stages of development, he declared: "In my opinion, the only two paths that can lead to a knowledge of the structure of the brain . . . are those of comparative anatomy and the anatomy of the fetus; for this labyrinth they are like the thread of Ariadne" (1826, p. 2). The principle of ontogeny was superbly illustrated by drawings of the developing olfactory system in the human fetus, published in the midnineteenth century by two French comparative anatomists, Leuret and Gratiolet (Fig. 2.5).

The concept that the evolution of a species is repeated during development of the individual embryo was formalized by Ernest Heinrich Haeckel (1834–1919), German zoologist and student of natural history; in 1866, he declared that "ontogeny recapitulates phylogeny." In Haeckel's

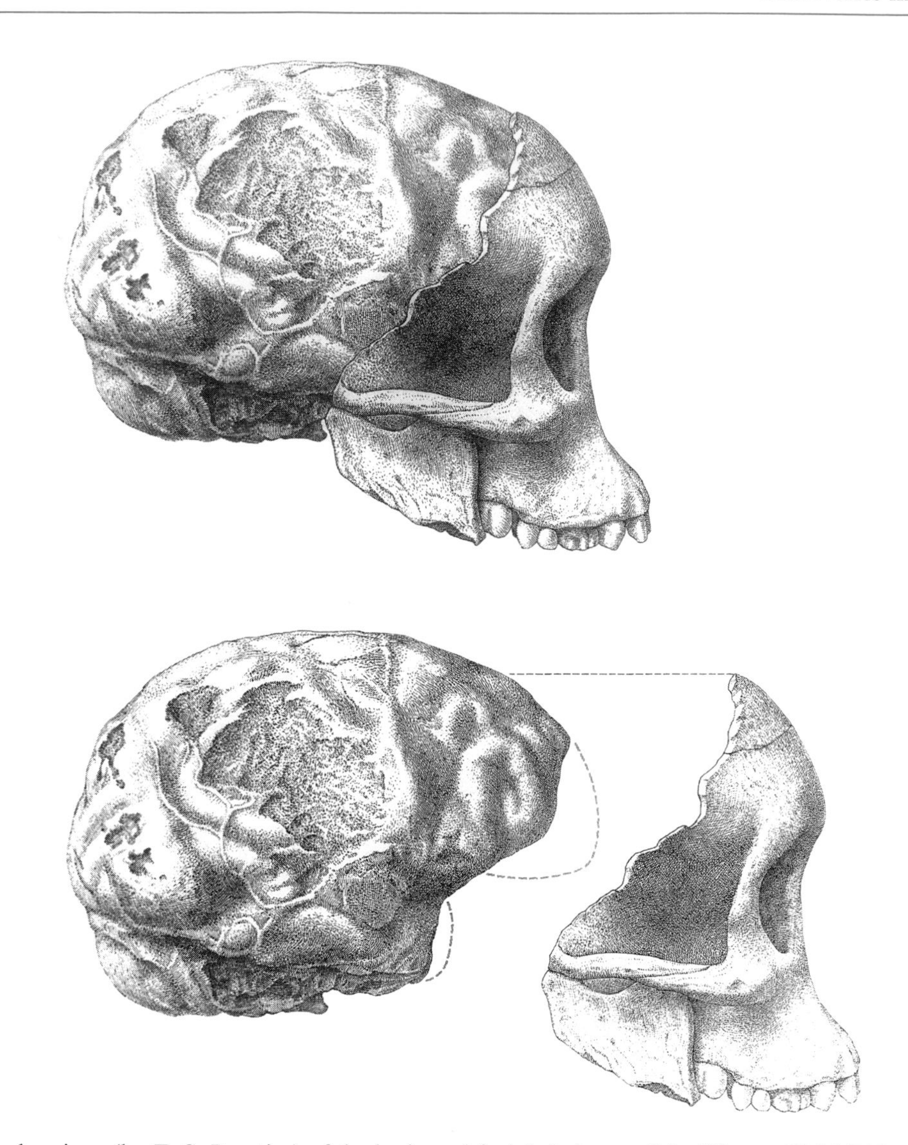

Fig. 2.8. Modern drawings (by T. S. Prentiss) of the brain and facial skeleton of the "Taung Child," identified by Raymond Dart in 1924, the first australopithecine fossil from South Africa to be discovered. Above—Natural cranial endocast that articulates perfectly with the skull from the same specimen, showing meningeal arteries, gyri, and sulci. Below—Cast and skull shown separately with outlines of parts not recovered. (From Holloway, 1974, p. 107.) The details emphasized Dart's view that the specimen's hominid features predominated, whereas the early rendering (*see* Fig. 2.3) leaned toward ape-like features.

view, ancestral adult stages were "accelerated" into the ontogeny of the descendants. The appeal of this concept was reflected in a contemporary German textbook: "The earliest characters of the embryonic head, or of its equivalent in all Vertebrates, point to its being phylogenetically, the most ancient portion of the body, and serve as a finger-post to the path of development of the Vertebrate body" (Gegenbaur, 1878, p. 413). Although Haeckel's "biogenetic law" was too sweeping and has been largely discarded (Ferris, 1922; Gould, 1977), his contribution was to combine the studies of vertebrate embryos at various stages of gestation (*see* Fig. 2.6, on p. 15, top) with their phylogenetic history and so to provide clues to evolutionary relationships among species.

A modern view holds that "[recapitulation] fell when it became unfashionable in practice, following the rise in experimental embryology, and untenable in theory, following scientific change in a related field (Mendelian genetics)" (Gould, 1977, p. 6). In its place, Gould proposes an approach to ontogeny that invokes a small genetic difference "with profound effects"—alterations in the regula-

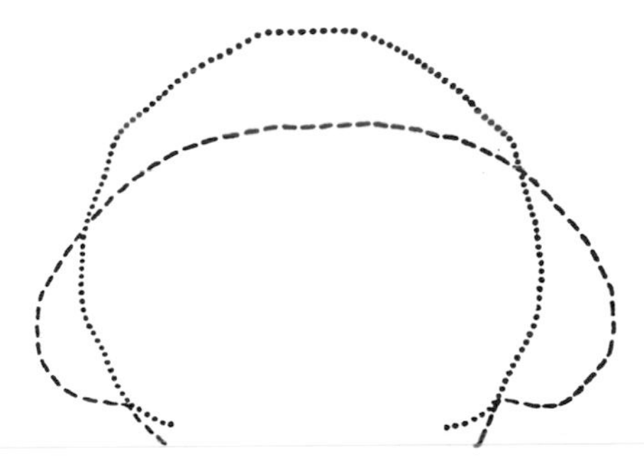

Fig. 2.9. Contour tracings of coronal sections through the widest part of the parietal regions of the endocranial casts in *Australopithicus* (dotted line) and in a modern gorilla (dashed line), showing that the ancient homonid had "set aside" a larger part of the cortex for storing information. (From Dart, 1925, p. 198, Fig. 6.)

tory system that slow, instead of accelerate, the rate of development in humans, called heterochrony, or differences in rate of timing of features already present (ibid., p. 9).

A traditional pictorial summary of phylogeny and ontogeny in the nervous system is the depiction of its functional unit, the neuron, from the work of the pre-eminent Spanish neurohistologist, Santiago Ramón y Cajal (*see* Chapter 7). In a single illustration, reproduced in Fig. 2.7 (*see* Fig. 2.7, on p. 15, bottom), this masterful observer set down the embryonic stages of a pyramidal cell (a neuron whose soma lies in the cortex, has apical and basal spiny dendrites, and with axons leaving the cortex for subcortical or other cortical areas; *see* DeFilipe and Jones, 1988, p. 565) and their morphologic development from very simple forms in fish to complex configurations in humans. In his Croonian Lecture of 1894, Cajal discussed the superiority of the mammalian pyramidal cells in the number of contacts among neurons they made possible and noted that:

> [A]s one descends the [phylogenetic] scale . . . the dendritic apparatus appears to be less differentiated, and the collaterals of the axons less numerous, less long, and less branched. . . .
>
> On the other hand, our research on the development of the embryonic nerve cells . . . has shown us that as the cerebral cortex increases [in size] the dendrites and the collaterals of the axons of the pyramids become proportionately longer and more branched. . . . (transl. in ibid., p. 86).

Cajal admitted that he was not proposing a new principle, but only providing it with "the positive facts of structure" (ibid., p. 87). His hypothesis regarding correlation of microscopic neuronal structures with intellect was to be echoed by new discoveries in the fossil record of overall brain size in the human lineage.

PALEONEUROLOGY IS INTRODUCED

A reliable documentation of the antecedents of the modern human brain was discovered in 1924 by Raymond A. Dart (1893–1988), Professor of anatomy at the University of Witwatersrand, Johannesburg, South Africa. Reuniting a natural endocast of a skull cavity with the piece of rock in which it had been embedded, he realized he was holding almost the entire facial skeleton of a three- or four-year-old child (Fig. 2.8) who had lived more than three million years ago. Referring to the skull and brain endocast from a species which he named *Australopithecus africanus*, Dart wrote "the specimen is of importance because it exhibits an extinct race of apes *intermediate between living anthropoids and man*" (1925, p. 195). One of the links in the long-sought connection between apes and humankind had been found and Dart exalted that this

> ultra-simian and pre-human stock . . . had profited beyond living anthropoids by setting aside a relatively much larger area of the cerebral cortex to serve as a storehouse of information concerning their objective environment as its details were simultaneously revealed to the

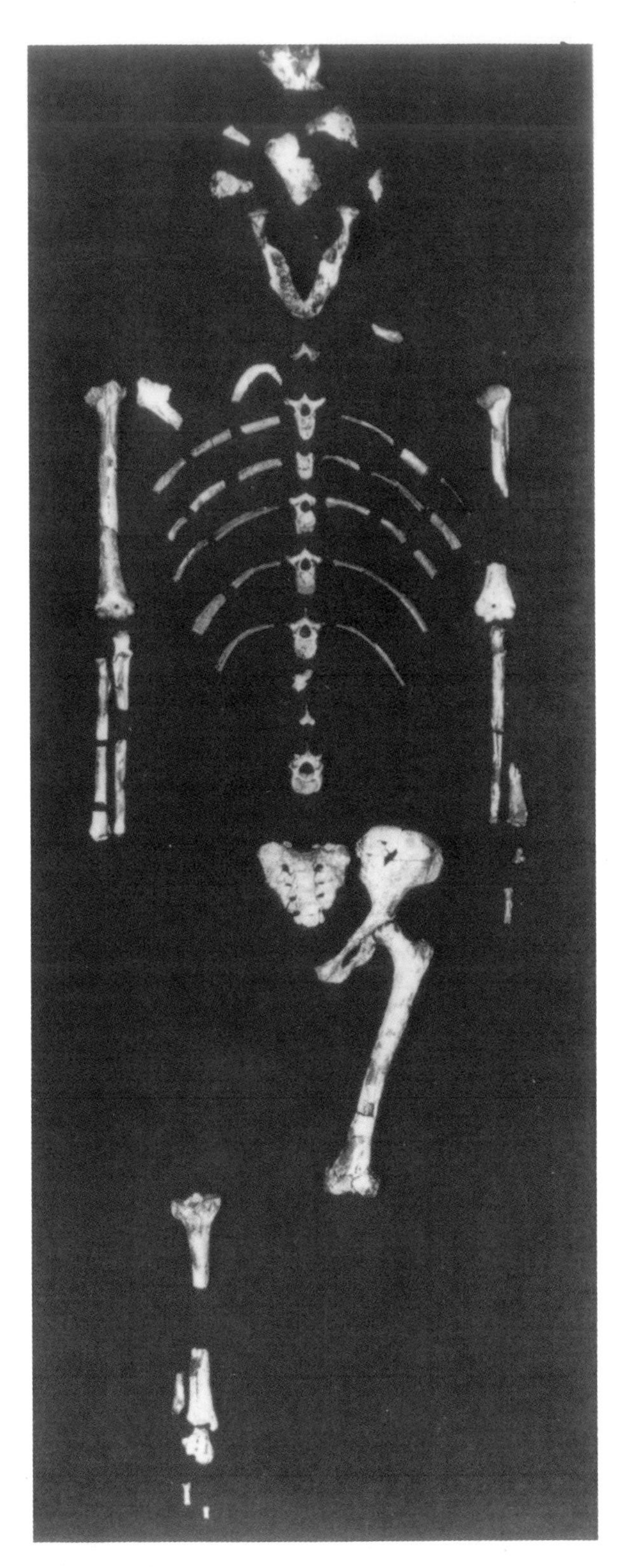

Fig. 2.10. Lucy's fossilized bones comprised the most complete hominid skeleton found before 1984. An australopithecine, Lucy lived ca. 3.18 million years ago. She was three feet tall, weighed about 60 pounds, and walked upright. The fossils have been returned to Ethiopia. (Impression from Johanson and Edey, 1981, p. 23.)

senses of vision and touch, and also of hearing. They possessed to a degree unappreciated by living anthropoids the use of their hands. . . . (ibid., p. 198).

The finding of *A. africanus* underscored the belief that in the course of evolution, the anthropoid apes became proficient in a specialized arboreal life, whereas the hominid sequence became adjusted for life on the ground with profound modifications of the skeleton for erect, bipedal locomotion. Transition from early hominid to human involved new adaptations of the hand, the pelvis, and the foot, accompanied by a reorganization and a substantial modification in the structure of the brain. Dart illustrated some of the evidence for such striking changes by superimposing the outline of the skull of a modern gorilla on that of the australopithecine he had identified (*see* Fig. 2.9, on p. 17).

A specimen of *Homo erectus*, a later hominid, was found in 1929 near modern Beijing by Canadian anatomist and anthropologist Davidson Black (1884–1934). The significance of Black's discovery, dubbed "Peking Man," was established by Franz Weidenreich (1873–1948), a German anthropologist who brought many endocasts back to the United States just before the Sino–Japanese war. In 1939–1940, the original skull and bones of Peking Man were packed in a foot locker and entrusted to a battalion of United States Marines for safe passage to the United States but disappeared en route (Bowers, 1972; Shapiro, 1974). The cranial capacity of Peking Man was 1000 cm^3, he had a sharply receding forehead, and is thought to have lived ca. 1.6 million years ago.

In 1974, Donald C. Johanson, American paleoanthropologist, unearthed in Ethiopia the most complete fossilized hominid skeleton found to that time. Forty percent complete and named "Lucy" for its female pelvic bone, she was diminutive and walked erect (Fig. 2.10). Lucy and other specimens indicate that *Australopithecus afarensis* had a small cranial volume of about 650 cm^3, roughly equivalent to that of modern apes. The skull, however, presents features not found in apes, which place it firmly among the Hominidae.

Other fossilized remains of the early hominids have been found in eastern parts of Africa, notably by Louis and Mary Leakey and their son, Richard. The elder Leakeys, in their passionate quest for

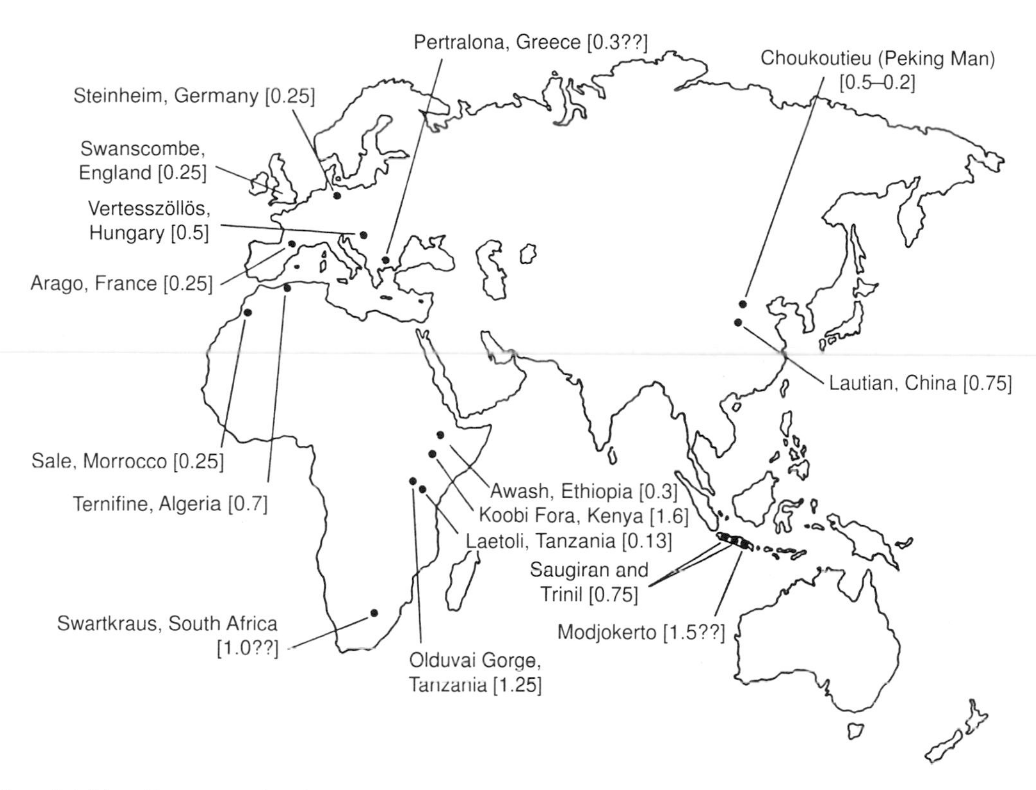

Fig. 2.11. Map of Africa, Europe, and Asia showing the sites where fossil remains of *Homo erectus* have been found and their probable age in millions of years. From these data, assumptions may be made of the time course and extent of *Homo*'s migrations. (Adapted from Lewin, 1984, p. 54.)

human fossils (cf. Morell, 1995), were early champions of the concept of multiple lines of hominoid species, in contrast to a single heritage. The stunning recovery in 1984–1985 of the scattered fossilized bones of an almost complete skeleton of a young boy was made near Lake Turkana in Kenya by a team directed by the Leakey's son, Richard, and Alan Walker. The size of the skeleton of this example of *Homo erectus* indicates that as an adult he could have attained a height of six feet, notably taller than other adult specimens found to that time. The Turkana bones predate Peking Man, suggesting the slow migration of *H. erectus* from an African cradle to far-flung settlements in Asia and Europe (Fig. 2.11).

The next known stage in human ancestry was *Homo sapiens*, which appeared between 250,000 and 100,000 years ago. An example was Neanderthal man, so named for the river in Germany where his remains were first discovered in 1857. His cranial capacity attained 1300 cm^3, and he had a well developed vertical forehead providing for the expansion of the frontal region of the neopallium. Cro-Magnon lived about 50,000 years ago in southern France and northern Spain, with a large cranial capacity measuring 1600 cm^3, an upright forehead, and well marked chin which conferred on him a definitely modern human countenance.

Comparison of cranial volumes, the most quantifiable measurement that can be made on prehistoric brains, albeit from endocasts, shows dramatic changes during evolutionary history, increasing about fourfold from *Australopithecus* to *Homo sapiens* (Tobias, 1971). Interpretation of the increase in absolute size was that the brain progressively acts and reacts to changes in neuronal reorganization, complexity of function, diversified behavior, and ultimately acculturation. Tobias (ibid., p. 151) perceived that two "elements of hominidization were alike in being long continued, sustained, and most marked: brain-size increase and cultural evolution." He proposed that it is in the middle of the chain that progress will be made in understanding the "tangled skein" that links the two ends: brain size and culture.

Fig. 2.12. A stone hand axe associated with *Homo erectus*. Its refined symmetry places it between 1.5 million and 200,000 years ago. (From Leakey, 1981, p. 134.)

Although the fossil record provides direct support for evolution of the Hominidae, evidence based on creatures living today, however extensive and compelling it may be, is only indirect. Studies of living species, however, have proved rewarding. The neurologic comparisons carried out by Lewin and by the Herricks, already mentioned, were based on living animal groups at different levels of phylogenetic advancement. Using endocasts, "the closest empirical 'windows' to the evolution of the hominid brain that we possess," Columbia University anthropologist Ralph L. Holloway, compared hominoid skulls for absolute and relative cerebral sizes, asymmetries, and lobar divisions, where possible, with those of modern skulls (1981, p. 155). With careful measurements to back his thesis, Holloway stated that "the early hominids already display evidence of reorganization [of the cerebral mass] toward a human pattern, particularly as regards the added development of the temporoparietal regions on the lateral surface" (1976, p. 346). This example of form adapting to function had been pointed out by C. Judson Herrick in 1926 (*see* p. 6, this volume).

EARLY EVIDENCE OF HANDEDNESS

The increasingly sophisticated use of the hand for creative and intellectual skills was evidenced at the Cro-Magnon sites by the presence of stone tools (Fig. 2.12) and small three-dimensional figurines found among fossilized skeletal remains. Of particular interest in the development of prehistoric art are the pictographs that were stumbled upon in caves of southwest France and northeast Spain. The 20,000-year-old drawings at Lascaux (Fig. 2.13) include frequent dots, lines, and geometric patterns, suggesting the use of symbols at that early period, although the first records of written language did not appear until about 5000 BC. Equally intriguing is the predominance of images of the left hand among the imprints or stencils of the hands of

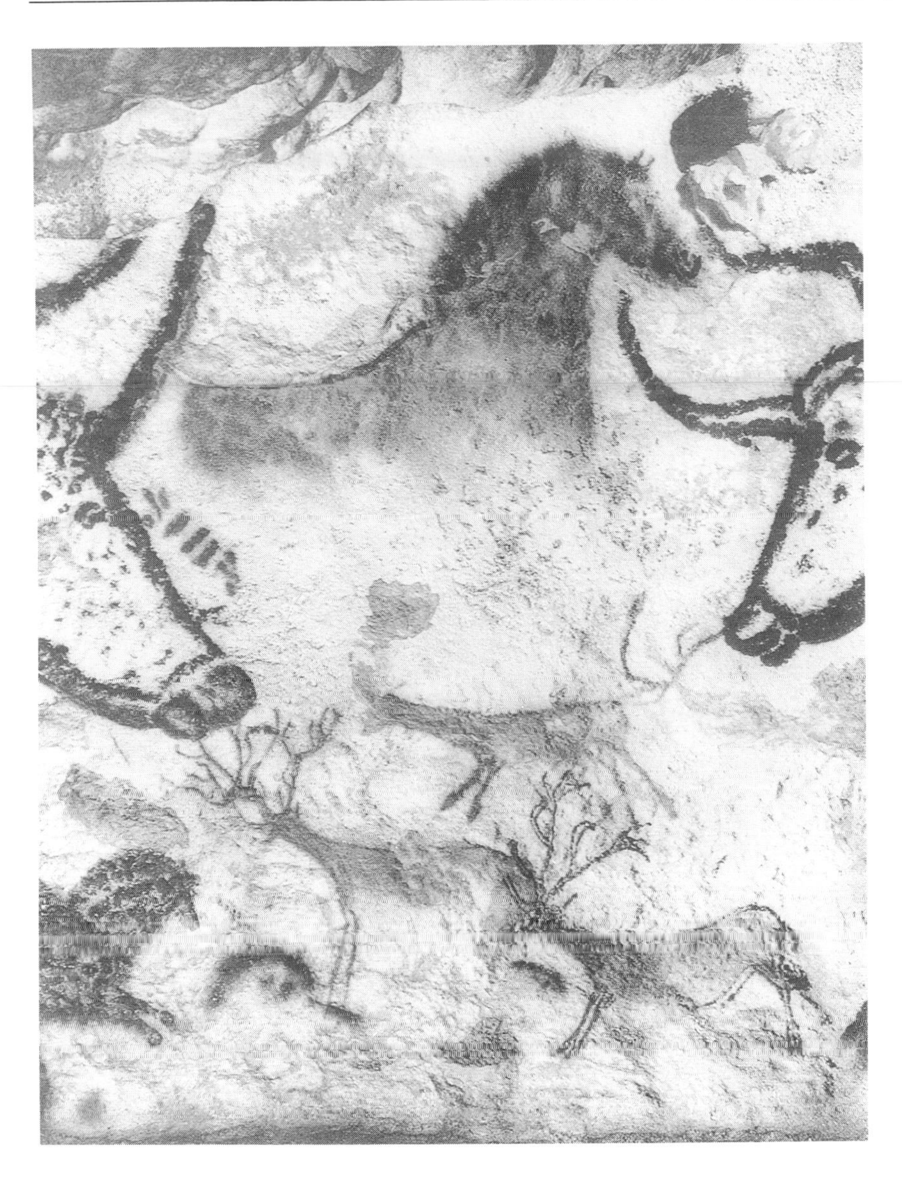

Fig. 2.13. Pictographs on walls of the Lascaux caves of southwestern France vividly document the artistry of people living during the last Ice Age: detail of bulls, horses, and deer at the entrance of the "Great Hall." (From Bataille, 1955, p. 52.)

prehistoric individuals (*see* Fig. 2.14, on p. 22). In one cave, 35 left and seven right hands were depicted, implying that the right hand of Ice Age people was more frequently utilized than the left for fine manipulative activities.

An early preference for the right hand is found again in the use of weapons depicted in ancient Egyptian and Assyrian carvings (ca. 3000 BC) of archers and spear-throwers (*see* Fig. 2.15, on p. 23). Moreover, in early illustrations of the act of writing, Egyptian and Sumerian scribes hold their stylus in the right hand (*see* Fig. 2.16, on p. 23). This characteristic practice was pictured in the medieval period as well, woven into the Bayeux tapestry story of combat between Norman and English warriors and seen repeatedly in illuminated manuscripts (*see* Fig. 2.17, on p. 24). Abundant evidence of the ubiquity of right-handedness occurs also among other old cultures, as in paintings by the Mayans found in the New World (*see* Fig. 2.18, on p. 25).

The exploratory daring of two twentieth-century neurosurgeons provided proof that the human brain accommodates selectively the intense and sentient use of the hand. Wilder Penfield and Theodore Rasmussen (1950) determined by electrical stimulation of the exposed brain of awake patients that a disproportionately large area of the motor cortex is

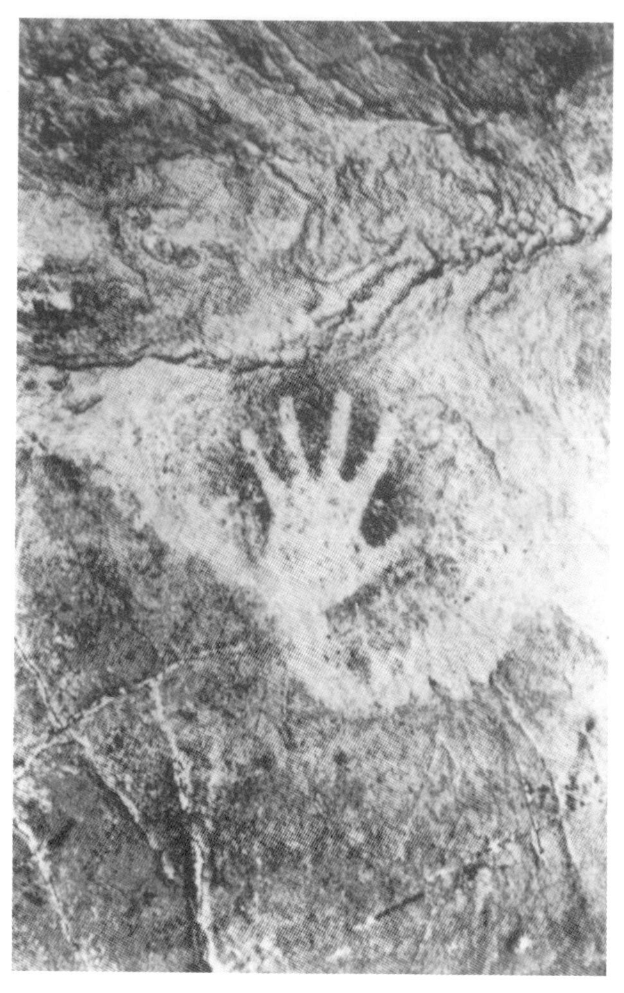

Fig. 2.14. Pictograph of a left human hand on a cave wall in southwestern France. Because outlines of more left than right hands were found, it was assumed that right-handedness was already characteristic of Ice Age people. (From Magoun, Darling, and Prost, 1960, p. 100, Fig. 48.)

devoted to representation of the hand, especially the thumb (*see* Fig. 2.19, on p. 26). Penfield's interest lay in preoperative mapping of the speech areas, and he persistently sought evidence of a relationship between aphasia, handedness, and cerebral hemispheric dominance. From descriptions of 522 cases in the literature (Penfield and Roberts, 1959), and his own additional cases, he concluded that "Man seems to have acquired language and to have become right-handed at about the same time in evolution. . . . Brain function and handedness may be unrelated except by disease" (ibid., p. 102). It remained for Norman Geschwind (1965) to correlate handedness with an anatomic substrate (*see* Chapter 6).

MODERN THEORIES OF EVOLUTIONARY NEUROLOGY

As the Ice Age was replaced by a more temperate environment and life became more conducive to social and cultural activities, the associated increase in brain size reflected its augmented complexity, particularly in cytological modifications. The cellular alterations included elaboration of additional dendritic processes, as already noted, as well as many departures from primitive reflex forms. The nature of those departures, as they were delineated by DeFilipe and Jones (1988) included establishment of new synapses among previously unconnected regions of the brain;

Fig. 2.15. The use of weapons in the hunt: Assurbanipal, the right-handed Assyrian warrior-king, from a bas-relief carved ca. 700 BC now in the British Museum.

feedback loops modulating neural input and output; neural facilitation or inhibition in new loci of influence on central activity; and the effect on current neuronal action of sensitivity to earlier neural activity. Returning to Ramón y Cajal's composite drawing (Fig. 2.7), he stated in 1894 that:

> The facts of observation . . . have suggested to us a hypothesis . . . as to why intelligence is acquired after a well-directed mental education, why intelligence is hereditary, why professional cerebral adaptations arise, and why certain artistic aptitudes are created.
>
> Cerebral gymnastics are not capable of improving the organization of the brain by increasing the number of cells . . . but it can be admitted as very probable that mental exercise leads to a greater development of the dendritic apparatus and of the system of axonal collaterals in the most utilized cerebral

Fig. 2.16. Two right-handed Egyptian scribes from "The delinquent tax-payers relief" in Saqqara (ca. 3000 BC) and now at the Oriental Institute of the University of Chicago. (Adapted from impression in Davidson, 1962, pp. 48–49.)

Fig. 2.17. Preference for the right hand was shown in images created during the medieval era. Above—In depictions of warfare woven into the Bayeux Tapestry (AD 1100). Below—An illustration of an author dictating to a right-handed scribe from an illuminated manuscript of Vincent de Beauvais, ca. AD 1234. (Mss. 240 Folio 8, the Pierpont Morgan Library.)

> regions (translated in DeFilipe and Jones, 1988, p. 86).

Cajal went on to say that his hypothesis was not a new principle, but, due to his findings, it was now based on "the positive facts of structure" (ibid., p. 87).

Fifty years after Cajal, C. Judson Herrick outlined in more general terms the possibilities for change:

> The cerebral cortex . . . has emerged from reflex centers . . . which are activated chiefly in response to what is going on outside the body—the exteroceptive apparatus . . . it

Fig. 2.18. Right-handedness in the Amerindians is illustrated in the Mayan "Mural of Bonampak" (ca. AD 850) in southern Mexico, depicting blood sacrifices to a displeased deity. (Adapted from Davidson, 1962, pp. 408–409.) Color plate for this figure appears as an insert after page 84.

> seems to serve . . . chiefly as an activator, reinforcing, inhibiting, or otherwise modifying and controlling the innate reflex patterns of the lower centers out of which it has grown (1926, p. 31).

Herrick and his short-lived elder brother, Clarence, were among the first true neuroscientists in that they stressed the integration of anatomical and behavioral evidence in the interpretation of the phylogenetic evolution of species. They were at the forefront of American naturalists in the late nineteenth century who were intent on exploring the native assets of the new nation represented by the abundance of fauna and flora. In so doing, they accumulated evidence of the brain's structural rearrangements to accommodate maturing behavioral patterns in the ascending phylogenetic scale.

A model of a structural–behavioral brain, the "triune brain," was proposed by an American neurologist, Paul D. MacLean, based on mankind's "oldest heritage," the reptilian brain, superimposed on which are the olfactory and limbic structures of the lower mammals, and surmounted in late evol-

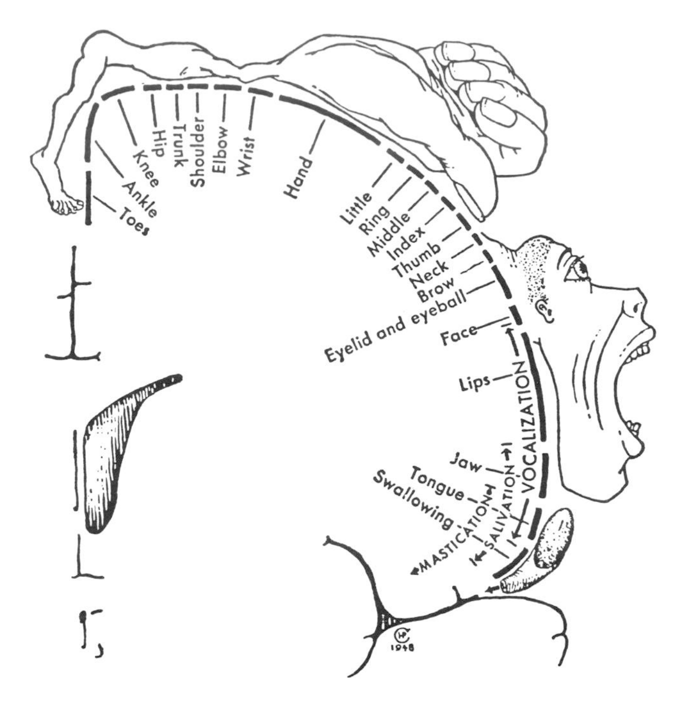

Fig. 2.19. A grotesque homunculus represents the extent of the areas of the motor cortex that control the muscles of the distorted part. Note that the area of cortex controlling the intricate action of the hand, especially the opposable thumb, is as large as that for the rest of the body and legs together. An equivalently large area controls the muscles of the face, in which the parts concerned with speech (mouth, tongue, and larynx) have the greatest representation. (From Penfield and Rasmussen, 1950, p. 57, Fig. 2.2.)

utionary time by the neocortex (1967; *see* Fig. 12.4). The triune brain model challenged the belief that precursor analogs of the three hierarchial brains have been present throughout phylogenetic history, waiting for time to draw them out in development of new species.

Knowledge explosively accumulating through studies in molecular neurobiology attests broadly and clearly to mankind's evolutionary heritage and the relationships among animal species. An early example is the study of patterns of neurofilament proteins, subunits of nonessential genes, which are known to contribute to the stability and diameter of some large axons and represent adaptation for unusually fast nerve conduction in appropriate circumstances. Among the living species studied, mammals possess three neurofilament protein subunits, reptiles two, the jawless fish (lampreys) one, and the arthropods (crabs) none (Lasek, Phillips, Katz, and Autilio-Gambetti, 1985). We may also cite the genetic differences among living species that indicate the branching of the gorilla–chimpanzee clade from the human lineage six to eight million years ago (*see* Gould, 1995, p. 138). Thus the evolutionary history of the human nervous system is being probed at the level of the gene, where it commenced.

OVERVIEW OF EARLY HUMANS

Darwin's *Origin of Species* promoted the ordering of the vast accumulation of descriptive, apparently unrelated, data on animal (and plant) species into a rational scheme, a phylogenetic scale. The advance of the concept of phylogeny is based on the proposition universally accepted by scientists that different species share a common ancestor. Because most determinations of evolutionary lineages are made on indirect evidence, they are cautiously speculative. Nonetheless, ontogeny suggests that the brain, or its representative, was the first animal organ to undergo a phylogenetic change. The unearthing of fossilized remains of hominoid and hominid specimens and their identification and dating by isotope and geological/artifactual strata have generated two sweeping theories of human evolution: a single line of human ancestry versus multiple hominid species with primitive human characteristics. The continuing addition of new findings from worldwide sites, together with uncertainties about where they fit into evolutionary history and whether the process was gradual or punctuate, as seems more likely (*see* Gould, 1995), have made paleoneurology a lively and adventuresome discipline.

The reliance on indirect measurements of prehistoric brain sizes has revealed a progression toward larger brains accompanying the ascent of the phylogenetic series, fourfold from *Australopithecus* to modern man. This trend mingles with increasingly complex behavior displayed in toolmaking and use, for example, and in early art work. Judging from the sculptures of ancient scribes and hunters and evidence from caves decorated during the Ice Age through the Mayan depiction of sacrificial ceremonies and medieval fighting, preference for use of the right hand that characterizes modern humankind was established early in human history. The connection between that behavior and a neural substrate was not elucidated until modern times, as will be described in Chapter 6.

3 The Ventricles and Their Functions

Contents

The cerebrum is the general motory as well as sensory organ and it is also the general laboratory of the fluid essences of its body.

(Swedenborg, trans. by Tafel 1882, vol. 1, p. 583)

The first documented reference to any part of the nervous system is found in the Edwin Smith Surgical Papyrus, originally compiled in Egypt ca. 3000 BC and known from a copy made ca. 1600 BC. Case Six of this surgical collection is an account of a head injury so severe as to expose the brain after penetrating the skull and meninges. The surgeon is directed to palpate the wound and feel the "corrugations which form in molden copper [the convolutions], (and) something therein throbbing (and) fluttering under thy fingers" (Breasted, 1930, vol. I, p. 166). Because the hieroglyphics for "brain" (Fig. 3.1) are usually followed by "in the skull," Breasted suggests "marrow of the skull" as the earliest designation of the brain.

The next recorded account of the human brain available to us is that of the Greek physician, Alcmaeon, who lived in southern Italy ca. 500 BC. Only the smallest fragments of his writings survive, but he can be quoted from Theophrastus, *De sensus:* "All the senses are related in some way to the brain; consequently [the senses] are incapable of action if the brain is disturbed" (Diels, 1952, p. 212; trans. O'Neill). These same ideas were expressed more fully in the Hippocratic book, *On the Sacred Disease:* "[From the brain] come joys, delights, laughter and sports, and sorrows, griefs, despondency, and lamentations. And by this, in an especial manner, we acquire wisdom and knowledge, and see and hear, and know what are foul and what are fair, . . ." (Adams, 1886, vol. II, p. 344).

Plato, like Alcmaeon and the Hippocratic physicians, supported the concept of the primacy of the brain in mental activities, placing therein the immortal part of the soul. His pupil, Aristotle, however, discounted the primacy of the brain, believing that it merely cooled the blood, and proclaimed the heart as the locus of thought and sensation. He described the brain as bilateral, with two membranes surrounding it and "in the great majority of animals, [the brain] has a small hollow in its centre" (Aristotle, 495[a]); that is probably the first reference to a cerebral ventricle. Although he stated that in proportion to body size, man has the largest brain of all animals and a woman's brain is slightly smaller, it is unlikely that he dissected a human brain (Clarke and O'Malley, 1968, p. 8). Aristotle's idea of the primacy of the heart was tenacious, however, and survived throughout the Middle Ages but with diminishing authority. The reason why this polymath could have been so wrong in considering the brain as secondary to the heart has been attributed to the fact that he was not a physician and, therefore, had no clinical data on which to

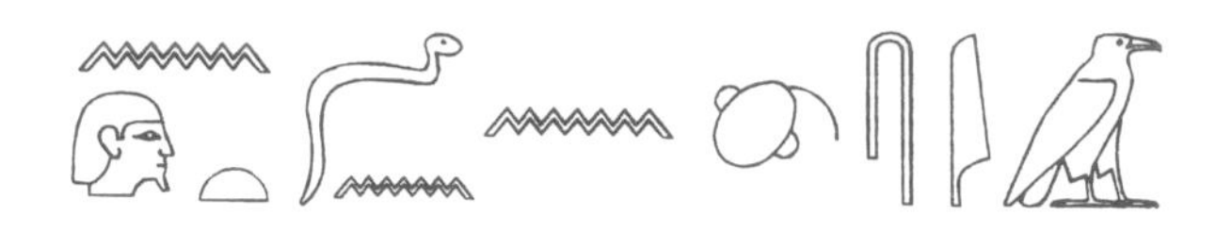

Fig. 3.1. The hieroglyphic transliteration from the Edwin Smith Surgical Papyrus (Egypt, ca. 1600 BC) meaning "marrow of the skull" or "brain." (Breasted, 1930, vol. I, Plate IIA; ×2.) [1]

judge (Gross, 1995). The culmination of this early Golden Age of investigation occurred at Alexandria during the third century BC. Founded the previous century by the young Alexander the Great, after his death the city came under the control of the Ptolemaic dynasty and developed a scientific sophistication that permitted extensive study of human anatomy. The writings of two Alexandrian investigators, Herophilus and Erasistratus, have not survived, even in fragments, but it is known from discussion of their work by Galen (Clarke and O'Malley, 1968, p. 11) that they believed the brain to be the organ of intelligence, in opposition to the heart as proclaimed by Aristotle. Herophilus seems also to have described the lateral ventricles and their horns, the third ventricle, and what he called the ventricle of the cerebellum, our fourth ventricle. The proximity of the fourth ventricle to the spinal cord and motor nerves led him to state that it contained the forces necessary for motor activities. Not only did he place the dominant principle of the soul (or mind) in the brain, but more specifically, he assigned it to the ventricular cavities, thus establishing an anatomic basis for the concept of localization of the mind within the cranium. His slightly younger contemporary, Erasistratus, continued those studies and described in detail the cerebral ventricles without mentioning their function, concentrating instead on the convolutions of the brain surface.

ANCIENT AND EARLY-MODERN VIEWS

The extension of Roman power, permeated with a new spiritual religion and philosophy, brought about the cessation of human dissection even in Alexandria. When Galen of Pergamum (AD 129–199) carried out his studies some 400 years after Erasistratus, the use of human material was interdicted. Nevertheless, both as investigator (with animal subjects) and as an elaborator of earlier writings, Galen extended the work of the Alexandrians with reference to the brain. He declared that the soul (or mind) is not in the ventricles as Herophilus was said to have written, but in the tissues of the brain. This belief was based on his clinical observation that cerebral lesions penetrating the ventricles deprived individuals of sensory or motor activities, but were not necessarily immediately fatal, as would be expected if the soul resided in the ventricular cavities. At the time of Galen's death, ca. AD 200, the early Greek belief in the brain as the body's central organ and locus of the soul had been reinstated, but not completely accepted. Of greater permanent significance, Galen's description of the nervous system as a functional unit—of the brain, spinal cord, and nerves as one system—had been introduced.

Until the sixteenth century, there was little further practical study of the human cerebrum. During that long eclipse, however, an associated development made headway: the localization of psychological functions within the ventricles . A Byzantine surgeon named Poseidonius, writing during the fourth century AD, probably from observations of head injuries, concluded that lesions at the front of the brain interfere with apprehension of all types of sensation, whereas trauma of the posterior part results in memory deficit, and damage to the middle ventricle produces a disturbance of reason. Those were the first written assignments of mental functions to specific cerebral ventricles.

Nemesius, Bishop of Emesa in Turkey (ca. AD 390), and a younger contemporary of Poseidonius, carried forward those ideas of ventricular function and proposed a schema which remained fairly standard for several centuries. Nemesius claimed that the anterior ventricles (lateral) accounted for the mixing of sensations and for imagination, the middle ventricle (our third) for cogitation and reason, and the posterior ventricle (our fourth) for memory, which proved to be a very stable concept. Without the benefit of first-hand observation, however, the first depictions of the brain were along geometric lines (Fig. 3.2). During the millennium from AD 400 to Vesalius, the artistry of drawings of the three-cell brain steadily improved (Magoun, 1958b), whereas knowledge of the cranial contents

[1] We are deeply grateful to Professor Stuart Tyson Smith of the Institute of Archaeology, University of California, Los Angeles for identifying the glyphs and pointing out that an alternative translation as "viscera of the head" was suggested by E. Iverson (1947).

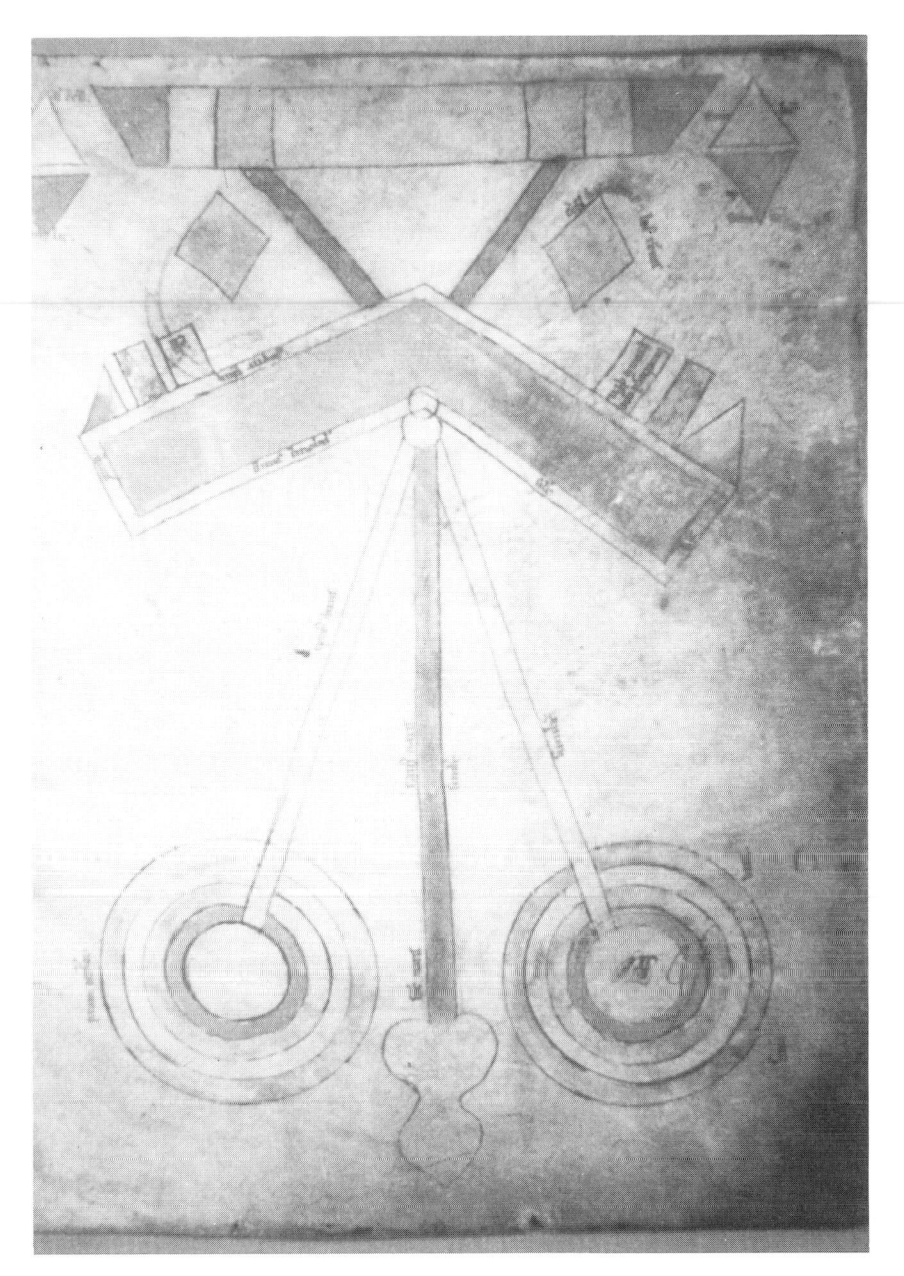

Fig. 3.2. A drawing from a Latin manuscript of the late twelfth to early thirteenth centuries. Above the two eyes and their converging nerves is the brain with its coverings and the attached "cells," only one of which is labeled. On the right, the diamond shape is inscribed "dwelling place of the brain or the place of reason." (M. R. James, 1907, vol. I, pp. 218, 219. From Gonville and Caius MS 190/223 f.br). Color plate for this figure appears as an insert after page 84.

lagged behind that of other parts of the body, the muscles and skeleton, for example. The 1503 drawing (Fig. 3.3) from the German encyclopedist, Gregor Reisch (ca. 1467–1525), assigned functions to each of the three cavities, a repetition of the original concept of Nemesius.

In the fifteenth century, the persistent medieval concepts began to give way to a more accurate depiction of the ventricles. The transition is clearly illustrated by Leonardo da Vinci (1452–1519), whose drawing executed ca. 1490 showed the conventional three cerebral cells and their relation to the optic nerve (*see* Fig. 3.4, on p. 30). About fifteen years later, he again drew the ventricles (of the ox) after injecting them with molten wax and removing the brain tissue when the wax had hardened, leaving a cast showing the true relationships (*see* Fig. 3.5, on p. 31). In what has been deemed a "most brilliantly innovative fashion . . . he discovered that the third ventricle did

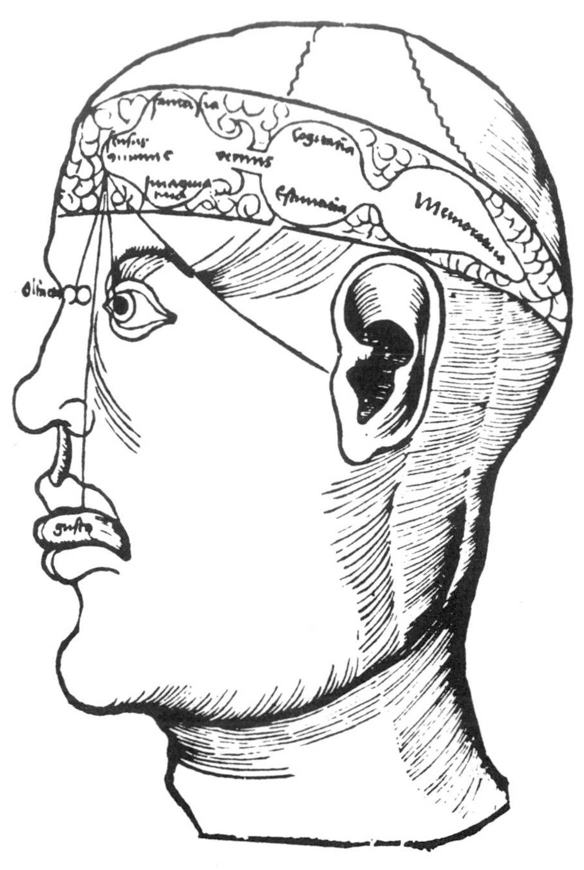

Fig. 3.3. An example of the durability of ideas, this popular drawing appeared in 1503 in Gregor Reisch's encyclopedia of grammar, science, and philosophy. Depicted are the three cells or ventricles containing the mental faculties: sensation, fantasy, imagination; cognition, judgment; and memory as had been suggested 11 centuries earlier. The cells are surrounded by what might be interpreted as convolutions. (From Reisch, 1503, Liber X, Tractus II, Fig. 18.)

indeed lie within the purview of the site for a *senso comune* inductively derived from his observations. . . . Wax casting was no accidental discovery. It was an extraordinarily inventive problem-solving response to a very real need at the time" (Todd, 1983, pp. 94, 95).

The origin of modern anatomy is traced to the midsixteenth century when Andreas Vesalius (1514–1564; *see* Fig. 3.6, on p. 32) was a dominant humanistic figure. An enthusiastic teacher, he supplemented his anatomy lectures at the University of Padua with demonstrations of human and animal dissections. To illustrate his publication describing those dissections, *De humani corporis fabrica libri septum* (1543), "an unprecedented blending of scientific exposition, art, and topography" (O'Malley, 1964, p. 139), he added drawings

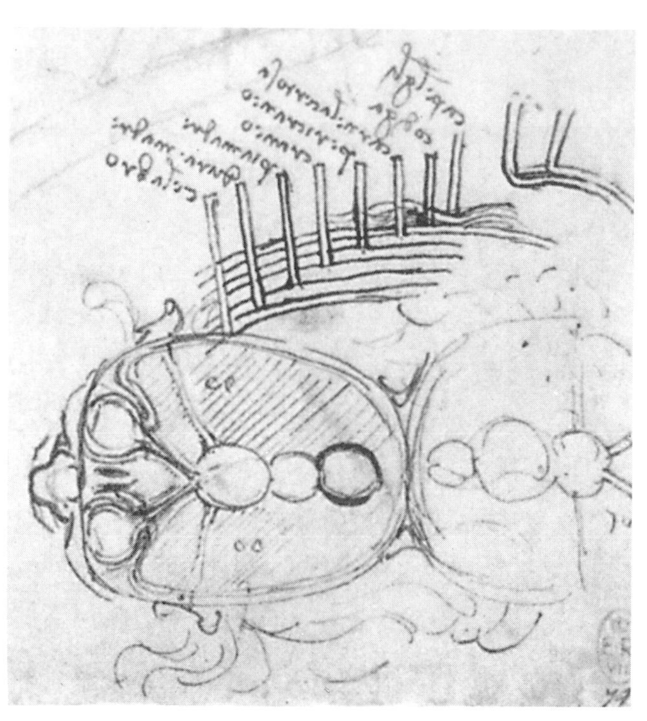

Fig. 3.4. Leonardo da Vinci's early drawing of the ventricles, executed about 1490 and now in the Windsor Castle Royal Library, shows the usual three connecting cells and their relation to the eyes. This configuration conformed to the accepted teachings of the Middle Ages. Todd (1983, 61 passim) provides a fascinating discussion of Leonardo's "mirror writing" and the important drawing of which this is a detail.

probably commissioned from draftsmen in Titian's workshop (ibid., p. 124; *see* Fig. 3.7, on p. 34). The cut wooden blocks were sent to Switzerland for printing under his supervision by a master scholar-publisher. Vesalius opposed the medieval theory that the soul resides in the ventricles, believing these to be mere channels for the passage of the fluid "anima" (*see* Fig. 3.8, on p. 35), and explicitly identified the brain as the main organ of intelligence, movement, and sensation.

One hundred years after the *Fabrica*, a role for the ventricles in formation and circulation of cerebrospinal fluid was proposed by the English neurologist, Thomas Willis (1621–1675; *see* Chapter 4). Striving to assign a function to the ventricles, which "result accidentally from the folding of the brain," Willis suggested that the animal spirits originate in the choroid plexuses of the ventricles which in addition serve as a conservator of heat from the blood, whereas "*the moderns* consider these places as vile and assert them to be merely sewers for the carrying away of excreted

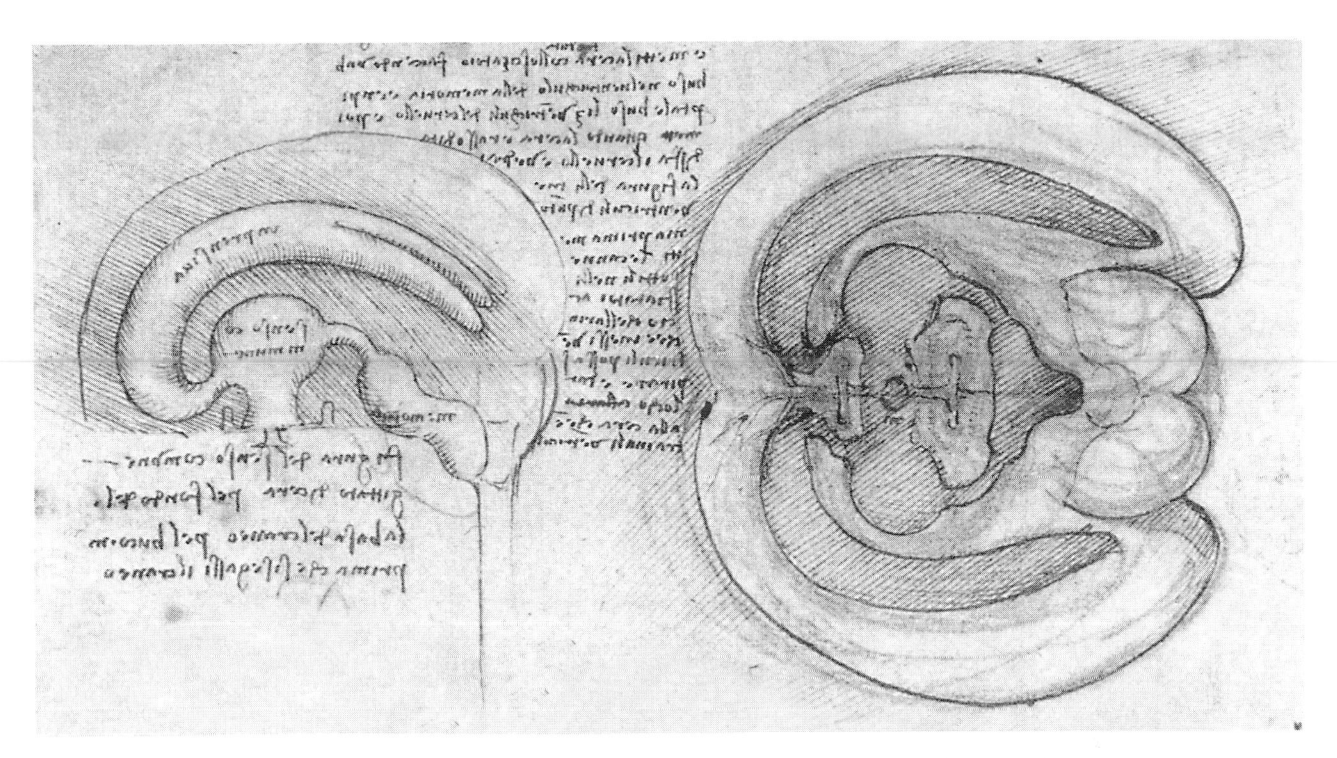

Fig. 3.5. Leonardo's later drawing, made ca. 1505 after he had injected molten wax into the ventricles of ox brains to produce casts shows the true anatomical relationship. Not willing to entirely discard the medieval doctrine of three cells, he labeled the ventricles in reverse writing (*see* Fig. 3.4) with their mental functions: perception, association, memory. (Clarke and Dewhurst, 1972, p. 51.)

matter" (1664, p. 74; translated in Clarke and O'Malley, 1968, pp. 724–725).

At the time when Willis wrote, the novel ideas about the world and its creatures propounded by the French mathematician and philosopher, René Descartes (1596–1650; *see* Fig. 3.9, on p. 36) were in full debate. Descartes's mechanistic explanation of all natural phenemona had created a scientific atmosphere in which either support or attack was required. Willis and most of his friends rejected the mechanistic doctrine and later he and Descartes were scolded by Steno (*see* Chapter 4) for accepting too many of the old ideas without some degree of skepticism.

One of those ideas centered around the pineal gland, which Descartes, just a year before he died in Sweden, had positioned within the ventricles:

> I seem to have plainly ascertained that the part of the body in which the soul immediately exercises its functions is . . . a certain extremely small gland, situated in the middle of the [brain] substance, and so suspended above the duct by which the spirits of its anterior cavities are in communication with those of the posterior that its slightest movements can greatly alter the course of these spirits and conversely the course of the spirits can greatly alter the movements of the gland (1649, Article 312, translated in Voss, 1989, p. 36).

In 1662, Descartes's major work was published posthumously in Latin and in French two years later. In *De homine (L'homme)*, he elaborated the mechanistic concept of human sensation and motion and retained the ancient idea of the soul (animal spirits) residing in the pineal gland, crudely depicting it from antecedent drawings rather than direct observation (*see* Fig. 3.10, on p. 37).

The belief that the soul resides in the ventricles ("sensorium commune"), and that the fluid in them can be animated, prevailed into the end of the eighteenth century; for example, it was upheld by the German *Naturphilosoph*, Samuel Thomas Sömmerring (1796; *see* Riese, 1946, for discussion). In fact, the idea that the soul could be anchored in a specific region was challenged by contemporary thinkers, notably Goethe and Kant among others,

Fig. 3.6. The work of the early anatomists culminated brilliantly with the teachings of Andreas Vesalius of Brussels. *De humani corporis fabrica libri septem*, first published in 1543, represented his conclusions from actual dissections carried out in Belgium, France, and Italy, and not only corrected Galen in many details but shifted the seat of brain function from the ventricles to the tissue surrounding them. This frontispiece in *Fabrica* is considered to be the only known authentic likeness of Vesalius (*see* O'Malley, 1964, p. 147).

and was gradually dispelled by nineteenth-century materialism. Having originated in the fifth or sixth century BC, "It must, therefore be one of the longest surviving aspects of biological thought," as Clarke and Dewhurst noted (1972, p. 85).

THE CEREBROSPINAL FLUID

The earliest writing specifically on the cerebrospinal fluid was by a Swedish "mystic," Emanuel Swedenborg (1688–1772). According to his translator and annotator, Swedenborg, in about 1736 or 1737, believed that blood from the common carotid and vertebral arteries mixed with the spirits from the brain in the choroid plexuses, "like a marriage union with its spouse" (Swedenborg, 1882, vol. I, p. 611). The passage continued: "The newly-born moisture . . . is discharged . . . through an equally great number of little emissary ducts and arteries . . . thus by little mouths it is distilled thence into the cavities of the ventricles" (ibid., p. 612). Swedenborg's translator, R. L. Tafel, actively championed his countryman: "No one, indeed, has poured such floods of light on the functions of the human body, and especially on those of the nervous system, as Swedenborg has done" (ibid., vol. II, p. vii). Tafel's two-volume tour de force consisted of quotations of all earlier, ancient, and modern, authors on a given subject, followed by Tafel's analysis, and then by his translation of Swedenborg's text on the same subject. Tafel pointed out that the cerebrospinal fluid was rediscovered by Cotugno in 1764, by Magendie in 1825, and by Key and Retzius in 1875.

A broader view was held by the Neapolitan physician, Domenico Felice Antonio Cotugno (1736–1822), which he described in *De ischiade nervosa commentarius:* "Whatever space exists between the sheath of dura mater and the spinal marrow, it is always *filled* . . . by a *water* similar to that which . . . fills the ventricular cavities of the brain, the labyrinth of the ear, or . . . the other cavities of the body to which the air has no entry" (1764, p. 12; translation from Clarke and O'Malley, 1968, p. 729). Cotugno observed that this fluid had not been noted by previous anatomists because of the "ridiculous method" usually followed in dissection, which was to cut off the head, allowing the fluid to drain away. He further declared: "It seems beyond all possible doubt that the spinal fluid, as well as that which humectifies all the other cavities of the body, constantly oozes from the extremities of the smallest arteries and, finally, is absorbed through very small inhaling veins, so that there is a continual state of renovation" (ibid., p. 731). As Clarke and O'Malley observed, although Cotugno seems not to have viewed the blood circulatory system as a closed circuit, he was right in assuming that the cerebrospinal fluid is continuously formed and drained away.

The cerebrospinal fluid claimed the attention of one of France's great physiologists of the early nineteenth century, François Jean Magendie (1783–1855), from his first mémoire on the subject, in 1825, to a final summarizing monograph in 1842. The most important of his contributions in this field include the discovery of the passageway between the fourth ventricle and the subarachnoidal space (the foramen of Magendie), the flow of fluid through it, which he named and characterized as a "natural fluid of the body" so useful that "it must be given first place in a list of these fluids" (Magendie, 1827, p. 79; translated in Clarke and O'Malley, 1968, p. 733), and his suggestion that secretion from the choroid plexuses is the chief agent of its formation. So great was Magendie's name in physiology that his *Précis élémentaire*. . . of 1824, a compendium for students, was published in Philadelphia in English the same year.

Ernst Faivre, in his dissertation (1857) at the University of Paris, attempted the first estimate of the surface area of the choroid plexus in man. Measuring only the lateral ventricles, Faivre reported that they averaged in man, 0.07 cm long and 2 cm wide, and consequently have a surface of 14 cm. Almost a century later, direct measurements of the plexuses were carried out by E. Voetmann in Copenhagen. He found (1949) that the epithelial cells of the plexus can be quantitatively detached from the underlying connective tissue and their cell size and density measured to yield an average 213 cm^2 in brains of adult male humans. As Voetmann pointed out, Faivre's calculations were inexplicably incorrect by an order of magnitude, and "My work, which *per se* is only a quantitative anatomical investigation, thus seems . . . additional evidence in support of the theory of secretion" (ibid., p. 54).

The question of the fate of the cerebrospinal fluid was addressed by two Swedish anatomists, Ernst Axel Hendrik Key (1832–1901) and Gustav Magnus Retzius (1842–1919). They provided the best evidence to that time (1875–1876) for drainage of the cerebrospinal fluid through the subarachnoidal villi, structures that had been described in 1705 by Antonio Pacchioni of Rome.

The origin and circulation of the cerebrospinal fluid were of vital importance to modern neurosurgeons, and two groups in the United States undertook experiments aimed at clarification of some of the existing uncertainties. At Johns Hopkins, Walter Edward Dandy (1886–1946) blocked flow through the cerebral ventricles of dogs and a condition of hydrocephalus developed, which did not appear if the choroid plexuses had been removed previously (Dandy and Blackfin, 1914). Believing he had entered into "a virgin experimental field" (Dandy, 1919, p. 129), although not entirely correct, Dandy wrote with emphasis: "From these experiments *we have the only absolute proof that cerebrospinal fluid is formed from the choroid plexus. Simultaneously it is proven that the ependyma lining the ventricles is not concerned in the production of cerebrospinal fluid*" (ibid., p. 134). From the experience gained from his experimental work, Dandy introduced diagnostic ventriculography to localize "intracranial affections" (1918, p. 5) by partial replacement of cerebrospinal fluid with air, a procedure that might be considered a latter-day extension of Da Vinci's substitution of molten wax for fluid.

That same year, similar experiments were carried out independently in the division of surgery of the University of Pennsylvania by Frazier and Peet (1914) and they wrote "It is now generally accepted that the cerebrospinal fluid is secreted by the choroid plexus of the various ventricles" (ibid., p. 269). Using chemical irritation to produce an obstructing inflammation or by injection of a thick substance, they confirmed that the choroid plexus is the major source, and with the dye phenolsulfonphthalein showed that absorption is largely by venous channels. Notwithstanding those conclusive experimental results and the corroborating clinical evidence regarding the formation of the cerebrospinal fluid, skepticism and uncertainty about its formation and physiology remained.

TWO "BARRIER" SYSTEMS: THE BLOOD–BRAIN BARRIER

Knowledge of the source of continuous renewal of the cerebrospinal fluid from the choroid plexuses and its drainage by way of the subarachnoid villi to the venous sinuses and hence into the systemic circulation did not solve the problem of why some substances in blood plasma were so slow to appear in the fluid. The wide difference in their concentrations in blood plasma and spinal fluid was epitomized by the puzzling distribution of the new industrial dyes when introduced into the bloodstream of experimental animals. An offshoot of the ascendancy of the dye industry in Germany at the close of the nineteenth century, that experimental work led to a massive, inconclusive body of medical literature. Conflicting results and misinterpretations abounded, not only because of uncertainties of bonding of dyes with proteins or other molecules, but also because of ignorance of the metabolic fate of dyes in tissues, and unreliable measurements, to mention a few obstacles (*see* Bradbury, 1979).

ANDREAE VESALII BRVXELLENSIS, INVIctissimi CAROLI V. Imperatoris medici, de Humani corporis fabrica Libri septem.

CVM CAESAREAE Maiest. Galliarum Regis, ac Senatus Veneti gratia & priuilegio, ut in diplomatis eorundem continetur.

BASILEAE, PER IOANNEM OPORINVM.

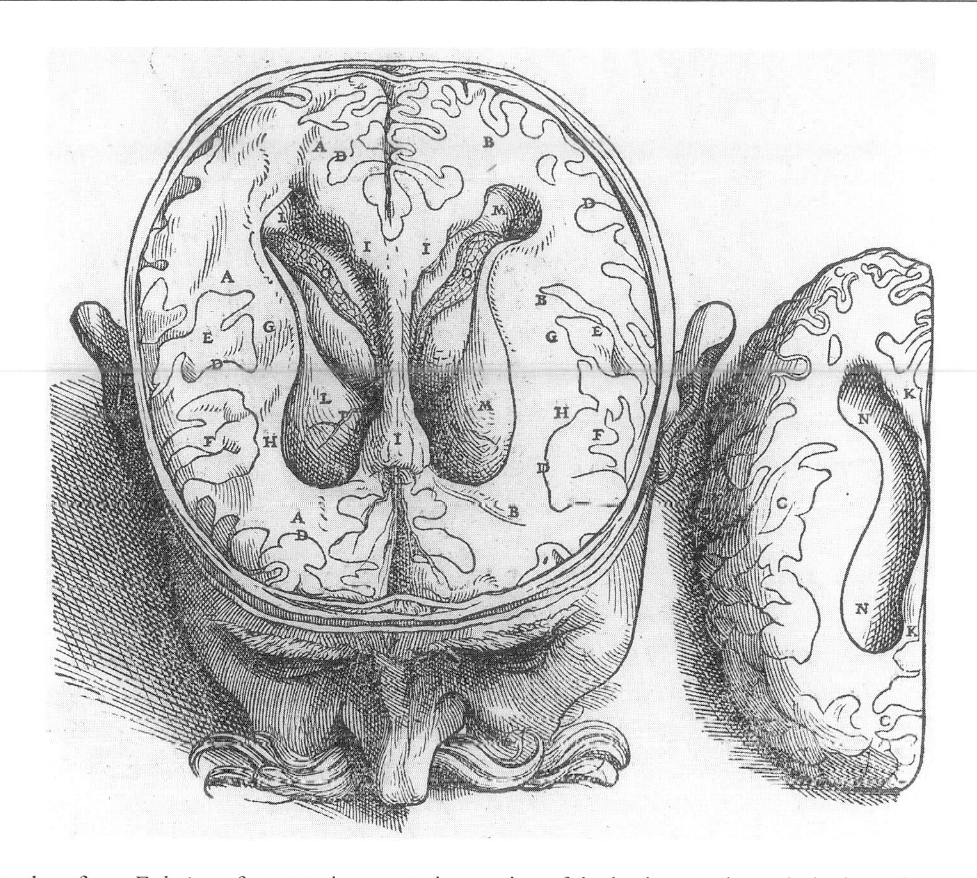

Fig. 3.8. This woodcut from *Fabrica* of an anterior posterior section of the brain cuts through the lateral ventricles (L, L, M, M) and exposes the choroid plexuses (o, o). It illustrates the detailed directions for the dissection, a practice found throughout the volume. (From Vesalius, 1543, Liber VII, Fig. IV, p. 608.) Vesalius had "the great negative virtue of refusing to see things through the eyes of another, and thus often avoids seeing the non-existent" (Singer, 1952, p. xvii).

The difficulties commenced with the observations of Paul Ehrlich (1854–1915), prominent German bacteriologist and discoverer of Salvarsan for the treatment of syphilis; with his associates, he devoted 25 years to the study of the relation of the chemical constitution of compounds, especially dyes, to their distribution and action. Ehrlich attributed staining inequalities between neural and nonneural tissues to differences in affinities of a dye for the respective tissues, an explanation that did not require the assumption of a special barrier. For example, although bilirubin is widely distributed, the brain tissue of patients with jaundice is not yellow because the brain lacks the proteins that attract the dye. In an address delivered in 1898, Ehrlich declared (1906, p. 420): "According to the views held by the majority of investigators and also by me this localization in certain organs depends in every instance on causes within the tissues and not on the vascular distribution."

The concept of a specialized barrier that isolates the cerebrospinal fluid from blood plasma was first clearly enunciated by Edwin E. Goldmann, a South African-German. In 1909, he showed that trypan blue, an acidic large-molecule vital dye, when administered intravenously stains the capillary-rich choroid plexuses but does not enter the cerebrospinal fluid, which he considered the carrier of nutrients and metabolic products respectively to and from the nerve cells. The corollary experiment—to inject the dye into the cerebrospinal fluid (Goldmann, 1913)—produces heavy staining of the brain. The only tenable answer is the presence of a barrier between the dye carried in the cerebral capillaries and brain tissue that is not present in the choroid plexuses, where "all substances" appear to

Fig. 3.7. *(opposite page)* The title page of *Fabrica*, or "Seven books on the structure of the human body," is a dynamic scene widely reproduced in the literature of medical history. It is discussed in detail by the modern expert on Vesalius's life and work, Charles Donald O'Malley (1907–1970).

Fig. 3.9. The only portrait of René Descartes known to have been painted by Franz Hals is an oil sketch dated 1649, in Copenhagen; the later copy reproduced above hangs in the Louvre, a measure of the great esteem held for Descartes during his lifetime as well as today.

have passage (*see* Fig. 3.11, on p. 38). Goldmann, with a pharmacologist's intuition, forecasted that: "Every practical attempt to affect the diseases of the central nervous system *chemotherapeutically* will have to rely upon the greater or lesser permeabil-ity of the epithelium of the plexus. . . ." (ibid., pp. 54–55; translated in Clarke and O'Malley, 1968, p. 750).

Goldmann's prediction was underscored two decades later by two other German physicians who studied the relation of some infectious diseases to the blood–brain barrier. On the basis of the results of their own experiments and those reported in the literature, Friedmann and Elkeles (1932), in a rather literal translation of their report, wrote: "[I]t can be stated that toxins differ among themselves in the facility with which they pass the blood-brain barrier" (1934, p. 723). Cobra venom and lamb-dysentery toxin pass, and the authors localized the barrier in the walls of the cerebral capillaries. They also noted that:

> There is another quality according to which the toxins under investigation fall into the same groups according to their ability to pass the B.B.B. All toxins which were shown to pass the B.B.B. are characterized by the lack of an incubation period and the rapidity of their action. Amongst the toxins investigated the ability to pass the blood-brain barrier is correlated to electrical charge. Diphtheria, tetanus and botulism toxins which do not pass . . . carry a negative charge at the pH of blood (ibid., pp. 724, 777).

The work of Goldmann with trypan blue was largely unknown until taken up, again in Germany, by H. Spatz (1933). He noted that various dyes injected intravenously not only do not stain the cerebral tissue, but also do not color the capillary wall, and so he argued that the cerebral capillary endothelium is the "essential" block to penetration of dye into the brain tissue. This lining of the fine cerebral vasculature was known to be a continuous single layer of cuboidal cells with closely adhering luminal and basal contacts to each other in tight junctions and joined by basal interdigitations between them. By midcentury, the "haemato-encephalic barrier" was generally accepted (Wislocki and Leduc, 1952) to consist of the capillary endothelium, the "chorioid" plexuses, meninges, and special regions in the brain, each with a different role in the formation and fate of the cerebrospinal fluid. Wislocki and his associates, from the results of their experiments showing differences in the penetrations of trypan blue and silver nitrate, suggested the possibility that there exists "a succession of thresholds at different levels of which substances could be held up depending upon the interplay of physical and chemical factors pertaining to both substance and threshold" (ibid., p. 389).

There was no general agreement, however, about the existence of cerebral extracellular spaces, let alone their possible significance for cerebrospinal fluid and neuronal dynamics. An English review (Woollans and Millen, 1954) attempted to sort out the conflicting histological evidence that the perivascular space in the central nervous system is artifactual—during histological processing the neuron shrinks more than its glial and capillary envelope. The authors argued that the neuronal ground substance has an important role and concluded that any real space existing around the neurons or cerebral capillaries has no relevance to the cerebrospinal fluid, but serves only in a physical

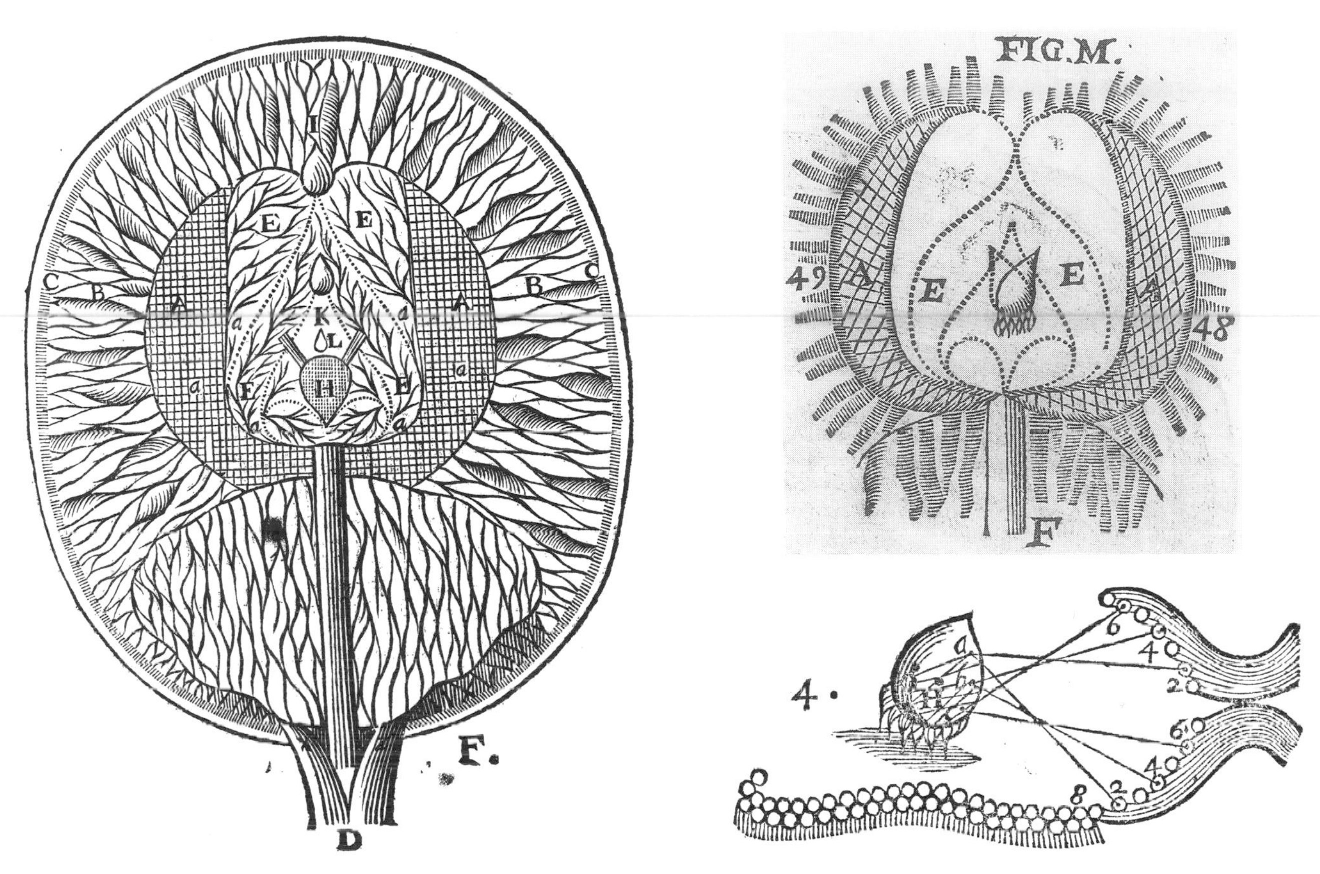

Fig. 3.10. Left—Descartes's diagram of the human brain showing the pineal gland (H) placed in the ventricles (EEEE), which can drain through I to the nose or into the nasopharynx through K and L. A, A represent the nerve tubes in cross section which are drawn longitudinally in B, B. The unlabeled cerebellum is at the bottom. (From the 1664 French edition, p. 62.) Right above—Another drawing illustrating how the pineal gland may turn "here and there" in the ventricles (E, E) (ibid., p. 64). Right below—Connections between the ventricular wall openings of the uncrossed optic nerve tubules and the pineal gland (H) (ibid., p. 65).

capacity as "cushions between the expansile vessels and the nerve cells" (ibid., p. 281).

Investigation of the anatomy of the blood–brain and cerebrospinal–brain barriers continued to hold center stage during midcentury, impelled by the introduction of a new tool, the electron microscope. The early submicroscopic visualizations of barrier elements (Van Breeman and Clemente, 1955; Maynard and Pease, 1955; Dempsey and Wislocki, 1955) revealed the "striking" proximity of neurons, glia, and glial processes and the close investment of brain capillaries by glia, precluding the presence of connective tissue as in "leaky" capillary walls of nonneural tissues (*see* Chapter 7 for further discussion of glia). The microscopists were groping for functional insights: "The anatomical relations strongly suggest that the astrocytes may serve as the principal transport system. . . . The expanded pericapillary feet might be expected to assist exchange with the blood. The other processes, pervasive in the extreme, reach all other parts of the neuropile" (Schultz, Maynard, and Pease, 1957, p. 384). As shown in Fig. 3.12, this concept envisioned an endfoot of the astrocyte in proximity to each of the fluid spaces of the two transfer systems. "An active homeostatic mechanism protecting the brain from acute changes in volume, operating continuously across the astrocytic-vascular membrane is suggested" (De Robertis and Gerschenfeld, 1961, pp. 60–61).

In another direction, the importance of the process of diffusion across the extracellular rather than

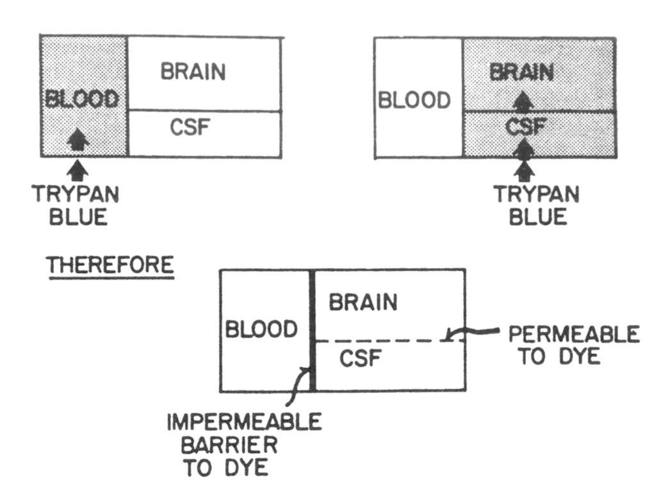

Fig. 3.11. Diagram of E. E. Goldmann's experiments in 1909 and 1913 which demonstrated the isolation of the cerebrospinal fluid from the blood plasma—the blood–brain barrier. (From Bradbury, 1972, p. 1025, Fig. 28.2.)

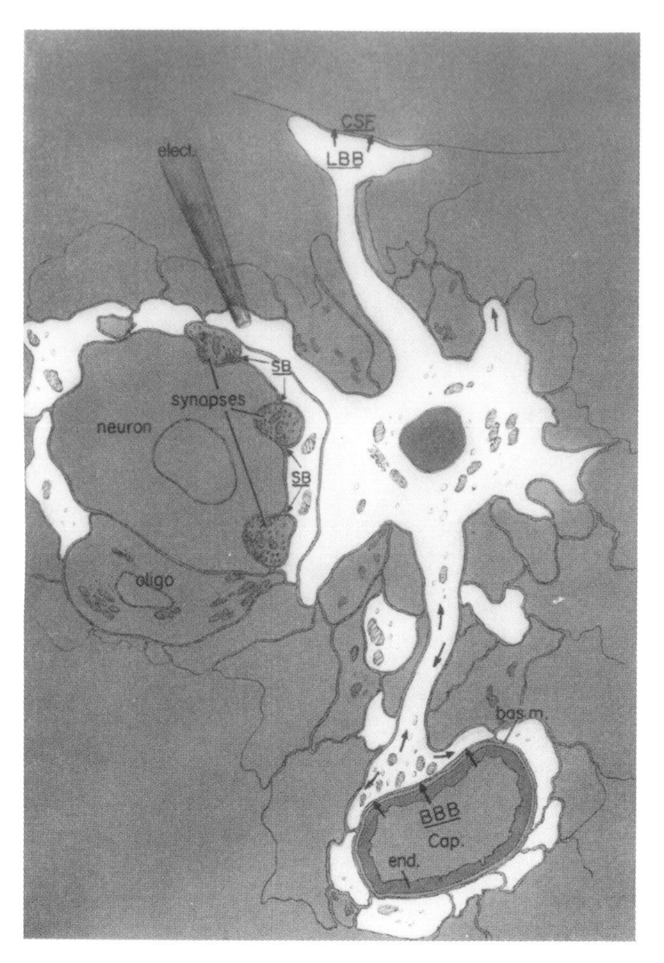

Fig. 3.12. A schema of the topographical relationships of an astrocyte in the gray matter. An endfoot surrounding a capillary (Cap.), endothelium (end.), and basement membrane (bas.m.) serves as the blood–brain barrier (BBB); SB—synaptic barrier, LBB—liquor–brain barrier denoting site of formation of the cerebrospinal fluid. (From De Robertis and Gerschenfeld, 1961, p. 17, Fig. 8; also in Davson, 1967, facing p. 97 and Bradbury, 1979, p. 9.)

by way of glia, was made clear by two neurophysiologists, the Hungarian-American, Stephen W. Kuffler and John G. Nicholls. In elegant perfusion experiments on leech single units, they showed (summarized in 1966) that diffusion could account for movement of solutes within the central nervous system including its maximal needs during neuronal activity. In their review, the authors present a diagram (Fig. 3.13) of structures then known to be involved in exchange of substances between vertebrate neurons and blood and cerebrospinal fluid.

Actual diffusion distances were calculated by Oldendorf and Davson (1967) based on data of sucrose transfer that focused on the role played by the extracellular space in brain tissue. That space is important because, as they pointed out, a steady state, not an equilibrium across membranes, is established with the cerebrospinal fluid acting as a sink. They calculated that "no point in human brain is more than about 2 cm from an ependymal or pial surface, most is within 1 cm, and most gray matter is within a few millimeters. This arrangement may be an indication of the distances over which extracellular diffusion transfer is effective" (ibid., p. 204). Those distances were later deemed a functional overstatement, however (Bradbury, 1979, p. 34), because the perivascular spaces presumably were not taken into account.

After the early application of the electron microscope to depiction of the substructures of the blood–brain and cerebrospinal fluid–brain barriers, higher resolution photomicrographs yielded increasing evidence of tight junctions between the ependymal cells making up the cerebral endothelium. With the evidence obtained in studies of various vertebrate species showing that juxtapositioned endothelial cells in general may have a girdle of attached surfaces at the luminal end of the junction (Bennett, Luft, and Hampton, 1959), the stage was set for the continued race to higher magnifications. Muir and Peters (1962) predicted that the five-membrane junctions they could see in various tissues, including the blood–brain barrier, would be found wherever a "sheet of cells" lies at the interface of

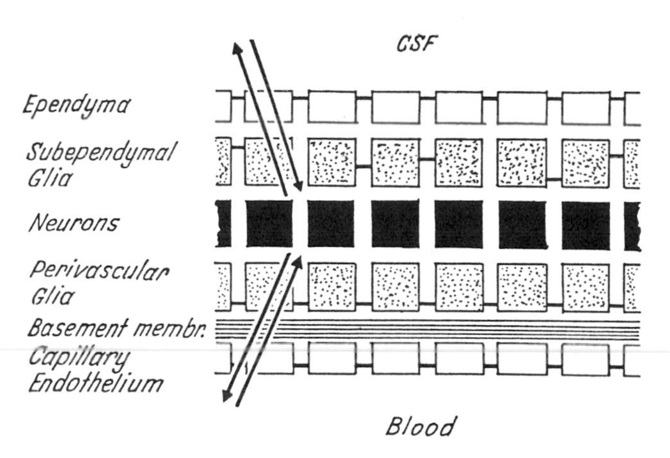

Fig. 3.13. A schematic drawing of the passageways involved in movement of solutes between neurons and the blood or cerebrospinal fluid of vertebrates, as known in the midtwentieth century. Short horizontal bars—tight junctions. (From Kuffler and Nicholls, 1966, p. 51, Fig. 27.)

differently constituted fluids, preventing the intercellular diffusion of ions and small molecules.

The fine details of the intercellular junctions in brain ependyma were worked out by Brightman and Palay (1963), who could see the five membranes to be so closely packed that the intercellular space in the "tight" zone obliterans, named by Farquhar and Palade (1963), was obliterated. The final proof of the impermeability of tight junctions, indeed of their very existence (*see* Crone, 1986), was provided by Reese and Karnovsky (1967). After systemic injection, horseradish peroxidase did not penetrate into brain tissue, whereas the enzyme passed easily through cardiac and skeletal muscle endothelium.

QUANTITATIVE PHYSIOLOGIC DATA

The advent in the midtwentieth century of a second indispensable tool, the radioisotope, into biomedical research was nowhere more significant than in the investigation of functional relationships between neuronal vital processes and the surrounding milieu. In a fresh physiological approach to the problem, in contrast to the "bewilderingly large . . . and not very informative" (Davson, 1955, p. 114) dye studies, the new markers provided precise analysis of the composition of the cerebrospinal fluid and could be relied on as indicative of function. The breach in the morphologists' and surgeons' dominance of cerebral ventricular studies was heralded by a publication of the British physiologist Hugh Davson, *Physiology of the Ocular and Cerebrospinal Fluids* (1956). Written in an acknowledged effort to balance the overwhelming abundance of structural studies that had focused research interest on neurons and neuroglia and inspired by the work of Ramón y Cajal and Rio del Hortega (*see* Chapter 7), the monograph is a combination of theoretical analysis, handbook, and review. Davson's mathematical treatment of the kinetics of solute transfer brought cerebral intracellular and extracellular fluids to the attention of biophysicists and general physiologists.

Davson had shown conclusively in 1955 that the concentrations of isotopes of Na, K, chloride, and other small molecules in rabbit cerebrospinal fluid did not match those in blood plasma, and therefore the formation of cerebrospinal fluid is not a simple filtration, but, rather, is due to an active transport of substances. Two decades of investigations by general physiologists was the foundation on which such a process could be demonstrated in the blood–brain barrier (Davson, 1967). A Canadian zoologist, Edward H. Craigie, in a comparative study of the cerebral capillaries in lower vertebrates, found a great deal of variation in the pattern of their embryonic development in different neuropils and wondered "why areas of neuropil with very few cell bodies are so frequently poorer in capillaries than areas with many cell bodies" (Craigie, 1938, p. 17). A possible explanation was provided by the work of the Viennese-American, Ernst Scharrer. Scharrer showed a direct relation between capillary density as outlined by intravenously injected India ink and the number of mitochondria present in the brains of some bony fishes, alligator, and opossum. Based on the large number of mitochondria found in highly vascular neuropil, Scharrer concluded: "Since mitochondria are carriers of respiratory enzymes, this relationship is probably significant and supports the view that capillary density indicates differences in local metabolic activity of the brain tissue" (1945, pp. 241–242).

Somewhat the same line of thinking led Vates, Bonting, and Oppelt (1964) to examine the role of respiratory enzymes in the choroid plexus of cats by observing the effects of inhibitors and activators of adenosine triphosphatase on the flow of cerebrospinal fluid from the aqueduct of Silvius. They concluded that the Na,K-activated ATPase system is of primary importance in cerebrospinal fluid formation, "presumably through the active secretion of Na ions into the ventricle" (ibid., p. 405). A decade later, supporting evidence was forthcoming in studies carried out by the American radiologist, William H. Oldendorf (1926–1993), which showed a comparative absence of mitochondria in the relatively permeable capillaries of nonneural tissues: "If the brain capillary wall performs an energy-dependent excretory function as a substantial part of its total metabolic work, it might be expected that brain capillary endothelial cells would contain more mitochondria than a tissue with non-specifically permeable capillaries, such as skeletal muscle" (Oldendorf and Brown, 1975, p. 736). Comparing counts of mitochondria in randomly selected electron micrographs of rat cerebellum and vastus medialis muscle, the hypothesis was confirmed, as shown in Fig. 3.14.

As already indicated, one of the most significant arguments against the primary role of the cerebrospinal fluid in transfer of substances to brain tissue is the slowness of their uptake and small concentration, compared with the situation in blood, a dynamic that is clearly inadequate to accommodate the high and rapid metabolism characteristic of brain tissue. Christian Crone in Stockholm had hypothesized that "facilitated diffusion" could account for the 25% of the normal blood glucose that crosses the blood–brain barrier (Crone, 1960), and that 10% leaves in a single passage through the brain of dogs (Crone, 1965), in contrast to only 3% to 5% of administered fructose glycerol. Those findings indicated the existence of a mechanism for carrier transport of glucose, but gave no evidence of the site at which it occurs. The author skirted any suggestion of a barrier and believed it "most reasonable to consider the possibility that the blood–brain barrier permeability reflects the permeability of the endothelial cells proper—in other words, that the slow exchange of material in the brain is due to low permeability of the cerebral capillaries" (ibid., p. 416).

The enigma of the delicate mechanisms responsible for the exquisite adjustments of the blood–brain and cerebrospinal–brain barriers to the needs of cerebral neurons for homeostasis simultaneously with fast transport has not yet been solved. Nevertheless, a teleologic argument has been made for their existence. Recalling the work of Craigie (1955) showing that ectodermal and mesodermal elements from the neural tube sur-

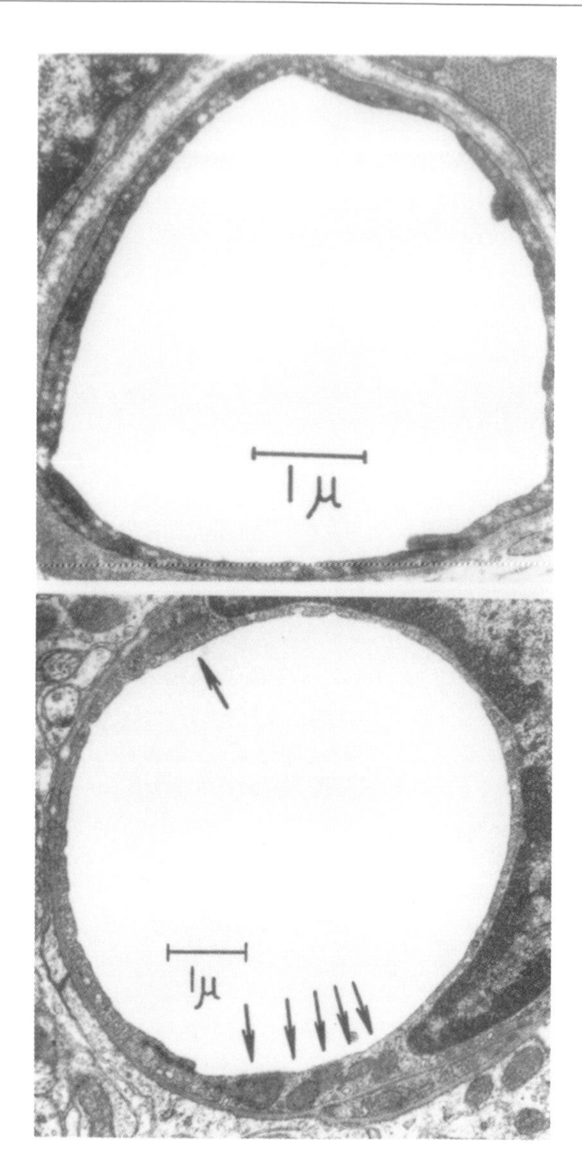

Fig. 3.14. Electron micrographs of rat capillaries. Above—In the vastus medialis muscle; note absence of mitochondria. Below—In cerebellum, with six mitochondria. (From Oldendorf and Brown, 1975, p. 737, Fig. 2.)

round the mature brain's larger blood vessels, M. W. B. Bradbury proposed in 1975 that the blood–brain barrier should be viewed as forming the interface between avascular ectodermal tissue invaded by mesodermal blood vessels. As he wrote subsequently:

> Ontogeny repeats phylogeny in that vascularization of the CNS anlage takes place in two main steps. . . . Firstly a superficial network of sinusoids surrounds the brain vescicles and the neural tube; so-called external vascularization. Subsequently sprouts from these vessels invade

the parenchyma by growing initially inwards and then dichotomizing to give branches running parallel to the surface—internal vascularization. . . . The blood-brain barrier itself may be regarded as an expression of this intimate connection between mesoderm and ectoderm (Bradbury, 1979, pp. 17, 18).

Another view considered the brain as an organ isolated from the other body organs by virtue of separate formations and circulations of blood plasma and cerebrospinal fluid. By applying a comparative approach, the phylogenetic advantage of the permeability modifications possible in a fluid isolated from plasma was pointed out by Cserr and Bundgaard (1984), who suggested that adaptive demands for increasingly complex nervous activity created a cerebral structure capable of assuring both equilibrium (homeostasis) and incredibly rapid adjustments as an advantage in the struggle for survival. Figure 3.15 reproduces their schematic diagram of the complex compartmentation of the fluid spaces in which the adjustment processes are thought to occur, as they are envisioned toward the end of the twentieth century. Comparison of this figure with Figs. 3.11 and 3.13 representing what was known at the beginning and middle of the century, respectively, demonstrates schematically the progress in our understanding of a complex and vital mechanism.

The use of cultured endothelial cells from brain, "sealed together by continuous, tight junctions" and containing few pinocytotic vesicles, as a model presaged faster progress (Goldstein and Betz, 1983) in the future than had been possible in the preceding few decades. Another expectation was the perfection of microperfusion techniques applied to single cerebral capillaries, to match the achievements of the microelectrode in revealing the details of ionic transfer in single cells.

OVERVIEW OF THE VENTRICULAR "CELLS"

Looking back on some of the discoveries relevant to the human cerebral ventricles, the earliest, as recorded in Grecian writings, proposed that the "hollow place" in the brain is the storage site for an "animus" and that it is the seat of motor and mental

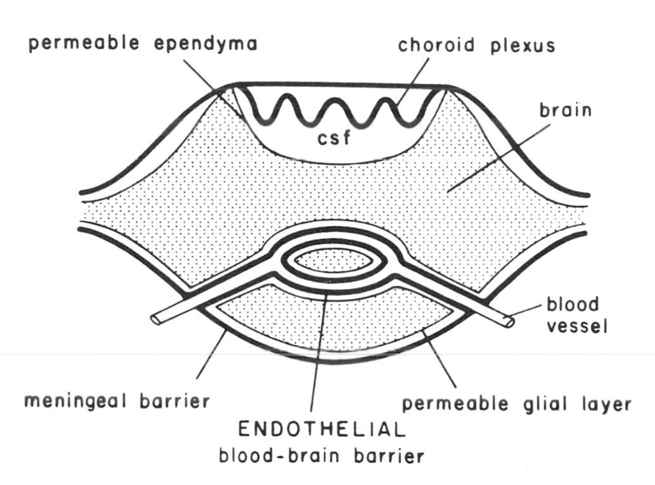

Fig. 3.15. Diagram of the dominant pattern in vertebrates of blood–brain and cerebrospinal fluid–brain interfaces as known in the late twentieth century. csf—Cerebrospinal fluid, heavy lines—selective barrier layers with tight junctions. (Adapted from Cserr and Bundgaard, 1984, R278, Fig. 1.)

activities. After Vesalius and Willis brought the animus ashore to reside in the brain substance, interest shifted to a structure thought to be within the ventricles—the pineal body. Descartes attributed to the pineal a central role in human sensation and action and by nature of his mathematical method and mechanistic interpretation of behavior, he introduced "one of the most powerful tools of all modern physiological research. This was the hypothetical model" (Crombie, 1959, p. 173). Although the role of the pineal gland may be far from what Descartes thought, he made the transition from the early philosophy to a new direction that was manifested by the Enlightenment.

The choroid plexuses within the ventricles became the center of inquiry during the nineteenth century with the realization that the cerebrospinal fluid is not a simple filtrate of blood plasma. At different times, three possible pathways between the ultimate neuronal microenvironment and blood plasma were identified and accepted as partial answers: passive diffusion, active transport, and secretion. With the substitution of reliable quantitative methods for the dyes that yielded at best only vague research data, a clearer understanding of how the delicate equilibrium of cerebral processes is sustained has become possible.

4 Surface Contours

Order or Chaos?

CONTENTS

This sulcus [of Rolando] . . . forms the surest point of departure for finding the convolutions.

(Ecker, 1873, p. 8)

The morphology of the mammalian spinal cord and brain-stem nuclei does not vary substantially among species except in size. In contrast, the surface of the cerebrum and cerebellum differ conspicuously from species to species and to a lesser degree among individuals of the same species. This variation in neocortical surface configuration is conferred by the arrangement of convolutions (gyri) separated by fissures (sulci). When it was recognized that the intelligent animal species possess complex brains with more convolutions and deeper sulci than do species of less intelligence, the question arose of how this came about. Was the convoluted folding a result of mechanical factors generated from fitting an enlarging cortical surface into a cranial box of relatively fixed size ("the most economical use of space"), or were there diverse functional associations that governed the directions in which the brain increased in size?

The German neurologist, Carl Wernicke (*see* Chapter 5), was among the first to address this question. In a seldom-remembered publication (1876), he asked whether the brain surface represents a complex of organs whose functions differ in extent and inner connections, "or can the arrangement of the convolutions be understood as merely an accidental result of mechanical growth relationships. . . . We . . . think that the development of a scientific phrenology is at hand, which will give great importance to deviations from the normal conventional pattern (Wernicke, 1876, p. 298; transl. by J. Ratclife and M. V. Anker).[1] The bulk of Wernicke's long paper consisted of a minute description of the primordial gyral system of the fetal human brain from which he conceded that he was unable to reach any conclusion about mankind's genealogy based on what he had learned from his studies of embryonic development.

The difficulties attending studies of contours in the human fetus was noted in 1890 by the Irish neurologist, Daniel John Cunningham (1850–1909), as being unduly complicated by the transitory nature of some nine furrows seen in the third to fifth month of gestation. The appearance and disappearance of these furrows, first reported by Meckel (1815), during ontogenesis made reliable measurements uncertain. Cunningham (1890) did not subscribe to the view held by some of his contemporaries that the occipital part of the brain is a secondary "budding off" from earlier structures.

[1]Wernicke here used the term "phrenology" in its broadest sense to mean separate mental faculties that are localized in different parts of the brain.

He claimed instead that the calcarine and parietal–occipital fissures, which appear early in primate embryogenesis, are not smoothed out in the succeeding period of "lack of harmony" between the rapid growth of the forebrain and the lagging skull and that consequently, what we may call cerebral tectonic forces come into play. The occurrence of those fissures that are permanent he attributed to a transitional phylogenetic period when the brains of quadrupeds evolved into the more complex organ of the primates.

Such mechanistic views recall the geologic analogy described in Chapter 1 of the primordial layering of the nervous system. As a result of the growth of the cerebral hemispheres within the confines of the skull, the only recourse for the expanding gyri is to turn downward at the occiput and then move forward along the base of the cranium, forming in aggregate the temporal lobe. Thus the temporal lobe—rather than the occipital—is actually the most caudal portion of the hemisphere of the brain, despite the fact that it incongruously lies beneath the frontal lobe in the anterior half of the cranium. This evolutionary configuration results in the third lower frontal convolution, identified at its posterior part as Broca's area, being contiguous with the first (upper) temporal convolution, both of them bordering directly on the most conspicuous fissure and the first-formed embryologically, the fissure of Sylvius.

Reinforcement of the mechanical notion was provided by microscopic observations (Bok, 1959) that revealed stretching and thinning of cortical layers at gyral crowns and compression and thickening in the depths of the sulci. More modern analysis, derived by modeling and with a "reasonable fit to available empirical data" (Prothero and Sundsten, 1984, p. 152), suggested that a gyral "window" exists through which all afferent and efferent fibers must pass, hereby casting a physical restriction on the number of nerve fibers that can occupy the space. The maximum area of the hypothetical window converts to a simulated brain of 2800 g (Welker, 1990).

THE CONVOLUTIONS OR GYRI

The coils of brain tissue that lie in what the ancients saw as a disorganized mass inside the human skull are believed to have been first noted by Praxagoras of Cos (ca. 300 BC) as convolutions or gyri (Steckerl, 1958, p. 55). Not long thereafter, Erasistratus (ca. 260 BC), is said to have recorded "an apt analogy between the appearance of the gyri and the coils of the small intestine seen when the abdomen is opened, a description used again and again by subsequent writers. [Erasistratus] also stated 'that man's brain is more convoluted than that of other animals because of his superior intellect' " (Clarke and O'Malley, 1968, p. 385). The opinions of the early Greeks regarding the formation of the cerebral convolutions as "a kind of excrescence put forth from the spinal cord" (Galen, 1854–1856, p. 561; translated in Clarke and O'Malley, 1968, p. 385) were scorned by Galen (*see* Chapter 1). His teleological interpretation of the subordination of form to function was popular and his tenets became accepted without questioning. From his nonhuman dissections, Galen made unwarranted inferences about human anatomy and expressed them as truths; ironically, his insistence on the importance of experimental studies was ignored and so some of his errors and speculations persisted without challenge into the Renaissance and beyond.

The spectacular end of the void in new knowledge of human anatomy after Galen was provided by Vesalius in the sixteenth century and included many discoveries pertaining to the nervous system. Vesalius complained that Galen had been too brief in describing the convolutions:

> . . . [H]e ought to have added that the convolutions reveal the great ingenuity of the Creator who formed them . . . for the nourishment of the substance of the brain. If the substance of the brain had been continuous without all that folding of membranous fibres, it would not have been sufficiently firm for the distribution through it of veins and arteries (Vesalius, 1543, p. 630; translated in Clarke and O'Malley, 1968, p. 386).

Vesalius's rough description of the convolutions (Fig. 4.1) heralded a gradual shift of interest from the fluid-filled ventricles to the brain substance surrounding them:

> [T]he gyri or convolutions . . . , which Erasistratus very nicely compared to the twistings of the thin intestines, are found with the same frequency over the entire surface of the brain. . . . I believe that they cannot be compared to any-

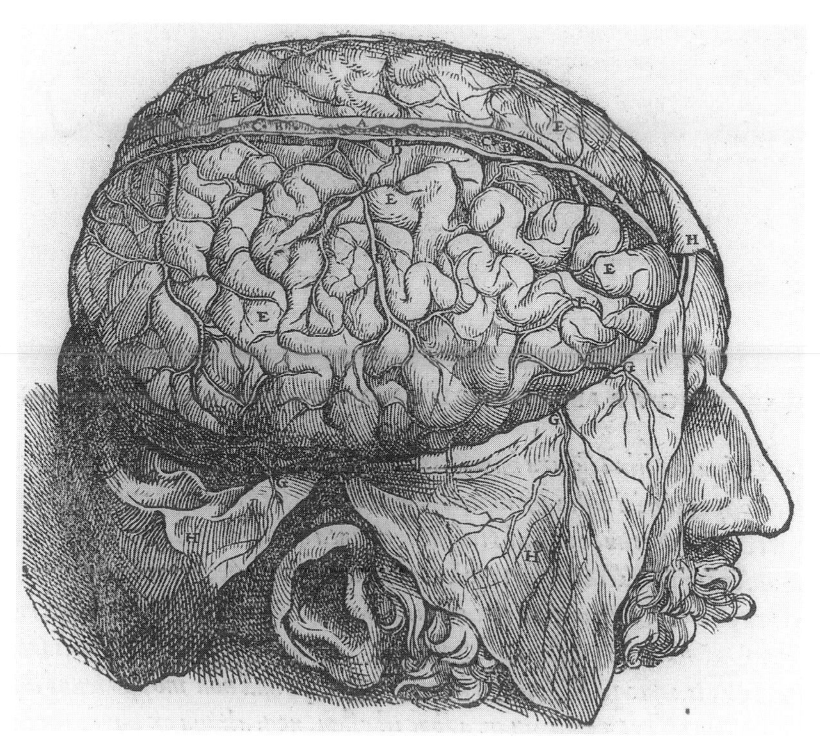

Fig. 4.1. The second drawing from a series by Vesalius depicting progressive dissections was published in *De humani corporis fabrica* (1543) and emphasized the intestine-like, random appearance of the convolutions. (From Liber VII, Tabula II, p. 606.)

> thing more happily than to clouds as they are usually delineated by either untrained art students or by schoolboys. . . . [T]here is nothing unusual about man's brain, and these convolutions appearing in its substance are also to be found in the brain of the ass, horse, ox, and other animals which I have examined (Vesalius, 1543, p. 130; translation in Clarke and O'Malley, 1968, p. 386).

Almost 150 years after publication of Vesalius's illustrations, the convolutions were still being depicted like coils of intestine without a pattern of arrangement. An extreme example was the 1684 illustration (Fig. 4.2) from Raymond de Vieussens (1641–1715), and like his predecessors, Vieussens did not assign a function to the convolutions. By the midseventeenth century, however, careful dissectors and observers had perceived something more than a chaotic mass of coils on the brain surface. Chief among them was the English physician, Thomas Willis (1621–1675; Fig. 4.3), whose pioneering and influential comparative studies of the gyri in many vertebrates, including man, drew attention to the brain's surface organization. The drawing in Willis's *Cerebri anatome* (1664) of the base of the human brain (Fig. 4.4) was executed by Christopher Wren, whose architectural prowess rendered an inferior naturalness compared with another illustration which appeared almost simultaneously in a second classic publication, *De homine* (1662) by Descartes (*see* Fig. 4.5, on p. 47), as Clarke and Dewhurst noted (1972, p. 68; 1996, p. 74).

Willis proposed a direct correlation between convolutional complexity and intelligence, as had Erasistratus, and maintained that memory is situated in the gyri. In his lectures, delivered about 1661–1662 after taking up his Sedeian professorship of natural philosophy at Oxford, Willis recognized that "movement is initiated in the cerebrum," and that "convolutions and gyrations . . . provide a more commodious area for [expansion of the animal spirits] in the use of memory and phantasy" (Dewhurst, 1980, pp. 55, 138). Most importantly, those lectures reinforced the belief that mental

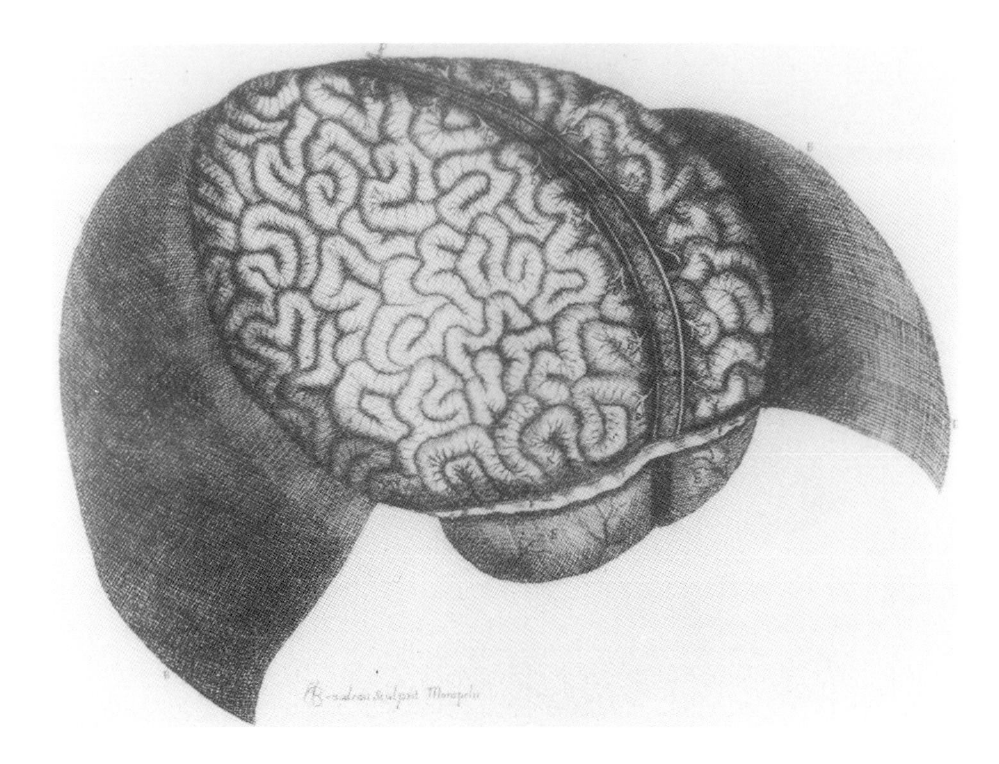

Fig. 4.2. This engraving by Beaudeau of Montpellier shows the meninges removed from the cerebrum and the cut ends of blood vessels. It is from *Neurographia universalis* (1685) by the French anatomist, Raymond de Vieussens (1685, Liber I, Tabula II, facing p. 11; X2/3). Almost 150 years after Vesalius, the convolutions were still depicted in an "enteroid" pattern.

activities are not carried out in the ventricles but in the brain substance itself, an idea that had been expressed by Sylvius in 1660 and by Erasistratus sixteen centuries earlier. In *Cerebri anatome* (1664), Willis provided a vivid picture of the human brain:

> [W]hen the gyri or folds have been laid open and separated from one another, the substance of the brain is seen to be . . . ploughed into furrows from which arise cliffs or ridges of uneven height, . . . and so the whole brain is variegated by a successive order of such inequalities. . . . [T]hese folds or convolutions are far more numerous and larger than in any other animal [than man] because of the variety and number of acts of the higher faculties. . . . Those gyri are fewer in quadrupeds, and in some such as the

Fig. 4.3. The frontispiece of Thomas Willis's *The Remaining Medical Works* . . . (1681) shows him at 45 years of age; D. Loggan was the artist. Willis is best known for his rediscovery of the arterial circle at the base of the brain that bears his name; he made many original contributions that earned him a reputation as the founder of modern neuroanatomy.

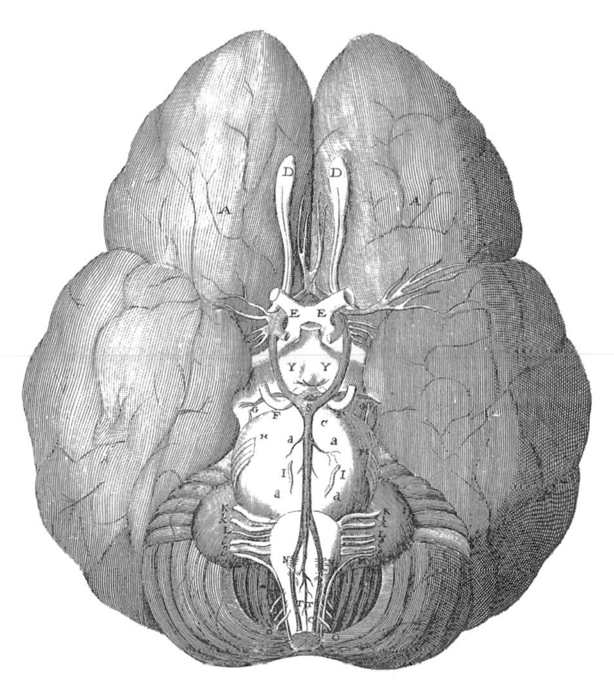

Fig. 4.4. Thomas Willis's careful dissections were closely observed by Christopher Wren, who drew most of the illustrations for his friend's *Cerebri anatome*, published in Latin in 1664. This engraving, from a drawing by Wren (Clarke and Dewhurst, 1972, p. 71), of the base of the human brain shows the olfactory bulbs (D) and the circle of Willis surrounding the optic chiasma (E). (From the 1664 octavo edition, facing p. 13.)

> cat, they are found to have a particular shape and arrangement so that this beast considers or recalls scarcely anything except what the instincts and demands of nature suggest (Willis, 1664, p. 65 passim; translation from Clarke and O'Malley, 1968, pp. 388–389).

Willis coined the term "neurologie" from the Greek (Skinner, 1949) and introduced it into English in 1681 (*see* Fig. 4.6 below, on p. 48). At first encompassing only the nerves, neurology was broadened subsequently to include the brain and spinal cord, and Willis is generally accorded the honor of having initiated this modern medical specialty.

During the Enlightenment of the late seventeenth and eighteenth centuries and the associated gradual increase in respect for evidence derived from

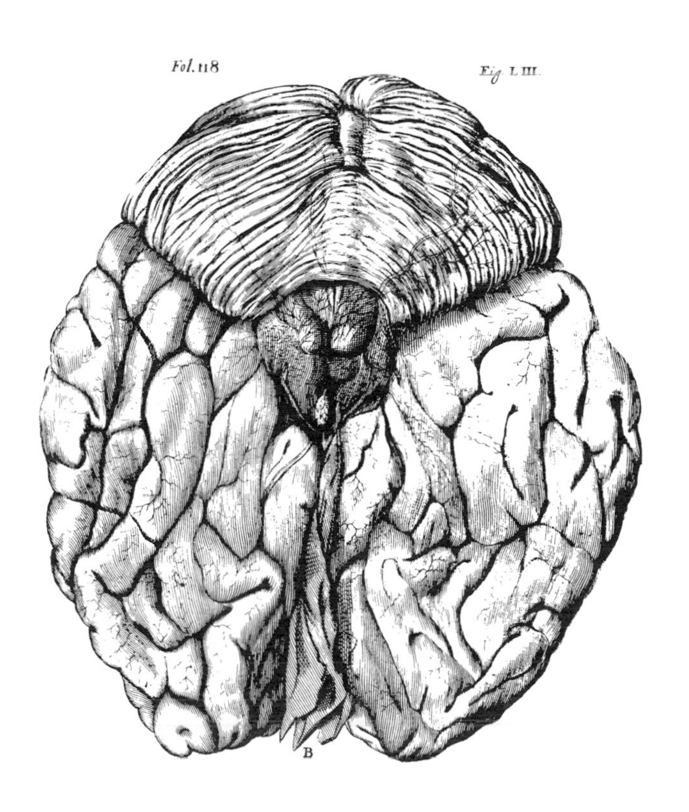

Fig. 4.5. The drawing from Descartes of the cerebral convolutions of the base of the human brain is surprisingly accurate for the time. The pineal gland is in the center of the figure and the cerebellum at the top. (From *De homine*, 1662, Fol. p. 118, Fig. LIII.)

THE
REMAINING
MEDICAL WORKS
OF THAT
FAMOUS and RENOWNED
PHYSICIAN
Dr Thomas Willis
OF
Chriſt-Church in OXFORD, and *Sidley* Profeſſor of
Natural Philoſophy in that famous UNIVERSITY.

VIZ.

I. Of Fermentation.
II. Of Feavours.
III. Of Urines.
IV. Of the Accenſion of the Bloud.
V. Of Muſculary Motion.
VI. Of the Anatomy of the Brain.
VII. Of the Deſcription and uſes of the Nerves.
VIII. Of Convulſive Diſeaſes.

The Firſt Part, though laſt Publiſhed.

With large Alphabetical Tables for the whole, and an Index for the Explaining all the hard and unuſual Words and terms of Art, derived from the Latine, Greek, or other Languages, for the benefit of the meer Engliſh Reader, and meaneſt capacity.

With Eighteen Copper Plates.

Engliſhed by *S. P.* Eſq;

To deſcribe all the ſeveral pairs of the ſpinal Nerves, and to rehearſe all their branchings, and to unfold the uſes and actions of them, would be a work of an immenſe labour and trouble: and as this *Neurologie* cannot be learned nor underſtood without an exact knowledge of the Muſcles, we may juſtly here forbear entring upon its particular inſtitution: but it may ſuffice concerning theſe nerves and their medullar beginning, that we advertiſe only in general what things may occur moſt notable and chiefly worth taking notice of.

Fig. 4.6. Thomas Willis coined the term "neurology" and used it at least four times, as on page 178 (above) of the text "Englished by S.P. Esq." [Samuel Pordage, Esquire] in 1681 from the Latin of the 1664 edition.

empirical research, the convolutions were viewed as not without some degree of order, and comparative studies became commonplace. In 1781, Felix Vicq d'Azyr (1748–1794) reported to the French Royal Academy his discovery that in the monkey the convolutions are sparse, symmetrical in both hemispheres, and their arrangement is constant among members of the same species. Later he found that in man (*see* Fig. 4.7, on p. 49), on the contrary, the convolutions are neither symmetrical nor alike among individuals (Vicq d'Azyr, 1786). The beautiful frontispiece of his atlas is reproduced in Fig. 4.8 (*see* Fig. 4.8, on p. 50) as an example of the lyricism and care lavished on scholarly publication during that era.

Although Vicq d'Azyr's European contemporary, Gasparo Ferdinando Felice Fontana (1730–1803), added nothing to the exact knowledge of human convolutionary patterns, he dissected countless human brains and produced extremely accu-

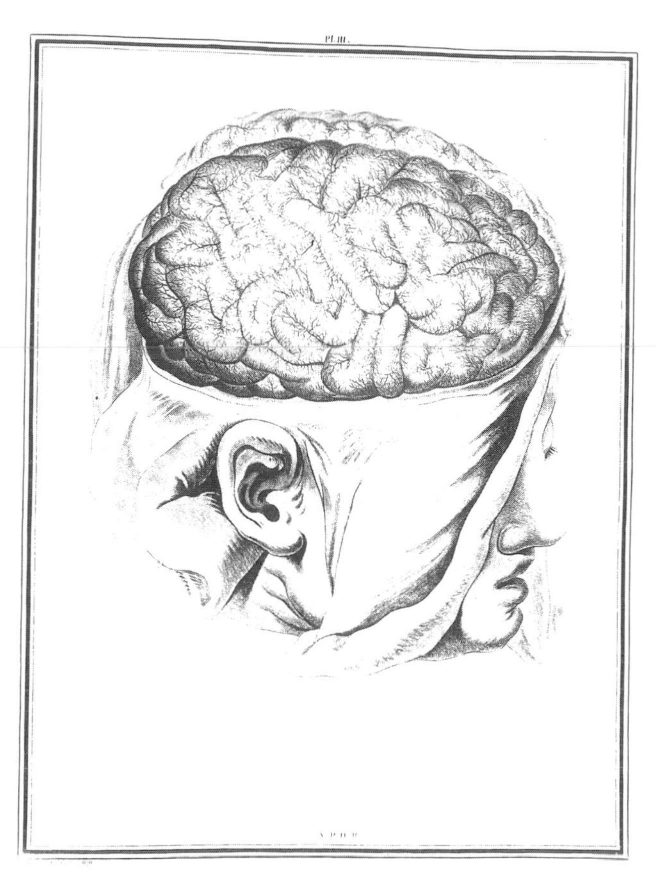

Fig. 4.7. Felix Vicq d'Azyr's depiction of the cerebral convolutions, drawn with the eye of an artist, shows a nascent perception of convolutional pattern. The dura mater of the right hemisphere has been deflected and the crest of the left brain appears at the top. Vicq d'Azyr wrote that the convolutions are not at all similar on the two sides of the brain. (From *Traité d anatomie et de physiologie*, [1786], Plate III.)

rate wax replicas of the cerebral convolutions. Fontana (*see* Fig. 4.9, on p. 51) was born in Austria and educated in Italy, where he taught histology, physiology, and anatomy as well as physics and chemistry. His "major interest, an intensely practical one . . . on the venom of the viper, began in Pisa in 1764. . . ." (Knoefel, 1984, p. 65). Called to Florence as experimental physicist to the court, the Duke of Tuscany commissioned him to establish a museum, later known as "Il Museo della Specola." Fontana created hard wax figures for teaching anatomy to physicians as well as to the artists who were attending human dissections to improve their knowledge of anatomy. His contributions to the museum included five, life-size, full figures demonstrating the nervous system and 55 additional focal pieces on the same scale (*see* Fig. 4.10, on p. 52). The collection of wax replicas so impressed the Emperor Joseph of Austria that he asked Fontana to replicate it, and by 1789 some 1200 pieces were transported by mule and barge to Vienna, where they are still on view in a beautiful building, Das Josephinum (*see* Fig. 4.11, on p. 53). As a French visitor touring Italy in 1792 wrote after seeing the wax collection, "What labor! What patience! But also what a splendid monument!" (ibid., p. 46). Ironically, the fame of Fontana's wax figures overshadowed his important studies of irritability of animal tissues that led him to suggest that "nerve electricity" produces muscle contraction (Brazier, 1970).

During the first half of the nineteenth century, anatomists began to systematically compare the brains of different animal species and discovered that as the cerebrum becomes enlarged, the surface pattern of the convolutions changes. The progres-

Fig. 4.8. The frontispiece of Felix Vicq d'Azyr's *Traité d'anatomie et de physiologie* (1786) foretells the delicacy and detail of this beautiful publication (*see* Fig. 4.7), only the first volume of which appeared owing to the author's early death. Color plate appears as an insert after page 84.

sion from the smooth brains of lower mammals to the densely convoluted cerebri of the human was elegantly illustrated in a French publication consisting of two volumes that appeared 20 years apart. In volume I, François Leuret (1797–1851; *see* Fig. 4.12, on p. 53) described in detail the brains of a great number of mammals:

> One ought to . . . understand thoroughly the cerebral convolutions of animals, to individualize them, so to speak, in order to compare them among themselves, . . . and to determine how they resemble and how they differ from those of man. . . . [T]his study has led me to a way of classifying animals according to the nature of their convolutions. . . . One thus sees how intelligence in these animals corresponds to the arrangement of their convolutions (1839, p. 368; transl. by M. V. Anker).

He found that, in general, the largest brains had the most numerous and most irregular convolutions and

Fig. 4.9. The unknown provenance of this portrait of Felice Fontana, exhibited in the Accademia degli Agiati at his birthplace, Rovereto, Italy, has cast doubt on its authenticity. Among Fontana's contributions were a description of the axis cylinder and experiments on irritability of nerve and muscle; the latter studies were forerunners of modern electrophysiology (*see* Brazier, 1970, p. 202).

identified 14 groups, from the smooth brain of rodents to the convolutional complexity of apes. The differences are illustrated in Fig. 4.13, on p. 54.

Leuret was born in Nancy, France, and although destined to the ministry, he soon found in science the necessary challenge for his eager mind. Having to rely on his own resources for subsistence, he enlisted in the army where he rushed through the heavy training program in order to pursue his studies, concentrating especially on intellectual functions and their pathology. At the Salpêtrière, the students remarked about the "scrawny appearance of the poor little soldier never missing one of the precious lectures by M. Esquirol. He even sold his piece of bread in order to buy a candle so he could study at night in the army barracks."[2]

After Leuret's untimely death, his student and friend, Pierre Gratiolet (1815–1865; *see* Fig. 4.14, on p. 55), pushed their investigations to a higher phylogenetic level, tracing the appearance and development of gyri in primates, from the relatively simple brain of the monkey to that of the chimpanzee, culminating in the highly convoluted and indented brain of *Homo sapiens* (*see* Fig. 4.15, on p. 56). He also followed the gyri in the embryo at different developmental stages, thus detecting those convolutions that could be construed as having appeared earliest in evolution. Gratiolet's most important observations were published in his *Mémoire sur les plis cérébraux de l'homme et des primates* (1854). He summarized them three years later in the second volume of Leuret and Gratiolet's *Anatomie comparée du système nerveux considéré dans ses rapports aver l'intélligence,* the subtitle of which is indicative of the scientific preoccupation of that place and time.

Early in the nineteenth century, Franz Joseph Gall (1758–1828) reintroduced the method of commencing dissections from below and following fiber tracts forward toward the cortex. That method had been used during the second half of the seventeenth century by Steno, Willis, and most effec-

[2]From Preface by M. Trélat in Leuret et Gratiolet, 1857, p. xvi.

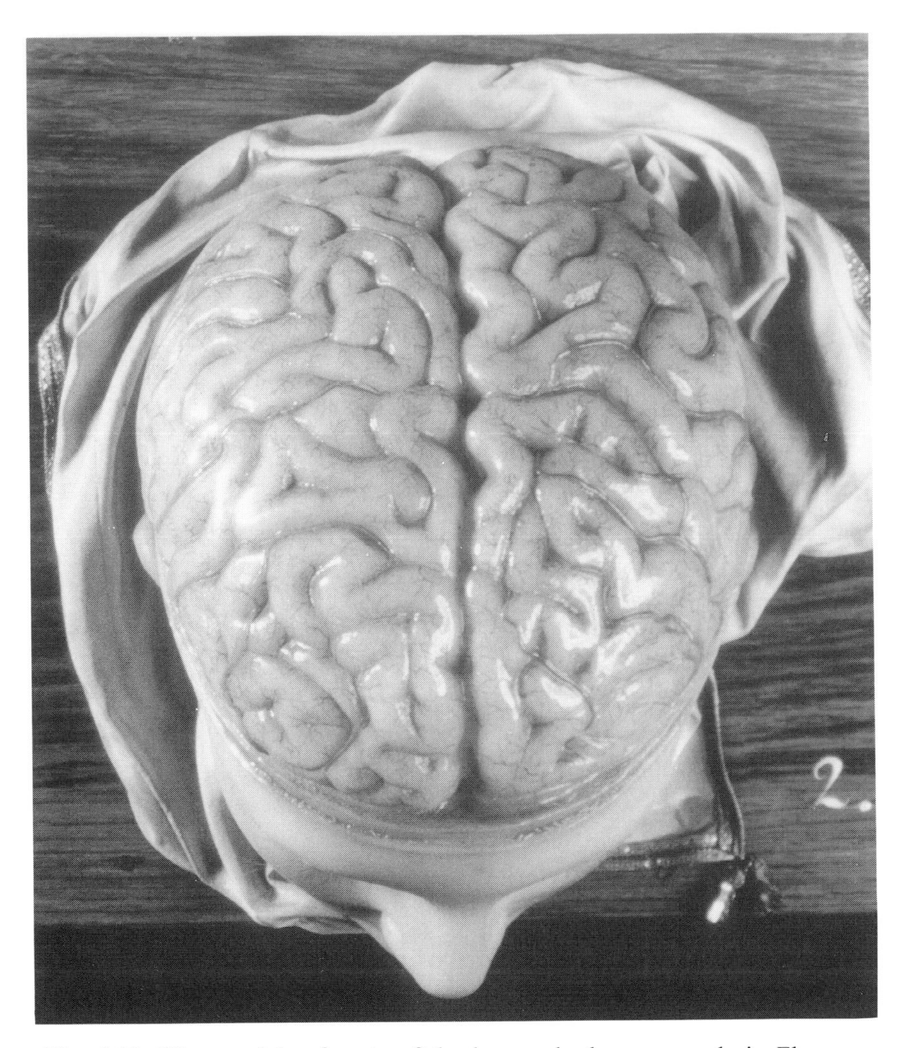

Fig. 4.10. Wax models of parts of the human body were made in Florence between 1775 and 1785 under Felice Fontana's direction. The photograph shows a model from the second set assembled (*see* text and Fig. 4.11).

tively by Vieussens. Gall's studies convinced him that the convolutions are the "expansion" of fiber bundles coming from all directions in the brain. He believed that the brain surface takes its shape according to the degree of localized functions it exercises and "these varieties of form . . . are developed on the surface of the cranium and head" (Gall, 1835, III, p. 2). His student, J. C. Spurtzheim, exploited this new cranial cartography, which became known as "phrenology" and spread to America (*see* Fig. 4.16, on p. 57).

Response of the general public to the concept of phrenology was mixed, as seen in this passage from a "letter" to "My esteemed Children" published in 1827[3] by a retired governess:

> Many particulars of the craniological system of Gall and Spurtzheim appear to be trifling and fantastical. Yet I have never been able to take a general view of a few scores of busts, and sculls [sic], without remarking that an ample development of the scull, from the orifice of the ear, to the prominent part of the forehead, is indicative of high intellect; and, on the other hand, that a great projection of the brain, from the ear to the lower and back part of the head, is connected

[3]Our thanks to Russell A. Johnson for calling this letter to our attention.

Fig. 4.11. The second set of Fontana's wax models, commissioned by Emperor Joseph II, is considered such a treasure that it is now on display in the Institute for the History of Medicine of the University of Vienna. (From a modern postcard.)

Fig. 4.12. François Leuret's persistence in pursuing science against great odds resulted in an outstandingly detailed study of the comparative anatomy of the nervous system in mammals. (From Noël, 1978, p. 1401.)

> with a strong tendency to the lower and brutal passions (Anon., 1827, p. 238).

The low esteem held for the assignment of character traits to specific loci on the cranium is implicit in a remark made by Napoleon Bonaparte and reported by Johannes Müller, according to that "erudite bibliothécaire," Soury (1899, p. 602): "What would become of the organ of theft if [the thief] had no property?" The insurmountable hurdle, however, to the scientific acceptance of craniology was the adverse opinion of Pierre Flourens (*see* Chapter 5), the internationally prominent French physiologist whose influence carried weight far beyond Paris; he republished several times (1846, 1863) an essay denouncing both the evidence put forth by Gall and Spurtzheim and the logic of their thinking. In London in 1882 a paper was read before the Casual Club facetiously titled "The Noble Forehead" (Clapham, 1881–1882). It attacked phrenology in general and claimed there was no proof for the belief that intelligence resides in the frontal part of the brain. Instead, five reasons for crediting the occipital convolutions with intelligence were advanced: they occur only in primates, are the last to form, are not occupied by motor areas, are undeveloped in idiots, and they atrophy in dementia. The Club's recording secretary added: "An interesting debate followed. . . ."(ibid., p. 624).

The cartography of Gall and Spurtzheim, although based on an erroneous assumption, called attention to the functional importance of the cortical convolutions and the sulci separating them. The

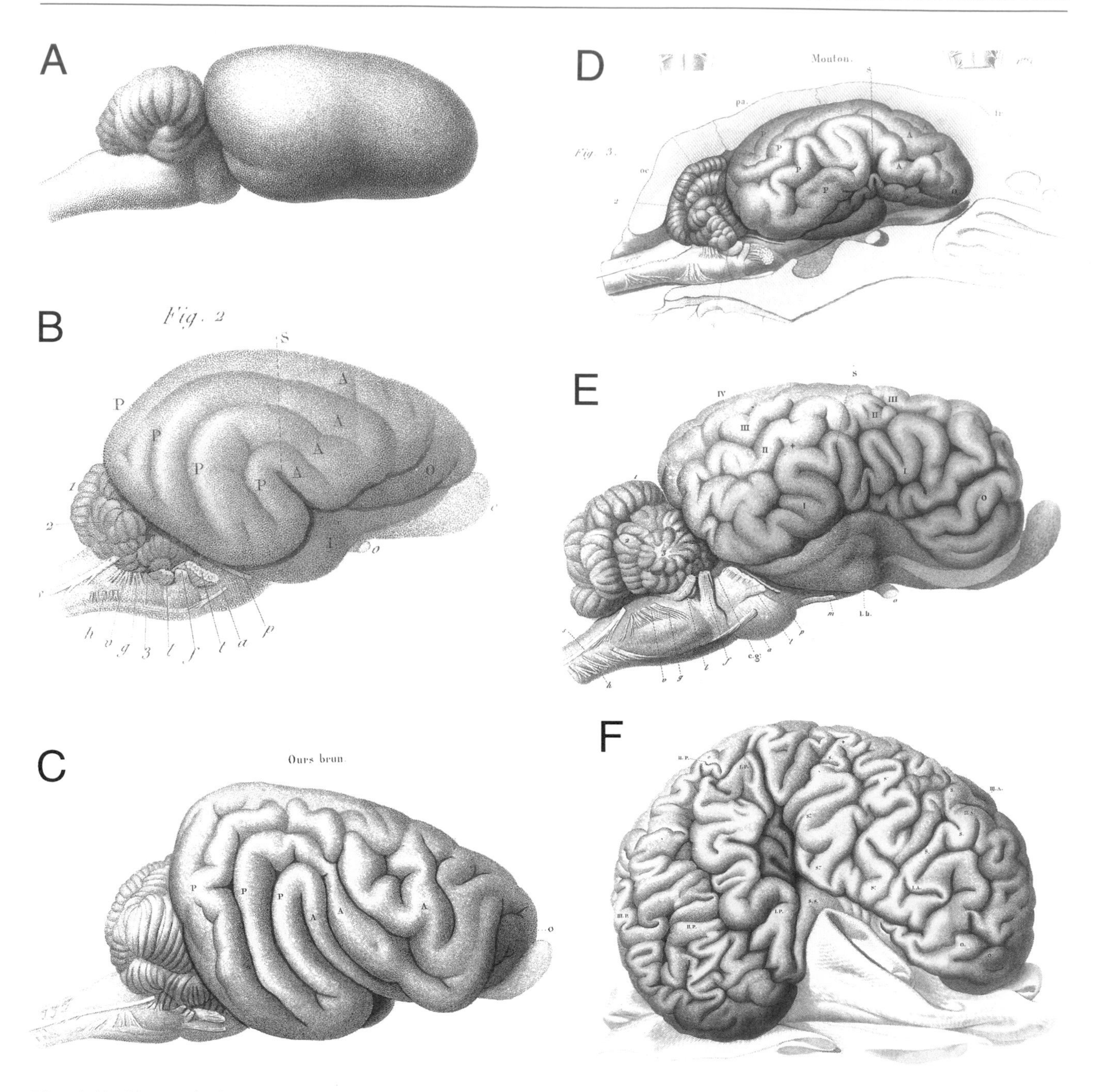

Fig. 4.13. Six engravings of mammalian brains from a series of 32 plates by François Leuret (1839) show the direct relationship of convolution complexity with higher phylogenetic position. A—In beaver, the cerebrum is smooth and cerebellum slightly furrowed. B—The fox brain has five smooth convolutions surrounding the sylvian fissure. C—In bear, the second convolution is enlarged anteriorly and indentations anticipate the human pattern. D—In sheep, the convolutions are more indented. E—The horse brain has more indentations than the sheep. F—In the elephant brain the sylvian (central) fissure divides frontal from caudal convolutions. (From Leuret et Gratiolet, 1859–1857 [sic], Plates III, IV, VI, VII, IX, and XIV, respectively; not in scale.)

plates for *Anatomie et Physiologie* (Gall was the sole author of the last two volumes) engraved in London under Gall's supervision and depicting the cortical convolutions with unprecedented accuracy (*see* Fig. 4.17, on p. 58), finally led to the demise of the misleading "intestinal-like coils"

Fig. 4.14. Louis Pierre Gratiolet is portrayed in his prime, wearing the rosette of the Société d'Anthropologie de Paris of which he was president the year before his untimely death at the age of 50. (From Broca, 1865a, frontispiece.)

analogy that had represented the cerebral cortex from antiquity. Even more significantly, Gall introduced the concept of the correlation of areas ("organs") of brain surface with specific faculties or functions. In his words: "In all organized beings, different phenomena suppose different apparatus; consequently, the various functions of the brain likewise suppose different organs" (Gall, 1835, vol. II, p. 254). Among the many localizations proposed in his craniology, at least one was on target: he placed language and speech in the frontal convolutions (ibid., vol. III, p. 55).

With Gall, we enter the new era of the functional cerebral cortex: the philosopher-psychologists recognized that the mental and organic realms are more than a geographic site and "physiological psychology was thus mobilized to fill a gap left by the decline in the respectability of phrenology" (Jacyna, 1982, p. 248). An even more significant role for Gall's ideas was as a bridge between Willis and the later localizationists such as Hughlings Jackson, Broca, Hitzig, and Ferrier. In trying to adjust skull form to the underlying convolutions, Gall's "organology, that dared to make an inventory of the mind and to find a pigeonhole for each item, was no doubt eighteenth century quixotic, but it casts a long shadow on twentieth century cybernetics" (Schiller, 1970c, p. 34).

THE FISSURES OR SULCI

Until the midnineteenth century, only two features of the cerebral cortex were named after individuals: the fissure of Sylvius in the seventeenth century and the fissure of Rolando in 1839. The sylvian fissure, the most conspicuous feature on the brain surface, was first depicted and so named by the Danish anatomist, Casper Bartholin (1585–1629), whose *Institutiones anatomicae* was pub-

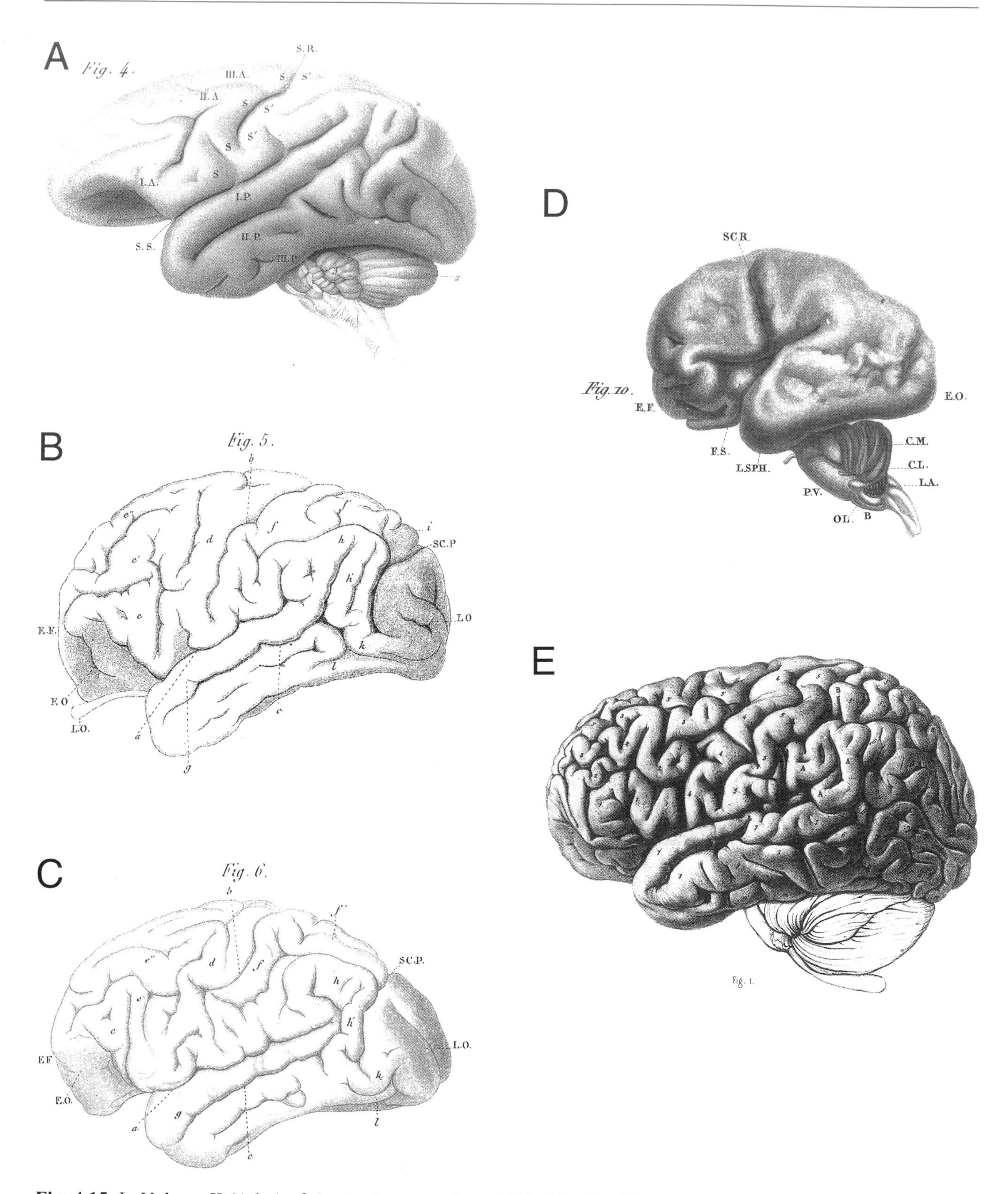

Fig. 4.15. In Volume II (Atlas) of *Anatomie comparée*, published by Gratiolet 14 years after its coauthor's death, the convolutions of the brains of subhuman primates show increasing elaboration. A—In the baboon brain the sylvian fissure (S.S.) lies obliquely instead of vertically, the convolutions are more indented, and a definite fissure (S.R., the rolandic) divides the frontal and parietal lobes. B—On the brain of the orang-utang the convolutions are more indented and a caudal

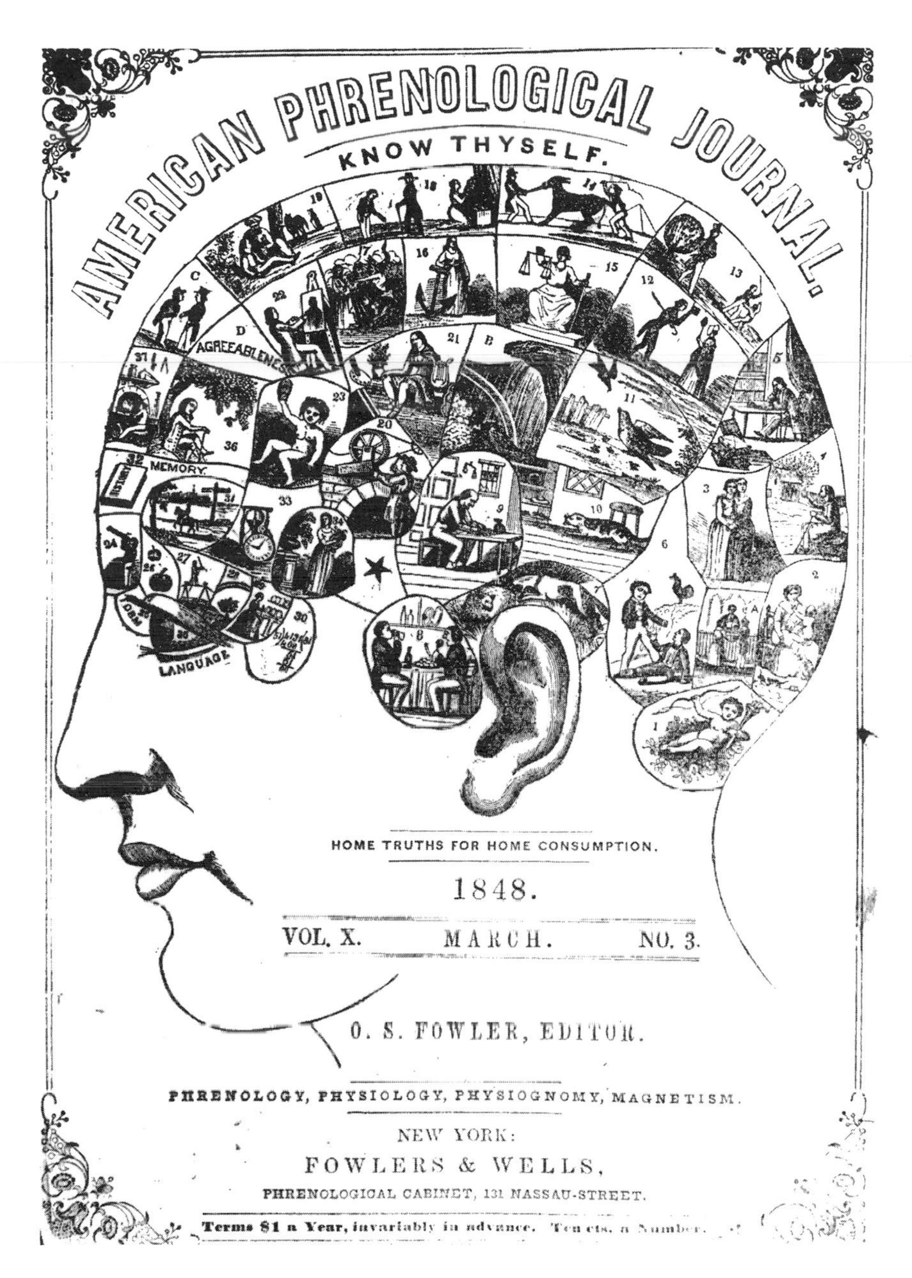

Fig. 4.16. The new cartography of F. J. Gall and J. C. Spurtzheim assigned specific functions to selected areas of the brain, popularizing the relationship of function to morphology. This "symbolical head" was published on the cover of the *American Phrenological Journal*, vol. X, no. 3, March 1848.

lished by his son Thomas (1616–1680) in 1641 (*see* Fig. 4.18, on p. 58). The volume contained an excellent rendering of the fissure in a lateral view of the hemisphere (*see* Fig. 4.19, on p. 58). Sylvius's (François de le Boë, 1614–1672; *see* Fig. 4.20, on p. 59) own description of this fissure was not published until 1663, five years after his appointment to the chair of medicine at Leyden:

> [The] entire external surface of the brain is covered with twistings (gyri) which are somewhat similar to convolutions of the small intestine. Particularly noticeable is the deep fissure or

fissure separates the temporal, parietal, and occipital lobes. C—The chimpanzee brain has three well defined temporal convolutions. D—From his extensive study of human fetal brains of known gestation, Gratiolet determined the order and precision of convolution development. This fetus was five-and-one-half-months-old and compares to the simplest animal brains. E—The brain of an adult man. (From Leuret et Gratiolet, 1857, vol. II [Atlas], Plates XV, XXIV, XXIX; and Gratiolet, 1854, Plate I, Fig. 1, respectively; not in scale.)

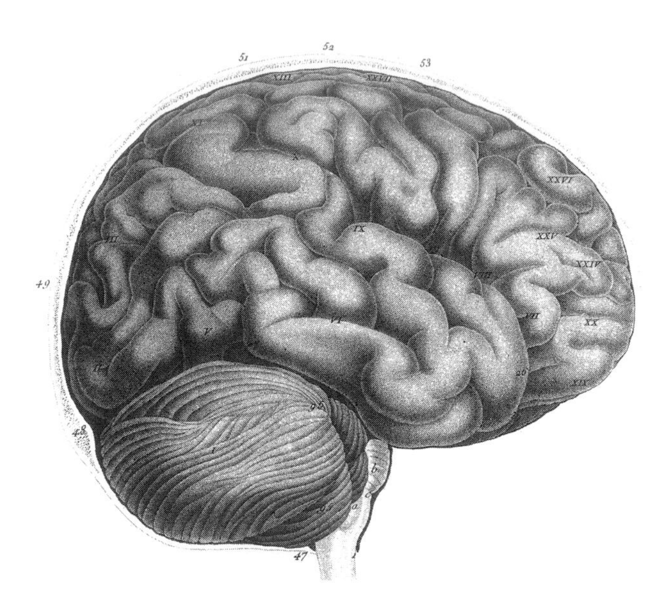

Fig. 4.17. Gall's plates for his and Spurtzheim's four-volume *Anatomie et physiologie*, published between 1810 and 1819, were true to nature and artistically drawn, with numerals on the cerebral surface to indicate the sites of personality traits. (From *Atlas*, vol. I, Plate IV.)

> hiatus which . . . begins at the roots of the eyes. . . . It runs posteriorly above the temples. . . . It divides the cerebrum into an upper, larger part and a lower, smaller part. Twistings occur along the fissure's length and depth (Le Boë, 1663, pp. 43–44; transl. Y. V. O'Neill).

In addition to the many French studies of gyri and sulci during the midnineteenth century, a German anatomist and embryologist, Émil Huschke (1797–1858), reported that the sylvian fissure does not lie in the same position in all mammals: It is relatively short and almost vertical in herbivores (sheep, ox, horse), longer and slightly slanted in the cat, panther, lion, fox, and dog, and occupies a middle position in the pig and elephant. The greatest length is attained in the apes and the greatest width and most horizontal position is found in humans. Huschke's findings showed great variability, and he was careful to point out that any correlation of intelligence with convolution complexity pertains to within-class comparisons (1854).

Alexander Ecker (1816–1887), anatomist and physiologist at the University of Basel, sought "a law for the formation of the convolutions . . . as a

Casp.
BARTHOLINI,
D. & Profes. Regii
INSTITVTIONES
ANATOMICÆ,
Novis Recentiorum opinionibus
& observationibus, quarum
innumeræ hactenus editæ
non sunt, figuris que.
Secundo auctæ.
ab Auctoris Filio
THOMA BARTHOLINO
MOVENDO
LVG. BATAVORVM,
Apud Franciscum Hackium
cIↃ IↃ CXLV

HIPPOCRATES
GALENVS
VESALIVS
RIOLANVS
C.BAVHINVS
SPIGELIVS
PAVIVS
O.HEVRNIVS

Fig. 4.18. The title page of *Institutiones anatomicae*, written by Caspar Bartholin (1585–1629) but not published until 1641, 12 years after his death, by his son Thomas. In this text, the conspicuous primary fissure of the human brain was named in honor of Sylvius; *see* Fig. 4.19.

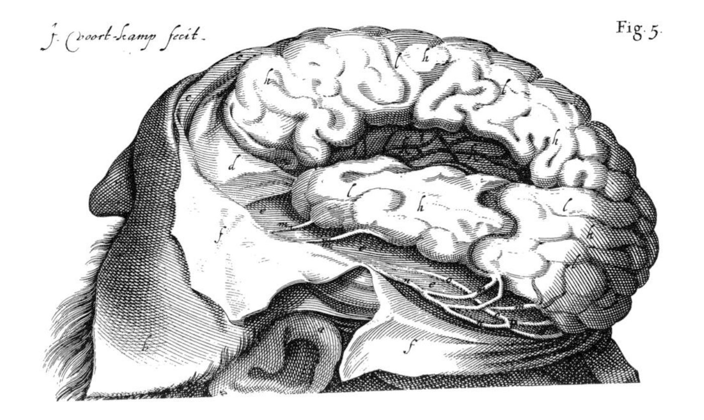

Fig. 4.19. Drawing of left cerebral hemisphere showing the fissure named after Sylvius in Bartholin's *Institutiones anatomicae* (1641, p. 261, Fig. 5); *see* Fig. 4.18, above.

Fig. 4.20. François de le Boë (Sylvius), famous teacher and medical chemist at Leyden, published in 1663 his own description of the fissure that had been named after him. His portrait was by the artist C. van Dalen, Junior; *see* Figs. 4.18 and 4.19.

necessary consequence of certain mechanical antecedents in the growth of the brain and skull" (1873, p. 4). He illustrated his idea by attributing the formation of the fissure of Sylvius to "the folding together of the entire hemisphere into an arch, having its concavity downwards" (ibid., p. 10). In the same context, Wernicke (1876, p. 303) described Leuret's three primordial longitudinal gyri (*see* below) as "forming concentric crescents around the Sylvian fissure" (Goldstein, 1970, p. 533). Actually, Foville in 1844 had already thus described the first gyrus in lower animals.

Turning now to the other great cerebral fissure, Luigi Rolando (1773–1831; *see* Fig. 4.21, on p. 60) was among the first to claim that the brain was not without form. This Italian anatomist and physiologist believed that the cerebral "enteroid processes," on the contrary, had regular and well determined shapes and positions (*see* Fig. 4.22, on p. 60). His recognition of order was manifest most notably by his description (1809) of what are now called the precentral and postcentral gyri, bordering the central sulcus which later Leuret named the "fissure of Rolando." In a recent paper on "The rise of the enteroid processes," Francis Schiller has further defined Rolando's role in establishing modern definitions:

> Two other [contributions of Rolando] are of abiding interest. Rolando showed what he considered to be a single convolution, i.e., that surrounding the Sylvian fissure. In his scheme of things, this was more fundamental than the [middle or central] vertical processes arising from it. The other . . . is the gyrus surrounding the corpus callosum. In a previous paper (1809), he had called it . . . the enteroid process of the crest. A little later, [in 1819], Burdach in

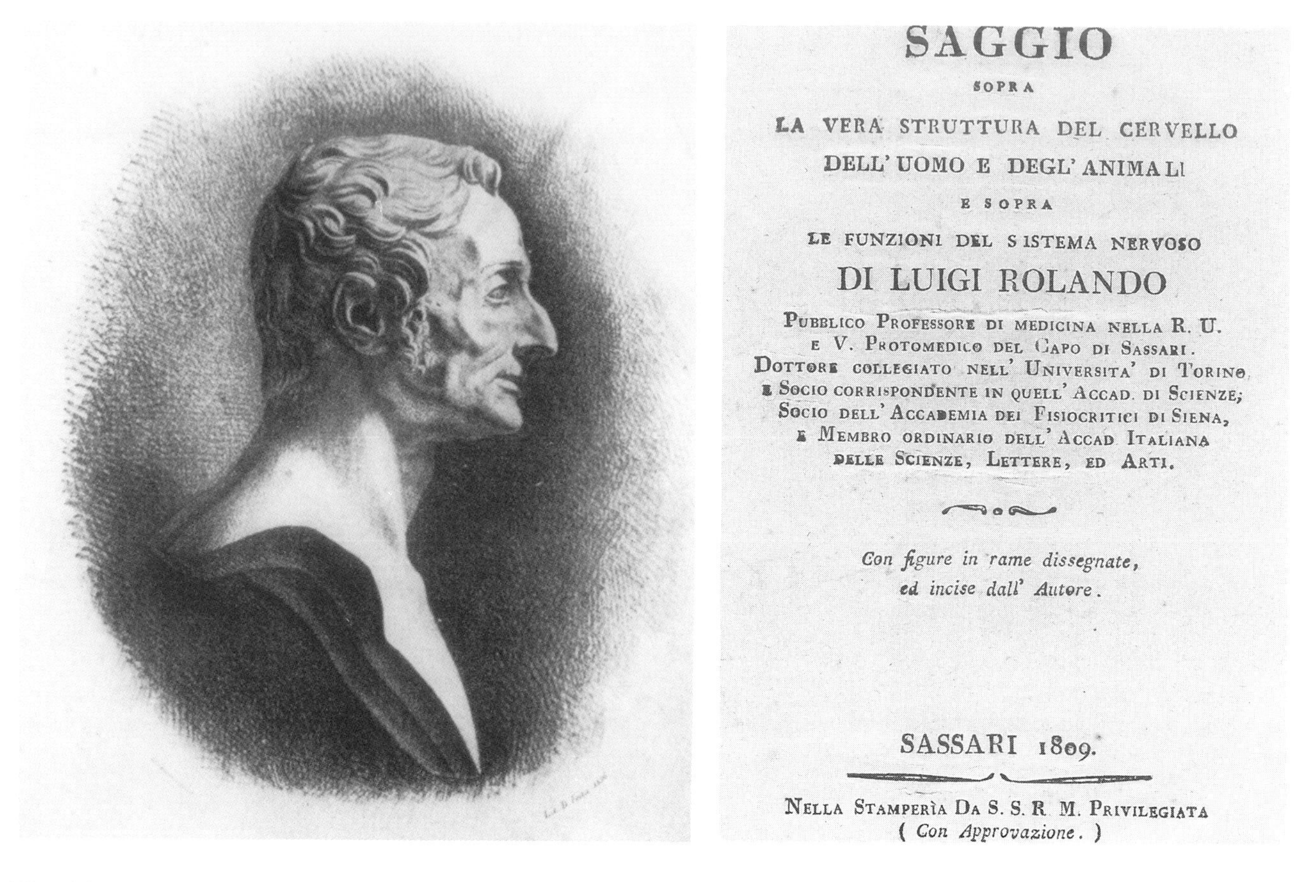

SAGGIO

SOPRA

LA VERA STRUTTURA DEL CERVELLO
DELL' UOMO E DEGL' ANIMALI

E SOPRA

LE FUNZIONI DEL SISTEMA NERVOSO

DI LUIGI ROLANDO

PUBBLICO PROFESSORE DI MEDICINA NELLA R. U.
E V. PROTOMEDICO DEL CAPO DI SASSARI.
DOTTORE COLLEGIATO NELL' UNIVERSITA' DI TORINO
E SOCIO CORRISPONDENTE IN QUELL' ACCAD. DI SCIENZE;
SOCIO DELL' ACCADEMIA DEI FISIOCRITICI DI SIENA,
E MEMBRO ORDINARIO DELL' ACCAD. ITALIANA
DELLE SCIENZE, LETTERE, ED ARTI.

Con figure in rame dissegnate,
ed incise dall' Autore.

SASSARI 1809.

NELLA STAMPERIA DA S. S. R. M. PRIVILEGIATA
(*Con Approvazione.*)

Fig. 4.21. Right—Luigi Rolando's best known publication, in 1809, was translated into French by Flourens in 1823 and included a description but no drawing of the central sulcus named for him by François Leuret in 1839. Left—The original engraving of Rolando's profile is at the Bibliotèque de l'Académie Nationale de Médecine, Paris.

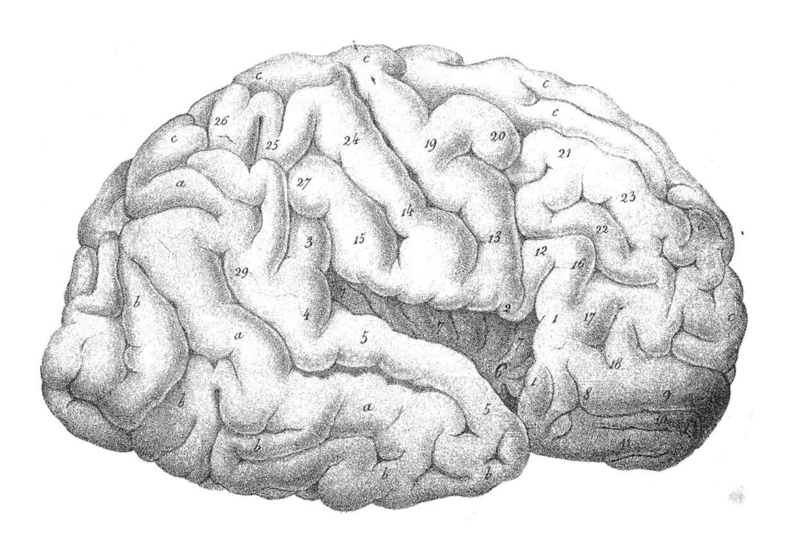

Fig. 4.22. Rolando's clear depiction of the central sulcus (named for him by Leuret in 1839) was published in 1829 and showed the precentral (13,19) and postcentral (14,24) gyri which not only separate the frontal and parietal lobes, but soon became associated roughly with motor and sensory activities, respectively (*see* Manni, 1973).

> Koenigsberg called it the 'gyrus cinguli.' It is . . . almost the whole of what, phylogenetically, remains in man of Broca's lower mammalian great limbic lobe (Schiller, 1965, p. 326).

About his naming of the fissure of Rolando, Leuret wrote in his description of the baboon's brain:

> The monkeys have three anterior and three posterior convolutions (running longitudinally across the lateral aspect of the hemisphere). . . . At a right angle across these convolutions is a fissure which separates them; it is as constant as the Sylvian fissure and I have called it the fissure of Rolando because it was this anatomist who first described it in man, where it is much more developed than in monkeys (Leuret, 1839, vol. I, 397–398; transl. M. V. Anker).

Leuret's observations on simians and their expansion to the primates by his friend and co-author, Gratiolet, promoted a great deal of comparative work with human brains. The high esteem for the work of those two French neuroanatomists was manifest in a biographical essay to introduce a reprinting of Broca's major works: "Based on this profound word of Leuret: the brain of a monkey, to a certain point, can be considered as a rough draft of that of man, Gratiolet untangled the chaos of the convolutions of the human brain" (Pozzi, 1888, p. vii).

OVERVIEW OF THE "COILS"

The early anatomists saw the surface convolutions of the brain lying *in situ* in the cranium as resembling the intestines in the opened abdomen, failing to note the relative stability of the deeper arrangement. Not until the midseventeenth century was there a growing recognition of order in the placement of the cerebral convolutions, most importantly by Willis. Comparisons among animals revealed patterns of arrangement according to animal classes and their position on the phylogenetic scale, which inevitably invited speculation about the relation of intelligence to the number of convolutions, notably by Leuret. Somewhat late in the history of the discovery of the convolutions, the question of what evolutionary pressure could have led to their formation was addressed. Wernicke was among the first to consider whether gyri and sulci are the result of mechanical forces or a response to functional associations. The physical form of the cerebral convolutions was understood long before assignment of function, which made a false start when Gall and Spurtzheim contended that confined areas of the brain surface were the locus of specific mental and moral faculties and were represented by the contours of the skull. In a broader sense, however, it was recognized by one of the important contemporary philosophers that by calling attention to the local representation of traits of character on the skull above the underlying brain, the possibility was opened to "a union of philosophy proper with physiology" (Laycock, 1860, vol. I, p. 65).

Lobes and Functional Localization

Contents

I do not believe in abrupt geographical localisations...

(Hughlings Jackson, 1874, p. 60)

Throughout neuroscience's history, the localization of functions to specific parts of the brain has been one of the most contentious and widely investigated puzzles. Traces of the centuries-old Galenic placement of animal spirits in the ventricular cavities, thus assigning there the source of man's intellect and mental functions, lasted into the seventeenth century, for reasons discussed in Chapter 1. The recognition that brain substance is the "material substrate of thought" had to emerge before real progress in localization could occur. As summarized elegantly in the translation of the historical monograph of Max Neuberger (1981), the localization of function first appeared in the Hippocratic writings, which assigned thought to the brain. In spite of the influence of Aristotle's views (i.e., the mind resides in the heart), the Galenists, too, placed the soul in the brain and gradually transposed "soul" to intellect, or "anima rationalis." Some 13 centuries elapsed before the great Italian Renaissance figures, Da Vinci and Vesalius, identified brain tissue, in contrast to the ventricular "cells," as the site of thought. Localization of specific functions as a concept was envisioned by "the founding father of experimental brain physiology" (ibid., p. 8), Thomas Willis (*see* Chapter 4). A century later, during "Haller's era" (1750–1770), belief in a more unitary substrate for thought was popular, to be superseded about 1830 by the ideas on localization of Gall and of Magendie. After it was established that convolutions are arranged in lobes with sulci between them, the roles of the separate eight lobes (four in each cerebral hemisphere) were gradually determined by observational and experimental studies.

Traditionally, the only acceptable method of studying human brain function had been observational—on the battlefield, or in conjunction with disease or accidental trauma. Here the key to new discoveries was the correlation of brain pathology with behavior. From scattered observations of the relation of postmortem findings with behavioral changes, a crude correlation at best, the specialty of pathological anatomy slowly took form. Its emergence was greatly speeded during the seventeenth century by the elegant dissections and astute observations of Thomas Willis and his contemporaries (Martensen 1992). One hundred years later, the Paduan anatomist–teacher, Giovanni Battista Morgagni (1682–1771), in *De sedibus et causis morborum* (1761), demonstrated a link between pathologic lesions in the brain and such clinical syndromes as the loss of speech.[1]

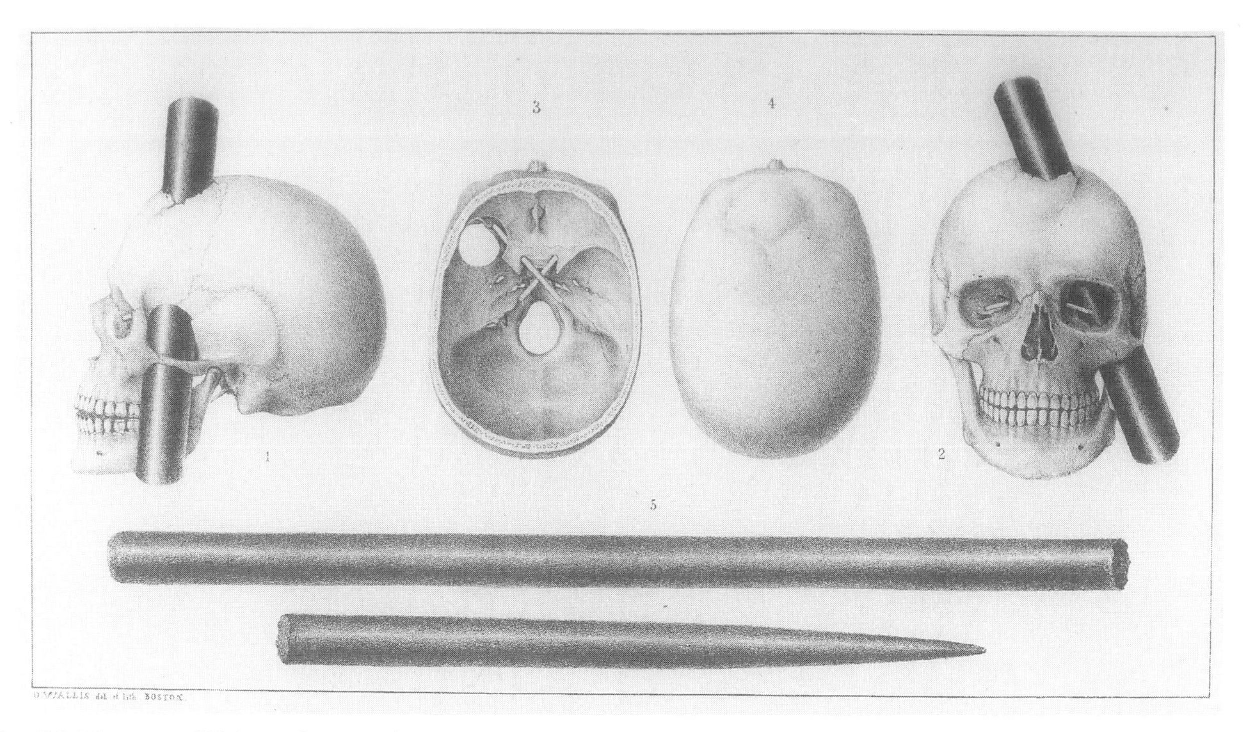

Fig. 5.1. The case of Phineas Gage's miraculous recovery, "perhaps unparalled in the annals of surgery" (Bigelow, 1850, p. 13), was published three times by John Harlow, the country doctor who treated him. The first two were written soon after the passage of a tamping iron through the brain, and the third 20 years after Gage's death and examination of the skull, thus verifying the early drawings. (Lithograph by G. Wallis from Bigelow, 1850, p. 12, Plate 1; Fig. 2 was reproduced by Ferrier in 1870 and modified spectacularly on the cover of *Science*, 20 May 1994.)

FRONTAL LOBE PATHOLOGY AND SPEECH: OBSERVATIONS

One of the most notorious modern examples of observational evidence involving the brain took place in 1848, in the well known case of the American laborer, Phineas Gage. Having survived the explosive passage of a tamping rod through his forebrain, he experienced a conspicuous psychological change from competent worker to semimoron. After his death, gross lesions were found (Fig. 5.1) in the left frontal lobe which impaired his "higher" mental functions rather than speech, movement, or sensation (Harlow, 1869).[2]

The French anatomist, Gratiolet, in his comparative studies of the convolutions of the brains of a variety of animals, described in Chapter 4, divided the hemispheres into well defined lobes by following the more or less deep indentations surrounding the gyri, noting their differences in development. He showed that the relative complexity of the lobes corresponds to the animal's position on the phylogenetic scale and he believed that in humans it depends also on race, age, and gender. In his conclusions, published in 1854 after more than 20 years of research, the *anterior frontal* lobes were honored as the seat of man's highest intellectual faculties, and sensation, "inclination," and passion were assigned to the *parietal* and *occipital* lobes. Gratiolet was quoted by his colleague, Paul Broca (1861, pp. 319–321), as saying: "It is in the frontal lobe that the majesty of the human brain in some measure resides." And a few years later, Broca stated (1865a, p. cxvi) that it was due to Gratiolet that "[t]he cerebral convolutions are no longer, as one could believe before, disorderly folds comparable to the sinuosities of the intestinal mass. The most constant order presides at their distribution." The tenacious metaphor of the "enteroid processes" was finally laid to rest, and neurologists found

[1]*See* Green (1985) for a fuller discussion of the rise of pathological anatomy.

[2]Sophisticated methods of reconstructing the wound were applied recently by Damasio and associates (1994).

Fig. 5.2. An early account of a speech disorder, the miracle of Zacharias as described by St. Luke, relates that when an angel appeared beside the altar to tell Zacharias that his wife, Elizabeth, would bear a child, Zacharias dropped his censer and exclaimed "That cannot be." Immediately, he was struck dumb for disbelieving Heaven's messenger, although his writing was unimpaired, as his companions indicate. (From the Gospel Book of Henry III, AD 1043–1045, Patrimonio Nacional, Madrid). Color plate appears as an insert after page 84.

attractive the exploration of functions associated with the separate lobes, thus entering a period of intense investigation of the extensive and deep interconnections and functional overlap characteristic of cerebral lobular arrangement.

Because dysfunctions of speech are readily recognized, they have been noted from early recorded history. In *Speech and Speech Disorders in Western Thought before 1660*, Y. V. O'Neill (1980) presented the first comprehensive account of the history of ideas about speech and its disorders during the pre-Renaissance era. One of the miracles, told in the First Chapter of the Gospel according to St. Luke, described Zacharias as being struck dumb by the Angel Gabriel for his disbelief (Fig. 5.2). He remained able to write, however, and subsequently his speech was restored when he recanted. This account is particularly interesting because it shows that as far back as biblical times, St. Luke, who had received medical training and was often referred to as "The Physician," recognized the relationship between speech and the brain.

French neurologists were the first to make significant contributions to the neuroanatomy of speech, beginning with Gall in the early nineteenth century, who assigned it to the frontal lobe, a correct concept derived from his mistaken belief that facility in language is correlated with protruding eyes.

Fig. 5.3. Jean-Baptiste Bouillaud with his son-in-law, E. Aubertin, supported Gall's ideas and offered 500 francs to anyone who could show him a lesion of the anterior lobule of the brain without some deficit of speech.

Jean-Baptist Bouillaud (1796–1875; Fig. 5.3), an advocate of some of Gall's ideas, in 1825 called the frontal lobe the "legislative organ of speech," distinguishing the ability to create speech from the faculty to articulate words, as had Gall. In what has been deemed "one of the most noteworthy outcomes of French clinicopathological correlations" (Clarke and Jacyna, 1987, p. 303), Bouillaud wrote:

> [It] is the brain which determines and regulates the muscular contractions in which these two functions [intellect and desire] are involved. . . . [It] is the same with all the organs charged with the execution of muscular motions . . . among others are the tongue and the eye, two admirable structures which play such important roles in the mechanism of intellectual functions. . . . I shall try to subsequently determine the site of the nervous center which directs the mechanism of the organ of speech. . . . It is not enough, however, that there exists in the brain a particular force destined to regulate and to *coordinate*, the marvelous movements by which man with his articulated voice communicates his thoughts, and expresses his ideas. . . . I believe that the nervous principle in question . . . resides in the anterior lobes of the brain (1825, pp. 25–30 passim; translation from Clarke and O'Malley, 1968, pp. 490–491).

Paul Broca (1824–1888; Fig. 5.4), French surgeon, neurologist, and anthropologist, defined language as the faculty to establish a constant relation between ideas and signs and cautioned that a function must be analyzed carefully before it can be successfully localized. From pre- and postmortem observations on a series of patients, he placed the ability to articulate language in the posterior part of the third frontal convolution of the left hemisphere.[3] In his 1863 report to the Société

[3] The determination of Broca's area has been recalled frequently, but in no instance more eloquently than by Francis Schiller in *Paul Broca*, a recent biography.

Fig. 5.4. Pierre Paul Broca was a precocious and prolific contributor to studies on modern anthropology, aneurysms, rickets, muscular dystrophy, hypnotism, and many other fields of medicine. Residing in Paris, a center of debate about cerebral localization, he deduced and, from study of a succession of patients, proved that motor aphasia is localized on the third left frontal convolution. This painting by an unknown artist hangs in the city hall of Sainte Foy, Broca's birthplace.

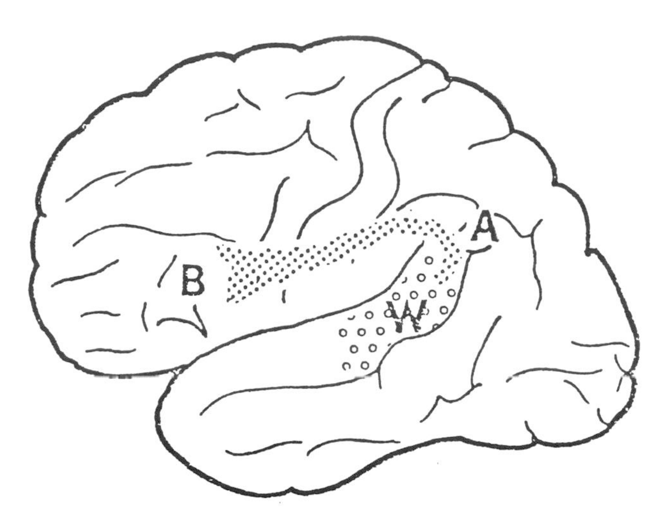

Fig. 5.5. The German neurologist Carl Wernicke was particularly interested in the three primordial convolutions which bend around the sylvian fissure in lower mammals. In the human brain the third convolution is formed by Wernicke's (W) and Broca's (B) areas joined by the arcuate fasciculus (A). (Portrait from Ziehen, 1906, frontispiece; photographer Max Glauer, Oppeln. Also in Haymaker and Schiller, 1970, p. 532.)

Anatomique de Paris, he described some of his own patients and one each of Charcot and Duchenne, declaring: "Here are eight instances in which the lesion was in the posterior third of the third frontal convolution. This figure appears to me sufficient to be strongly presumptive. And, remarkably, in all patients the lesion was on the left side. I do not dare draw from that a conclusion and I await new facts" (1863, p. 202; transl. M. V. Anker).

Some 40 additional cases came from an unexpected source that same year when a manuscript written by a physician from a small town in the south of France was deposited with the Academy of Medicine (*see* Joynt and Benton, 1964). The unpublished paper of Marc Dax, written the year before his death, had been languishing 27 years in Montpelier and was brought to Paris by the son, who added 140 more cases from his practice and the literature. The father had concluded indecisively that: ". . . not every affection of the left hemisphere will necessarily lead to a change in the memory for words; but if this memory is altered by any disease of the brain, one must look for the cause in the left hemisphere, and look for it there even if both hemispheres are affected" (quoted from Schiller, 1979, p. 193). The son claimed priority on the part of Dax père for speech localization in the left hemisphere, but Broca remained above any controversy. The introduction of the new-found manuscript both invigorated an argument that had become somewhat extended and attracted new interest on the part of peripheral on-lookers. Broca's attention became more focused and after Charcot found two additional cases, Broca declared:

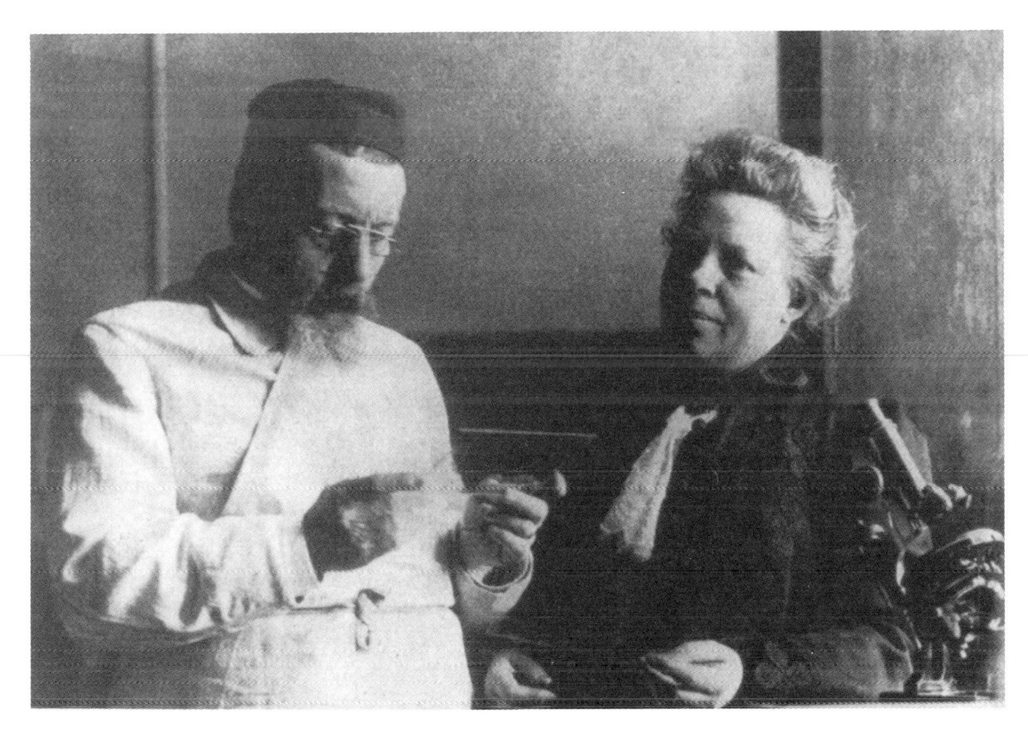

Fig. 5.6. Educated in Paris, where he spent his professional career, the Swiss-born Joseph Jules Dejerine was renowned for his psychotherapeutic success and his localization (with Vialet) of word blindness. He collaborated with his wife, Augusta, the first woman "interne des Hôpitaux" in Paris. (From Thomas, 1951, p. 466.)

"I persist in thinking, until informed to the contrary, that true aphémie, that is to say loss of speech without paralysis of the organs of articulation and without loss of intelligence, is lodged in lesions of the third frontal convolution. . . . [Furthermore, I note] the singular predilection of the lesions . . . for the left hemisphere of the brain" (1865b, p. 378).

The identification of the seat of motor speech was followed slightly more than a decade later by the complementary discovery—the locus of cerebral injury responsible for sensory aphasia—by Carl Wernicke (1848–1904; Fig. 5.5) in Breslau. Wernicke spent six months in Theodor Meynert's Viennese laboratory, and was familiar with the latter's dissection of the arcuate fasciculus arching around the sylvian fissure with some fibers going to the caudal portion of the upper temporal gyrus in the left hemisphere (Fig. 5.5, right). The small book from which Wernicke's international fame quickly sprang, *Der aphasische Symptomencomplex* (1874; *see* Eggert, 1977 for a translation), was a notable accomplishment for the 26-year-old, for whom the sensory speech area is named. Wernicke's study (1876) of the primordial gyrus system (*see* Chapter 4) shed new light on understanding aphasia problems, offering a theory that resolved the apparently conflicting clinical symptoms of the disorder and opening a new direction for aphasia research.

Another major historical step was taken by Joseph Jules Dejerine (1848–1917; Fig. 5.6), whose work was greatly aided by his wife, Augusta Marie Klumpke (1859–1917). With Vialet (1893), he identified the cortical lesions associated with alexia in the supramarginal and angular gyri of the parietal lobe. That discovery interrelated the written aspects of communication in the occipital cortex with auditory aspects in Wernicke's area in the temporal lobe and motor expression in Broca's area in the frontal lobe adjacent to the laryngeal area of the motor cortex.

The triangle of these three "areas" naturally appealed to those with geometrical mind-sets. To a modern historian, "[a] diagram is a very helpful device indeed. . . . A diagram of the nervous system is intended mainly to symbolize the road along which travels the still hypothetical nervous current. . . ." (Riese, 1959, p. 19). At the time of the

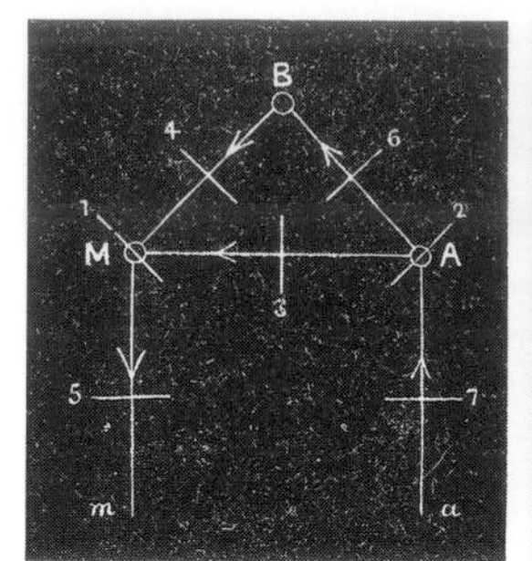

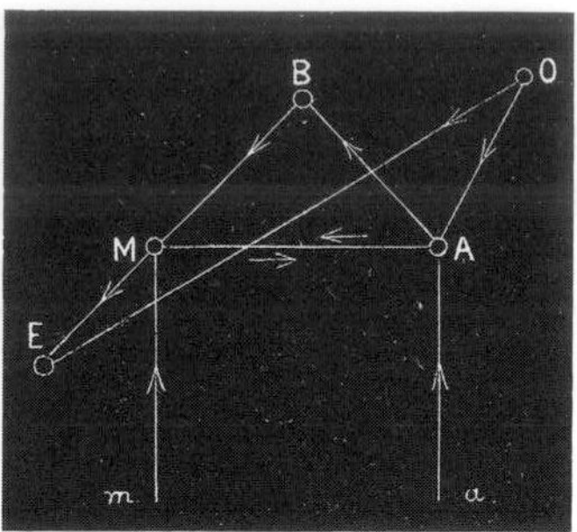

Fig. 5.7. Characteristic diagrams drawn in the late nineteenth century by a neurologist to explain the largely theoretical connections of "brain centers" associated with aphasia. A—Center of auditory images; M—center of motor images; B—concept elaboration; E—center for writing; O—center for vision. (From Lichtheim, 1885, pp. 436, 437.)

"localizationists," however, when an ensuing school of "diagram makers," arose which included in Great Britain the neurologist Henry Charlton Bastian (1837–1915) and other eminent physicians in Germany especially, they were alluded to disparagingly for their "evil influence" by Henry Head (1861–1940; *see* Fig. 10.4 left), one of the chief representatives of the British neurological establishment at the turn of the century. In Head's view, the diagram makers had twisted the clinical findings to conform with schemata of centers and tracts for speech that did not exist (Fig. 5.7). The British and German clinicians had been entrapped by the neatness with which the diagrams fitted their interpretations of the clinical signs they observed. Even Wernicke published (1903) "A case of isolated agraphia" in which the "clinical obtuseness [and] want of theoretical insight" were puzzling (Head, 1926, p. 397).

Head "is particularly significant . . . for his role in resurrecting and promoting the theories of Hughlings Jackson" (Harrington, 1987, p. 266). How this came about was described autobiographically:

> We believed that every sign and symptom could be deduced from a local lesion in some cortical centre or injury to the parts between them.
>
> But . . . later . . . I lost this robust faith. Moreover, I could not make my own observations fit the phenomena which should have been present. . . . For example, a lesion of the cortex produced results which could only be classed under the heading of defective recognition of certain relations between sensory stimuli and the perception to which they give rise (Head, 1926, pp. vii–viii).

His studies of war casualties and a rereading of Hughlings Jackson's verbose papers on aphasia, representing some 30 years of observation and diagnosis by that gifted clinician which generally had been overlooked in obscure journals, had moved Head (1915) to republish them. By making them more accessible, Head had hoped to help clear the air. He has been severely castigated, however, for omitting Jackson's physiological views of consciousness, a failure "bordering on intellectual dishonesty" (Harrington, 1987, p. 266).

Regarded as the "Grand Doge" of neurology at the National Hospital, Queen Square, London and looking the part, Hughlings Jackson (1835–1911; Fig. 5.8) theorized that the degree of dysfunction after brain damage depends on the complexity of the task involved: A hierarchy of function exists by which simple automatic patterns are retained after more complicated capabilities are lost. In his words:

> . . . I say that the highest cerebral centres are the least organised. . . although they are the most complex, whereas the lowest centres are the most organised, although the least complex. . . . If the highest centres were not modifiable, we should be very simple machines; we should make no new acquirements. If the lowest ("vital") centres were to become as modifiable as the highest ones, life would cease (Jackson, 1889, p. 356; Taylor, vol. 2, 1958, p. 395).

According to the historian R. M. Young (1970, p. 204), "Jackson based his argument on two theo-

Fig. 5.8. Although small, the National Hospital, Queen Square, London (depicted in 1866) during Hughlings Jackson's 45-year tenure (1862–1907) became the world center for study and treatment of neurological disorders. (Impressions from Holmes, 1954, frontispiece and facing p. 33, respectively.) Hughlings Jackson's portrait, showing a dreamy-eyed conceptualist, was painted about 1894 by Lance Calkin, and hangs in the Royal College of Physicians, London.

ries: [Thomas] Laycock's hypothesis of Reflex Cerebral Action and Spencer's hypothesis of 'Nervous Evolution,' " and made his most important contribution by extending the sensory-motor reflex into the highest cerebral centers.

A poor speaker, Hughlings Jackson lost an argument when he ineffectively presented his early views against rigid localization of function at a meeting arranged by the British Association for the Advancement of Science in 1868, billed as the "opportunity of immediately comparing the best English and the best French views on the pathology of this remarkable disease" (*Lancet* II, 1868, p. 226). Broca, the central figure in the localization issue, had been invited to speak on aphasia and as his paper was "clear, definite, and precise" (Head, 1920, p. 395), it won support for his views on strict localization centers. A few years after the "debate," Jackson stated: "Whilst I believe that the hinder part of the left third frontal convolution is the part most often damaged, I do not localise speech in any such small part of the brain. To locate the damage which destroys speech and to locate speech are two different things" (Jackson, 1874; also in Taylor, 1958, vol. 2, pp. 129–130).

The confusion and conflict over interpretation of the aphasias continued at an international level well into the twentieth century. In France, Pierre Marie (1853–1940) in 1906 reassessed the early studies and added new material from his Bicêtre clinic. As the problem continued to excite heated disputation, in 1908 the Société de Neurologie de Paris devoted three meetings to aphasia in an attempt to resolve the clinical and pathological aspects. Like their European peers at the end of the nineteenth century, some prominent American physicians were thinking and writing on aphasia. Two neurologists and a psychiatrist may be singled out (Fig. 5.9). In 1890, Moses Allen Starr (1854–1932) published 50 cases of sensory aphasia in which Broca's area was not diseased. Charles Karsner Mills (1845–1931) was regarded at the time as "dean" of American neurology, having served an unprecedented two terms heading the American Neurological Association as its president, in 1886 and 1924. He wrote, with W. G. Stiller, an important paper on aphasia (1907) in relation to lesions of the lenticular zone. The Swiss-American, Adolf Meyer (1866–1950), firmly believed in the neural basis of behavioral disorders; his work on aphasia culminated in a Harvey Lecture on "The present status of aphasia and apraxia," in 1910.

After this period of uncertainty and contention, the diagrams and arguments remained relatively neglected until a generation or so later, when an American school re-examined the early work, led by Norman Geschwind (1926–1984), a Boston neurologist who reintroduced Wernicke's concept of the "disconnection syndrome" (1965). With the gradual acceptance of the idea that speech depends on several groups of cortical neurons connected by association pathways, it followed that dysfunctions could be caused by disruption of those connections at many points. By coupling careful clinical observation (in the manner of Hughlings Jackson) with the latest neuroanatomic and physiologic knowledge, Geschwind placed behavioral neurology on a firm foundation and substantiated his findings with solid evidence of asymmetry related to handedness, as discussed in Chapter 6.

The revival of the disconnection syndrome had been preceded by several significant discoveries by American investigators. The first related work was the report by Wilder Penfield and Theodore Rasmussen (1949) that ongoing speech is arrested by electrical stimulation of the exposed cerebral cortex in patients under local anesthesia. Those authors not only confirmed the localizations of both Broca's and Wernicke's areas, but they also identified a third speech-related region in the supplementary motor cortex (Fig. 5.10). Ten years later, G. A. Wada and Rasmussen (1960) reported the results of injecting sodium amytal into the left carotid artery to unilaterally anesthetize the speech area, producing muteness. Somewhat later, G. A. Ojemann and C. Mateer (1979) extended the electrical stimulation technique to a more elaborate study of verbally expressed cortical faculties, indicating that the arrest of naming and reading is due to a deficit in output from short-term memory (*see* Ojemann, 1982).

Sorting out the components of speech and their anatomic correlates within the brain is far from complete. What can be declared is that the long story of discovery of human brain function continues from astute observations of speech behavior and neuroanatomy. We now turn to the early history of discoveries of functions of the frontal cortex that used different methodologies.

FRONTAL LOBE EXPERIMENTS BY STIMULATION

As the enlightened age of inquiry deepened in the eighteenth century, experimentation in the sciences, and more focally, in the relation of brain to behavior, supplemented the relatively constricted observational studies in humans. With experimental (i.e., subhuman) animals, two methodologies were available: stimulation (usually electrical but also chemical) of the brain surface and surgical ablation or damage to various parts of the brain. The experimentalists were not slow to point out the advantage of using different approaches to solving a problem, as exemplified a century ago in a study of eye movements in dogs: "[I]t is of unmistakable value, that the conformity of the experiments by both methods [stimulation and ablation] warrants the assumption that there is some certainty in the knowledge acquired" (Munk, 1890, p. 64). Again, some of the most effective work was carried out in France. Indicative of the general emphasis on experimentation was a contest provided by the Académie Royale de Chirurgie, which offered annual "prize questions" related to problems encountered in its members' practices. The question for 1768 (Neuburger, 1981, p. 172) elicited a series of studies by Louis Sebastian Saucerotte (1741–1814) on 28 dogs. From the results obtained by ablation, Saucerotte concluded that not only did the pathways controlling movement originate in the cerebral hemispheres, but also that their removal afflicted muscles on the opposite side of the body. In addition, he found that the anterior part of the cerebrum controlled movement in the lower limbs and the lateral part the upper limbs. Another Frenchman, François Pourfour du Petit (1664–1741), had earlier investigated the contralateral innervation of the motor and visual systems (1710);

Fig. 5.9. Left—Moses Allen Starr, professor of nervous diseases at the College of Physicians and Surgeons of Columbia University (1888–1918), is considered an American pioneer in localization. (From Pearce Bailey, 1975, p. 112.) Center—Charles Karsner Mills was the first professor of neurology at the University of Pennsylvania (1893–1915) and helped develop the clinicopathological method of research on disease. (From McHenry, 1969, p. 330.) Right—Like Freud, Adolf Meyer had a substantial interest in neurology at the beginning of his long career as a psychiatrist, most of it at Johns Hopkins School of Medicine (1908–1950).

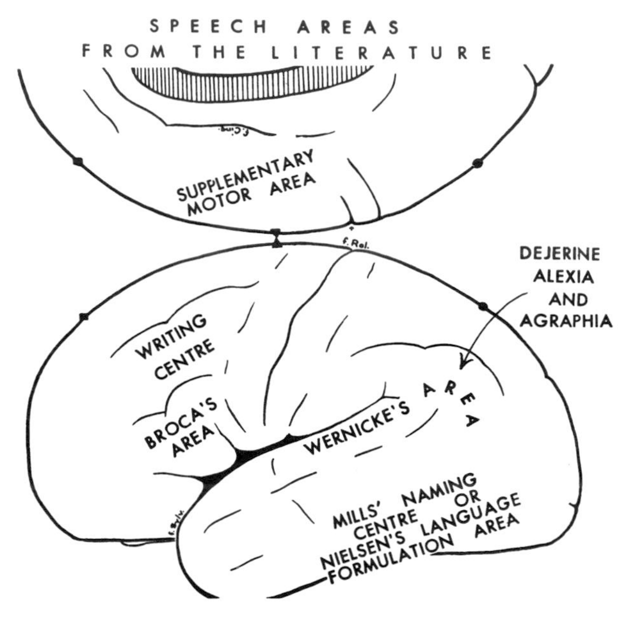

Fig. 5.10. Above—Informal photograph of Herbert Jasper (left), who carried out the electroencephalographic studies on Wilder Penfield's (right) patients at the Montreal Neurological Institute. Below—Speech areas from the midtwentieth-century literature. The center for writing was first described by Wernicke; Nielsen's language area refers to Johan M. Nielsen (1890–1969), the contemporary major figure in aphasia studies on the West Coast. (From Penfield and Roberts, 1959, p. 195.)

unfortunately his findings were ignored until the midnineteenth century. Clarke and Jacyna (1987, p. 425, n.10) suggest that Pourfour du Petit's studies may have been the basis for the "remarkable statement" of Emanuel Swedenborg in 1745 describing the motor homunculus in general as we know it today.

As discussed in the preceding chapter, Gall's assignment of psychological functions to a mosaic of cerebral localization, propounded early in the nineteenth century, marked the beginning of serious consideration of a correlation between brain anatomy and behavior.[4] During that same period, Rolando in Italy reported the effects of a strong electrical current applied to the brains of various farm animals, but those early studies, although demonstrating cerebral control of motor activity, were not widely known and made little impression until they were translated and disseminated by Pierre Flourens in France.

The most influential physiologist of his day, Marie Jean Pierre Flourens (1794–1867; Fig. 5.11), carried out experiments somewhat similar to Rolando's with the same results, but concluded that motor control resided in the cerebellum. Although Flourens's ideas, even when wrong, outshone and outlasted those of his contemporaries, surprisingly, his experiments (reported in 1824) were mostly on birds, and his stimulation by needle pricks was far from precise. Further, he seems not to have heeded Vicq d'Azyr's statement that the brains of birds differ from those of mammals having little or no neocortex (Clarke and Jacyna, 1987, p. 254).

Progress in cerebral localization slowed during the middle decades of the nineteenth century, in spite of the contributions of clinical observers such as Broca, until unequivocal evidence of motor localization was forthcoming. Like a rocket burst, corroboration appeared from Germany in the work of Eduard Hitzig (1838–1907; Fig. 5.12), whom Clarke and O'Malley (1968, p. 508) deemed a typical example of the pure experimental physiologist. Encouraged by movements of the eyes in human subjects "easily" produced "by conducting constant galvanic currents through the posterior part of the head" (von Bonin, 1960, p. 79), and one experiment on a rabbit, Hitzig proceeded to "a definitive solution to the problem."

Assisted by his more experienced friend, Gustav Theodor Fritsch (1838–1927), stimulation and ablation experiments were carried out on the dog's cerebrum (Fritsch and Hitzig, 1870). Their drawing of the loci they identified showed the fairly discrete points at which mild electrical shocks,

[4]*See* Clarke and Jacyna (1987) for a detailed revaluation of Gall in the context of his times.

Fig. 5.11. The brilliant and formidable Pierre Flourens recognized that parts of the brain have specific functions, yet he maintained that sensations are represented equally throughout cortex. Right or wrong, his ideas had great influence because of his exalted academic position at the Collège de France and as permanent secretary of the Académie des Sciences.

passed through a glass capillary 1.5 mm in diameter and filled with physiologic saline (0.7% NaCl), reliably produced contraction of muscles on the opposite side of the body. In two dogs, the response was abolished by careful ablation of the sites of stimulation.

The partnership broke up after those experiments, and Fritsch pursued a career in anatomy, including publication of studies on the brains of fish and electric eels. In contrast, Hitzig, although busy as neurologist and psychiatrist at the University of Halle, continued his experimental work on localization and confirmed and extended the early findings to the monkey, accumulating evidence that disproved Flourens's holistic concept of cerebral equivalence. From firsthand accounts we have two characterizations of these men whose single collaborative paper opened a field in neuroscience still active today. Clarence Herrick wrote (1892a, p. 87) "[I] will not soon forget an accidental meeting with Professors Fritsch and Hitzig, both of whom are splendid specimens of physical development and of German culture at its best." To von Bonin (1960, p. xii), however, "Hitzig was a psychiatrist of renown, a stern and forbidding character of incorrigible conceit and vanity complicated by Prussianism."

The announcement of electrically excitable cortical sites by the young and unknown Germans, Hitzig and Fritsch, was immediately tested by

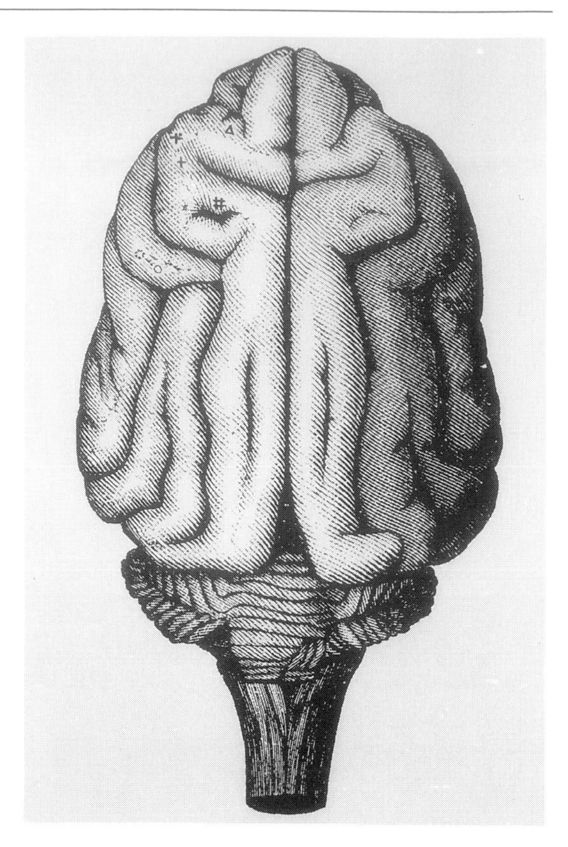

Fig. 5.12. Eduard Hitzig (left) with Theodor Fritsch fulfilled Hughlings Jackson's prediction that movement is localized on the cortex. (Right) In their report, the dorsal view of the dog's brain (Fritsch and Hitzig, 1870, p. 313) showed stimulation points for the muscles of the neck (triangle), anterior leg (+), posterior leg (#), and facial nerve (o-o). (Portrait from Haymaker and Schiller, 1970, p. 230.)

British investigators. David Ferrier (1843–1928; Fig. 5.13), a Scottish physician practicing in London, not only confirmed and extended the German findings (Ferrier, 1873), but also anthropomorphized them: he superimposed the excitable points found in monkeys onto a map of the human brain (1876). In his meticulous experiments, Ferrier used faradic instead of galvanic current, and a bipolar electrode.

During the last decade of the nineteenth century, the same types of experiments were carried out on animals phylogenetically closer to humans. Victor Alexander Haden Horsley (1857–1916; *see* Fig. 5.14, p. 78), London's famous general surgeon who made his mark in neurosurgery, recognized the advantage of cerebral localization of bodily movements in determining where the diseased brain should be operated. With E. A. Schäfer (1888), he reported finding excitable points both in front of and behind the central sulcus in the orang-utang. In search of more precise localization, a few years later C. E. Beevor and Horsley (1890) explored the exposed left hemisphere of an anesthetized orang-utang in 2-mm steps with threshold shocks using an inductorium and platinum wires applied parallel to the surface. When the animal's blood pressure commenced to fail, they opened the right skull and repeated the mapping, noting only a "trifling difference" between the two hemispheres in the positions of responding sites. Their published map is shown in Fig. 5.15 (*see* p. 79); because there was only one animal, they drew no conclusions. Perhaps Horsley's most significant contribution to neuroscience was made 15 years later, when he inspired another collaborator, Robert Clarke, to design a stereotaxic instrument, as will be described in Chapter 9.

Among the host of investigators attracted to the serious study of localization of function, Charles Scott Sherrington (1857–1952; *see* Fig. 5.16, p. 80), "[t]he great English physiologist of the truncated

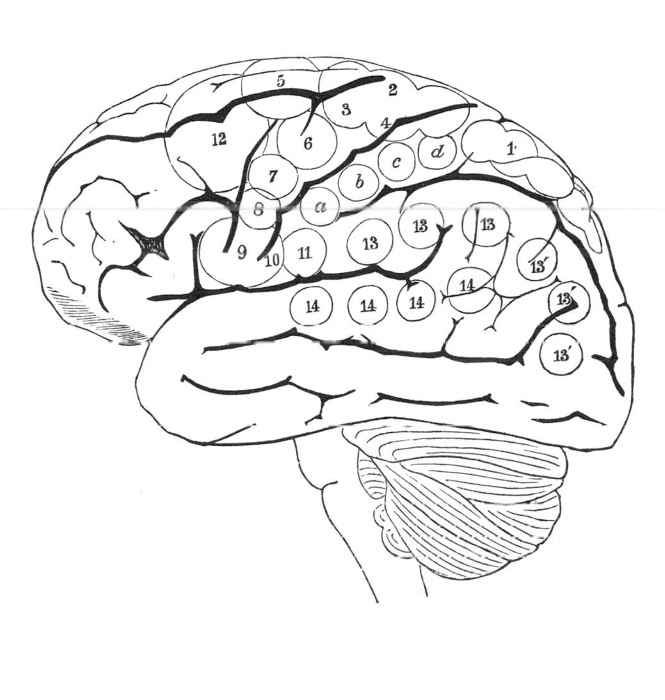

Fig. 5.13. The British school of localization was led by David Ferrier, who initiated experiments in monkeys at the West Riding Lunatic Asylum that confirmed Fritsch and Hitzig's results. Ferrier then boldly superimposed the sites stimulated on the monkey brain onto an outline of the human brain (1876, p. 304).

nervous system of decerebrate animals" (Magoun, 1977, p. 880), quantified the work on anthropoid apes, taking advantage of a month-long visit of the young American surgeon, Harvey Cushing in 1901.[5] After observing Sherrington's crude technique in an experiment on a chimpanzee, Cushing took over and "opened the skull" of a gorilla and an orang-utang, making precise sketches of the cortical surface. Some 15 years later, Sherrington expressed his gratitude to Cushing in a paper summarizing his faradic stimulation and ablation experiments on 22 chimpanzees, three gorillas, and three orang-utangs (Leyton and Sherrington, 1917). The delay in publication was occasioned by Sherrington's earlier "brush" with Horsley: An increasingly sharp exchange of letters between them had appeared in the British medical weekly *The Lancet* (March 10, 1894, p. 642) over an unrelated priority misunderstanding and Sherrington tactfully chose to wait until after Horsley died.

In the late nineteenth century, surgeons who were acquainted with the animal experiments were understandably tempted to discover if the human brain could be similarly stimulated. Such an opportunity was presented to Roberts Bartholow (1831–1904; *see* Fig. 5.17, p. 81), then a practicing physician in Cincinnati and professor in the Medical College of Ohio, in a patient with an eroding tumor of the scalp and skull. Although today we may doubt how informed was her consent—she was reported to be feebleminded—with her permission, wires were inserted just through the dura over the precentral gyrus; stimulation led to contralateral muscular contraction (Bartholow, 1874).

[5] *See* the perceptive biography of Cushing by Elizabeth H. Thomson (1981) for more on this Continental "grand tour."

Fig. 5.14. Victor Horsley (center) with a succession of collaborators worked in a small room at University College, London. This photograph of Sir Victor dictating during an experiment was taken in 1913. (From Paget, 1919, p. 200.)

The physician was soundly castigated for carrying out such an extreme procedure. Within 15 years, six additional human cases were itemized (Beevor and Horsley, 1890) and within another 20 years stimulation of the human brain had become relatively common. A group of 142 cases was reported in 1912 by a German surgical neurologist, Fedor Krause (1856–1937; see Fig. 5.18, p. 82). With monopolar stimulation through a saline-filled glass pipet tapered to a 1.5-mm diameter, he mapped in great detail the precentral sites where stimulation produced muscular contractions (*see* Fig. 5.19, p. 83). He admitted that sometimes he had permitted the anesthesia to lapse (Krause, 1912, p. 293), thus foreshadowing Penfield's famous explorations. Krause found unexcitable areas of cortex, as had Beevor and Horsley; he also noted that cortical excitability to faradization is easily fatigued.

Wilder Graves Penfield (1891–1976; *see* Fig. 5.10, right), the much-traveled American-Canadian neurosurgeon, had seen in Breslau the human electrocorticograms taken from awake patients in Otfrid Foerster's clinic. When Penfield returned to North America in 1928, he was able to carry out explorations of his own at McGill University and later at the Montreal Neurological Institute, which he planned and directed. During his long and productive career (*see* the authoritative biography by his grandson, Jefferson Lewis, 1981), he worked with many collaborators, including the surgeon and neuropathologist, William Cone, and an electroencephalographer, Herbert Jasper. Prior to surgical removal of diseased foci, Penfield identified cortical areas from which mild stimulation elicited motor or sensory responses, changes in speech and vocalization (described earlier), memory of past experiences, and visual and auditory effects. Penfield apparently believed that all surgeons with similar opportunities had an obligation to add scientific data to the knowledge of brain function (Penfield, 1937). He utilized his findings to formulate a "centrencephalic system" to serve both sides of the brain, where integration of function could occur. That hypothesis has not been supported; rather, Penfield is remembered chiefly for his contributions to the knowledge of epilepsy and for his homunculi (*see* Fig. 2.19, p. 26).

Another method of stimulation, called chemical neuronography, although no longer practiced, had a brief but productive vogue. It was brought

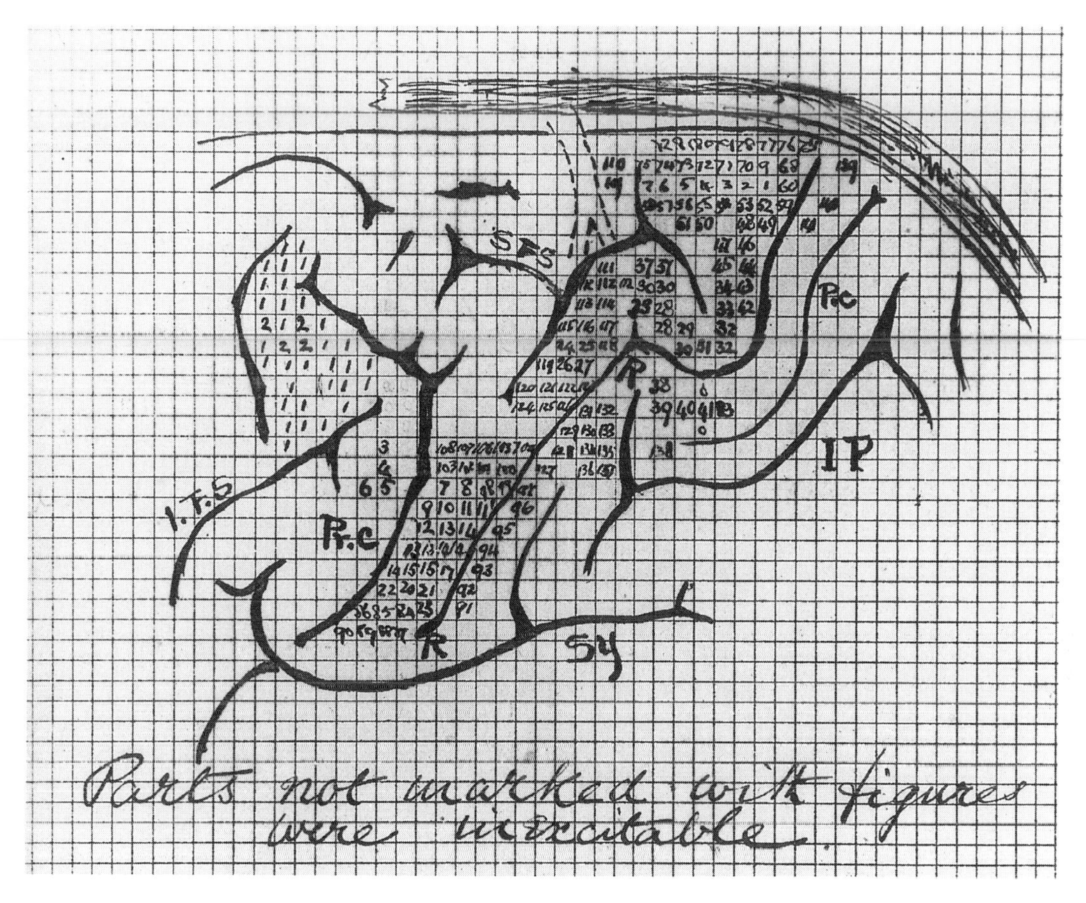

Fig. 5.15. The remarkable precision of Beevor and Horsley's (1890, p. 156) mapping of the orang-utang cortex is shown in this "Photograph of the original drawing of the portion of the cortex examined, made at the time of the experiment."

first to England by a Dutch neurophysiologist, Johannes Gregorius Dusser de Barenne (1885–1940), and then to Yale University in 1930. Magendie had used strychnine (1822) to substantiate his findings of the motor function of the dorsal spine nerve roots. In Dusser's hands,the procedure entailed application of small pieces of absorbent paper soaked in a solution of strychnine to 1-mm^2 areas on the cat's exposed spinal cord or cortex, which rendered the associated cutaneous and deep pressure receptors hypersensitive, thus "localizing" the function of the small poisoned area of the cerebral cortex (Dusser de Barenne, 1916). In later experiments (1924), the electrical disturbance produced by the simultaneous firing of the underlying cells, which did not cross synapses and could be picked up at the nerve fibers' terminals, were recorded. This method tested physiologically the connections found by the anatomists' dissections of dead brains, but divulged nothing about the pathways traversed from origin to destination; its major usefulness was in demonstrating the multiplicity of corticocortical connections (*see* Fig. 5.20, p. 83).

After Dusser's death in 1940, the remaining members of the team, neurophysiologist Warren Sturgis McCulloch (1898–1969), and neuroanatomist Gerhardt von Bonin (1890–1979), were lured to the Illinois Neuropsychiatric Institute at Chicago by Percival Bailey (1892–1973), neurologist and director of the INI (*see* Fig. 5.21, p. 84). Their imaginative work was augmented by a stream of bright young postdoctoral students and added to the luster of Chicago's strength in neuroscience research in the 1930s and 1940s, when laboratories at Northwestern University Medical School (S. Walter Ranson) and the University of Chicago (Ralph W. Gerard) were at their peak. The method of "strychninization" was briefly useful in showing neuronal connections before the advent of more precise methods based on silver staining, axonal flow, and the electrophysiologic evoked potential.

Fig. 5.16. Charles Scott Sherrington was the nucleus of a school of neurophysiology at Liverpool, subsequently at Oxford; among many other achievements, he mapped the cerebrum of a gorilla and an orang-utang. (Portrait by R. G. Eves, 1927, in Woodward Biomedical Library, University of British Columbia, Vancouver, Canada.)

FRONTAL LOBE EXPERIMENTS BY ABLATION

Many early experimentalists who stimulated the cerebral surface also carried out the complementary experiment of removing parts of the brain as a second approach in their search for localization of function. Indeed, during an era (the late nineteenth century) when experimental ablations were most widely carried out, the attempts of Freidrich Leopold Goltz (1834–1902), a German physiologist, were particularly influential in establishing current principles of cerebral localization. His first work (1869) with decapitated ("spinal") frogs demonstrated a range of activities preserved through spinal reflexes (e.g., jumping when stimulated, swimming) and were followed by studies on mammals. On the evidence of his carefully tended decerebrate or decorticate dogs (1888) Goltz concluded that the doctrine of constricted centers was untenable, thus foreshadowing the holistic theories of brain function. Not surprisingly, the frontal lobes, easily accessible from the top of the skull, were widely subjected to this method of study. In the human condition, Hitzig (1874, p. 261; quoted in translation by Halstead, 1947, p. 25), declared that the frontal lobe is the seat of abstract intelligence, a somewhat expansive statement as he used dogs and monkeys for his experiments. The frontal lobe "guides and integrates the personality," according to Flechsig (translated in von Bonin 1960, p. 16), an observation supported by Phineas Gage's change in character after extensive frontal damage, mentioned above.

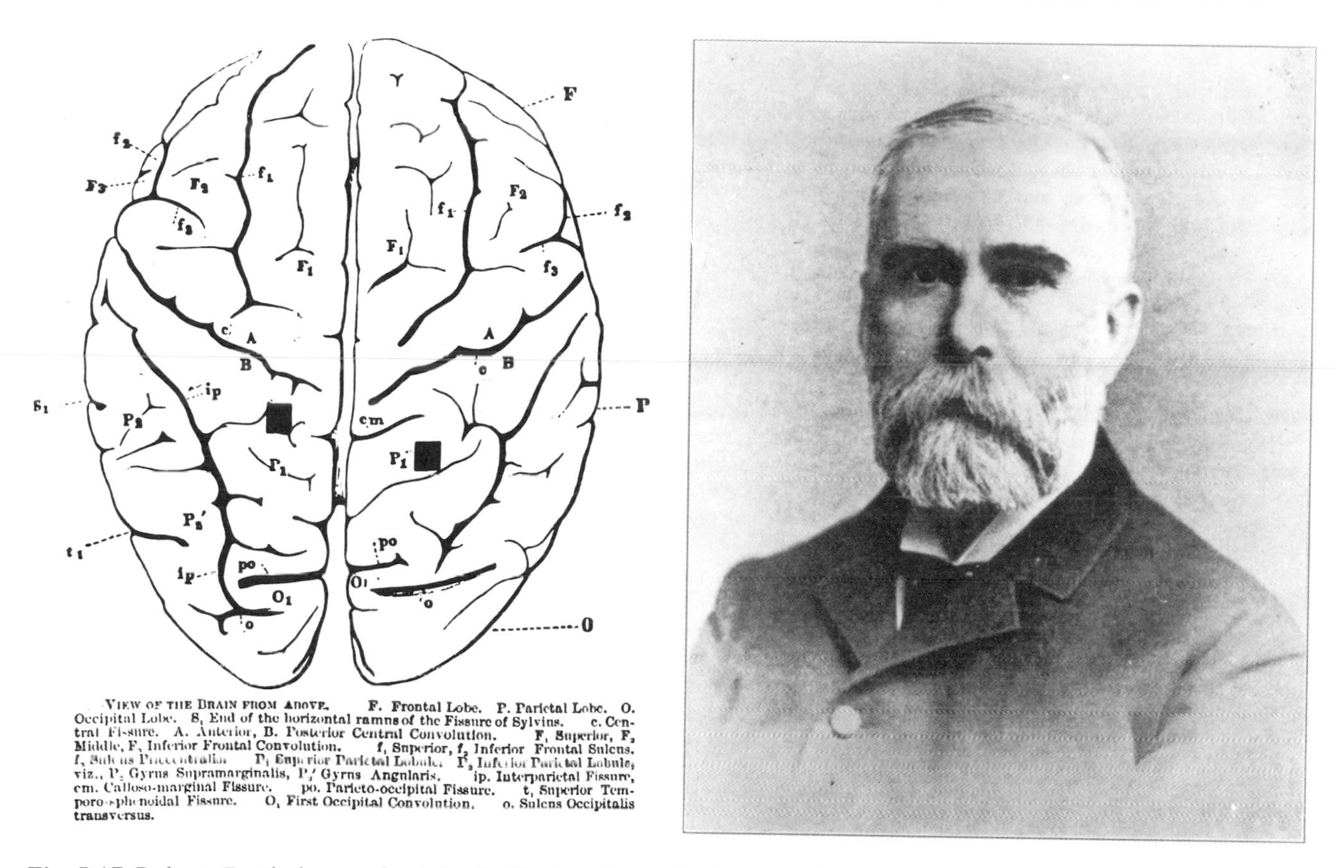

Fig. 5.17. Roberts Bartholow, a physician in Cincinnati, was the first American to stimulate the living human cortex. The points of stimulation (black squares) as described by Bartholow (1874) were superimposed on a sketch by Ecker (1873, p. 9) of the human cerebral hemispheres (×1.5).

Laboratory observations of frontal lobectomized monkeys by Leonardo Bianchi (1895) at Naples, Italy, were of a similar nature:

> While intact animals let nothing escape them and are constantly poking about and taking stock of their environment, mutilated monkeys allow a number of things to pass unobserved. . . . Memory becomes enormously reduced. . . . The mutilated monkey does not use past experience, but persists in repeating the same actions, without profiting from their futility so as to arrive at a determined object (translation in Bianchi, 1922, pp. 71, 73).

Bianchi's monograph, *The Mechanisms of the Brain and Function of the Frontal Lobes* (1922), opened with a chapter on the "History and Evolution of the Frontal Lobes," followed by a summary of his many observations of animals and man with frontal lobe injuries. The frontals were the "organ of intellect" as anyone could see that his dogs rendered afrontal were "weak-minded." A cautionary note was added by John Fulton who perceptively remarked: "Both Ferrier and Bianchi were good observers, but the behavioral changes which they described were difficult to appraise because at the time they worked objective methods had not yet been developed for analyzing behavioral disturbances in quantitative terms" (1951, p. 51).

Not everyone agreed that the frontal lobes were so necessary, however. Students learned from the famous English physiology textbook edited by Schäfer (1898, p. 772) that "complete severance" of the prefrontal region did not produce obvious symptoms in monkeys. Relatedly, Jacques Loeb (1859–1924), a German émigré to the United States where he became a prominent experimental biologist, wrote: "There is perhaps no operation which is so harmless for a dog as the removal of the frontal lobes" (1900, p. 275).

Fig. 5.18. Fedor Krause, German surgeon–neurologist, in 1908 published a textbook, later translated into English as *Surgery of the Brain and Spinal Cord*, that appeared in 60 reprintings/new editions within 60 years.

More detailed American investigations, however, were disclosing grave deficits, as Sherrington noted in his *Integrative Action of the Nervous System*: "New methods of promise seem to me those lately followed by Franz, Thorndyke [sic], Yerkes, and others [which permit study of] the influence of experimental lesions of the cortex on skilled actions recently. . . acquired" (1906, p. 307). Shepherd Ivory Franz (1874–1933; Fig. 5.22), an experimental psychologist whose extensive work on the cerebrum of lower primates has been neglected, resolved the apparent paradox. After completing his doctorate at Columbia in 1899, Franz pursued postdoctoral research in Bowditch's physiological laboratory at Harvard Medical School where he extended the earlier European work by combining frontal lobectomies in cats with Thorndike's "puzzle box" studies of animal learning. In 1904, Franz established a laboratory for physiological psychology at McLean Hospital at Waverly, Massachusetts, and commenced to elaborate the concept of "re-education" of behavior, initially impaired or abolished by experimental or clinical injuries of the frontal lobes. Moving to Washington, DC, in 1907 as director of laboratories at the Government Hospital for the Insane (St. Elizabeths Hospital), Franz extended his studies of frontal lobectomy from cats to monkeys, with essentially similar results. He differentiated between the unchanged emotional condition of the animal after removal of the influence of the frontal lobes and subtle behavioral changes, writing that "the associational loss . . . could not be determined by simple observational methods" (1907, p. 63). Franz's valu-

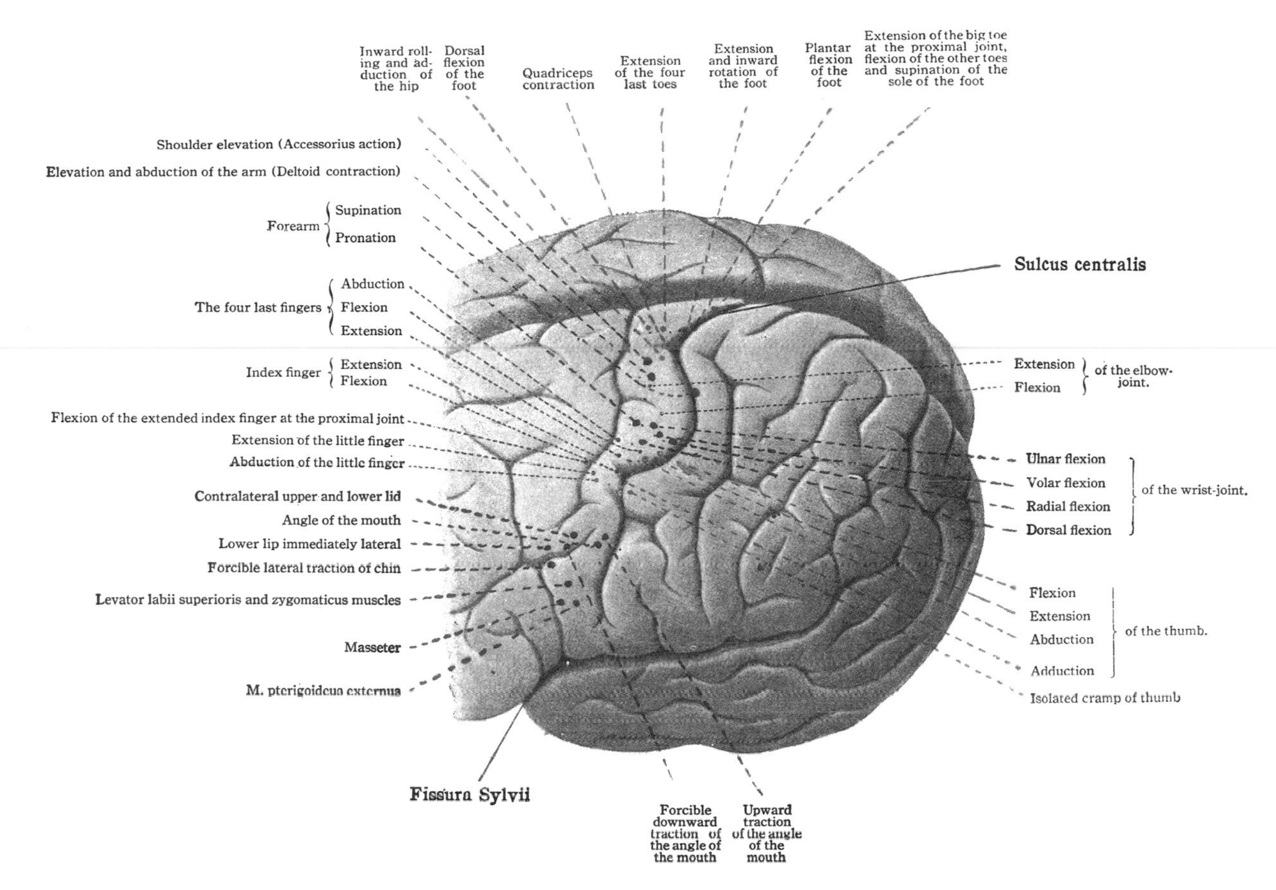

Fig. 5.19. Fedor Krause's map showed the left hemisphere with "results obtained by stimulation" (caption of his Fig. 66, 1902). There is doubt that all points were from human patients, some data from subhuman primates having been incorporated (*see* Clarke and Dewhurst, 1972, p. 116, Fig. 140).

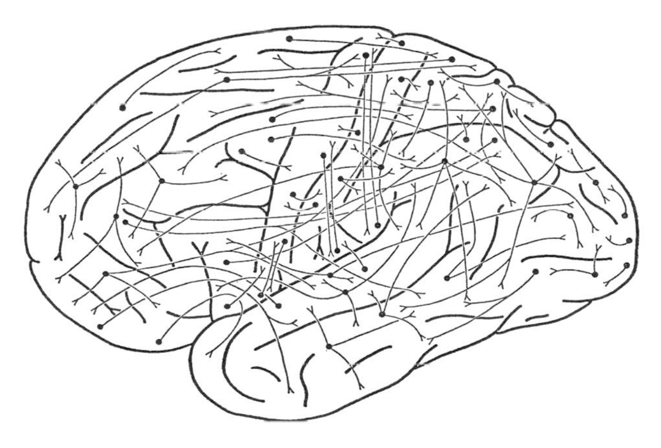

Fig. 5.20. "Corticocortical connections as revealed by physiological neuronography in the chimpanzee" (caption in Bailey and von Bonin, 1951, p. 239, Fig. 112). *See* text for explanation and Fig. 5.21.

able contribution was the application of systematic and sophisticated psychological testing before and after operation. He declared: "[T]he cerebrum as a whole may be looked at as a very labile organ because of the numerous possibilities of connections which may be made" (1915, p. 158). Experiencing enormous difficulties without clear conclusions from motor cortex stimulation in the higher anthropoids, Leyton and Sherrington (1917, p. 144) heartily agreed with Franz.

After a two-decade lapse, the revival of behavioral and physiological studies of effects that follow lesions of the prefrontal cortex in lower primates was initiated by Carlyle Jacobsen in John Fulton's Laboratory of Physiology at Yale in 1930. Jacobsen (1934) was interested in the retention of skilled movements after motor and premotor lesions. As a consequence of Harvey Cushing's presence (*see* Fig. 5.23, p. 85) at "The Laboratory" (1933–1937), operative technique, record-keeping, and postoperative care of the animals were

Fig. 5.21. In the basement research laboratory at the Illinois Neuropsychiatric Institute, Warren McCulloch, Percival Bailey, and Gerhardt von Bonin (from left) for a decade used strychnine neuronography to study interconnections at the cortical and subcortical levels.

Fig. 5.22. Shepherd Ivory Franz, one of the pioneer physiological psychologists in the United States, during the early twentieth century helped lead psychology in the direction of the objective quantification of behavior.

superb, which contributed greatly to the decisiveness of the results. Fulton presented the effects of lesions of the primate prefrontal cortex in two chimpanzees, "Becky" and "Lucy," at the International Neurological Congress in London in 1935, thereby launching the idea of frontal lobectomy as a therapy for certain mental illnesses, with far-reaching consequences.

António Caetano de Abreau Freire De Egas Moniz (1875–1955; *see* Fig. 5.24, p. 86), neuropsychiatrist at the University of Lisbon, Portugal, was in the audience and was so impressed that on his return home he arranged for an associate neurosurgeon, P. Almeida Lima, to perform frontal lobectomies in mentally ill patients whose disease did not respond to treatment. With publication of the apparently favorable consequences, therapeutic frontal lobectomy swept rapidly through Western medicine. From his first 20 patients in whom prefrontal connections had been severed, reported as operative trials ("tentatives operationes"), Moniz concluded that the frontal lobes can substitute for each other in psychic integration and that their bilateral destruction does not produce a total disruption of psychic life (Moniz, 1936).

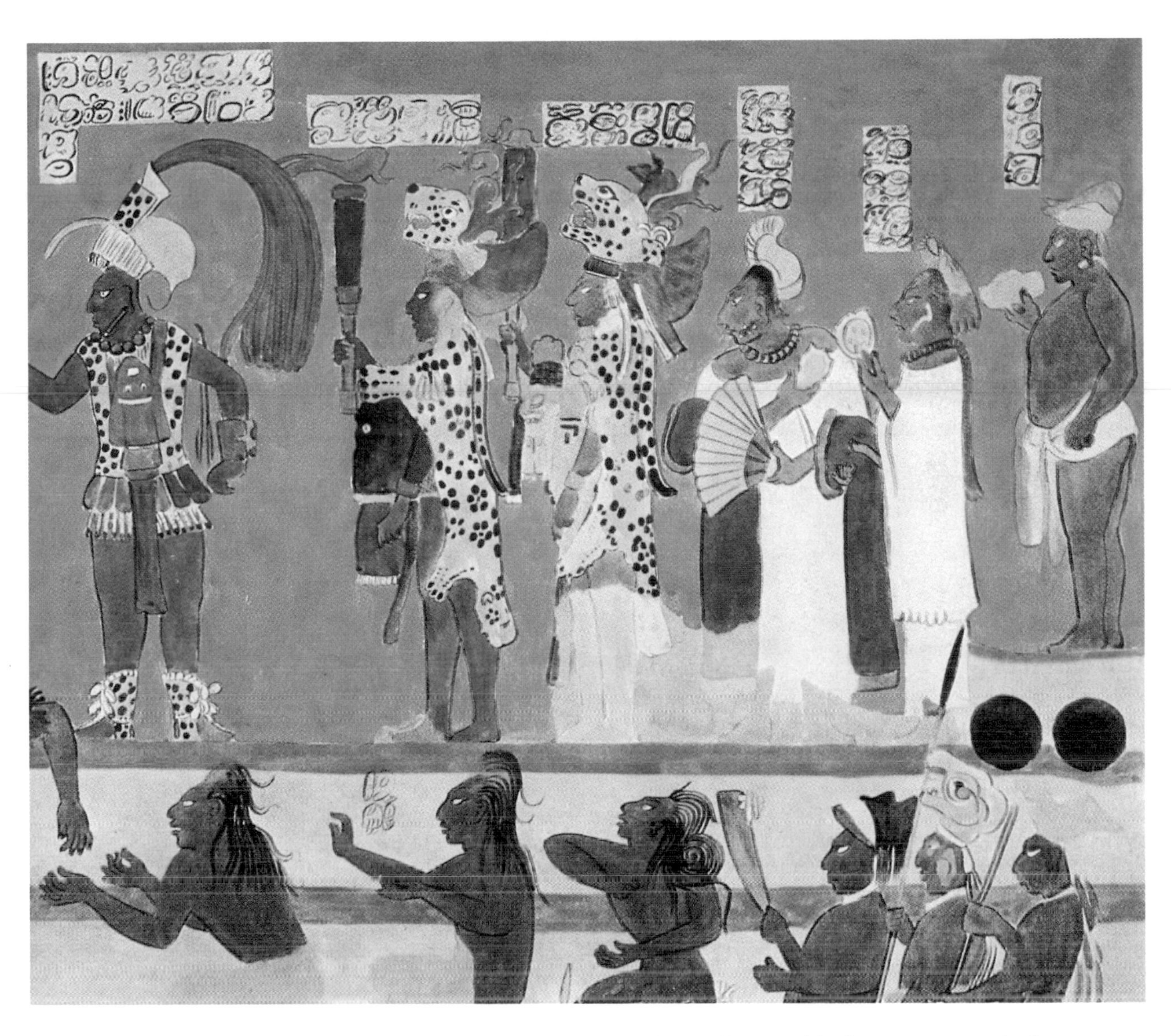

Plate 1 (Fig. 2.18 from Chapter 2). Right-handedness in the Amerindians is illustrated in the Mayan "Mural of Bonampak" (ca. AD 850) in southern Mexico, depicting blood sacrifices to a displeased deity. (Adapted from Davidson, 1962, pp. 408–409.) *See* discussion in Chapter 2.

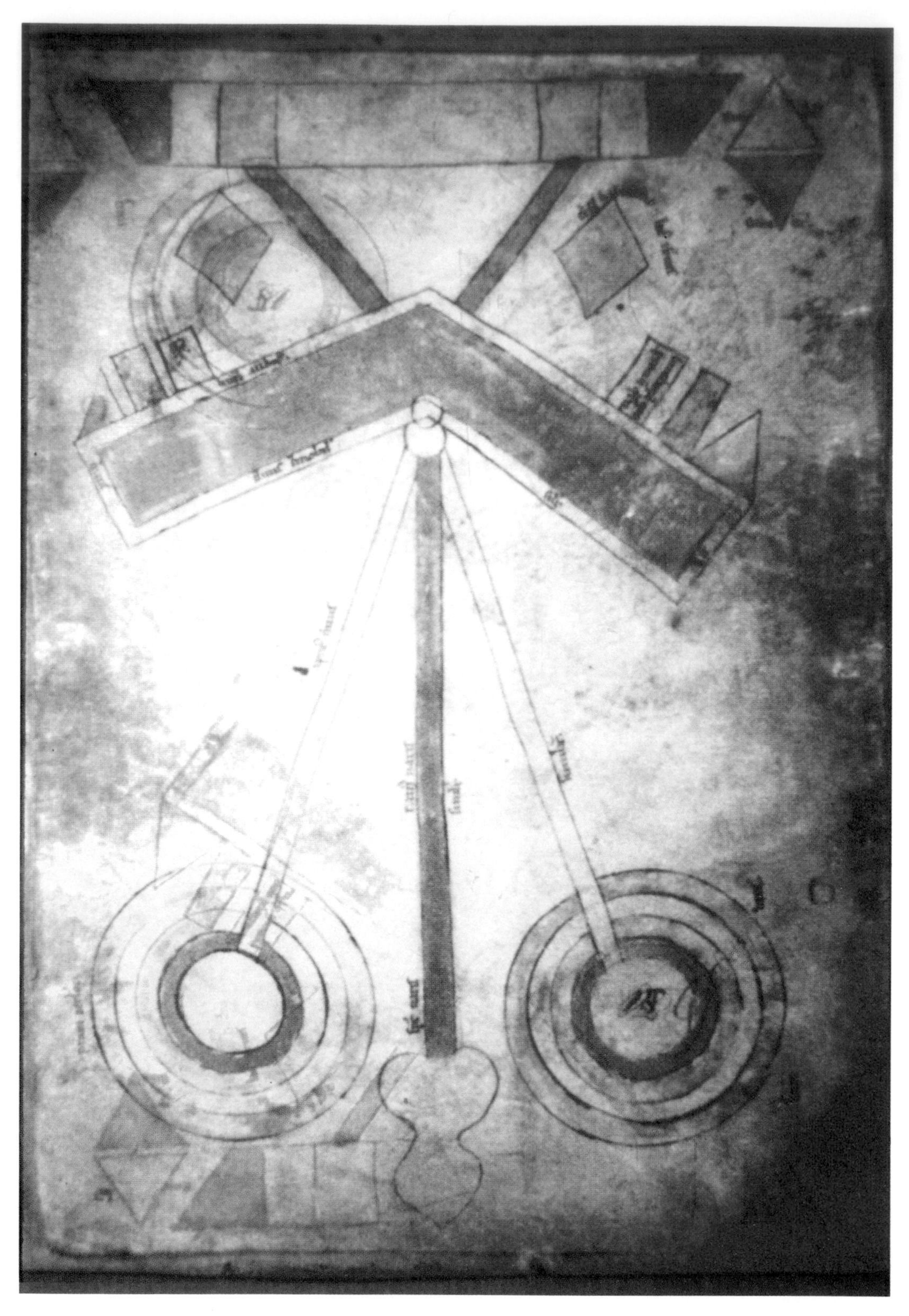

Plate 2 (Fig. 3.2 from Chapter 3). A drawing from a Latin manuscript of the late twelfth to early thirteenth centuries. Above the two eyes and their converging nerves is the brain with its coverings and the attached "cells," only one of which is labeled. On the right, the diamond shape is inscribed "dwelling place of the brain or the place of reason." (M. R. James, 1907, vol. I, pp. 218, 219. From Goncille and Caisu MS 190/223 f.br.) *See* discussion in Chapter 3.

Plate 3 (Fig. 4.8 from Chapter 4). The frontispiece of Felix Vicq d'Azyr's *Traité d'anatomie et de physiologie* (1786) foretells the delicacy and detail of this beautiful publication (*see* Fig. 4.7), only the first volume of which appeared owing to the author's early death. *See* discussion in Chapter 4.

Plate 4 (Fig. 5.2 from Chapter 5). An early account of a speech disorder, the miracle of Zacharias as described by St. Luke, relates that when an angel appeared beside the altar to tell Zacharias that his wife, Elizabeth, would bear a child, Zacharias dropped his censer and exclaimed "That cannot be." Immediately, he was struck dumb for disbelieving Heaven's messenger, although his writing was unimpaired, as his companions indicate. (From the Gospel Book of Henry III, AD 1043–1045, Patrimonio Nacional, Madrid.) *See* discussion in Chapter 5.

Fig. 5.23. Cushing is shown with another guest of the Harvey Cushing Society (now the American Association of Neurological Surgeons) meeting in New Haven in May, 1935. (From Thomson, 1981, facing p. 270.)

In the United States, Walter Freeman and James Watts, professors of neurology and neurosurgery at George Washington University in the District of Colombia, initiated a major program in *Psychosurgery in the Treatment of Mental Disorders and Intractable Pain*, the title of their report published in 1942. Except for the difference in the instruments involved, the similarities between this "operation" and that suffered accidentally by Phineas Gage in 1848 are obvious. Moreover, as the postoperative periods became extended, an increasing number of patients presented behavioral and cognitive debilities similar to those displayed by Gage. A quantitative study of 50 lobectomies by Ward C. Halstead led him to write: "The frontal lobes . . . are the portion of brain most essential to biological intelligence" (1947, p. 149). Disturbing reports by Milton Greenblatt in Boston, Fred Mettler in New York, and others, gradually discredited frontal lobectomy. The Moniz–Freeman–Watts procedure was finally abandoned due in part to the advent of psychopharmacological methods for treating mental illness and the introduction of other operative techniques.

An alternative approach to the imprecise destruction and aftereffects of prefrontal lobectomies became possible with identification of connections between that part of the cortex and the dorsomedial

Fig. 5.24. World reknown was bestowed on Portuguese neurosurgeon Egas Moniz for two achievements: visualization of the cerebral vasculature and the first human frontal lobotomies, which he claimed had favorable results. Portugal honored him philatelically in 1967.

nuclei of the thalami (*see* Clark, 1948). In 1947, E. A. Spiegel, H. T. Wycis, M. Marks, and A. J. Lee published in *Science* their version of a stereotaxic apparatus for humans with which precise lesions in targeted nuclei could be placed, thereby creating a new branch of neurosurgery.

Uncertainties about the motor function of the frontal lobes, generally localized to the more posterior gyri, had been banished by an overwhelming body of evidence that commenced with the work of Hitzig and Fritsch, as described. The more anterior frontal lobes, the so-called prefrontals, however, remained mysterious. Since Schiff's work in 1875, their direct stimulation was known to produce visceral changes. The anatomic connections between the prefrontal cortex and deeper structures were demonstrated in 1895 by Constantin von Monakow (1853–1930), who described pathways from the thalamus to the prefrontal regions. The abundance of corticocortical fibers disclosed by chemical neuronography and other methods has been mentioned.

The fact that the prefrontal lobes have so many connections has given rise to a host of proposals regarding their function. The most thorough modern study is the ongoing work of Joaquin M. Fuster, who has proposed a "conceptual outline" of his own. Through detailed experimental analysis of behavioral performance, motility, and emotional reactions in animals with precisely placed prefrontal lesions, he has pinpointed "processes of attention, short-term memory, and suppression of interference from internal and external sources" (1980, pp. 143–144). Fuster believes those processes are organized temporally by the prefrontal cortex to make possible purposive sequences of cognitive and motor acts, a hypothesis requiring proof by prolonged and sophisticated experimentation on the part of many investigators.

THE PARIETAL LOBES AND EVOKED POTENTIALS

During evolution, the human parietal lobes pushed upward behind the frontals, which moved forward, whereas the visual and auditory cortical areas could move only back and downward, respectively. Constituting about one-fifth of the total cortical area in man, only about one-third of the parietal surface is exposed, the remainder being buried in sulci. According to the English neurologist, Macdonald Critchley, who was Senior Physician at Queen Square and from whose comprehensive book, *The Parietal Lobes*, the foregoing information is derived, disorder here causes a greater variety of clinical signs than it does in any other cerebral region (1953, p. 411).

After Meynert (1868) divided the cerebrum into anterior, motor quarters and posterior, sensory quarters on the basis of cell morphology and after Flechsig concluded (1896) that the parietal lobe was an associational area because in his hands it was electrically nonexcitable, parietal lobe function remained uncertain. Ferrier (1873), as mentioned, located motor areas in the monkey not only in the usual precentral gyrus but also in the postcentral and superior temporal gyri, a configuration recalling the schematic diagrams of the neural substrates of speech. The experiments of Horsley and

Schäfer (1888) did not help, for they identified areas in both precentral and postcentral gyri in monkey and orang-utang where stimulation elicited contraction in contralateral muscles. The viewpoint of "the stern and sterling" Hermann Munk (1839–1912), that the motor cortex was also sensory, was upheld by Dusser de Barenne, whose chemical neuronography demonstrating the wide-ranging corticocortical connections of the frontal cortex has been described. Dusser found that strychnine placed on either side of the central (rolandic) sulcus greatly exacerbated sensitivity to somatic sensory stimulation. Shown first in cat (1916) and, after the disruption of the Second World War, in monkey, he outlined the first major subdivisions in the sensory cortex (1924). Again, the absence of clear-cut answers created confusion about interpretation as well as doubt about method. The time had come for the new, direct technique waiting in the wings, the sensory evoked potential.

The investigations of electrical activity in the brain during the midnineteenth century commenced with studies of the centrifugal activity of the brain. We can recall Du Bois-Reymond's discovery that the contraction of muscle following nerve stimulation was accompanied by centrifugal current along the nerve to the muscle and that Fritsch and Hitzig applied electrical shocks to the dog cortex to produce limb movement. After those novel and spectacular beginnings, the centrifugal phenomena were studied widely. In 1875 the correlary centripetal experiment was announced by a young and inquiring physiologist at the Liverpool Royal Infirmary. Like Matteucci some 30 years earlier, Richard Caton (1842–1926; Fig. 5.25) looked for electrical currents in the brain. Whereas the Italian found the demarcation current between cut and intact brain surface, the Englishman saw what is now called the evoked potential (a term claimed to have been coined by Gerard, 1975, p. 468) in response to sensory stimulation. Caton reported finding action potentials picked up from the frontal lobe of an anesthetized rabbit when, for example, light was directed to the contralateral retina (1875). Similar findings, independent and 15 years later, appeared in the doctoral thesis of Adolf Beck (1863–1942) at Cracow. Not having cameras, both he and Caton had to draw the action potentials they observed. The first photographs of evoked potentials were published in 1913 by a Ukrainian physiologist, Vladimir Vladimirovich Prawdicz-Neminsky (1879–1952). The international fencing over priority of discovery of evoked potentials and of the variation in the underlying "spontaneous" cerebral action currents (the electrocorticogram) that followed Beck's publication of his observations was apparently ended in 1891 by Caton's dignified letter to the editor of *Centralblatt für Physiologie*, citing his prior publications.

"Spontaneous" background electrical activity was eventually recorded through the intact scalp and skull of humans as "brain waves" by Hans Berger (1873–1941) in Germany (1929; Fig. 5.26). He described "checking" of the alpha rhythm by opening the eyes or by a sudden sound. There were skeptics, however (Berger was relatively unknown), until Adrian and Matthews (1934a,b) confirmed and extended the finding to a variety of species, joining a flurry of investigations. Ten years after Berger, the transient negative deflections evoked by sensory stimuli were noted on the electroencephalogram by Pauline Davis (1939), who saw not only checking in the EEG, but also noted the "anticipatory potential" that in some individuals precedes the on-effects of sensory stimuli. The anticipatory on-effect is diffuse, variable, and subject to the "psychological set" of the anticipation. Those findings may be said to have foreshadowed the contingent negative variation described by Grey Walter as the "slow potential waves in the human brain associated with expectancy, attention and decision," the title of his paper in 1964.

The full possibilities of the sensory evoked potential as a research tool for cerebral studies was utilized initially only in acute experiments on laboratory animals. When Wade Hampton Marshall (1907–1972; Fig. 5.27) shifted his graduate field at the University of Chicago from biophysics to physiology, he brought to Gerard's laboratory the skill and knowledge necessary to design and construct amplifier and stimulator circuits specifically attuned to the study, first, of peripheral nerve action potentials (Gerard and Marshall, 1933; Marshall and Gerard, 1933) and then moved to the higher, more complex level of evoked potentials in the central nervous system (Gerard, Marshall, and Saul, 1936). They recorded both "spontaneous" and evoked electrical activity from auditory, visual, and somesthetic stimuli, feeding the amplified signal through a loudspeaker, a more sensitive device for

Fig. 5.25. The first demonstration of spontaneous cerebral electrical activity and its modification after sensory stimulation was made by an English surgeon, Richard Caton, in 1875. With electrodes placed directly on the cortex of anesthetized animals and shining a light on the retina, he visualized the passage of current with a mirror galvanometer. (From Brazier, 1959, p. 50.)

distinguishing potential changes than the faint trace on the cathode-ray tube of the early oscillograph. The electrode was manipulated by a modified Horsley–Clarke stereotaxic apparatus built locally to their specifications. To locate the position of the electrode tip, Gerard borrowed photographs of stained serial sections of the cat brain from Ranson at Northwestern University, superimposed grids on them, and claimed priority in producing the first atlas of the cat's brain. The arrangement made it possible to correlate precisely the anatomic structures with the electrical activity generated by specific sensory stimuli. As Gerard wrote many years later: "We followed evoked potentials . . . from sight, sound, touch, and proprioception into all sorts of regions where these sensory impulses were not supposed to go. . . ." (1975, p. 468).

Later, at Johns Hopkins Medical School, Marshall inaugurated a second program of cortical mapping, using a state-of-the-art amplifier built during a summer at Harvard in cooperation with Albert Grass in Hallowell Davis's laboratory. With cat or macaque sufficiently anesthetized to minimize the Berger rhythm, and cotton thread electrodes moistened with Ringer's solution positioned on the pial surface, they carried out "observations with the facility comparable to that enjoyed in studies of the axon potentials of isolated nerve trunks" (Marshall, Woolsey, and Bard, 1937, p. 389). With discrete tactile stimulation (movement of a few hairs or touch of a von Frey hair on bare skin), they established that the responses appear only on the parietal side of the central sulcus, that they are stable, can be inhibited by another nearby similar stimulus, and correspond roughly with the topography of the motor precentral points, a feature that Caton had noted. In the more detailed publication (1941), studies on additional animals verified the surface-positive wave in response to discrete stimuli on the contralateral postcentral gyrus and central sulcus (*see* Fig. 5.28, p. 91), and demonstrated an overlap of submaximal responses. In England, Adrian and Moruzzi (1939) examined the strychnine-evoked potential, with loudspeaker effects dramatically described by Moruzzi many years later (*see* L. H. Marshall, 1987, p. 228). When Wade Marshall's interest shifted to vision, Clinton N. Woolsey brilliantly continued the mapping studies both at Hopkins and later at the University of Wisconsin. He and Rose (Woolsey, 1943; Rose and Woolsey, 1943) identified a second somatosensory area (S II) in the parietal lobe.

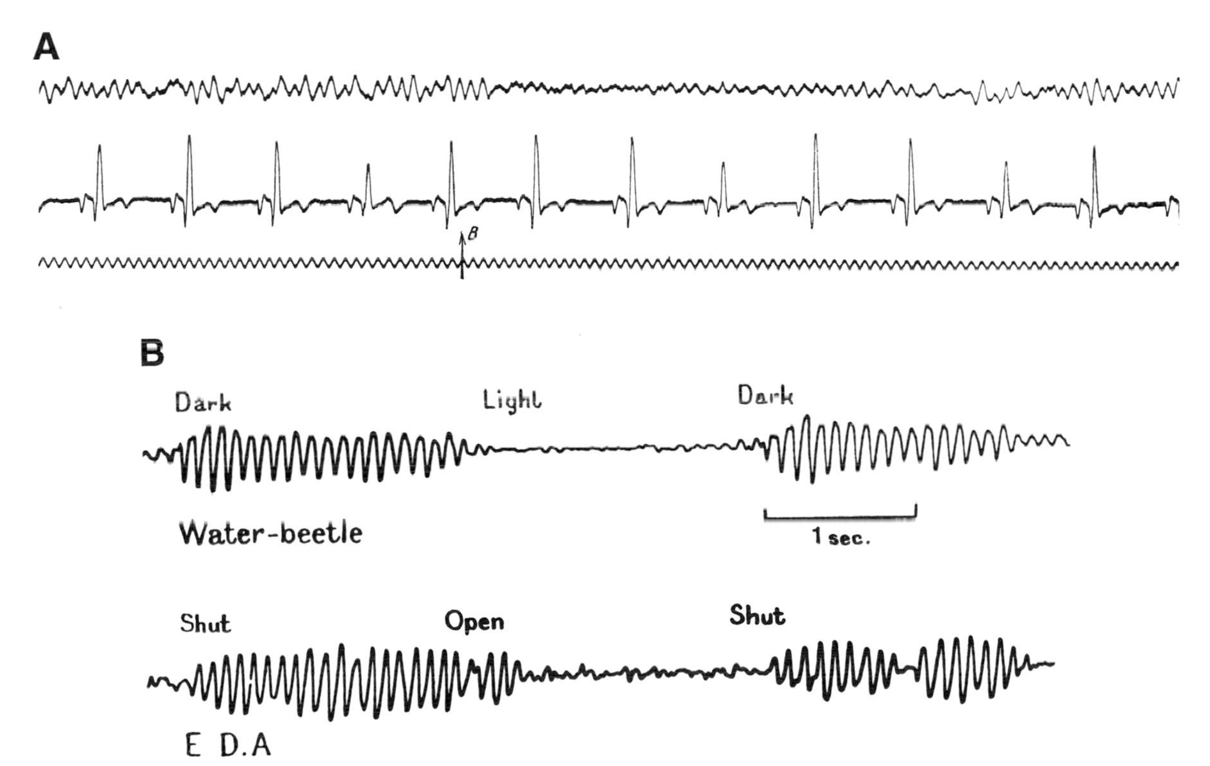

Fig. 5.26. Johannes (Hans) Berger, a "founder of psychophysiology," recorded in 1924 the first electrical potentials obtained from the intact skull of man. (Portrait signed in 1933, from Haymaker and Schiller, 1970, p. 172.) A—The first recording in man of alpha-wave blocking by sensory stimulation. Top—EEG from forehead and occiput. Middle—Electrocardiogram. Bottom—Time in 0.1 s, *B* marks stroking subject's right hand. (From Berger, 1929.) B—Recordings from man (E.D.A.) and beetle demonstrating the ubiquity of alpha waves and their susceptilibity to visual input. (From Adrian and Matthews, 1934b, p. 373, Fig. 15.)

From their first detection, oscillations in the electrical currents picked up on the cerebral cortex or on the scalp were being scrutinized for what they represented in terms of their underlying electrical components as well as what they revealed about the activity of the responsible neurons. In 1934, Edgar Douglas Adrian (1889–1977; Fig. 5.29) at Cambridge University, already the author of two monographs on fundamentals of nervous system activity, with Brian H. C. Matthews published experimental

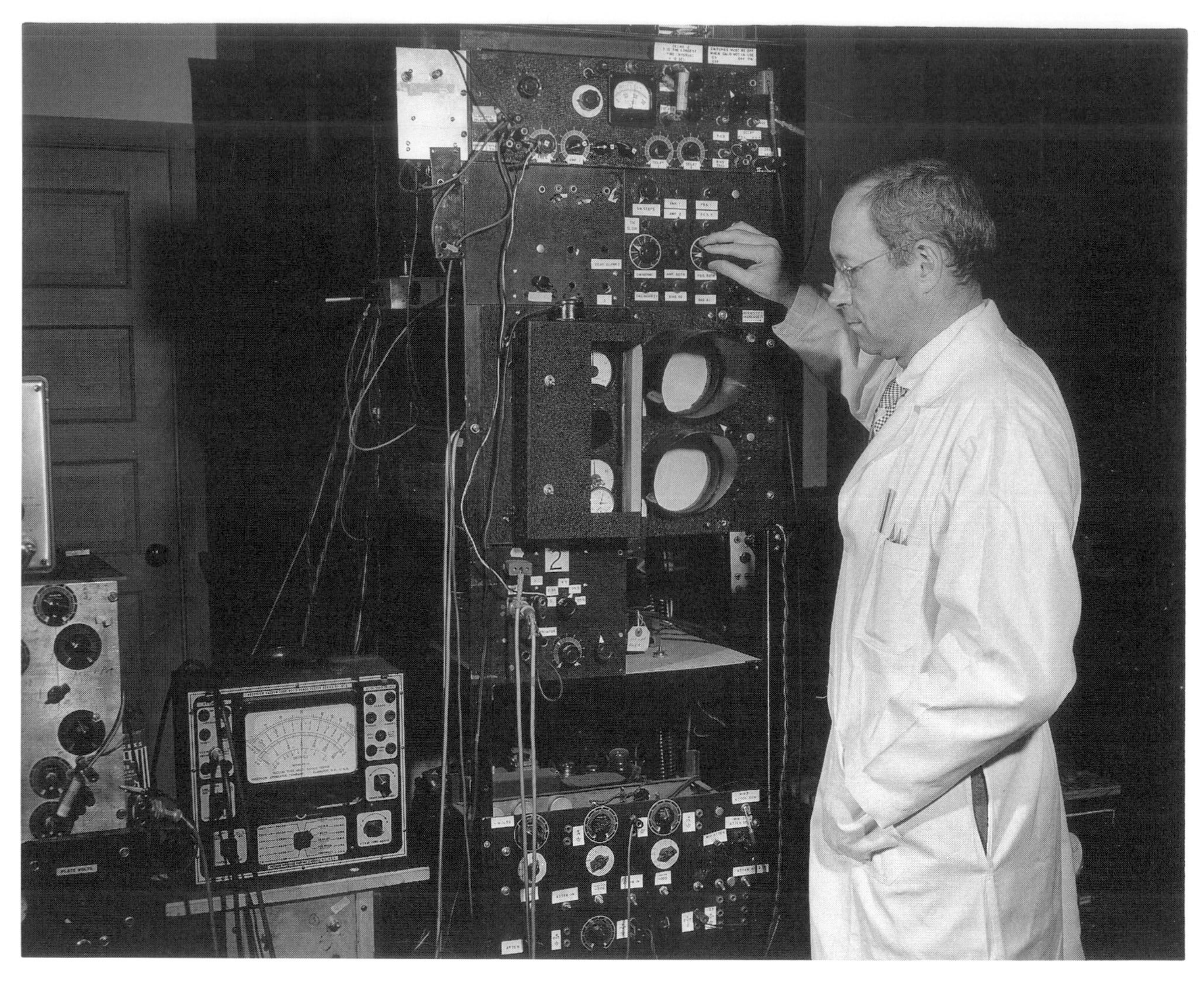

Fig. 5.27. The first laboratory of neurophysiology at the National Institute of Mental Health, Bethesda, Maryland, was established by Wade Hampton Marshall in 1947. The photograph shows the electronic gear in use shortly thereafter. Compare with the complex set-up used 20 years earlier at Washington University (Fig. 5.40).

results showing that the evoked cortical potentials resembled Gasser and Erlanger's evoked potentials in peripheral nerve. Adrian and Matthews also noted that the slow background waves "gave way" (were checked in Berger's term) after a noxious stimulus. Adrian had the capacity to follow his experiments where they led him (Moruzzi, 1980, p. 17), and shortly after the detailed paper of Marshall, Woolsey, and Bard (1941) appeared, Adrian (1941) published his own experiments using stimuli of touch, pressure, and movement on four species of anesthetized animals. He, too, found the loudspeaker more satisfactory than the visual cathode-ray tube display; he could easily distinguish potentials of individual cells ("units") from the background noise when he pushed the wire electrode through the layers of cortical cells. Adrian reported finding the "after-discharge" from touch stimuli in cat, dog, and monkey but not in rabbit and attributed it to neurons activated by the primary afferent volley arriving at the thalamus.

Concurrent with the animal experimentation, the application of the electroencephalograph to clinical studies made great strides in diagnosis and treatment of epilepsy but without resolving debates about its underlying neurophysiology. The human EEG could not contribute much more than clues, however, until a method was available to distinguish reliably between a large, irregular background activity and the small signals from an

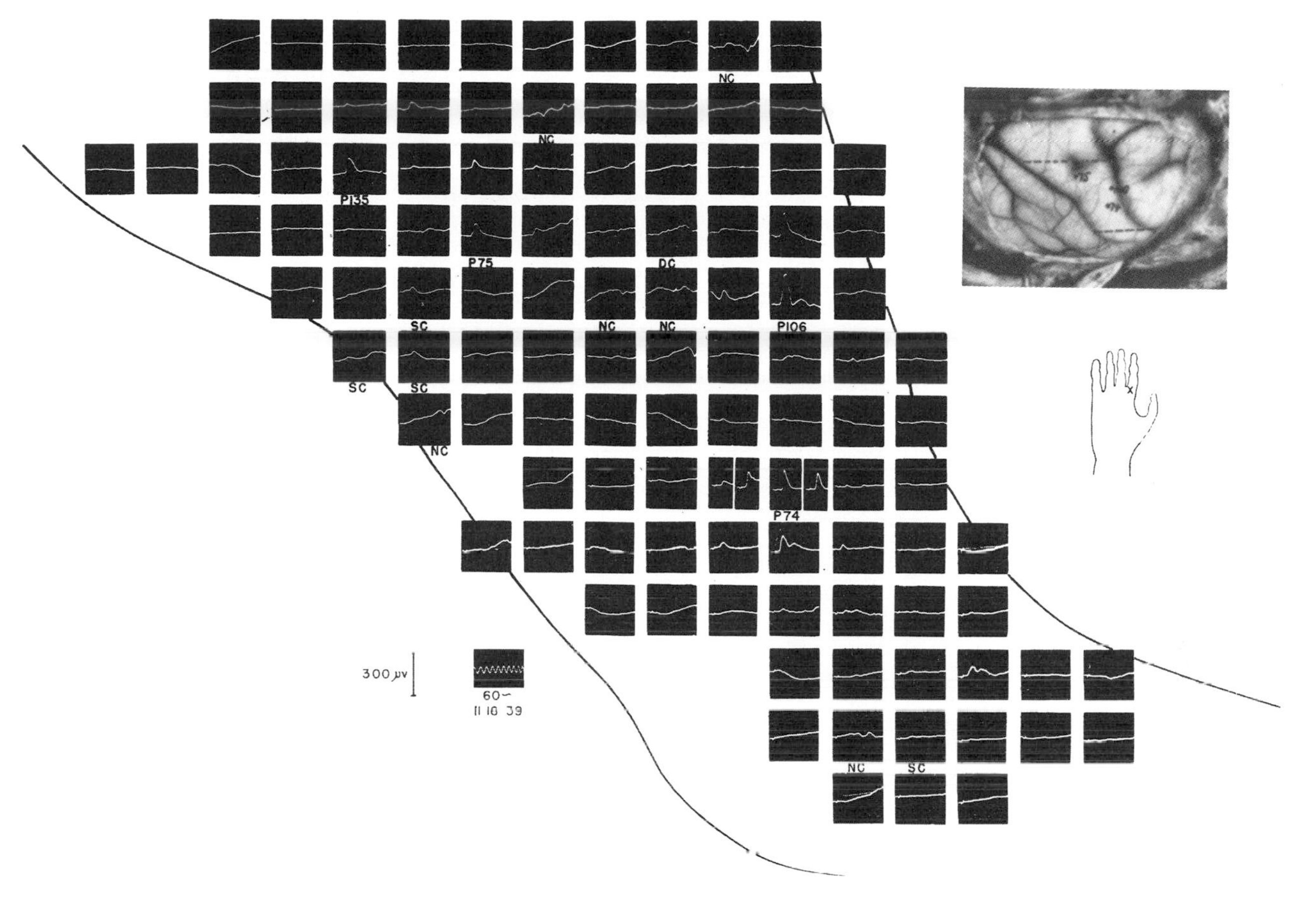

Fig. 5.28. An early map of the distribution of the action potential response to a tactile stimulus in the monkey. (From Marshall, Woolsey, and Bard, 1941, p. 8.)

evoked potential. This problem was solved by George Duncan Dawson (1912–1983; *see* Fig. 5.30, p. 93), English engineer and inventor, using a noninvasive method of photographic superimposition of faint oscilloscopic traces (Dawson, 1947). In 1951, he demonstrated an improvement, his electromechanical averager, which summed small evoked potentials, and two years later he published schema and photographs of the apparatus, circuit diagrams, and illustrations of control experiments (Dawson, 1954). From that beginning, averaged sensory evoked potentials became routine in contemporary clinical laboratories (Giblin, 1986).

The averaging technique made it possible for William Grey Walter (1910–1977), who had worked briefly with Dawson, to discover the contingent negative variation mentioned earlier. Walter (1964) found that when a warning or conditional stimulus is presented before the "imperative" stimulus (*see* Fig. 5.31, D, p. 94), a slow negative wave appears over wide areas of the frontal nonspecific cortex, which he named the E-wave for expectancy. He also found that the strength of the E-wave is a measure of the significance to the subject of the forthcoming signal.

The study of evoked potentials, and their more modern appellation, the "event-related potential," had a limited scope, as Fulton pointed out (1949, p. 377). They could establish the existence of cortical representation (of S II, for example) but could contribute nothing to an understanding of function. The record of the event-related potential has many components (*see* Fig. 5.32, p. 94) as revealed by changing its temporal registration but none of its underlying neurophysiology was elucidated. What was understood at that time was that the human parietal cortex is almost entirely, if not fully, somatosensory in function.

The contribution of a small group of military casualties had enormous impact on understanding

Fig. 5.29. "A close inspection," showing two of the titans of electrophysiology in the 1930–1950 era, Edgar Douglas Lord Adrian (left) and Alexander Forbes. Each contributed to the solution of many problems in neurophysiology and was held in the highest regard by his contemporaries, at home (Cambridge and Harvard Universities, respectively) and abroad.

the functions of the parietal lobes. The detailed examination of six patients with bilateral parietal lobe trauma sustained during the First World War by the eminent British neurologist, Gordon Holmes, revealed the existence of a disorder of visual space perception (Holmes, 1918)[6] and indicated that normal spatiovisual perception requires coordination of information from retina, oculomotor muscles, and neck muscles. The startling effect on space perception of damage to one parietal lobe was described in 1941 by W. R. Brain in the journal that he edited. Patterson and Zangwill (1944) cited a statement in the *Handbuch der Neurologie* that the right parietal lobe is important in the disorder in patients with left-brain dominance and added to the literature two additional patients with parieto-occipital trauma, stressing the relevance of their findings to the functions of the nondominant hemisphere. Further testing on patients and experimental work in the last 40 years have implicated attentional and motivational factors in parietal functions, in addition to the known convergence there of inputs from the spatial relationships of the surroundings (*see* J. C. Lynch, 1980).

Recalling Hughlings Jackson's abhorrence of strict boundaries between functional cortical areas, a suggestion was made that the visual "neglect" characteristic of patients with parietal damage, evidenced in their drawings and recall of visual experiences, may be explained by the lack of a sharp

[6]For a description of the signs and symptoms of parietal lobe dysfunction, *see* Gross (1995).

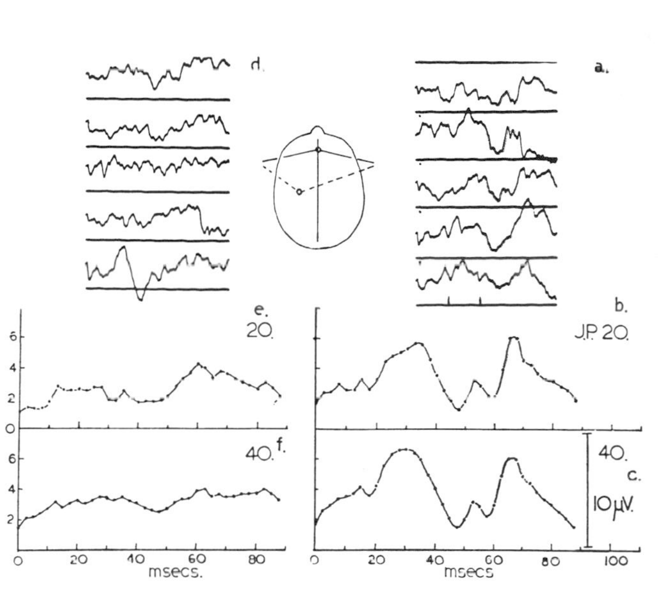

Fig. 5.30. George Duncan Dawson invented an electromechanical averager with which he demonstrated the first published averaged electroencephalograms. a—Five responses from the left scalp to a stimulus applied to the right arm which are indistinguishable from the spontaneous activity. b,c The averages of 20 and 40 records, respectively. d, e, f Control records without stimulus. (From Dawson, 1954, p. 68; portrait from Giblin, 1986, p. 2.)

boundary between the occipital and parietal lobes (Hyvärinen, 1982). Hyvärinen's summary relates functional dominance and form:

> At first glance, it might seem that there is no common functional feature for the whole parietal lobe. However, a certain degree of coherence can be seen in its various functions. Anteriorly, the somatosensory cortex mediates accurate information concerning cutaneous stimuli and joint positions. . . . Posteriorly, this information is used in area 5 [posterior superior lobule] for somatic sensory guidance of larger-scale movements. . . . The inferior parietal lobule receives visual input in addition to the somatic one. This visual input is related to positions in the extrapersonal space. . . . All parts of the parietal lobe are thus linked with a system that serves perception of the body and its parts to external space. . . . [A] gradual shift between the occipital and the parietal functions is natural and is perhaps reflected in the lack of a sharp anatomical boundary between the parietal and occipital lobes of the human brain (ibid, p. 177).

THE TEMPORAL LOBES

A decade after Broca consigned motor aphasia to the left frontal lobe, thereby introducing evidence of cerebral specialization and functional asymmetry simultaneously, identification of the "sensory aphasia" area in the first temporal convolution was made by the young Carl Wernicke, as noted above. In addition to Wernicke's area con-

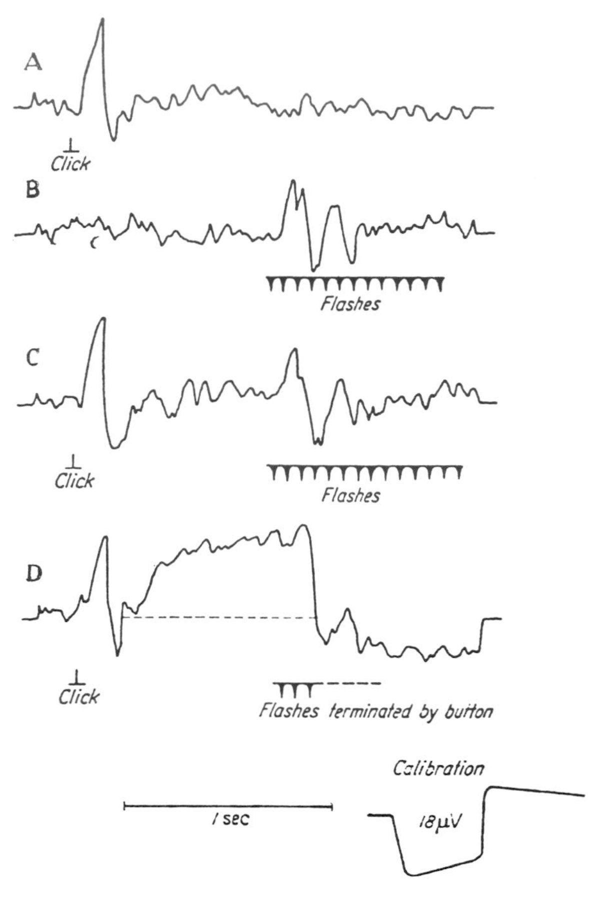

Fig. 5.31. Electroencephalograms illustrating the effect of expectancy (E-wave) on brain electrical activity. (From Walter et al., 1964, p. 305. *See* text for explanation.)

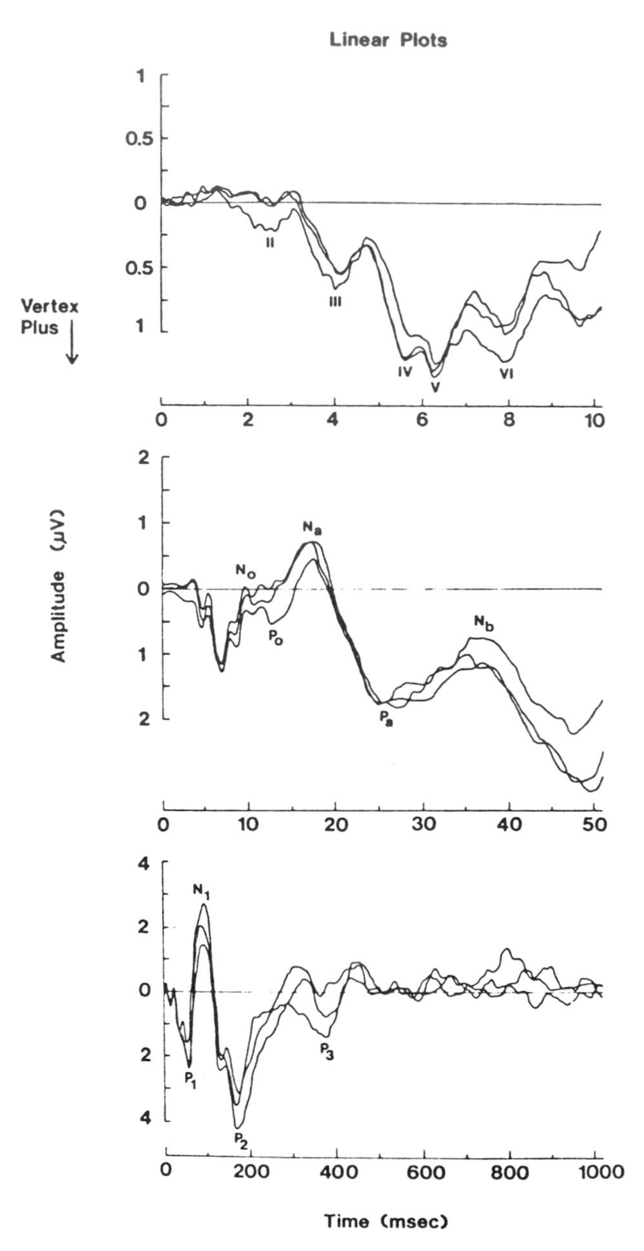

Fig. 5.32. Auditory evoked potentials from the human scalp (vertex-right mastoid electrodes) recorded at three speeds, illustrating the increase in detail of electrical activity with temporal increases in recording. (From Galambos, 1976, p. 382.)

cerned with speech comprehension, Ferrier's opinion that the auditory sense was centered in the superior temporal gyrus had gained currency. Schäfer found the evidence "insufficient," and with the help of an American working in London, Sanger Brown, he repeated (Brown and Schaefer, 1888) the bilateral destruction of the superior temporal gyrus in six monkeys. "In all six cases," he reported, "[h]earing was not only not permanently abolished, it was not perceptibly affected" (1888a, pp. 373–374). He suggested that Ferrier's single monkey had been deaf before the operation, an insult to which Ferrier made a blistering reply in the next issue of *Brain* (1888); Schäfer's immediate rebuttal was equally acrid (1888b).

The agitated exchange concerning the "deafness" of Ferrier's monkey enlivened the debate about the cerebral center for hearing but did not detract from the reality of its location in the temporal lobes, where both Ferrier and Hermann Munk (1881) placed it. Potentials evoked by sound, whether recorded at the exposed cortex or from the scalp, have had an unusually useful application to clinical situations and have provided clarification of some fundamental neuronal processes as well.

The discovery of the cochlear microphonic potential by E. G. Wever and C. W. Bray (1930a,b) demonstrated that evoked potentials of the special senses are more complex than the somatosensory evoked potentials recorded from the parietal cortex.

A Hungarian-born experimental physicist, Georg von Békésy (1899–1972; Fig. 5.33), who completed his career holding a professorship in sensory sciences, "provided us with the knowledge of the physical events at all strategically important points in the transmission system of the ear" (Bernhard, 1964, p. 720), including the transduced frequency code that goes forward to the higher nerve centers. Békésy's great skill in micromanipulation was conducive to his success in understanding the psychophysics of the special senses and neural inhibition (Békésy, 1967).[7]

Hallowell Davis (1896–1992; Fig. 5.34) developed the basic science of the auditory evoked (or "electrical" as Davis preferred) response (AER) into a useful otological tool. As student and teacher at Harvard, Davis's research interest in the auditory nervous system did not waver during a brief excursion into electroencephalography in the late 1930s, a period when he maintained that the AER represented an alteration in the EEG, not a generator source *per se* (Davis, Davis, Loomis, Harvey, and Hobart, 1939). With collaborators, Davis was one of the first to study computer-summed AERs, and at the Central Institute for the Deaf in St. Louis brought his extraordinary knowledge of the auditory nervous system to bear on human hearing problems.

In contrast to the historical arguments about the results of damage to one gyrus of the frontal lobe, as seen in speech deficits, the results of removal of the entire temporal lobe were certain and striking. In 1939, Heinrich Klüver (1897–1979), eminent physiological psychologist, and Paul Clancy Bucy (1904–1992), neurological surgeon, described the results of bilateral temporal lobectomy at the informal neurology club that met monthly in Chicago. The symptoms in the macaque included behavior indicative of the "psychic blindness" described earlier by Munk, strong oral tendencies in examining objects, profound emotional changes, and increased sexual activity. Anticlimactically and with chagrin, a year later, Bucy described the symptoms to the Chicago Neurological Society, concluding with these words:

Fig. 5.33. Georg von Békésy, eminent psychophysicist, admired art as well as science, and during his long career collected a great many art objects, which he bequeathed to the Nobel Foundation in Stockholm. (From Herrlinger, 1971.)

> It is extraordinarily appropriate that this subject should be presented [here]. Just over half a century ago, in 1888, Sanger Brown, one of the founders and the first president of this Society . . . removed both temporal lobes. . . .
>
> There is no doubt that the symptoms we have currently observed were first reported in 1888 by Sanger Brown and Schaefer. Unfortunately, however, they did nothing toward elucidating the mechanisms involved (Bucy and Klüver, 1940, p. 1144).

The role of the temporal lobes in epilepsy is associated with the name of J. Hughlings Jackson in Jacksonian epilepsy. Although he did little experimental work, neurology is indebted to him for the thoroughness of his clinical observations and "for a general philosophy [of cortical localization]" (Schiller, 1970a, p. 247). Hughlings Jackson

[7]Békésy and Rosenblith (1948) published a detailed and comprehensive history of hearing, which is especially helpful in placing it in the context of other sciences and the development of instrumentation.

Fig. 5.34. The career of Hallowell Davis (photographed in 1929) at Harvard in the Psycho-Acoustic Laboratory and at the Central Institute for the Deaf in St. Louis was instrumental in maintaining the association of American audiology primarily with psychoacoustics and only secondarily with otology and neurophysiology.

noted that not all "fits" occur with a motor component; there are also sensory discharges, which can take the "shape of reminders of common sensations. . . . The special senses may also participate. . . . Jackson . . . gave them a habitation, in the temporal lobe towards its inner side. . . ." (G. Jefferson, 1935, not paginated).

Another dimension of temporal lobe specialization was contributed by Penfield's conscious patients. In 1931, in his pre-excision exploration of the surface of the exposed left temporal lobe, he included by chance, stimulation of the upper convolution and evoked the patient's recall of a long-past experience. By 1959, he had accumulated records from 1000 such patients which he reviewed for the first Alexander Welsh Trust Lecture delivered before the Royal College of Surgeons of Edinburgh. After his lyrical introduction praising the four Monros who had graced the University of Edinburgh's chair of anatomy for 126 years, Penfield described a few selected cases, and concluded "there is hidden away in the brain, a record of the stream of consciousness" (Penfield, 1960, p. 181). He proposed that the area on the left temporal lobe, where stimulation produces this experience of recall, be designated the "comparative interpretative" area, but such a descriptive term was not widely accepted.

Functions of the human temporal cortex revealed by unilateral or bilateral lesions, and by stimulus-response trials have been elucidated also by human commissurotomy, described in Chapter 6. Historically, the roles of the auditory and language functions of the human brain were discovered in advance of the more complex associational role in perception and memory of auditory, visual, and olfactory input. The fascinating asymmetry of function originally demonstrated by Broca, Wernicke, and Penfield and earlier interpreted as dominance of one hemisphere of the brain over the other, is now regarded as complementary specialization (Milner, 1974).

OCCIPITAL LOBES AND VISION

This most posterior lobe of the modern human cerebral hemisphere was probably the first cortical lobe for which a physiologic function was clearly established. The assignment of vision to the occipital lobe was not arrived at directly, however. In the early nineteenth century, Flourens (1824) found a decorticated hen to be blind and therefore assigned vision (together with hearing) to the cerebral cortex as a whole. In Germany, Karl Friedrich Burdach decided, on the basis of postmortem findings, that the thalamus had something to do with seeing (1826), but conceded that the pathways involved were obscure. Definite evidence of how the visual impressions travel from the eyes to the brain was provided by an Italian anatomist, Bartholomeo Panizza (1785–1867; Fig. 5.35). In 1855, he wrote a classic paper on the optic nerve for the journal of the Lombardy Institute of Science, Letters, and Art, in which he described his dissection method of gentle scraping (in the manner of Gall) so as not to damage the fibers of the optic tracts and traced the pathways in birds and fishes to their terminations in the occipital lobes. He determined the crossing of optic fibers at the chiasma and showed that a lesion on one side of the brain affects the eye of the oppo-

Fig. 5.35. In 1834, a series of letters was published by Bartholomeo Panizza, describing his work on the nerves. Twenty-one years later, his *Osservazioni sul nervo ottico* appeared in which he reported dissection and ablation experiments that showed vision to be localized in the occipital cortex, thus becoming the first to identify a cerebral lobe with a specific sensory function.

Fig. 5.36. The austere Hermann Munk's role in providing the correct evidence for the localization of the great sensory modalities is often overlooked (*see* Schiller, 1970a, p. 247). He established "the facts of cortical blindness and mind blindness" and uncovered hemiopia.

site side (Soury, 1899, p. 1446). Panizza also demonstrated experimentally that the occipital lobe is essential for vision and extended his findings to the human, noting the occurrence of an occipital or deep hemispheric lesion in patients with visual defects that did not involve the eye's lens system (J. R. Green, 1985, p. 19). Panizza's work was recognized in his time, for he rose to become provost at the University of Pavia and to hold membership in the Académie Française des Sciences.

In experiments confirming and extending Fritsch and Hitzig's cortical stimulations, Ferrier (1874) stumbled on the site for producing conjugate eye movements. He was primarily an experimental ablationist, however, and believed that removal of the angular gyrus in the posterior parietal lobe produces blindness (1873). The criticism of Ferrier's work by Hermann Munk (1839–1912; Fig. 5.36), Prussian physiologist appointed professor at the Berlin Veterinary School, was direct and unflattering. In a series of lectures in Berlin, one-third of which were on vision (1881), Munk explained his omission of reference to Ferrier's experiment because "there was nothing good to be said about it" (from translation in von Bonin, 1960, p. 104). Munk claimed that Ferrier's "worthless and gratuitous constructions" were based on insufficient examination carried out when the brain was still suffering the trauma of the operation.

Munk's bilateral removal of a small (10- to 15-mm) area in the dog's occipital cortex produced "the situation of the [puppy] whose eyes have just opened" (ibid., p. 98), which he called psychic or mind blindness: The animal can see but not recognize objects. If the entire occipital cortex is removed, a condition of irreversible cortical blindness results. Munk's contribution was not only the differentiation of cortical and mind blindness but also, in line with Hughlings Jackson's beliefs, his emphasis on the nonexistence of sharp boundaries around cerebral functional areas. As cytoarchitectionists were also discovering, the function and morphology of one area is a continuum with its bordering areas, each tapering into the other.

Thirty years after Panizza had followed the optic tracts to the cortex by careful gross dissection, the German psychiatrist, Bernard Aloys von Gudden (1824–1886) followed their course microscopically with newly available methods for tissue staining. He used functional ablations, noting the disappearance of nerve centers and their connections after cutting cranial nerves or removal of sense organs from young animals. He showed that the large fibers of the optic tracts terminate in the lateral geniculate nuclei which, together with the superior colliculi, form part of the visual system (von Gudden, 1870).

In 1876 von Gudden, who was then director of an asylum, was visited by a young Russian-Swiss, Constantin von Monakow (1853–1930), with whom he shared his fine brain sections that clearly showed atrophy in the superior colliculus region of the visual pathway after neonatal removal of an eye. Continuing that line of research on rabbits, Monakow proved that the lateral geniculate nucleus degenerates, whereas the remainder of the thalamus is normal after enucleation; he summarized his work in Monakov (1914). In the words of an affectionate and perceptive biographer, "von Monakow laid the foundation upon which the present knowledge of developmental and functional unity of the thalamus and cortex securely rests" (Yakovlev, 1970, p. 487).

One of the many eminent graduates of the University of Bologna, Luigi Luciani (1840–1919), was an expert in many aspects of physiology. His five-volume textbook, *Fisiologia dell'Uomo*, was published in five editions and several translations, including English. His presentation (1884–1885) of overlapping areas of sensory representation on the cortex (Fig. 5.37) again fitted well with Hughlings Jackson's concept of nonabrupt delineations. Luciani disproved Munk's idea of the "cortical retina," or specific retinal segments projecting to specific cortical segments.

Toward the end of the nineteenth century, members of the great school of English neurologists, deeply engaged in localization studies through both clinical observation and animal experimentation, turned from their earlier involvement with speech functions to resolving the cortical representations of the senses. In studying sensation, however, experiments on laboratory animals were necessarily limited because of inability to interpret their perceptions. Schäfer, in his 1889 paper, "Experiments on the electrical excitation of the visual area of the cerebral cortex in the monkey," which confirmed that removal of both occipital lobes renders the animal permanently blind, refuted Ferrier's claim that occipital cortex was not excitable by electrical stimuli, and concluded there was a need to look for the effects in man. That opportunity was provided the London group by a Swedish medical practitioner from Upsala, Salomon Eberhard Henschen (1847–1930; Fig. 5.38). He had devoted his career in ophthalmology to looking for effects of disease that were confirmed postmortem, and reported 40 such cases at the Congress of Experimental Physiology in London, 1892. The first paragraph of his paper, although published almost a century ago, contains an unusually comprehensive view of brain and behavior and expresses the concept of present-day neuroscience:

> In order that modern psychology may possess a definite starting point and foundation, we must obtain an exact idea of the anatomical construction and physiological function of the paths and cortical areas of our senses, as well as their connections with other parts of the brain. In consequence of the law of evolution, there exists in man the highest development of the different

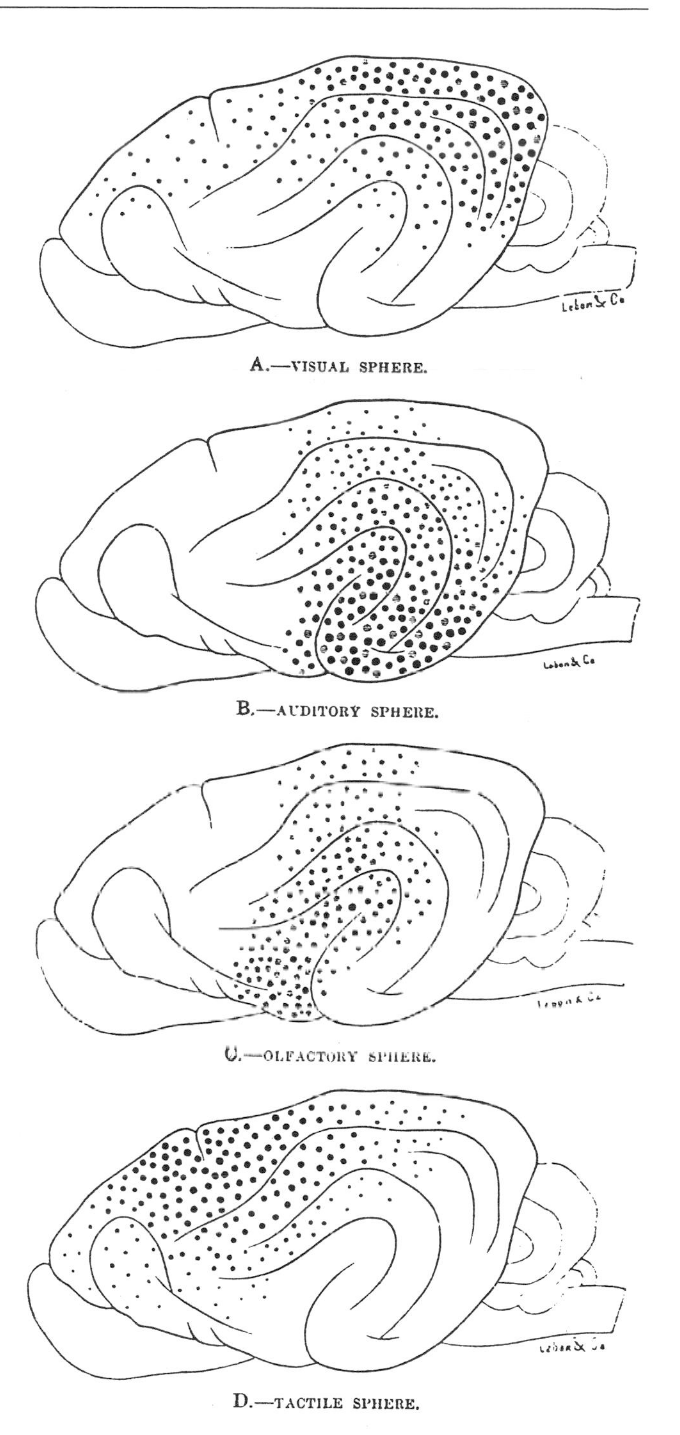

Fig. 5.37. Schematic figures of left cerebral cortex of the dog, on which the centers for (top down) the visual, auditory, olfactory, and tactile spheres are represented. By superimposing the schemas "we discover a remarkable fact: each sphere possesses a territory of its own, and in addition has a common territory, the parietal lobe, representing 'the centre of centres' where 'engrenage' or partial fusion of the sensations occurs." (From the English edition of Luciani, 1911, pp. 159, 160.)

Fig. 5.38. The Swedish internist Salomon Eberhard Henschen spent most of his career defending his deduction from patient observations and from the literature that the site of acute vision is in the calcarine fissure of the occipital lobe. This photograph was taken in his study in 1917.

> centres, and psychical life. For this reason we must principally study the paths and centres of the senses in *him*, whilst comparing the results obtained by experiments on animals (Henschen, 1893, p. 170).

Henschen's specific goal was to define more precisely Munk's visual center or macula, from observations on his own patients with hemianopia and from cases in the literature. He singled out the pulvinar region of the thalamus in the optic pathway and designated where the macula was represented in the optic radiation and calcarine fissure of the occipital lobe, labeling it a "cortical retina," as had Munk. Henschen's contributions were significant, but his writings were punctuated by expressions of bitterness at their lack of recognition.

The experimental psychologists, too, were quietly contributing interesting findings about perception and sensation, chiefly in the visual modality. At St. Elizabeths, Franz (1911) tested color vision in monkeys and found they can discriminate colors if presented together with an object to which they attend. Further, ablation of areas of occipital cortex caused a disturbance of movements that depend on visual coordination, not due to visual deficit, but to the loss of the afferent input from the eye muscles. However, "One animal in which the cortex surrounding the calcarine fissure was destroyed showed a true visual disturbance corresponding with those described in man" (Franz, 1911, p. 112).

Again, the First World War presented opportunities for study and Gordon Holmes and W. T. Lister (1916) reinvestigated the cortical representation of the retina in systematic detail. The area striata, they found, represents the upper retina dorsally and the lower retina ventrally, with macular vision caudal and peripheral vision rostral. Other investigators also had made use of war casualties, among them Pierre Marie in France (Marie and Chatelin, 1914–1915), and Tatsuji Inouye (1909) during the Russo–Japanese war; their localizations of macular vision were confirmed by Holmes and Lister. In the United States, Hans-Lukas Teuber (1916–1977), German–American physiological psychologist, studied the acute and later chronic, stable effects on vision of certain head wounds from the Second World War. He and associates showed that the macular fibers do not form a single bundle and the corresponding elements of the visual system are imperfectly aligned (1960).

Henschen's "cortical retina" became the experimental model of necessity when George Holman Bishop (1889–1973; Fig. 5.39) was catapulted from Erlanger's department of physiology[8] at Washington University, St. Louis (Fig. 5.40), into the ophthalmology department after a misunderstanding with his chief concerning channels of information (*see* L. Marshall, 1983b). In 1933, Bishop and his collaborators took the challenge of applying their extensive experience with peripheral axons to the infinitely more complex central nervous system and inaugurated a series of studies of the electrical activity of the optic system's various components. The initial experiments were descriptively titled "The cortical response to stimulation of the optic nerve in the rabbit" (Bartley and Bishop, 1933). With neurologist James L. O'Leary's expertise in Golgi techniques added to the team's talents, work was extended to the cat's occipital cortex and they studied in detail the shape, amplitude, and timing relations of responses at various levels of the visual pathway evoked by electrical stimulation of the optic nerve. In later work, Bishop sought to explain the recorded action potential in terms of the electrical events in neural soma lying in specific cortical layers (Bishop and Clare, 1952), a monumental undertaking which contributed to Bishop's reputation, in the estimation of Lorente de Nó (1976), as having produced order out of a chaotic state of electrophysiology.

The occipital cortex has a position of prominence in the history of the human brain—anatomically, physiologically, and behaviorally—because it is the primary cortical projection of the most extensively studied of the senses. In 1934, the opinion

Fig. 5.39. George Holman Bishop's (shown ca. 1950) long and brilliant career in neurophysiology was at Washington University, St. Louis. His education in basic sciences, his intellectual powers, and his ability to let the experiment lead him greatly advanced the knowledge of central and peripheral nervous system activity.

was stated that vision "is, I think, the function which in the higher mammals has become most corticalized" (Dusser de Barenne, 1934, p. 103). Dusser based his opinion on the fact that vision has a distinct and stable localization and little chance of reparation when it is damaged, in contrast to other functions which are more diffuse and may show marked repair of lost function. His statement was made prior to discovery of many additional cortical centers concerned with vision whose relation to the phylogenetic response to evolutionary influence is still not understood (*see* Fig. 2.4, p. 13).

An additional reason for the preponderance of research on the visual system compared with the other senses is the relative ease with which it can be studied using the methods long applied in the neu-

[8] A photograph of Erlanger's well shielded and stabilized basement laboratory of the department of physiology is shown in Fig. 5.40.

Fig. 5.40. The "plant" in the basement laboratory of the department of physiology at Washington University, St. Louis, with the sturdy platform in the foreground to support the sensitive string galvanometer. The dispersion of the equipment mandated the presence of two experimenters; compare with Fig. 5.27 of apparatus in use about 20 years later.

romuscular system. The system's accessibility, together with the biological application of radioisotopes after the Second World War, combined to rapidly advance the relationship of structure and function during the middle decades of the century. A prime example, presented in Chapter 7, is the investigation in exquisite detail of the retinogeniculostriate pathways in cat and monkey by David Hubel and Thorsten Wiesel and their collaborators.

OVERVIEW OF THE CORTICAL HEMISPHERES

The cortical convolutions (Chapter 4) and their aggregation into four major hemispheric lobes constituted the setting for the second stage of brain anatomy to be addressed historically, after the ventricles, especially in regard to function. In the seventeenth century, Thomas Willis believed that the "cerebral folds and convolutions are for the storage of sensory impressions as though in various closets and storehouses, and to enable these impressions to be evoked for any given occasion" (Willis, 1681, p. 34). The perception of an orderly arrangement of the convolutions into lobes and the recognition of a species-specific pattern engrossed eighteenth century men like Vicq d'Azyr in France and Felix Fontana in Italy. The most influential figure to appear in the next century was Gall, who believed the convolutions to be the very origin of behavior.

In the early nineteenth century, human brain history progressed slowly and quietly, compared with the stormy situation that developed in the later decades. In 1836, Marc Dax wrote in his private memoir that in some patients loss of speech correlated with left cerebral trauma. The manuscript was brought to attention by his son coincidentally with Broca's first cautious proposal of a left-side dominance for speech in 1863, and led to the accumulation of illustrative cases and counterarguments. The amorphous state of knowledge was somewhat clarified by a set of principles from one of the great neurologists of any period, Hughlings Jackson. As a younger colleague pointed out, Jackson's "doctrine that 'destructive lesions never cause positive effects, but induce a negative condition which permits positive symptoms to appear,' has become one of the hallmarks of English neurology" (Head,

1915, p. 3). Another admirer more recently said of Hughlings Jackson: "He took an astronomical view of neurons' functions . . . rather than the brave niggling over small points that is the best that most of us can do" (G. Jefferson, 1935, not paginated). Hughlings Jackson would have been comfortable with the warning of Dusser de Barenne (1934) about the parochialism of designating "localization" for a function that so patently takes place in many parts and levels of brain; Dusser preferred "corticalization" as the more accurate term. On many occasions, Hughlings Jackson countered the strict localization of function, an idea which was in ascendancy, and with his contrast of mental and nervous states, separated psychiatry from neurology (Brown, 1977).

Perhaps the greatest beneficiaries of the discovery of functional localization may have been the neurosurgeons, for the concept bestowed a scientific certitude on their placement of brain pathology and hence the locus of subsequent ablation. The year 1886 saw the first application of the principle of cerebral localization to successful surgical practice, according to Dasheiff (1994): Horsley (1886) in London, and Park (*see* Park, 1913) in Buffalo. The discovery by Berger and the use of electrical potentials recorded through the scalp added an immense arsenal to the surgical treatment of some forms of brain pathology. On the research front, the electrically evoked potentials so exclusively used by the axonologists in the 1930s, during the next decade was carried into the depths of the central nervous system largely initiated by the work of W. H. Marshall. The unexpected richness of surface and depth responses to a variety of sensory stimuli opened a wide door to sensory localization studies and facilitated the subsequent analysis of single-unit activity. Five decades later, the preciseness of noninvasive localization within brain tissues and clusters of specific cells promises a focus on behavioral effects with far greater resolution than was possible only a few years ago.

6 Cerebral Asymmetry and Behavioral Laterality

Contents

Is the brain, which is notably double in structure, a double organ, 'seeming parted, but yet a union in partition'?

(Maudsley, 1889, p. 161)

Maudsley's quotation above reformulated an earlier pronouncement of Sir Isaac Newton in which the remarkable overall exterior symmetry of *Homo sapiens* was credited to the deity (Oppenheimer, 1977). Whereas the symmetrical design of the human body is apparent from the outside, with two each of eyes, nostrils, ears, and limbs, the interior is recognized as asymmetrically arranged, most obviously in the thorax and abdomen where the organs are single. Asymmetry of the cerebral hemispheres is less discernible, nonetheless the early anatomists knew that the two cerebra are not mirror images in their gross structure. The more subtle absences of cerebral symmetry are revealed only piecemeal as discoveries of anatomic, electrophysiologic, and chemical differences between the two sides of the brain allude to differences in function and cerebral asymmetries and behavioral lateralities alternate in yielding to the search for scientific truth.

As discussed in Chapter 1, modern primitive invertebrates provide ample evidence for processes that occurred early in evolutionary biology. The simple brain of flatworms is considered a representative example of the onset of anatomic asymmetry, but whether or not there is also lateralization of function at that primordial stage is debatable: "No longer radially symmetrical, the larger flatworm brains show considerable lateral development, with side lobes. Although this might be considered an early example of lateralization in the most general sense, there is no evidence of any asymmetry of function in the two sides" (Hodgson, 1977, pp. 23–24).

The view of the phylogenetically ancient brain as a cluster of cells was defended by the organizer of a conference on brain evolution and lateralization: "[M]any are prepared to accept that these early creatures [had] a nervous system little more than a straight tube. The whole history of lateralization of the brain shouts out against such a view" (Dimond, 1977, p. 480). The recognition of the physical occurrence of asymmetries in the brain is ancient, but the discoveries of functional differences—execution of a function predominantly by one side of the brain—were delayed until the functions could be assigned to a specific region, or "localized."

HEMISPHERIC DIFFERENCES

A chief physician to Marie Antoinette and a talented artist, Vicq d'Azyr, in describing the convolutions in 1786 (*see* Chapter 4), stated that not only are the cerebral hemispheres of man not similar, they also possess more anomalies, i.e., asymmetries, than do those of any other mammal. The general acceptance of these facts was assured by their incorporation into a textbook written by an eminent French neurophysiologist and translated and published in Philadelphia:

> The brain of man is that which offers the most numerous *circumvolutions*, and the deepest *sinuosities*. The number, the volume, the disposition, of the circumvolutions are variable; in some brains they are very large; in others they are less and more numerous. They are differently disposed in every individual; those of the right side are not disposed like those of the left. It would be an interesting research to endeavor to discover if there exist any relation between the number of *circumvolutions* and the perfection, or imperfection, of the intellectual faculties—between the modifications of the mind and the individual disposition of the cerebral *circumvolutions* (Magendie, 1824a, p. 104).

As in France, English confirmation of brain asymmetries was from an elevated source. Henry Holland (1788–1873), physician-in-ordinary to Britain's royal couple, wrote in his *Medical Notes and Reflections:*

> Though the nervous system . . . is subject to fewer anomalies than any other organs of the body, yet are these deviations more frequent in man than in many of the mammalia most nearly approaching him in structure. . . . It is further to be noticed, as an anatomical fact, that in the brain and spinal marrow, the external parts on the two sides are less exactly symmetrical than those within; the surface of the brain showing this perhaps more distinctly than any other part (Holland, 1839, p. 98, footnote).

Hemispheric asymmetry was recognized by another British neurologist, Arthur Ladbrooke Wigan (1785 baptized–1847): "That it is an error to suppose the two sides of the cranium to be always alike, that on the contrary, it is rarely found that the two halves of the exterior surface exactly correspond. . . ." (1844, p. 29). That statement was one of 20 "propositions" offered as proof of the separation of psychic from neural phenomena put forth in his attempt to establish the concept of the "duality of the mind," the subtitle of his treatise on the cause of insanity. And as Harrington (1987) pointed out, in another proposition Wigan added physiology to interpretation of the dual brain by suggesting that the ability to synchronize the activity of the two brains is an acquired skill.

Major support for those views had already come from the French physician–neurologists who first proposed and later vehemently debated the "seating" of a function in specific brain foci. Leuret in his 1837 classification of mammalian brains (*see* Chapter 4, this volume) showed that the arrangement of the left convolutions does not resemble that on the right. His collaborator, Gratiolet, in 1854 recognized that the left and right sides of the brain control the right and left sides of the body, respectively. Their friend Broca, after focusing medical attention on functional asymmetries by his demonstration that speech articulation is localized in the left third frontal convolution (*see* Chapter 5), claimed in addition that the frontal gyri are more numerous on the left than on the right, and "que tout au contraire le lobe occipital droit est plus riche en circonvolutions," (quoted by Hughlings Jackson, 1868, p. 358, fn a; also in Taylor, 1958, vol. 2, p. 143). Later, Broca remarked that lesions on one side of the brain—the nondominant right in most patients—will not affect the homologous part in the opposite hemisphere, thereby recognizing the functional difference between right and left hemispheres.

One of the factors thought to contribute to cerebral asymmetry was an unequal rate of development. In the second volume of their famous *Anatomie* (Leuret et Gratiolet, 1859, p. 241), Gratiolet pointed out that the two hemispheres do not develop symmetrically: the frontal gyri are formed faster, i.e., earlier in fetal life, on the left than on the right, whereas in the occipital-sphenoidal (parietal) area the reverse occurs. Those observations were among the earliest, which, together with Flechsig's myelinogenesis theory, suggested a graded maturational element in brain development, a concept figuring prominently in twentieth-century theories of developmental disabilities. Hughlings Jackson paid tribute to Gratiolet's findings with a neat analogy: "M. Baillarger quotes from Gratiolet a statement to the effect that the frontal convolutions on the left side are in advance of those on the right in their development. Hence, if this be so, the left side of the brain is sooner ready for learning. It is the elder brother" (Hughlings Jackson, 1866, p. 661). The

Fig. 6.1. The well-traveled Édouard Brown-Séquard was a late achiever, not finding the niche he deserved until 61 years of age, when he succeeded Claude Bernard at the Collège de France. Brown-Séquard's international reputation was bolstered by appointments at The National Hospital, Queen Square, London; Harvard Medical School; and the Medical College of Virginia. The topics of his extensive writings and lectures included spinal cord hemisection, vasomotor constriction, epilepsy, and adrenalectomy. (From Fulton, 1966, Plate 63, between pp. 310 and 311.)

implications of this statement are twofold: the "elder brother" is the dominant hemisphere, and the right (less utilized) hemisphere is capable of being educated. Those were the themes of Brown-Séquard's (Fig. 6.1) Toner Lectures in 1874 delivered during his third visit to the United States. He emphasized the distinct differences in the capabilities of the two hemispheres and strongly advocated educating both sides so that the right might learn to substitute for the left rather than remain "useless." The well attended lectures, delivered by an authoritative scientist from the Old World to a receptive audience in the New World, greatly hastened the public acceptance of the idea of cerebral asymmetry and the notion that the cerebra are "educated" by the functions they carry out, a concept held by Broca himself.

Those admonitions jibed with the nineteenth century's popular belief in the role of education for moral and material benefit and one wonders how many parents and educators inflicted strange regimes at home and in school in the name of better education of the "useless" hemisphere. Indeed, J. Liberty Tadd, director of the Public School of

Industrial Art in Philadelphia, published in 1899 his theories and practices in a hardcover book bearing the imprint of a young student in pinafore drawing ellipses with both hands and the quotation "Be ye transformed by the renewing of the mind." But all training should be taken in moderate doses, as a neurolinguist recently commented about a group of graduate students at the California Institute of Technology who were at chance level in space-relation tests:

> They invariably named the blocks and their representations as the basis for making a match. When probed as to whether they ever used other methods, they [said] there was no other method. It was pointed out that they could have simply visualized the whole stimulus, at which they manifested surprise and said that this strategy had not occurred to them. One gets the suspicion that 18 or so years of formal schooling in the sciences may functionally ablate the right hemisphere. However, it must be kept in mind that the descriptions these students have of their mental function were controlled by the left hemisphere (Levy, 1974, p. 154).

The overdevelopment of the analytical left brain was a theme popular with some psychologists who championed the idea of a shift in dominance according to which mode of mental processing is active. The neglect of the "allegedly intuitive mystical right brain" (Harrington, 1987, p. 283) has been expressed in more general terms: "With a recognition of the physiological basis of the dual specializations of consciousness, we may be able to redress the balance in science and psychology, a balance which has in recent years swung a bit too far to the right, into a strict insistence on verbal logic that has left cortex and perspective undeveloped" (Ornstein, 1972, p. 69).

In the human brain, the most striking asymmetry appears in the temporal lobes, the "seat" of auditory functions (*see* Witelson, 1977). Differences there had been noted by Luys in 1879 and were confirmed by Constantin von Economo (1876–1931), Austrian neurologist and aeronaut (Fig. 6.2), who described in 1930 (with L. Horn) the upper temporal lobe on the left as usually larger than that on the right. Six years later, the German investigator R. A. Pfeifer (1936) reported that same asymmetry and in addition confirmed Broca's observation, made many years before, that the temporal lobe on the right side had a greater number of gyri than that on the left.

With the exception of isolated studies such as those just described, however, during the early twentieth century there was relatively little sustained interest among neurologists in the structural differences of the cerebral halves. Instead, the exciting fossil finds being uncovered in Africa presented the possibility of comparative studies of the crania of prehistoric man. Especially appealing was examination of the endocasts of fossilized skulls for clues to the evolutionary origin of language and handedness. An American neurologist, Frederick Tilney (1875–1938), "carried to perfection the method of studying endocranial casts" (Oppenheimer, 1977, p. 8). After pointing out that because the cranium begins embryologically as cartilage or membrane, it is susceptible to pressure from the developing brain, Tilney presented careful and objective measurements of several prehistoric hominid brains to show that the left frontal lobe and the left inferior frontal convolution are larger than on the right. Those data, according to Tilney (1927) constituted evidence that early man had language and was right-handed, a conclusion that fitted what he was looking for, as revealed by the subtitle of his paper: "The psychologic foundations of human progress."

Other distractions that precluded much interest in brain asymmetries included the rise of psychoanalysis and ideas of the psychic brain not based on anatomy, from hysteria to schizophrenia, and even Lashley's holistic view of the organ (Harrington, 1987, p. 261), which recalled that held by Flourens a century earlier (ibid., p. 269). Interest was rekindled during the midtwentieth century by Norman Geschwind's (*see* Fig. 6.3, p. 110) reintroduction of Wernicke's "disconnexion syndrome" (1963), as related in Chapter 5. He and Levitsky (1968) verified the early studies by careful postmortem examination of 100 adult human brains and found that the planum temporale, a cortical area involved in speech lying behind the primary auditory region, is larger on the left than on the right in about two-thirds of the brains examined (Fig. 6.3). Additionally, the tilt of the sylvian fissures in hardened brains of human and subhuman primates differ (*see* Fig. 6.4, p. 111, top), as Huschke had shown in 1854 (*see* p. 58, this volume). Substantiating evidence was obtained in living brains using a noninvasive technique to visu-

Fig. 6.2. In addition to obtaining his medical degree, conducting research, and acting as assistant to several clinical directors, Baron Constantin von Economo was the first Austrian to obtain the international pilot's diploma and served aviation with distinction. Six months before he died in 1931, he delivered the honorary president's address at the 30th anniversary of the Aero Club, which he had helped found. This dashing photograph was taken in 1910. (From Economo, 1937, p. 101.)

alize the flow of blood (arteriography). Judging from the course of the large arteries (Fig. 6.4, bottom), the left parietal operculum was larger in 38 of 44 right-handed patients and the same size in 15 of 18 left-handed patients (LeMay and Culebras, 1972). A larger planum temporale on the left in most subjects examined was also found in fetal brains (Wada, Clarke, and Homm, 1975), thus weakening the theory that "education" of the right hand was the evolutionary basis of left-brain specialization. As those authors declared, "[T]he human brain possesses a predetermined morphological and functional capacity for the development of lateralized, hemispheric functions for speech and language" (ibid., p. 245).

HANDEDNESS AND SPEECH AND OTHER ASYMMETRIES

The possible association of handedness with the cerebral area for speech insinuated itself into the arguments that were the hallmark of the French neurological school during the 1860s (*see* Chapter 5). A linkage between hemispheric differences and human handedness was alluded to vaguely by Bouillaud, a supporter of certain of Gall's ideas, in a presentation to the French Imperial Academy of Medicine in April of 1865 in which he separated articulation of speech from its comprehension. A few months later, in a paper read to the anthropologists (1865b), Broca specifically stated his convic-

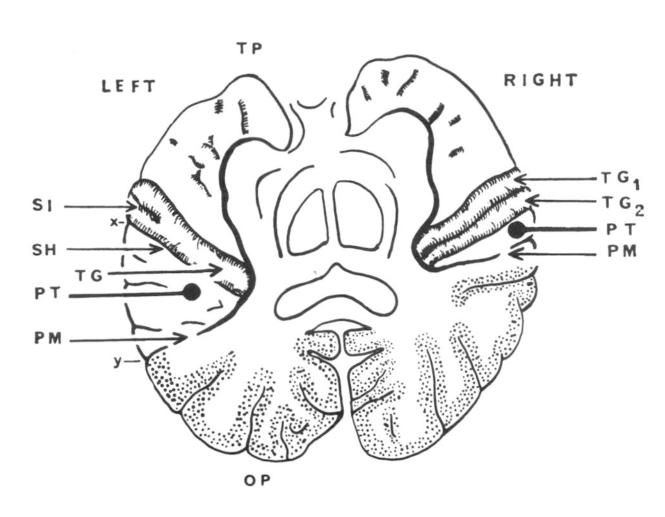

Fig. 6.3. Norman Geschwind, American neurologist, reopened the study of the aphasias and discovered the neural correlate of speech and handedness. The diagram shows the asymmetry of the superior temporal lobes in human brains: the left planum temporale (PT) is several times larger than that on the right. (Diagram from Geschwind and Levitsky, 1968, p. 186.)

tion of an association between right-handedness and the left-brain site of speech. The preference throughout civilization for the use of the right hand (*see* Chapter 2) seemed to correlate with the greater mass of the left cerebrum.

Modern corroboration of the link between handedness and hemispheric asymmetry in living brains became possible with computerized axial tomography. Using CAT scans, Galaburda, LeMay, Kemper, and Geschwind (1978) verified the century-old reports of brain asymmetries by Broca (1863), Eberstaller (1884), and Cunningham (1892), who was the first to note that the posterior sylvian fissure of the adult human brain is usually higher on the right than on the left. Comparing right- and left-handed subjects, Galaburda and associates found the numbers roughly matched the left-right distribution of their measurements of planum temporale as mentioned above. They also showed that in most right-handers the frontal lobe of the right hemisphere is wider and occasionally extends farther forward than the left frontal lobe, whereas the left occipital lobe is wider and extends farther to the rear (*see* Fig. 6.5, p. 112).

Having established the existence of a gross anatomic substrate for laterality of speech and the possible relationship of functional handedness with external brain asymmetries, the search was on for additional examples of asymmetry in the internal arrangements of the central nervous system. An unequal "degree of crossing" of the pyramidal tracts was reported in Leipzig by Paul Flechsig

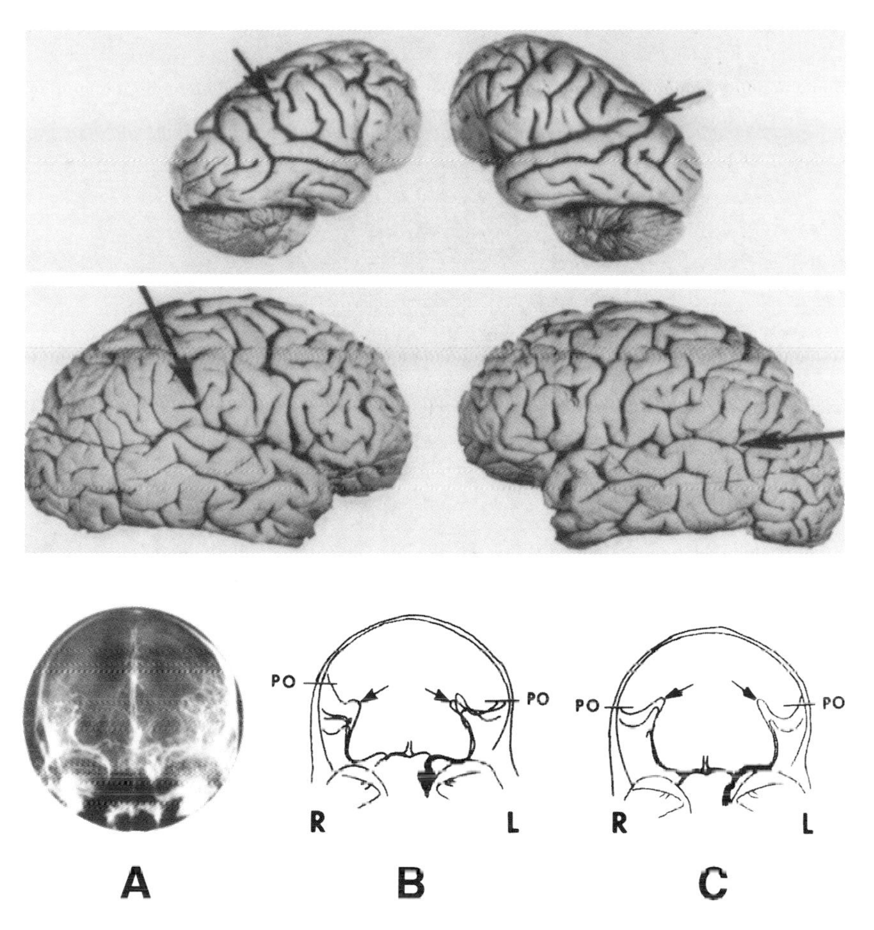

Fig. 6.4. Above—The well known right–left asymmetry of the sylvian fissure (arrows) in both human (bottom pair) and ape (top pair) brains (LeMay and Geschwind, 1975, p. 50), was found in living brains (**A**, below) by carotid arteriography (LeMay and Culebras, 1972, p. 169). Drawings of arteriograms of right-handed **(B)** and left-handed **(C)** subjects. In **(B)**, note the narrower (more acute) angle of arteries leaving the posterior sylvian fissure on the left caused by the larger left parietal operculum (PO).

(1876, p. 382) and was extended by G. Elliot Smith. Examining human brains in Cairo, Elliot Smith was puzzled by an "aberrant grouping of the pyramidal fibres" predominantly around the left olive in the medulla oblongata (*see* Fig. 6.6, p. 113). He speculated that: "Perhaps there is a greater tendency for fibres to be 'crowded out' of the pyramidal tract which comes from the demisphere preeminently concerned with the more skilled movements" (G. E. Smith, 1904, p. 382). Fleschig's and Smith's observations were reinforced almost a century later by Yakovlev and Rakič (1966) in a comprehensive study of neonatal and fetal brains from the famous Yakovlev collection of specimens, then at Harvard University. In 87 of 100 brains, they found that more pyramidal tract fibers cross from the left cerebrum than from the right, more of the latter remaining uncrossed so that below the decussation the right side of the spinal cord carries more pyramidal fibers than does the left.

The ventricles did not escape the search for gross cerebral hemispheric differences. A study of 87 right-handed neurological patients (McRae, Branch, and Milner, 1968) revealed that in 30% the occipital horns of the lateral ventricles were the same size in each hemisphere; in 57%, the left horn was longer; and in 13%, the right was longer. Related to handedness, right-handed patients were five times more likely to possess longer left horns, whereas in nonright-handers the chances were equal of the horns being the same size.

The early evidence for behavioral lateralization had been obtained from patients with unilateral brain loss leading to changes in sensory functions,

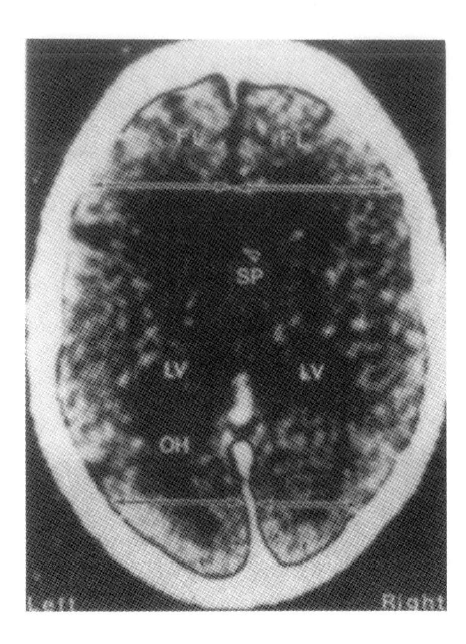

Fig. 6.5. Computerized axial tomogram showing usual pattern of hemispheric asymmetry of the brain of most right-handers. Note wider frontal lobe (FL) on right, wider left occipital lobe (lower arrows), and more prominent left occipital horn (OH) of the lateral ventricles (LV). (From Galaburda, LeMay, Kemper, and Geschwind, 1978, p. 855, ×1.5.)

but the method was necessarily limited. The use of evoked potentials to study interhemispheric differences in normal humans "is a challenging and exciting prospect. . . . Studying sensory losses caused by unilateral brain damage . . . [is] not a replacement for . . . the normally functioning brain" (Regan, 1972, p. 131). Before the mid-twentieth century, attention of neurolinguists and neuropsychologists was directed largely to the aspects of functional asymmetries related to speech comprehension and production. The shift of research to the auditory centers, aided by the psychophysical discoveries of Wever and Bray, von Békésy, H. Davis, and others, mentioned earlier, accelerated the application of the evoked potential method to the question of brain functional asymmetry and led to a series of neurolinguistic studies that distinguished the effects of verbal from nonverbal stimuli.

Using auditory evoked potentials (AEPs), the Canadian investigator Doreen Kimura found that with dichotic listening, in left-hemisphere-dominant patients, speech is comprehended more accurately in the right ear and music in the left. In her words: "It appears that when different verbal stimuli are presented to the two ears, those stimuli which arrive at the ear opposite the dominant hemisphere are more efficiently recognized" (Kimura, 1961, p. 169). The right-ear advantage for speech was believed to be due to the fact that stimuli traveling to the left auditory primary receiving area have a short, more direct path to the left language areas (Krashen, 1976). Evidence for this hypothesis came from hemispherectomized subjects and those with commissurectomies. So, too, verbal materials evoke AEPs of greater amplitude when recorded from the left scalp, whereas clicks are initially processed primarily in the right brain (Cohn, 1971). In most infants less than a year old, Molfese, Freeman, and Palermo (1975) demonstrated that AEPs are lateralized to the left with nonverbal stimulation, suggesting a "biological origin" of laterality. A further evidence of biological laterality in humans is the report that the alpha rhythm recorded over the right cerebrum precedes that recorded from the left in most healthy subjects (Liske, Hughes, and Stowe, 1967).

Other electrophysiologic signs of functional asymmetries in the human brain were found in the visual system. "Asymmetric cerebral functions underlying evaluation of visual stimuli are reflected in human evoked cortical potentials," Buchsbaum and Fedio concluded (1970, p. 209). Using verbal and nonverbal stimulation of left and right visual fields of right-handed subjects, they found increased perceptual accuracy and a more variable waveform from occipital electrodes when information was transmitted uncrossed to the dominant verbal or nonverbal hemisphere. Wada may have carried the concept of cerebral dominance to an extreme edge when he reported at the Tenth International Congress of Neurology (1973) that satisfactory meditation is the result of shifting the larger-amplitude cerebral evoked response from the "speaking" hemisphere to the "silent" hemisphere.

Although historians have ridiculed the relevance of animal studies of laterality to the situation in humans (Oppenheimer, 1977), nonetheless some fascinating discoveries continue to be made utilizing nonhuman subjects. The psychologist S. I. Franz (1913) attempted to determine the hand preferred by six monkeys in taking food. The results were inconclusive (1913), and his plea for gathering more data was accepted by Karl Lashley. Using two of Franz's monkeys, Lashley reported (1917)

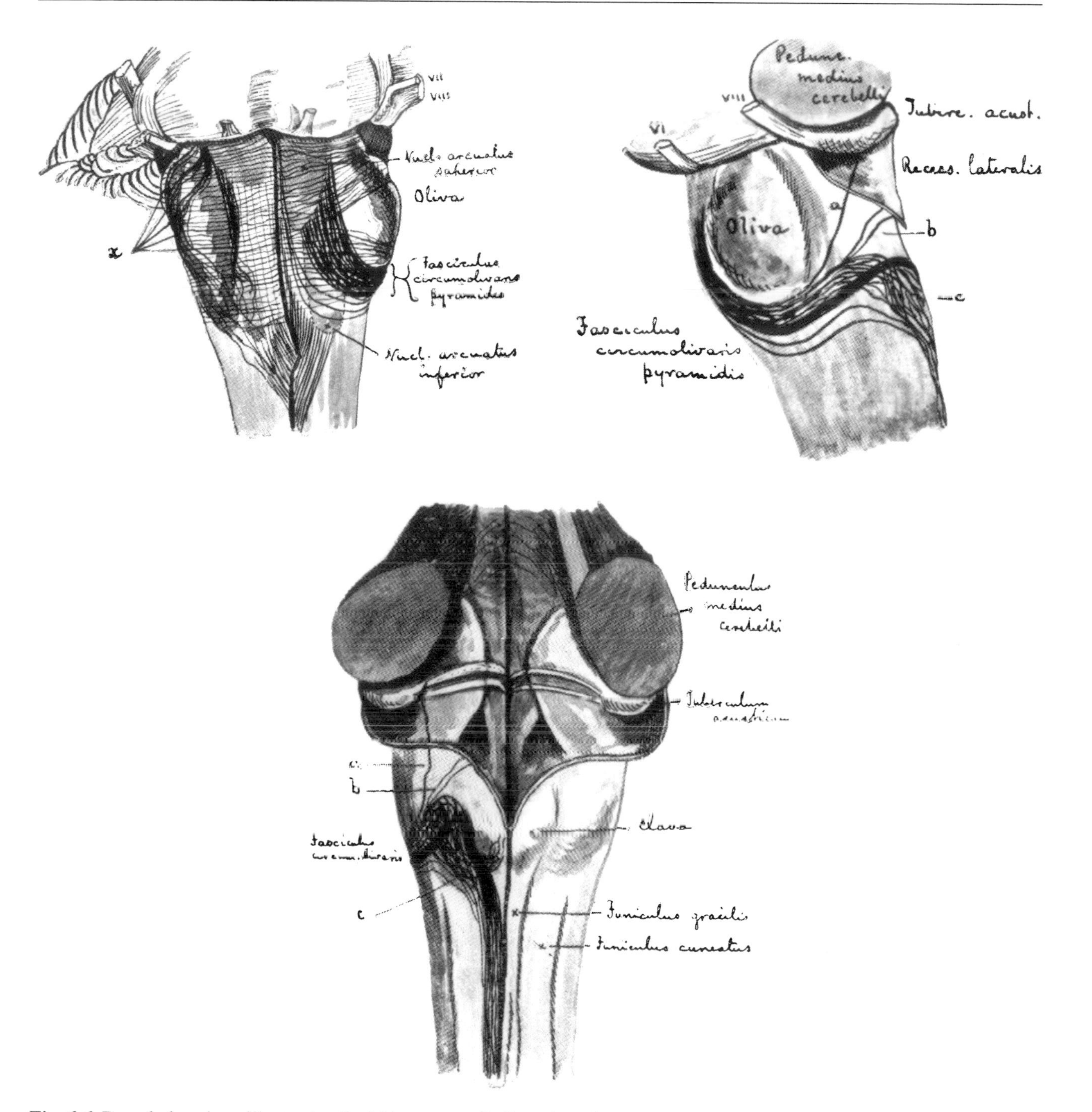

Fig. 6.6. Rough drawings illustrating Smith's autopsy findings in 15 human brains in which the "fasciculus circumolivaris pyramidis" in every case was larger on the left side. Clockwise from left—ventral, left lateral, and dorsal medulla oblongata. (From G. E. Smith, 1904, facing pp. 378, 381, Figs. 2–4, ×1.)

how easily modified the hand preference is and cast doubt on the value of studies on handedness even in children.

Again in nonhuman primates, the radiologist-neurologist team of LeMay and Geschwind (1975) found that brains of the great apes show some of the same anatomic asymmetries as do human brains. In fact, the orang-utang's brain is particularly human-like, and the authors suggested that studies of handedness and language should be conducted in that

Fig. 6.7. The preferred direction of turning of adult rats was reported to be reliably predicted by their tail positions as one-day-old newborns. (From Ross, Glick, and Meibach, 1981, p. 1959, Fig. 2, ×1.)

species rather than with the usual chimpanzees. Hamilton and Gazzaniga (1964) saw evidence of laterality in discrimination of visual tasks in nonhuman primates. Those early studies laid the groundwork for recent experiments by others on split-brain monkeys carried out at Caltech, the home base of such experiments, which again provided strong support for the "biological origin" of functional asymmetries. The animals were trained to discriminate visual stimuli that are known to involve lateralized processing in humans; the data showed that on the visual tasks used, nonhuman primates possess significant hemispheric specialization (Hamilton and Vermiere, 1988).

The laboratory rat has participated extensively in the quest for demonstrations of biological handedness. In a study that was part of a 30-year inquiry into the control of forelimb preference of the rat, it was found that paw-preference depends on the activity of a small number of cells in motor cortex of the frontal lobe (Peterson and Devine, 1963). The complete lateralization of this center and thus the possibility of using tissue from the unaltered side of the rat brain as its own control was exploited by Křivánek and Burešova (1972) in studies of acetylcholinesterase in brain tissue; they found a 10% increase in enzyme activity in cortex contralateral to the overtrained preferred paw. Relatedly, it was found that direction of turning can be controlled by contralateral striatal injections of amphetamine, a dopamine-depleter (Glick, Jerussi, Waters, and Green, 1974), and that female rats prefer to make dextral turns (Glick, 1974). In a curious tally, tails of newborn rats were reported to have a dominant direction of curvature according to gender (Ross, Glick, and Meibach, 1981; Fig. 6.7). More recently, asymmetries were reported in the postmortem distribution of neurotransmitters in the human brain: the concentration of dopamine (and other transmitters) was higher in tissue of the nigrastriatal path contralateral to the presumed (no handedness data were available) side of preferred handedness (Glick, Ross, and Hough, 1982).

One of the most provocative situations reported in lower animals came from the laboratory of psychologist Victor Denenberg, who has pursued a career-long interest in the effects of early experience on subsequent behavior. He and his associates found that routine handling influenced the hemispheric laterality of rats, presumably by enrichment of the environment. The results of those and other experiments led to a theory of innate functional asymmetry and generated much discussion in an innovative journal, *Behavior and Brain Science*, in which selected papers are followed by thoughtful, signed commentaries: Denenberg's 21-page article (1981) merited an additional 28 pages of commentary from 18 authors, plus the author's response.

Another example of functional asymmetry was documented by a series of studies with birds. Frederico Nottebohm of Rockefeller University in 1971 published "Neural lateralization of vocal control in a passerine bird," a report of experiments demonstrating that humans are not unique in their hemispheric specialization for vocalization, as Penfield and Roberts (1959) believed. Nottebohm and colleagues discovered that transection of the left but not the right hypoglossus nerve in the adult chaffinch produces permanent deficits in their song and suggested, because both hypoglossal nerves are present and the effect is restricted to the left, this lateralized specialization at the hypoglossal level resembles handedness and speech functions at the human cerebral level.

In the 1980s, it became known that anatomic asymmetries extend to the level of individual neurons. Because it is the dendrites that receive incoming information from other neurons, a greater dendritic complexity implies more connections. Counts of the number of dendritic branchings (*see* Fig. 6.8, left) of layer III pyramidal cells in tissue blocks taken postmortem from the top (motor) and

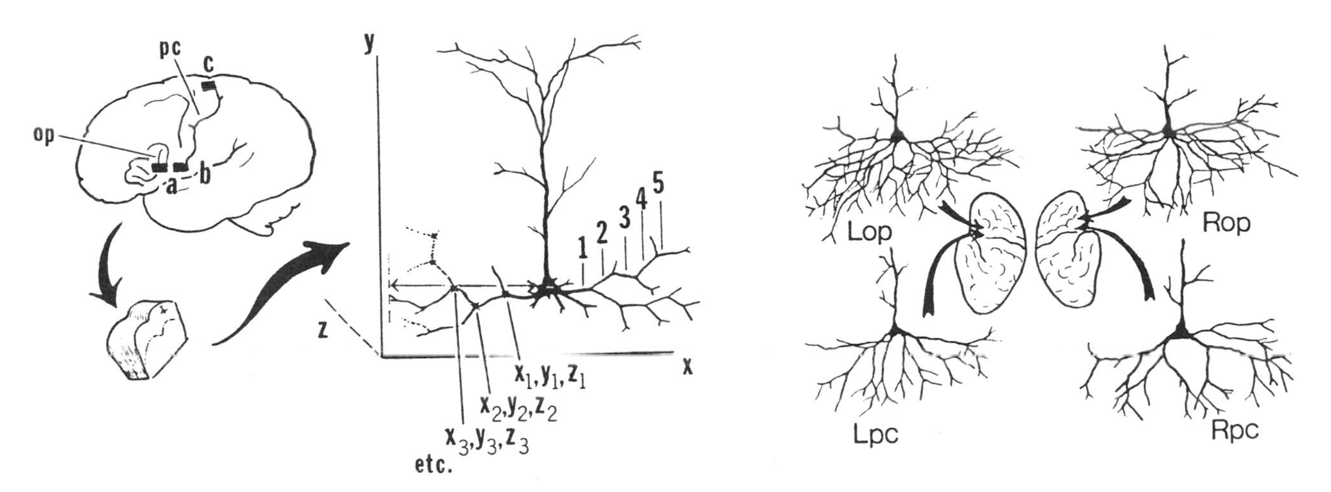

Fig. 6.8. Left—Diagram showing method of analysis of dendritic branching. Tissue blocks were taken from Broca's speech area (a, op) and the foot (b) and top (c) of the precentral gyrus (pc). After histological preparation, basilar dendritic segments (1–5) and bifurcation points (x,y,z) were measured. Right—drawing of four representative layer III pyramidal neurons, from the left and right speech (Lop,Rop) and left and right motor (Lpc,Rpc) areas. Note the relatively complex branching in the speech areas, especially on the left. (From Scheibel, 1984, pp. 45, 50.)

the foot (language) and Broca's area (articulation) by A. B. Scheibel and coworkers (1984) showed increased branching in the language area and longer dendrites in the motor areas (Fig. 6.8, right), perhaps in association with a greater complexity in processing of information than in motor activity, as the authors suggest. Such findings could be portents of future discoveries of increased density of brain receptor elements for specialized hand activities.

A decade earlier, brain asymmetries related to gender had begun to replace the earlier comparisons based on handedness, function, or early experience. Although gender dimorphism in the distribution of the massa intermedia, presumably conferring to women greater thalamic connectivity, was reported from Germany by Rabi (1958), it was almost 20 years before additional examples were announced. As mentioned above, greater asymmetry in the male human planum temporale was noted by Wada, Clarke, and Homm (1975). In experimental animals gender dimorphism was reported in synaptic organization (Raisman and Field, 1971) and in cell volume (Pfaff, 1966; Nottebohm and Arnold, 1976; and Gorski, Gordon, Shryne, and Southam, 1978). A more recent study of postmortem human brains seemed to demonstrate gender dimorphism in the size of the corpus callosum (De LaCoste-Utamsing and Holloway, 1982). Those pioneer studies formed a springboard to the eager pursuit of a biologic substrate associated with gender orientation of humans that is taking place during the last decade of the twentieth century

GROWTH OF KNOWLEDGE OF THE CORPUS CALLOSUM

The conspicuousness of the bundle of nerve fibers on which many of the cerebral asymmetries depend exposed the corpus callosum to early scrutiny and eventually it, too, was found to be asymmetric. In the first century BC, Galen believed that the function of the great band of white fibers joining the cerebra was to support the two halves of the brain. A millennium-and-a-half later, Vesalius (1543) named the pale and firm substance the corpus callosum and wrote that it "is observed to be in the middle of the brain . . . and comes into view when the brain has been separated," as in Fig. 6.9. Vesalius believed it not only "relates the right side of the cerebrum to the left; then it produces and supports the septum of the right and left ventricles; finally, through that septum it supports and props the body formed like a tortoise [fornix] so that it may not collapse. . . ." (1543, Lib. VII, Cap. V, p. 633; translated in Clarke and O'Malley, 1968, p. 579). After passage of another century, the corpus callosum was assigned a slightly less mechanical function by Willis, who placed the imagination

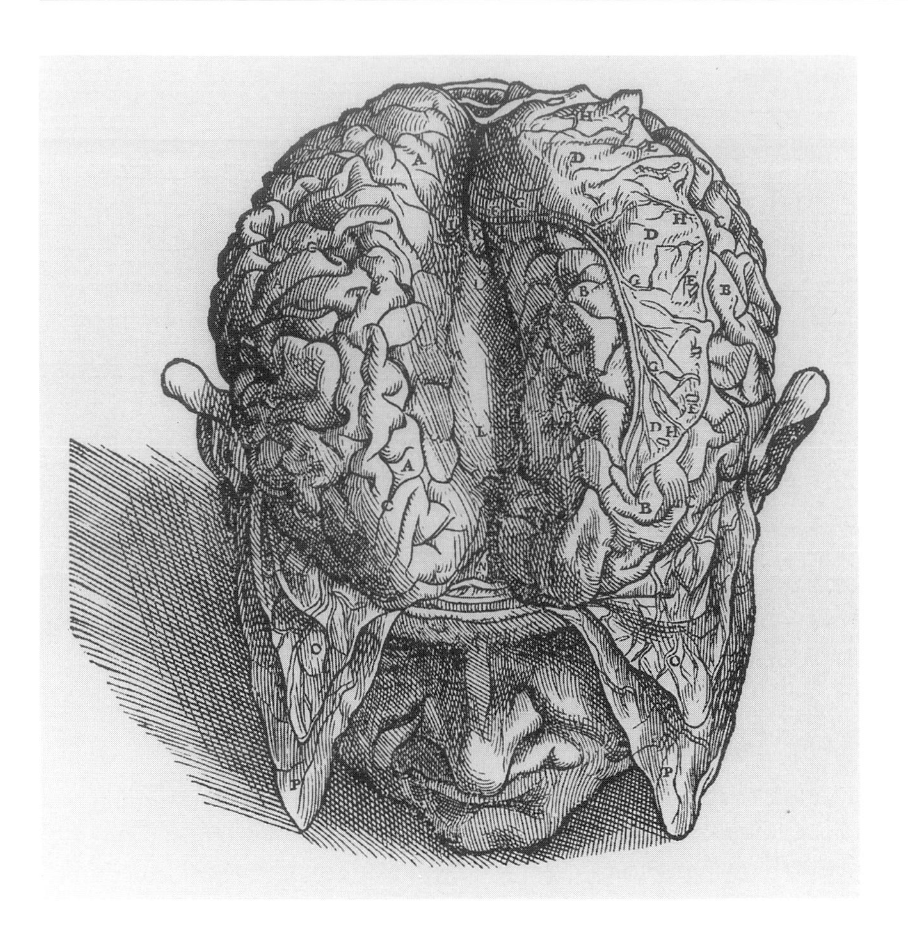

Fig. 6.9. To illustrate his description of stages in a progressive dissection of the human brain, Andreas Vesalius published this view of the corpus callosum as it appears when the cerebral hemispheres are drawn apart. (From 1543, p. 620, Fig. 15.)

there: "[W]hen a sensory impulse of the [animal] spirits is carried like a surge of water farther into *the corpus callosum* and from there into *the cortex of the brain* . . . [t]his is immediately succeeded by *imagination*. . . ." (1664, p. 64; translated in Clarke and O'Malley, 1968, p. 335).

Steno introduced midsagittal sectioning of the brain to observe the corpus callosum (*see* Figs. 8.2 and 8.3) and other central neural structures, but when he lectured in Paris he confessed: "We know so little of the true structure of the Corpus Callosum, that a man of a tolerable Genius may say about it, whatever he pleases" (Steno, 1669, p. 9). Publication in 1796 of the beautiful drawings of the dissections by the German anatomist, Samuel Thomas Sömmerring (1755–1830; Fig 6.10) and those of Gall and Spurtzheim (1809, 1810) executed before their preoccupation with "cranioscopy" (*see* Fig. 6.11, p. 118) introduced a number of outstanding depictions of the corpus callosum and other cerebral structures that appeared in the first half of the nineteenth century in France (*see* Fig. 6.12, p. 119; and *see* Fig. 6.13, p. 120). This flowering of anatomic illustration, an accompaniment of the refinement of experimental science during the Enlightenment, was associated with speculations regarding function, but not much progress was made toward uncovering reliable facts, for good reasons.

The experimental study of the corpus callosum commenced with Johann Gottfried Zinn (1749) in Göttingen; however, after transections of the tract his dogs displayed no clear-cut signs except lethargy. The experiment was replicated by all the great investigators of the time, but because of gross infection and trauma, there were no consistent changes in mobility or "psychic" behavior that could be attributed solely to the transection (*see* G. A. Russell, 1979). Gradually, however, the results of stimulation of the exposed corpus callosum (Brown-Séquard, 1877) and the careful fiber degeneration studies of Gall and of Meynert in the 1870s, led to the important realization that, in general, the commissures carry homotopical fibers:

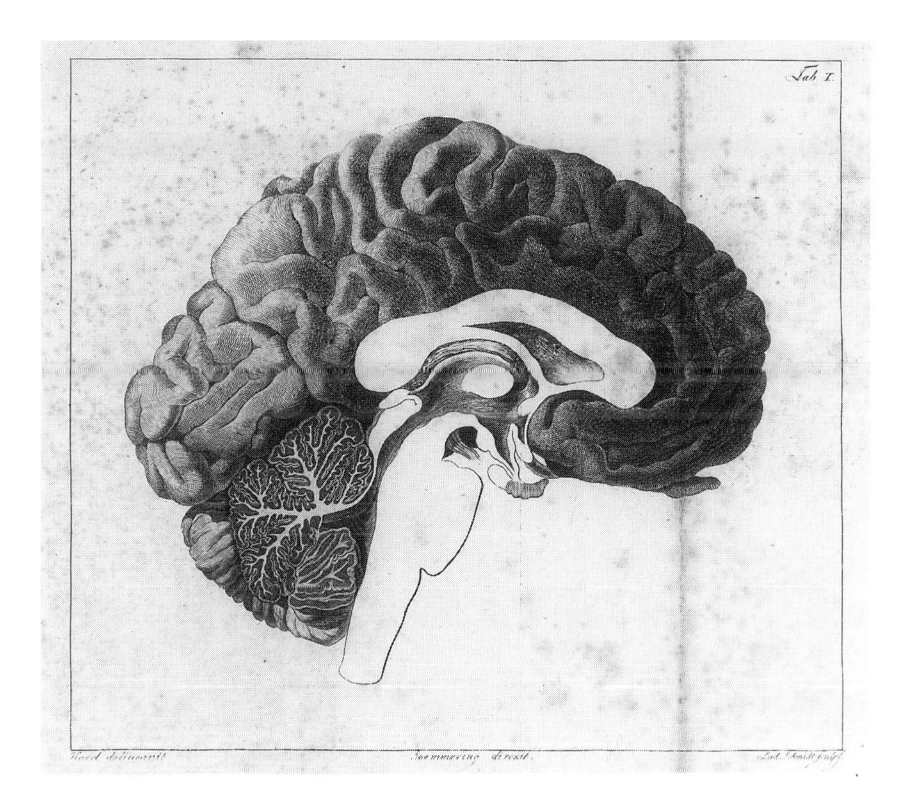

Fig. 6.10. Samuel Thomas Sömmerring "followed nature" in his illustrations and closely supervised the artist (C. Koeck) and engraver (L. Schmidt) of this outstandingly accurate depiction of a medial section of a human brain. (From Sömmerring, 1796, Tab. I, ×2/3.)

"The corpus callosum bundles . . . unite *identical parts of the cortex* of the two halves of the *cerebral hemispheres* (Meynert, 1872b, p. 698; translated in Clarke and O'Malley, 1968, p. 604).

A decade later, Meynert's beliefs were challenged by David J. Hamilton of Aberdeen, regarded by some writers (Ironside and Guttmacher, 1929) as one of the more significant investigators of the corpus callosum. His careful studies of adult (1884–1885) and embryo (1885) human brains convinced him that the great commissure consists chiefly of decussating cortical fibers which course downward to the basal ganglia. The idea of a "crossed callosal tract" was so contrary to the accepted wisdom that Hamilton reviewed what Willis, Foville, and Gratiolet had written and thought he could find evidence that they "foreshadowed the truth." As for Meynert's statement that he had followed single fibers continuously across the callosum from one cortex to the other, Hamilton believed that to be impossible because under the microscope a fiber does not remain in the same visual plane.

Hamilton's challenge to orthodoxy provoked a response from Beevor in a Preliminary Note: "The [conclusions] are so different to what has usually been considered to exist, and the results to clinical medicine would be of such great importance, that I do not think the subject can be approached without systematically working the question out" (1885, pp. 377–378). And so Beevor spent a year tracing fibers in brain sections. He concluded that it "is easy to see how a decussation may be imagined to exist, but I must maintain that no general decussation can actually be demonstrated" (ibid., p. 379). Both men hardened their specimens in bichromate of potash and used one of Weigert's stains (Beevor, hematoxylin; Hamilton, acid fuchsin) and boasted of the beauty of the results, but Hamilton had subjected the tissue to treatment with gelatin which stretched it one-third to one-half normal size and

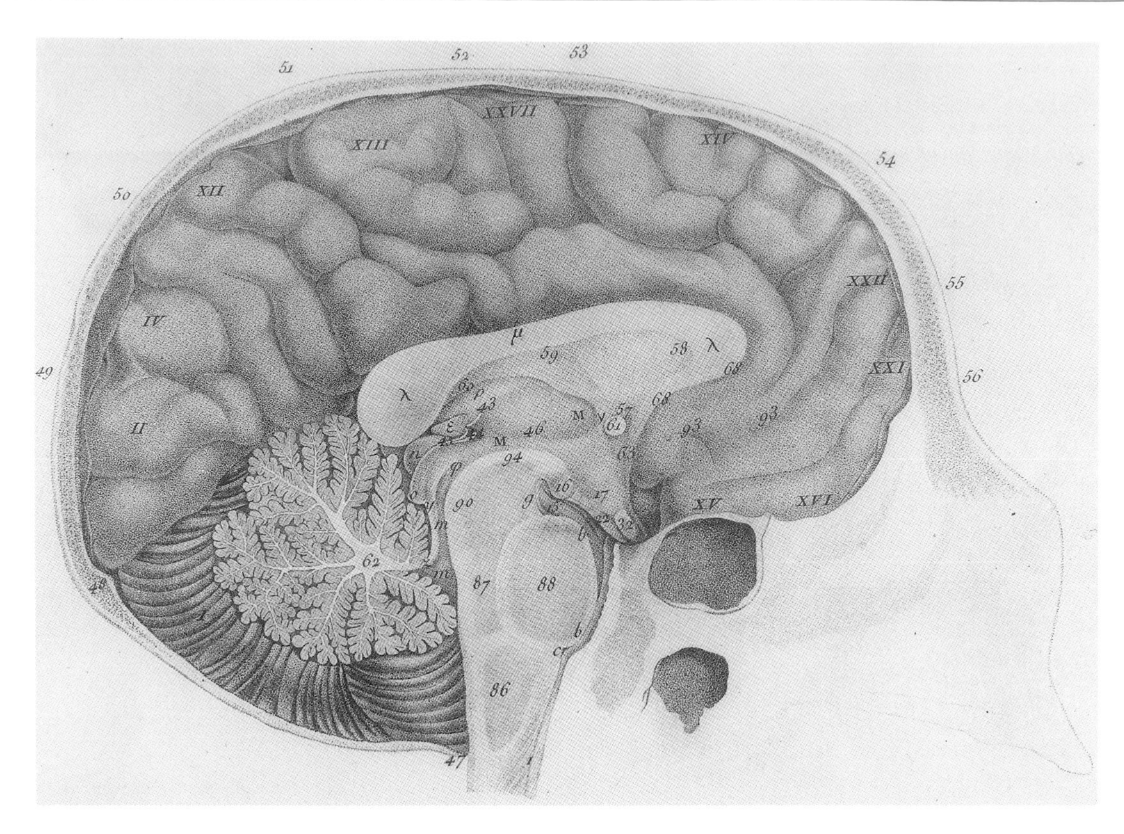

Fig. 6.11. Franz Joseph Gall's drawing of the midsagittal view of the human brain was engraved in England under his personal supervision. The Roman numerals correspond to phrenological areas; the corpus callosum is marked λ, μ, λ. (From Gall and Spurtzheim, 1810, vol. 2, Plate XI, ×1.)

thereby improved the processing. Yet subsequent observations from an entirely novel approach to the problem of callosal function rendered Hamilton's view a lost cause.

The use of experimental conditioned reflexes reinforced the concept of the existence of corresponding points on the two hemispheres that had been demonstrated by the early ablation/degeneration studies and also provided proof that the phenomenon depends on the corpus callosum. From England, a Russian expatriate, Gleb Vassilievitch von Anrep (1891–1955), reported (1923) decrement-free irradiation of stimulated points to the opposite side, using easily quantitated salivation in dogs as the classically conditioned response. In Mother Russia, K. M. Bykov, a young assistant at the Pavlov Physiological Laboratories, the cradle of conditioning, resolved to continue the study of the "paired work of the hemispheres, . . . [o]n the recommendation of my dear teacher Ivan Petrowich Pavlov." He wrote, in translation: "For the solution of these questions it was necessary first of all to sever the two hemispheres by cutting the corpus callosum, and on such animal, by the method of conditioned reflexes, to trace the work of all sections of the cortex" (1925, p. 9). Bykov is credited with being among the first to utilize evoked potentials experimentally (discrimination of sound direction by dogs) and to have demonstrated definitively that information exchange between the cerebral hemispheres depends on the intact corpus callosum: "In summary, these experiments [of Bykov] are of great importance in showing that the bilateral synergic activity of the hemispheres may be dependent on the corpus callosum, and indicating the part which the corpus callosum plays in the development of symmetrical reproduction of function in the hemispheres" (Ironside and Guttmacher, 1929, p. 453). The point-to-point connection of one side of the cortex to the other by way of the corpus callosum was finally proven by direct electrical stimulation: Curtis and Bard (1939) showed that a single shock on one hemisphere is picked up maximally at the corresponding point on the other hemi-

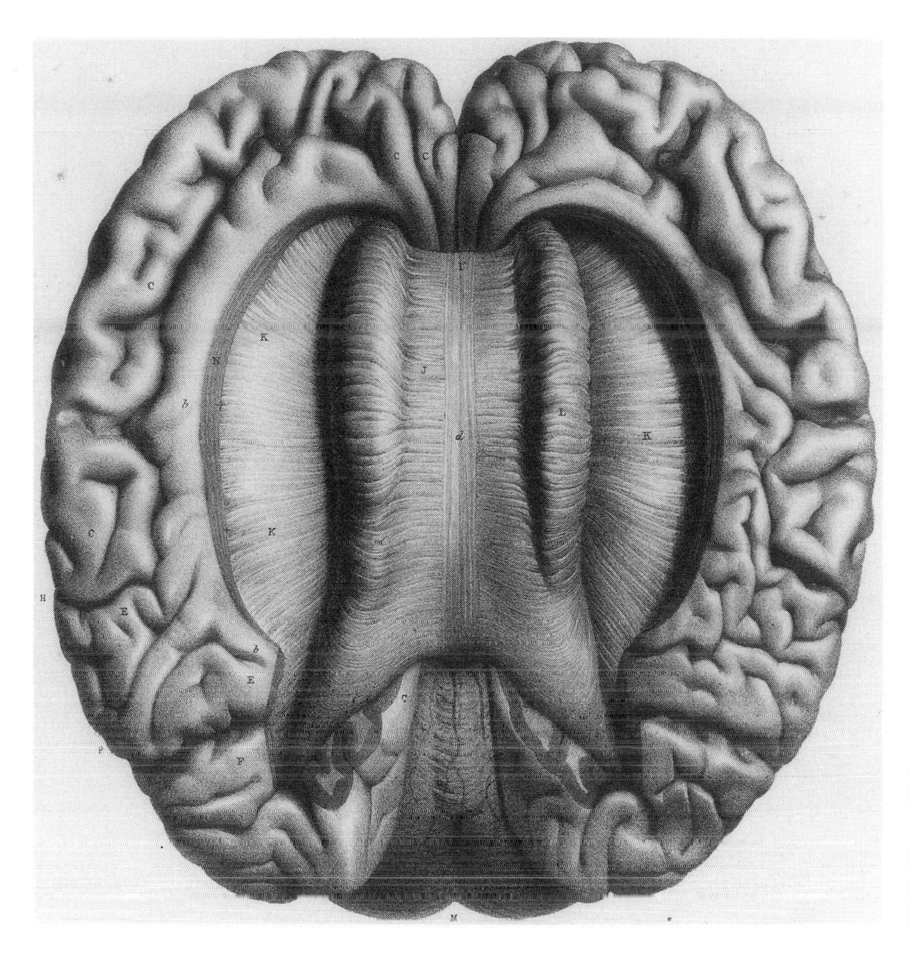

Fig. 6.12. The corpus callosum from the atlas of Louis Achille Foville's *Traité complet* (1844) depicted his careful dissection from above; the artist was E. Beau and the engraver F. Bion. (Plate VII, Fig. 1, ×1.)

sphere, a finding confirmed by McCulloch using strychnine (1944).

At Yale, H.-T. Chang set about to analyze the electrical activity of the callosal fibers as a possible clue to cortical activity. His study in the cat (1952), in which the constituent parts of the evoked potential were correlated with their origin, was predicated on the fact that the corpus callosum consists only of axons arising in the cortical deep layers and terminating in the superficial layers of the opposite side (except for ipsilateral olfactory fibers). Chang's results seemed to show that callosal fibers do not participate in the corticothalamic reverberating circuit pattern (*see* Chapter 11).

Turning now to the state of knowledge about the function of the corpus callosum in the human, clues had come from isolated cases presented by neurologists such as Dejerine and others (*see* Joynt's lucid discussion, 1974, p. 120 passim), as well as by deductions from the histological studies of Meynert and of Ramón y Cajal, but they were largely ignored. It was a generality that extensive damage in one hemisphere would be accompanied in time by atrophy of the corpus callosum, but evidence was lacking about which condition was the primary abnormality. From observation and reports of cases of callosal trauma, the Dutch neurologist, C. T. Van Valkenburg, (1913, p. 122) wrote: "I know of no degeneration investigations on the human brain which have led to any positive results."

In the United States, two neurosurgeons were sufficiently comfortable with the absence of proven callosal function to hazard a trespass on what "has always been looked on as hallowed ground" (Van Wagenen and Herren, 1940, p. 741). They carried out partial or complete commissurotomies on 10 patients with intractable epilepsy in an effort to prevent spread of the epileptogenic impulses to the second hemisphere. There were "no untoward effects," and pre- and postoperative testing revealed few abnormalities (K. W. Smith and Atelaites, 1942).

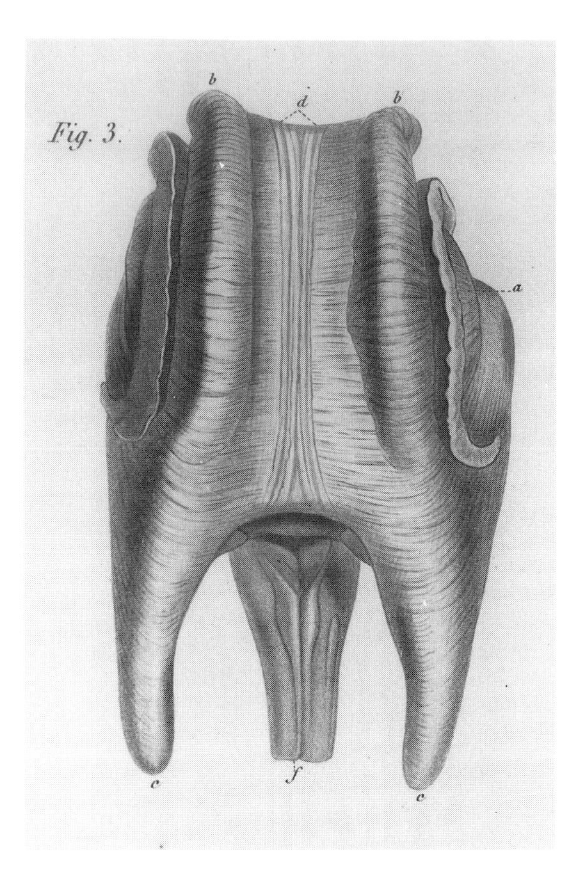

Fig. 6.13. This rendering of the upper surface of the human corpus callosum dissected out with some cerebral nuclei appeared in the atlas accompanying the *Anatomie comparée* of Leuret et Gratiolet, vol. II, 1857. a—external corpus striatum; b,b—anterior end of corpus callosum covering frontal horns of lateral ventricle; c,c—occipital horns of lateral ventricle covered by posterior fibers of corpus callosum; d—longitudinal nerves of Lancisi coursing in front of the medial valley of the corpus callosum. (Vol. 2, Plate XXV, Fig. 3, ×1.)

Conclusions of that kind prompted one of France's most eminent neurophysiologists to exclaim: "[The] section of important associative white tracts such as corpus callosum does not seem to affect mental performances. . . . These results are so disturbing that one may be tempted to admit the irrational statement that a heterogenous system of activities in the nervous system could form a whole in the absence of any identified liaison" (Fessard, 1954, p. 208).

The discrepancy between the relatively massive size of the corpus callosum, estimated to contain about one million fibers in the human (Bailey and von Bonin, 1951, p. 301), coupled with its prominent position and the slight effect on ordinary behavior of its transection "posed for many years one of the more intriguing and challenging enigmas of brain function" (Sperry, 1961, p. 1749). Ronald E. Myers and his mentor, Roger W. Sperry (1914–1994), were the first to demonstrate convincingly the independent functions of the cerebral hemispheres. After they observed (Myers and Sperry, 1956) in cats with a severed optic chiasma a mnemonic carryover via the corpus callosum in the hemisphere not receiving information, the next step was to repeat the tests in animals that were also commissurotomized. First at the University of Chicago, then at Caltech, their experiments pointed compellingly to the functional asymmetries of the cerebral hemispheres. In the early 1960s, that work was brilliantly applied to the human situation by the chance presentation of a patient in the emergency room of the White Memorial Hospital in Los Angeles, where one of the young neurosurgical residents, Joseph E. Bogen, admitted "W. J." in status epilepticus. Following his initial care of W. J., Bogen began to puzzle over the possibility of commissurotomy, a procedure he had first discussed with Myers at Caltech and that had been used apparently successfully on a few patients 20 years earlier (*see* above). Bogen continued to take care of W. J. and even arranged for him to spend six weeks at the National Institutes of Health in Bethesda, Maryland, where he was told his situation could not be remedied.

At that point, Bogen broached the ideas he had been harboring: What is a human being with a split brain; and could we get an answer by surgically treating patients whose epilepsy was otherwise intractable? First he persuaded W. J. and his family to consent, then he approached a prominent neurosurgeon, Philip J. Vogel, who agreed after extracting Bogen's promise to find cadavers for practicing the tricky procedure. Sperry had been consulted and his graduate student, Michael Gazzaniga, put W. J. through sophisticated psychological tests preoperatively.

After the operation, W. J.'s convalescence was stormy, but within a few months he was not having seizures, was leading a normal life, and saying that he hadn't felt so good in years. In 1962, W. J.'s case was published in two articles, one by Bogen and Vogel, "Cerebral Commissurotomy in Man" and the other by Gazzaniga, Bogen, and Sperry, "Some Functional Effects of Sectioning the Cerebral Commissures in Man." In 1969, when the number of split-brain patients had increased to eight, Bogen published a series of three papers under the general title "The Other Side of the Brain."

> Results accumulated over a period of 6 years demonstrated that the cortical commissures were critical to the inter-hemispheric integration of perceptual and motor function. These studies also revealed that the mute right hemisphere was specialized for certain functions that deal with nonverbal processes, while, not surprisingly, the left hemisphere was dominant for language. For the first time in the history of brain science the specialized functions of each hemisphere could be positively demonstrated as a function of which hemisphere was asked to respond (Gazzaniga, 1981, p. 517).

With his detailed appreciation of the perceptual and cognitive capabilities of the two disconnected sides of the brain, and his recognition that "two hemispheres united are better than two hemispheres divided," Sperry (1968, p. 135) found that his interpretation of consciousness was gradually modified (*see* p. 207, this volume).

Studies on the topographic distribution of fibers that pass through the corpus callosum for obvious reasons have been carried out almost exclusively on experimental animals. Some of the first involved monkey brains studied after constricted transections of the corpus callosum had been made earlier (Pandya, Karol, and Heilbronn, 1971) and showed that cutting the rostral half disturbs frontal lobe fibers and cutting the caudal half disturbs the fibers to the parietal, temporal, and occipital lobes. In addition to the experimental work, an occasional opportunity to test callosal function in the human presents itself. In what they "believe to be the first observation of interhemispheric availability of higher-order information in the absence of sensory transfer" (Sidtis, Volpe, Holtzman, Wilson, and Gazzaniga, 1981, p. 344), the authors tested a patient before and after transection of the posterior half of the corpus callosum. When his seizures returned after 10 weeks, the anterior half was cut and the testing repeated. The behavioral results are summarized in Fig. 6.14, p. 122.

Partial or complete agenesis of the corpus callosum in humans presents an additional opportunity to learn more about how the two hemispheres interact. That condition has been reported since at least 1812 (Reil) from autopsy observations. Introduction of the ventriculogram technique by Dandy (*see* p. 33, this volume) made it possible to identify agenesis of commissural fibers during life, as in five patients at "The Neuro" (Hyndman and Penfield, 1937). That individuals with this anomaly can lead a productive, normal life was evidenced by Sperry's subject whose callosal agenesis was not revealed until she was examined for headache during her successful college career (Saul and Sperry, 1968).

In an effort to summarize the vast amount of research on cerebral hemispheric differences—anatomic, behavioral, and electrophysiologic—only a very small part of which has been mentioned here, we may quote an investigator who has a grasp of the large picture as seen today:

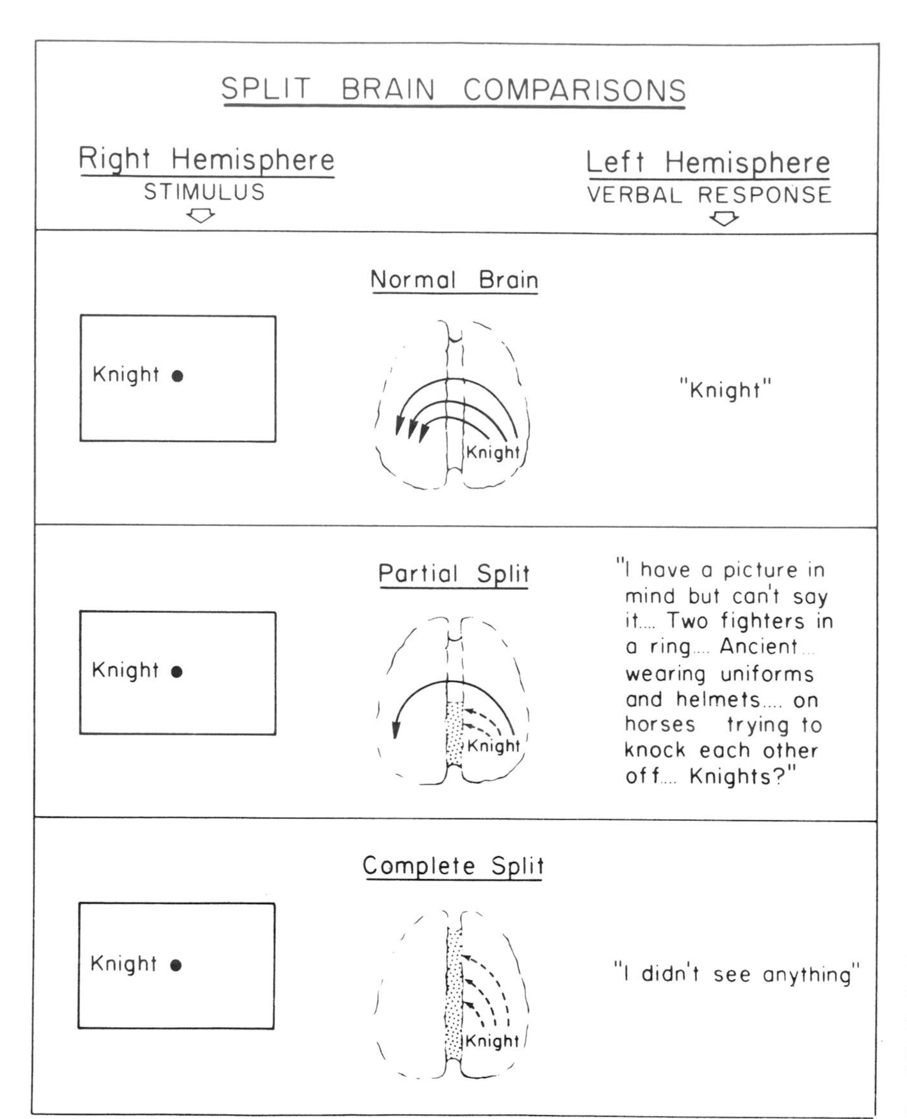

Fig. 6.14. Schema of a patient's naming ability before and after each stage of a two-stage commissurotomy, showing cognitive transfer without sensory information. (From Sidtis, Volpe, Holtzman, Wilson, and Gazzaniga, 1981, p. 345, Fig. 1.)

> [T]he human cerebral hemispheres exist in a symbiotic relationship in which both the capacities and motivations to act are complementary. Each side of the brain is able to perform and chooses to perform a certain set of cognitive tasks which the other side finds difficult or distasteful or both. In considering the nature of the two sets of functions, it appears that they may be logically incompatible. The right hemisphere synthesizes over space. The left hemisphere analyses over time. The right hemisphere notes visual similarities to the exclusion of conceptual similarities. The left hemisphere does the opposite. The right hemisphere perceives form, the left hemisphere, detail. The right hemisphere codes sensory input in terms of images, the left hemisphere in terms of linguistic descriptions. The right hemisphere lacks a phonological analyzer; the left hemisphere lacks a Gestalt synthesizer (Levy, 1974, p. 167).

In a lighter vein, it is appropriate to close this chapter on asymmetry from a distinctly global point of view, epitomized by the logo (Fig. 6.15) prepared in 1982 for the International Brain Research Organization (IBRO). As described by a former editor of *IBRO News*

> I drew the design to incorporate the brain and the world together to symbolize international brain research. On looking at it later I realized that

Fig. 6.15. The logo of the International Brain Research Organization (IBRO), incorporated temporarily in the masthead of its newsletter by the editor, M. I. Phillips, was not intended to suggest asymmetry (*see* text).

there are some visual puns which were not intended. Why is the brain part the left hemisphere? Is the left hemisphere of the world more verbal and more dominant? Is the frontal pole always north? Is the northern hemisphere more motor and the southern hemisphere more sensory? I disclaim any intention to make these inferences—but that, of course, is my left hemisphere speaking (Phillips, M. I./personal communication to H. W. Magoun).

OVERVIEW OF CEREBRAL DIFFERENCES

The "double brain" probably arose from symmetrical bulging in the leading end of primordial forms represented by the modern flatworms (*see* Chapter 2). At some era in prehistory, doubleness became a "basic, abiding and persistent quality" (Dimond, 1977, p. 477) of all brains, simple or complex. Among the reasons that make a double brain essential are the necessity to monitor a vast amount of external space, to control limbs on separate sides of the body, and to provide a more diverse perceptual and mental capacity. As the outgrowths became true bilateral brains, they probably controlled mostly the ipsilateral activities of the organism. The two sides also probably maintained contact through development of a "ladder-type" system as found in annelids and arthropods (ibid., p. 478), thus foreshadowing a corpus callosum. The problem with the phylogenetic development of two hemispheres was how to synchronize the action of the two sides, and in the midnineteenth century, the solution was thought to rest in the corpus callosum, where "there is certainly a sufficient apparatus of union to unitize the cerebral action," with "one supreme legislative power for both hemispheres" (Buchanan, 1849–1850, pp. 515, 519).

Differences in the number, size, and position of gyri and sulci between the two sides of the brain were noted long before their functions were accurately described. The first functional asymmetry was revealed in the midnineteenth century in the astute observations, by Dax and Broca, of patients who had impairment of speech and frontal lobe damage. The evidence became associated with the notion of preferential "education" to produce a picture of dominance of one side of the brain over the other. That concept held sway for half a century and was difficult to change because when first proposed, it was not challenged and became part of the psychological wisdom without the usual scientific scrutiny. Geschwind and his associates re-established interest in behavioral differences related to the dual brain by identifying the correlation of unequal size of the planum temporale with handedness, opening an era of new discoveries of asymmetries at anatomic, physiologic, and biochemical levels.

The independence of the cerebral hemispheres was demonstrated by interruption of their main connection, the corpus callosum. Sophisticated psychological testing by Sperry and his group reversed the initial claim of no observable effect and began the detailed elucidation of the different roles of the two halves of the brain. The usefulness of such an arrangement seems obvious:

> The ability to know left from right in space, and first from second from third in time, might be very difficult or even impossible with a symmetrical brain. In other words, the perception and memory of directionality in both space and time, important in human thought, may necessitate neuroanatomic asymmetry. In this framework one might suggest that one of the unique features of *Homo sapiens*, perhaps basic to others, may be in the marked left-right neuroanatomic asymmetry of the cortex (Witelson, 1977, p. 351).

The work of Sperry and his collaborators rescued the cerebral hemispheres from the threat of one side dominating the other and initiated a new era of discoveries in brain specialization.

7 The Anatomic Substrate

Cerebral Fine Structure

CONTENTS

I threw myself into the task with sure faith that in that dark thicket where so many explorers have been lost, I should capture, if not lions and tigers, at least some modest game scorned by the great hunters.

(Ramón y Cajal, quoted by Courville, 1970, p. 150)

The nineteenth century is considered the great age of cellular biology because it witnessed fundamental insights gained from detailed studies of life processes and the underlying structural elements on which they rest. The names of Johannes Müller and Theodor Schwann are prominent here, representing the Continental scientists who were endeavoring to answer specific questions about the basic form of both plant and animal tissues. Great progress in knowledge of the *macro*anatomy of the nervous system had been made, yet there was no understanding of its basic structural elements, that is, the neural tissue (histology), and more specifically the nerve cell itself (cytology). Progress in the study of fiber-tracts and pathways coursing to and from the main motor and sensory regions in the brain depended on the development of new knowledge at the *micro*anatomic level and histological description became the research focus for investigators such as Meynert in Vienna and Betz in Russia during the midcentury. With the introduction of the compound microscope and microtome and improved fixation (hardening) and staining techniques, those histological approaches prompted great refinements in cytological research and forecast our present reductionist era of molecular biology.

CONNECTEDNESS OF CELL BODY AND FIBER

Before the widespread use of the achromatic microscope (ca. 1830), there were few reliable observations of the appearance of neural tissue. Nerve fibers in the periphery were understandably the first to be looked at: they are firm and accessible to manipulation (Clarke and O'Malley, 1968, pp. 30–31). Combining his skill in slow grinding of perfect lenses with a superior visual acuity, the Dutchman, Anthony van Leeuwenhoek (1632–1723; Fig. 7.1) examined a great many life forms including the cross section of an ox's peripheral nerve (Fig. 7.2, left), a drawing of which he sent to the Royal Society, then 44 years later published a much less crude rendering (Fig. 7.2, right). The magnifying power[1] and the techniques used for tissue preparation were indeed very poor at the time; the microscopes suffered from chromatic and spherical aberration and the tissues were pliable and unstained. By the late eighteenth century, Fontana

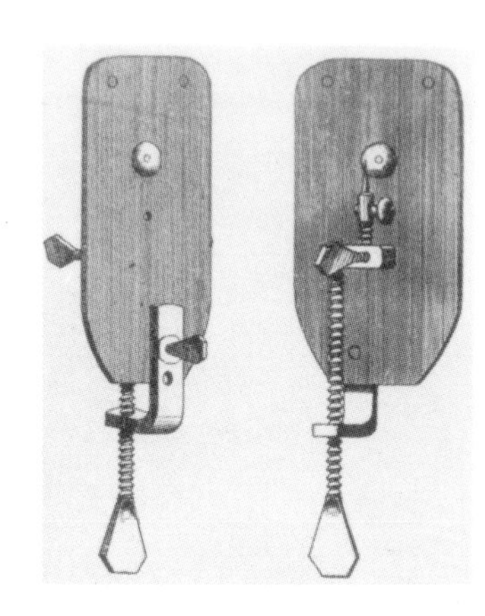

Fig. 7.1. Anthony van Leeuwenhoek's portrait shows him as a member of the Royal Society of London in his early adulthood and appears in his publication *Epistolae Physiologicae* (1719). The drawing of both sides of one of his original microscopes shows the glass lens at the top and the vertical adjustment screw with its support and the needle on which the specimen was impaled, ×1.

had been able to distinguish separate fibers in the optic nerves (Fig. 7.3). His observations of 1781 (translated into English six years later), convey a sense of the contemporary knowledge of nerve histology:

> I wish to know what the primitive [elementary] structure of the nerves is, that is to say, whether it is composed of channels or simple threads; whether it merely consists of globules, or contains a non-organic, irregular, spongy matter. . . . [Anatomists] have disputed [this] for more than three thousand years . . . and seem during this time to have done nothing more than multiply doubts and hypotheses (translation quoted in Knoefel, 1984, p. 235).

Progress in histological studies accelerated when achromatic, nondistorting microscopes became available, about a decade before fixation in chromic acid was introduced in 1840 by Adolph Hannover (1814–1894). Considered the first Danish microscopist, he studied tissues of both normal and abnormal nervous systems in an endeavor to associate structural derangement with dysfunction, during the same period when the French (Dax père, Baillarger, and later Broca) at the macro level were correlating damage in the left cerebral hemisphere with speech disorders.

An additional impetus stemmed from the enduring debate about whether or not the fibers are attached to the cell bodies. The challenge of demonstrating such a connection with the new compound microscope created an exciting period of claims and counterclaims. The first obstacle was the misconception that tissues, including brain tissue, consist of "globules." That bit of traditional wisdom was banished by two Englishmen, Thomas Hodgkin and J. J. Lister, who used the latter's compound microscope to look at "sufficiently stretched" brain tissue. They reported in 1827 that "one sees instead of globules a multitude of very small particles, which are most irregular in shape and size. . . ." (Hodgkin and Lister, 1827, p. 137).

Further progress was documented a few years later by C. G. Ehrenberg in Germany. Using the centripetal method of dissection reminiscent of Gall's practice, Ehrenberg wrote (1833, p. 455) that he traced the motor nerves passing into the brain substance and proved to himself that they are

[1]For a discussion of the early microscopes' resolving power, *see* Hughes (1959, p. 12ff); extended accounts of Leeuwenhoek's revolutionary new technique appeared in (Brazier, 1982; 1984).

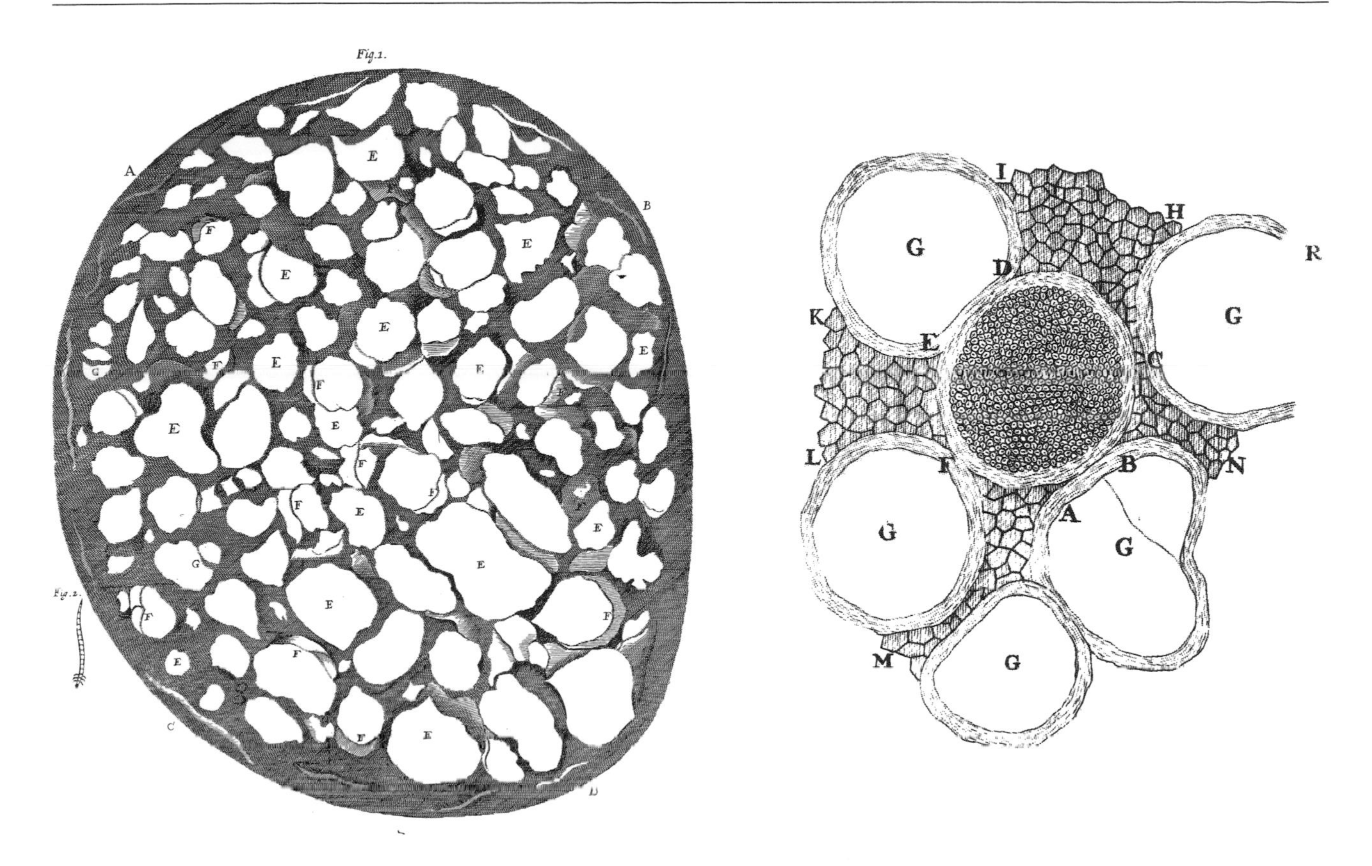

Fig 7.2. Left—The original sketch of a section of dried optic nerve of a cow sent by Leeuwenhoek to the Royal Society in 1675 and published as Fig. 1 in its *Philosophical Transactions*, ix–x, opposite p. 416, 1676–1678. A,B,C,D—Optic nerve, E—places left white and lucid from which soft globules have been extruded, F—transparent globules, G—globule particles sometimes lying across a cavity (1675, facing p. 416), ×1. Note Fig. 2, showing one of the "wondrous small creatures" the microscope revealed. Right—Extensively redrawn cross-section of a peripheral nerve from Leeuwenhoek's published letter to a friend (1719, facing p. 312). The "nervule" in the center, in which the tiny individual lines indicate the cavity of each of the "vessels," is surrounded by additional nerves (G) and fatty parts, ×1. An enlargement graces the cover of current issues of *Neuroscience*.

extensions of it. Although Ehrenberg clearly placed the fibers within the brain substance, it was several years before they were seen to be connected with the brain's "particles."

Some of the delay in making that discovery may be attributed to Gabriel Gustav Valentin (1810–1883), considered by some historians to have been one of four "forgotten leaders" of nineteenth century neuroscience (Kisch, 1954). Trained in Breslau, Valentin worked closely with his Czech teacher, Purkyně, and in 1836 published some of their views: "On the course and fine endings of the nerves." That paper is considered a milestone in nervous system histology because it identified ground elements common to all parts of the nervous system (Kisch, 1954). Clarke and Jacyna

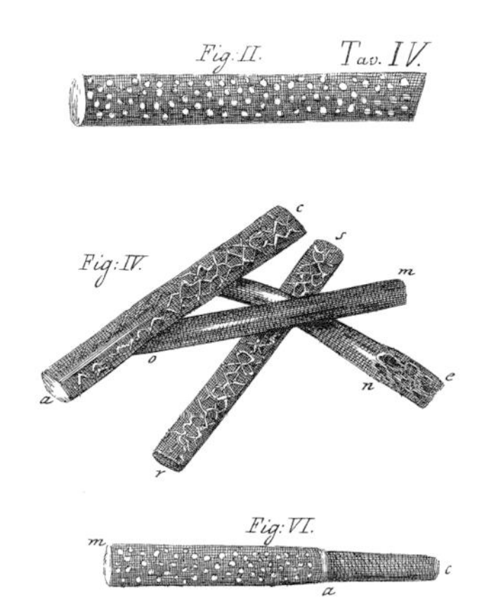

Fig. 7.3. Fontana's depictions of the "primitive nerve cylinder," illustrates what could be seen with the microscopes available during the late eighteenth century. The bottom figure probably shows the cylinder surrounded by neurilemma (*see* Clarke and O'Malley, 1968, p. 36). (From Fontana, 1781, Plate IV.)

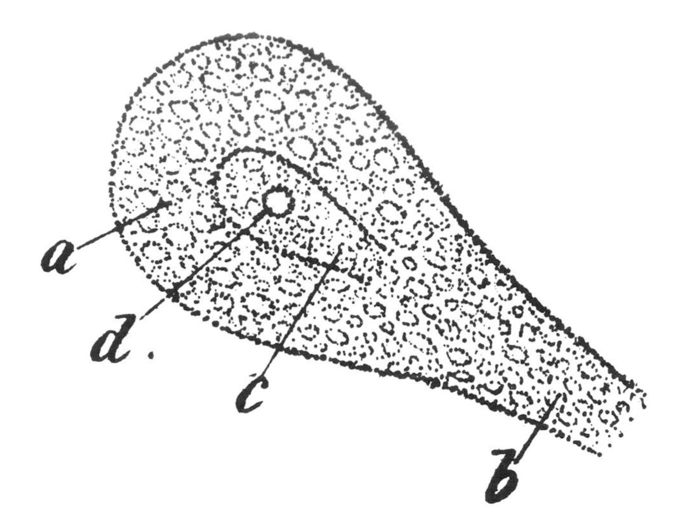

Fig. 7.4. This drawing of a cell from human cerebellar cortex is from Valentin's 1836 publication (Plate VII, Fig. 54). a—Parenchyma, a term which Valentin introduced; b—an appendage; c—nucleus; d—nucleolus; enlarged.

(1987, p. 83) argue that Valentin and Purkyně were striving to find analogous examples of the unity of nature in keeping with the romantic and ideological concepts of *Naturphilosophie*, which "... provided the framework that guided their microscopical researches." The analogy of the three structures of the ovum (parenchyma, nucleus, nucleolus) to the three elements seen in the nerve cell (Fig. 7.4) was persuasive to those latter-day romanticists. Their unitary point of view is seen in Valentin's statement, which brought fiber and cell body close together in the nervous system but did not join them:

> The finer structure of the nervous systems in man and vertebrates . . . has an admirable simplicity such as the human mind could hardly understand until now. . . . The entire nervous system is made up of two elementary basic substances, namely the isolated globules of the covering substance [cortex] and the isolated, continuous primitive fibres (Valentin, 1836, p. 157, translated in Clarke and O'Malley, 1968, p. 45).

A year after Valentin's paper appeared, Jan Evangelista Purkyně (1787–1869; Fig. 7.5) published his own work on the nervous system (1837). Among his illustrations are drawings of cells (since named for him) in the human cerebellum (see Fig. 7.6, p. 130). Purkyně could find "nothing definite" to prove a connection between cell bodies and fibers, although his statement is interpreted as showing a belief that such a connection exists (Clarke and O'Malley, 1968, p. 53).

A stronger champion of the idea that nerve fiber and cell body are united was Robert Remak (1815–1865; *see* Fig. 7.7, p. 131). This Polish neurocytologist made several pivotal contributions (Kisch, 1954): He is credited as the first to depict the microscopic structure of the cerebral cortex, to have clearly described the axon, and, most importantly, to have postulated and found evidence of a connection between nerve fibers and cells. Remak was straightforward in his statement: "*The organic* [nonmyelinated] *fibres originate from the very substance of the nucleated globules* [nerve cells], and this observation, although it is very difficult . . . is nevertheless so clear that it cannot be doubted" (Remak, 1838, p. 9, translation from Clarke and O'Malley, 1968, p. 52).

The uncertainties and contradictions were laid to rest by Rudolph Albert von Kölliker (1817–1907), the great Swiss histologist considered to have declared (1849) the unity of the fiber and cell-body almost 50 years before Waldeyer's publication (Bonin, 1970). Hannover's chromic acid fixative enabled Kölliker to work successfully on the cerebral cortex and, in 1849, he reported two distinct fiber networks: a fine matrix formed by dendrites and seeming to connect with the cell bodies, and a coarser network of axons. He regarded the cortical substance as the place of origin of all

Fig. 7.5. A pen drawing by Max Švabinský from the frontispiece of the Memorial Address at the 200th Anniversary of Jan Evangelista Purkyně, held in 1987 at Charles University, Prague. (With kind permission of Dr. Vilém Kuthan.)

nerve fibers of the cerebral hemispheres and also of those in the corpus callosum. Kölliker should be appreciated for his *Handbuch* (1852) of human histology written expressly for artists and students and put into English two years later.

The unequivocal proof that neural fibers are connected to cell bodies was delayed until the beginning of the twentieth century. Ross Granville Harrison (1870–1959; *see* Fig. 7.8, p. 132), then a biologist at Johns Hopkins University, devised a method of hanging from a cover glass a drop of lymph serum in which he suspended embryonic nerve tissue. For as long as four weeks, he could see processes growing out into the coagulated serum (1907; Fig. 7.8); with Harrison's demonstration, cytophysiology caught up with the more advanced microanatomy. Many years later, the sequel was provided by Rita Levi-Montalcini, who observed embryonic nerve cells forming processes in the presence of added nerve growth factor isolated from mouse salivary gland (1964; *see* Fig. 7.9, p. 133). Her first experiments were carried out in Italy under the extremely difficult conditions of the Fascist regime, as she describes in an autobiographical account (1975).

CELLULAR ELEMENTS

Before individual cells in the cerebral cortex were seen clearly, a striped pattern of layers was discernible; "lamination is one of the hallmarks of the cerebral cortex" (Jones, 1984a, p. 521). As noted, the first observations on the cortex were made on fresh tissues unfixed, or poorly fixed in alcohol; staining was not yet a general practice. Such were the tissues examined by Vicq d'Azyr, who reported (in translation) that he saw in the posterior lobes a "white linear tract that follows all the contours . . . and gives . . . the appearance of a streaked ribbon" (1781, p. 511). Looking at sections of a frozen brain, an Italian medical student, Francesco Gennari (1752–1779), also described (1782) the pale line, which today bears his name, seen most prominently in the grey cortical sub-

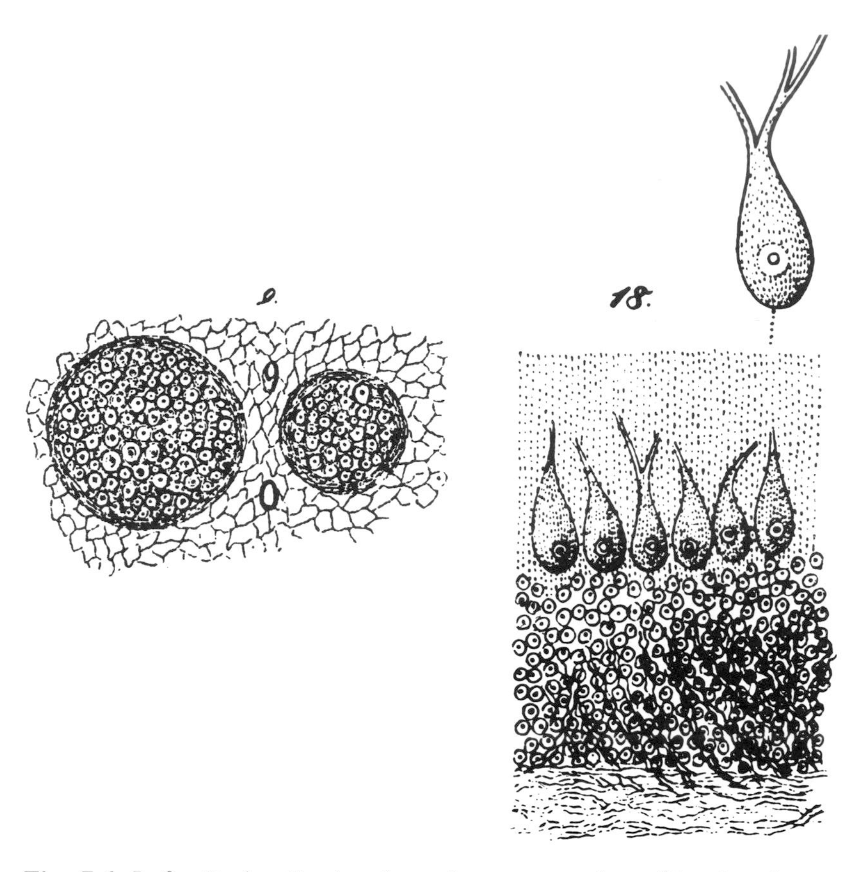

Fig. 7.6. Left—Purkyně's drawing of a cross-section of hardened nerve (unspecified) showing the inner canals (axons) of the nerve fibers. Note its resemblance to that published more than 100 years earlier by Leeuwenhoek (*see* Fig. 7.2). Right—one of Purkyně's famous drawings of cerebellar cortex showing what could be discerned before the Golgi silver impregnation technique came into use. Purkyně was among the first to depict these cells, subsequently named for him, in their context rather than as isolated elements. (From Purkyně, 1837, facing p. 174, Plate 11, Fig. 9, ×1/3; Fig. 18, ×2.)

stance of the occipital lobe. The next significant observation was made by Jules Gabriel François Baillarger (1809–1890; *see* Fig. 7.10, p. 134), in Paris. He found the cerebral cortex to be made up of layers and that fibers connect it with the internal white matter. In an attempt to correlate morphologic variations with functional elements such as mental disease, he examined thin sections of patients' cerebral cortex postmortem placed between two flat pieces of glass. "The researches which I have undertaken have made me recognize that the grey cortical substance of the convolutions of the brain is thus formed of *six layers* . . . grey and white alternately, from within out" (Baillarger, 1840, p. 151; transl. M. V. Anker). Other prominent neuroanatomists mentioned layers only briefly, for example, Ehrenberg in 1833 and Remak in 1844. The description of cells within the layers came in 1858, when Rudolf Berlin (1833–1897), looking at sections cut from cerebral material fixed in potassium dichromate and stained with ammoniacal carmine, distinguished not only an arrangement of layers in the cortex but also several typeforms of cells and suggested their classification into pyramidal, small and irregular (granular), and spindle-shape.

One of the great schools of neurology of his time was established by Theodor Hermann Meynert (1833–1892; *see* Fig. 7.11, p. 135) in Vienna. A neurohistologist–psychiatrist, Meynert's skillful

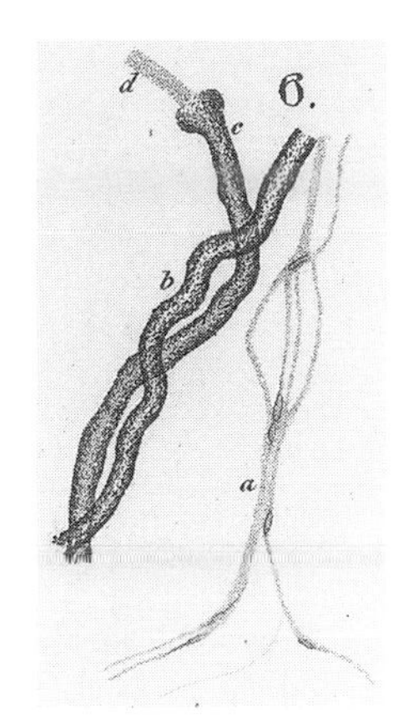

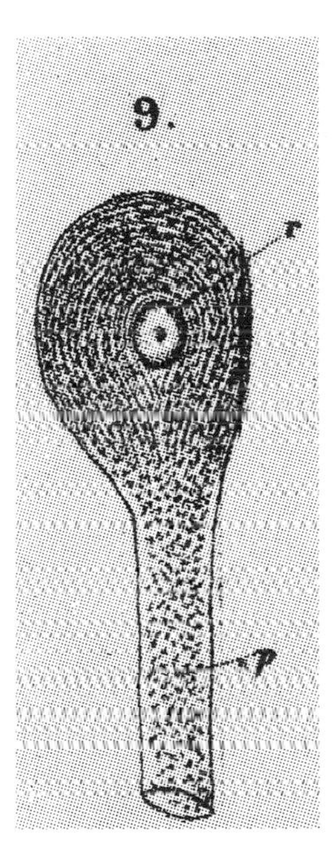

Fig. 7.7. Left—Robert Remak in 1838 clearly depicted myelinated (medullated, "primitive tube") and nonmyelinated ("organic") fibers and described the axon. Right, top—Myelinated (b) and nonmyelinated fibers (a) from sheep dorsal lumbar root. (From Remak, 1838, Plate I, Fig. 6.) Right, below—Crawfish nerve cell. (From Remak, 1844, Plate XII, Fig. 9; enlarged 2× to show the parallel-oriented lines representing neurofibrillae.)

studies on brains of mammals, especially the human, provided the first systematic description of the cellular elements of the cerebral cortex (1868). Using Berlin's staining methods, he found that the commonest type of cortical stratification, covering the surface of much of the human brain, consists of five layers (*see* Fig. 7.12, p. 136): an outer "neurological layer" containing a few angular nerve cells, a second layer of small pyramidal cells, a third of large pyramidal cells, a fourth of multiform or granular cells, and a fifth layer composed of large, squat pyramids and more deeply lying spindle-shape cells. Meynert also traced what he called projection and association fibers to and from the cortex, respectively, and theorized on their functions in the brain considered as an organ. He insisted that conscience, intelligence, and memory depend on cortical mechanisms representing, so to speak, perceptions. In his words, "we may argue that the cerebral cortex is the surface upon which the entire body is projected by these nerves" (1885, p. 36). Meynert's ideas carried

Fig. 7.8. Left—The first indisputable evidence that mammalian nerve-cell soma and nerve fibers are connected was presented by Ross Harrison (1908, p. 407, Fig. 22, ×2/3), using isolated embryonic spinal cord cells. Sketches 50 minutes apart. Right—Harrison photographed in 1911.

great weight; from that time, research in histologic organization of the cortex became the groundwork for further studies into the vital and especially the higher intellectual processes of the brain.

The details of cortical organization preoccupied the Russian anatomist, Vladimir Aleksandrovich Betz (1834–1894; *see* Fig. 7.13, p. 137, left), particularly the quality and quantity of the histologic elements in man and ape. He prepared the cerebral hemispheres with an innovative hardening procedure (alcohol, iodine, and potassium dichromate) for thin sectioning and staining with carmine to reveal the cellular components and their processes. After collecting 4000 histologic preparations, he announced in a German publication (1874) his discovery of the giant pyramidal cells (*see* Fig. 7.13, right, p. 137), later named for him, in human precentral cortex and described their development and distribution in fetal brains.

On the 125th anniversary of Betz's birth, his work was featured by M. S. Spirov, head of Betz's old department of anatomy at Kiev. In the following passage, Betz (1874) described his findings:

> The anterior part of the hemisphere consists of cells, discovered by myself and not yet seen by anybody. These cells are the largest in the central nervous system—I will call them callosal pyramids—measuring from 0.03 to 0.06 mm in width and from 0.04 to 0.12 mm in length, not including processes. All of them possess 2 major processes and 7 to 15 auxiliary protoplasmic processes, branching into smaller [ones].
>
> The cells of the anterior [motor] center do not form a continuous layer but cluster in nests, 2, or 3 or more in one place. . . . Sometimes the nests consist of 5 cells of different size. . . . (quoted by Spirov, 1962, p. 109).

Betz, like Meynert, described five cortical layers, but went further, recognizing 11 basic regions of the cortex: "Every part of the cortex differs structurally from other parts of the brain. . . . This struc-

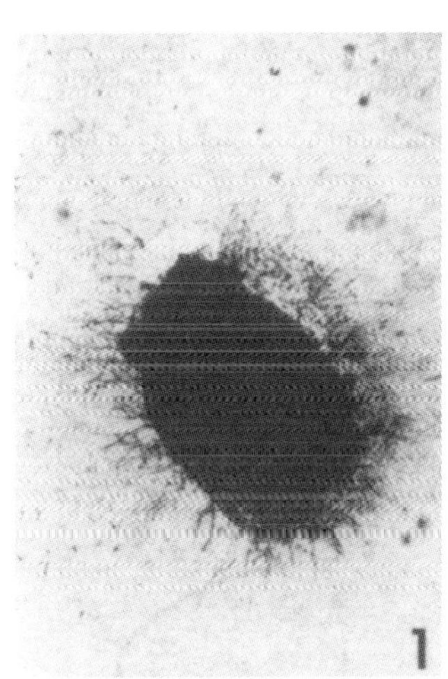

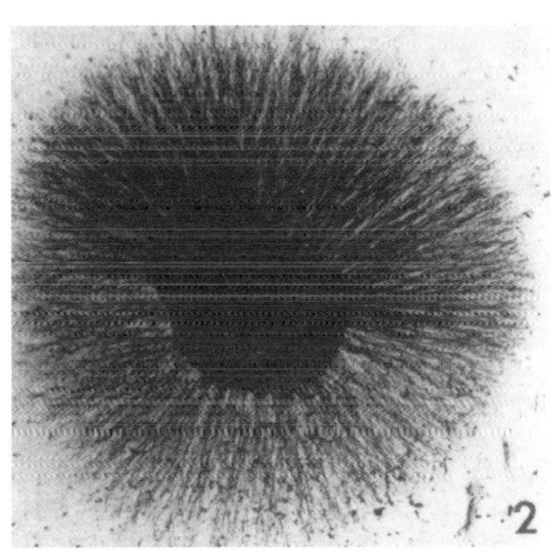

Fig. 7.9. Above—Portrait of Rita Levi-Montalcini by G. Merighi (1988), one of a series of medals honoring Italian Nobel laureates; she was awarded the prize in 1986. Below—(2) Photomicrograph of "halo effect" of seven-day embryonic chick sensory ganglion after 24-hour incubation in serum containing nerve growth factor compared with ganglion in control medium (1). (From Levi-Montalcini, 1964, p. 109.)

tural differentiation . . . is the expression of the localized functions in the cortex" (ibid., p. 109). From the original German text of 1874, we learn that Betz found fewer and smaller cells in brains of young individuals, and that the cells were more numerous and larger in the *right* hemisphere than in the left (*see* also Chapter 6). In the same paper, Betz also put forward the important concept that motor functions are represented anterior to the central sulcus (of Rolando) and sensory functions posterior to it. In another prodigious work (1881), dedicated to Broca, Betz analyzed 5000 microscopic preparations ranging from human fetuses to 14-year-old children.

History has not been kind to the significant work of the Australian investigator, Alfred Walter

Fig. 7.10. François Baillarger combined his interest in psychiatric disorders with a curiosity about the underlying neural substrates, the cerebral white and grey substances. He was able to show that some disorders are correlated with alterations in cerebral tissue.

Campbell (1868–1937; Fig. 7.14, p. 138), an unappreciated pioneer in cortical histology. After receiving his medical degree from Edinburgh and completing postdoctoral studies in pathology on the Continent, notably with Krafft-Ebing, successor to Meynert in Vienna, Campbell served outstandingly as neurologist, alienist, pathologist, researcher, and writer, during the two decades he was abroad. His facility in French, German, and Italian placed him in great demand as a reviewer of foreign publications for British journals.

During his appointment as resident medical officer and director of the pathology laboratory at Rainhill Asylum, Liverpool, Campbell moved in the sphere of Sherrington and his group at the University of Liverpool, where cerebral localization was being studied in anthropoid apes. In line with his ongoing studies on human brains, Campbell worked up several of Sherrington's experimental brains,

> . . . with the object of ascertaining whether the cortex of the parts which responded to electrical excitation could be differentiated from the "silent" parts, by the possession of any distinctive histological structure. . . . I think I was able to prove to Professor Sherrington and Doctor Grünbaum's satisfaction that it is just as possible to define the motor area on the histological bench, as on the operating table (Campbell, 1905, p. 20).

The proof had been published in 1903 as an appendix to Grünbaum and Sherrington's paper.

With the aim to "further the establishment of a correlation between physiologic function and histologic structure" (Campbell, 1903, p. 488), Campbell undertook what a fellow Australian has called "the most comprehensive and detailed studies ever . . . of the whole of the cerebral cortex"

Fig. 7.11. During Theodor Meynert's long professional life in Vienna, he contributed crucially to the neurohistology of the cerebral cortex as well as to the study of psychiatric disorders. (From Kolle, Vol. 2, 1959, facing p. 98.)

(Porter, 1988, p. 1). The studies were based on alternate sections stained for cells and for fibers, and in Porter's view they "provided the major and most essential background to Penfield's classic observations on the human brain" (ibid., p. 4). Campbell's monumental two volumes were published in 1905, the year he returned to Australia and completed his career with private practice and writing. Campbell recognized about a dozen regions in the human brain (*see* Fig. 7.15, p. 139) and provided more accurate information of the extent and limits of functionally differentiated areas than had previous investigators using physiologic methods.

Three years after Campbell's cartography appeared, the German neuroanatomist, Korbinian Brodmann (1868–1918) presented the results of his own microscopic analysis of cortical areas, without mention of Campbell's prior work. Brodmann, and later, Cécil and Oskar Vogt in the early twentieth century refined and redefined Campbell's cortical areas and added to them detailed parcelling with sharp borders based on minute cellular differences. Ironically, today it is Brodmann's system of numbering the parcels that prevails and he is generally credited with establishing the field of comparative cytoarchitectonics. With very detailed "histotopographical" work, Brodmann showed (1908) that the increasing differentiation of cortical areas is associated with specific functional patterns in animal evolution. Although Brodmann believed most localization boundaries to be sharp, he proposed a "Principle of relative localization":

> In addition to the sharply circumscribed organs, we have also found in the cerebral cortex structural areas with variable boundaries. Their tectonic features are blended more or less

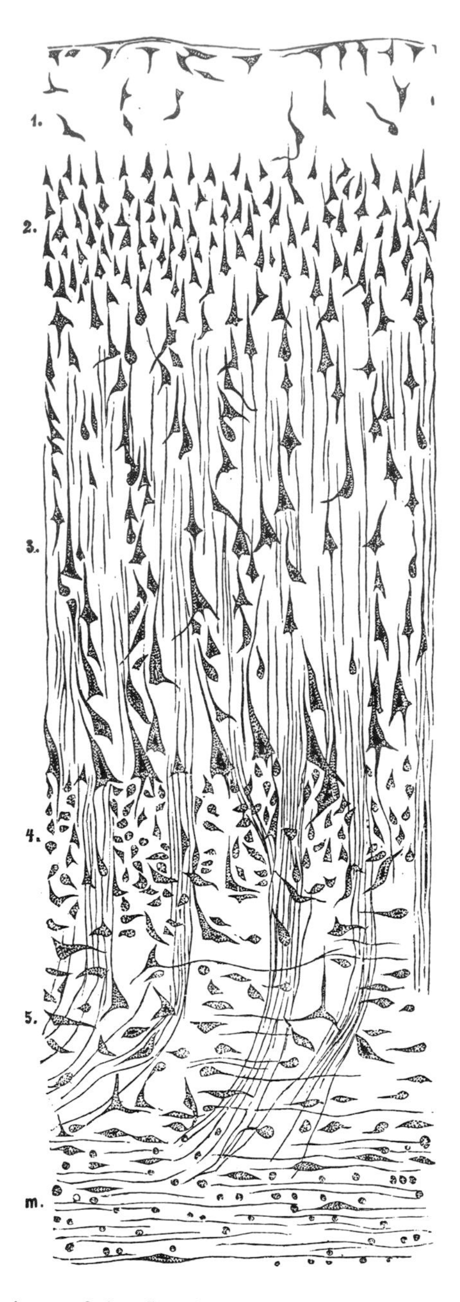

Fig. 7.12. Drawings of the five layers of cells (*see* text) in cerebral cortex observed by Theodor Meynert and determined by him to be representative throughout the human brain. (From Meynert, 1885, p. 57, Fig. 24, ×1.) Compare with Ramón y Cajal's drawing in Fig. 7.19.

with those of the neighboring areas with which they partially overlap. This fact suggests the possibility of a more or less overlapping of some functions in the cerebral cortex (Brodmann, 1909; translated in Clarke and O'Malley, 1968, p. 555; also in von Bonin, 1960).

THE INDEPENDENT NEURON

The ambiguous sightings and intense arguments concerning the connection of nerve fiber with cell body, described above, were mild compared with the agitation engendered by the corollary question of whether neurons are organized as a network

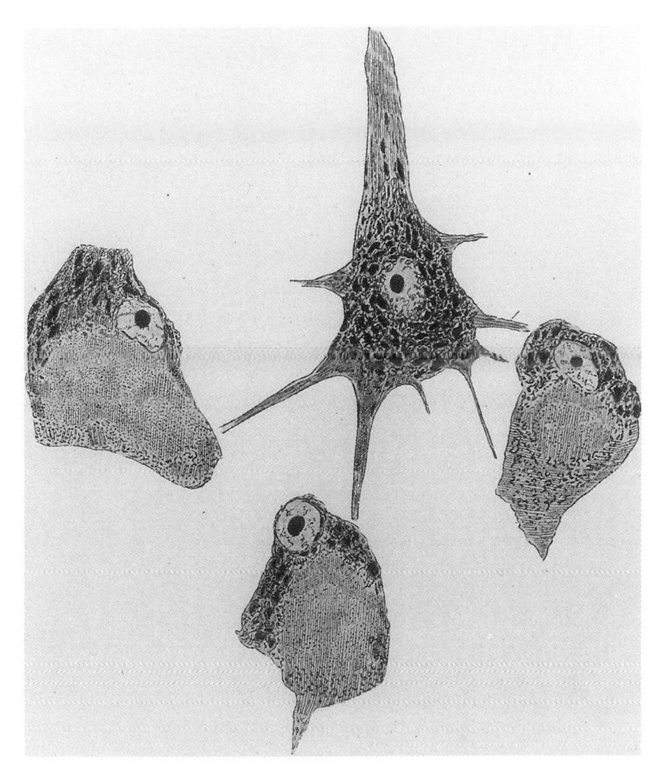

Fig. 7.13. Left—Vladimir Betz found the giant cells that bear his name in Meynert's fourth layer of the precentral cortex. (Portrait from Spirov, 1962, p. 107.) Right—Camera-lucida drawing of a normal human Betz cell (center) for comparison with three Betz cells undergoing chromatolysis after amputation of the contralateral leg. (From Campbell, 1905, p. 52, Fig. 3, ×1.)

(reticulum) or are in contact with each other without continuity of protoplasm. The lines drawn extended beyond the primary participants, the anatomists and some physiologists, although at that time the latter for the most part were still concentrating on nerve–muscle relationships. In addition, the biologists had been arguing for half-a-century over whether or not the nervous system could be fitted into the great cell theory enunciated in 1839 by Theodore Schwann, who proposed that all living tissues are composed of cells (*see* Shepherd, 1991, p. 4). The major reticularists in the middle nineteenth century were Deiters, Gerlach, Kölliker, and Golgi, whereas the neuronists had His, Forel, Nansen, Ramón y Cajal, and Waldeyer among their ranks. Deiters (1865) drew the first clear anatomical distinction between axons and dendrites and described the fine terminal fibers of the former on the latter to create a misleading picture of a network. Gerlach, with the new carmine (the first aniline dye to be used on nerve, 1858), showed "networks" in the spinal cord and proposed a "double origin" theory of motor and sensory axons. Kölliker, too, distinguished motor and sensory axons connecting through their dendritic nets. And Golgi (*see* below) saw axons with collateral fibers, which he speculated formed anastomoses in networks, and dendrites which served a nutritive, nonconducting function.

The supporters of the independent neuron were led by the embryologist Wilhelm His, who relied on what he could see—the outgrowth of the spinal cord axons precedes that of the dendrites and thus ontogeny proves they are separate. And August Forel said that after cutting a nerve, the lack of spread of degenerative signs to adjacent fibers demonstrated the absence of anastomoses. There was also the keen Norwegian neuroanatomist Fridjhof Nansen, whose work before he became absorbed by exploration of the Arctic, bridged the Golgi network and the neuron doctrine. Finally, Ramón y Cajal's investigations seemed to settle the disputed question, as described below.

Acceptance of the neuron doctrine was popularized by Wilhelm Waldeyer (1836–1921) in 1891 in a review that appeared serially in a Ger-

Fig. 7.14. Alfred Walter Campbell, Australian anatomist and physician, was trained in Scotland and on the Continent, and became fluent in several languages. He was a champion of the correlation of structure and function (*see* Fig. 7.15).

man journal. He stated four "tenets" regarding what constitutes a neuron and summarized the ideas of the "neuronists" as follows:

> The nervous system consists of numerous nerve units (neurons) connected with one another neither anatomically nor genetically. Each nerve unit is composed of three parts: the nerve cell, the nerve fibre, and the terminal arborizations. The physiological process of conduction can take place in the direction from cell to end arborizations, as well as in the reverse direction. Motor conduction takes place only in the direction from cell to terminal arborization, sensory conduction runs now in one, now in the other direction (Waldeyer, 1891; translation from Clarke and O'Malley, 1968, p. 115; also in Shepherd, 1991, p. 183).

The "enormous difficulty of demonstrating three-dimensionally the interrelations of neurons in the contemporary absence of adequate histologic techniques" (Palay, 1977, p. 3) was partially remedied by C. Fromman, who reported in 1864 a silver stain for nerve tissue which adheres to the surface of the nerve cell (Hughes, 1959). The improved chemical reaction of potassium dichromate with silver nitrate to form silver chromate (which penetrates the nerve fiber) enabled Camillo Golgi (1843–1926; *see* Fig. 7.16, p. 140), an Italian physician, to develop *la reazione nera* for impregnating selectively the nonmyelinated neurons in a tissue slice, rendering soma and processes visible.[2] His most famous publication, *Studi sulla fina anatomia degli organi centrali del sistema nervoso* (1886), elaborated his earlier findings of two types (*see* Fig. 7.17, p. 141, right) of cerebral cortical cells—those with long processes that travel to the white matter and subcortex (type I) and others with processes confined within the cortex (type II). Strictly from their morphology, he decided that type I cells are motor and type II sensory. Of lasting importance was his finding that axis cylinders (axons) are invested with col-lateral branches but he again, in error, arranged them into anastomos-

[2]For concise and lucid discussions of neuroanatomical research techniques, see the articles by Peters, pp. 764–766, and by Heimer, pp. 766–769 in the *Encyclopedia of Neuroscience* (1987).

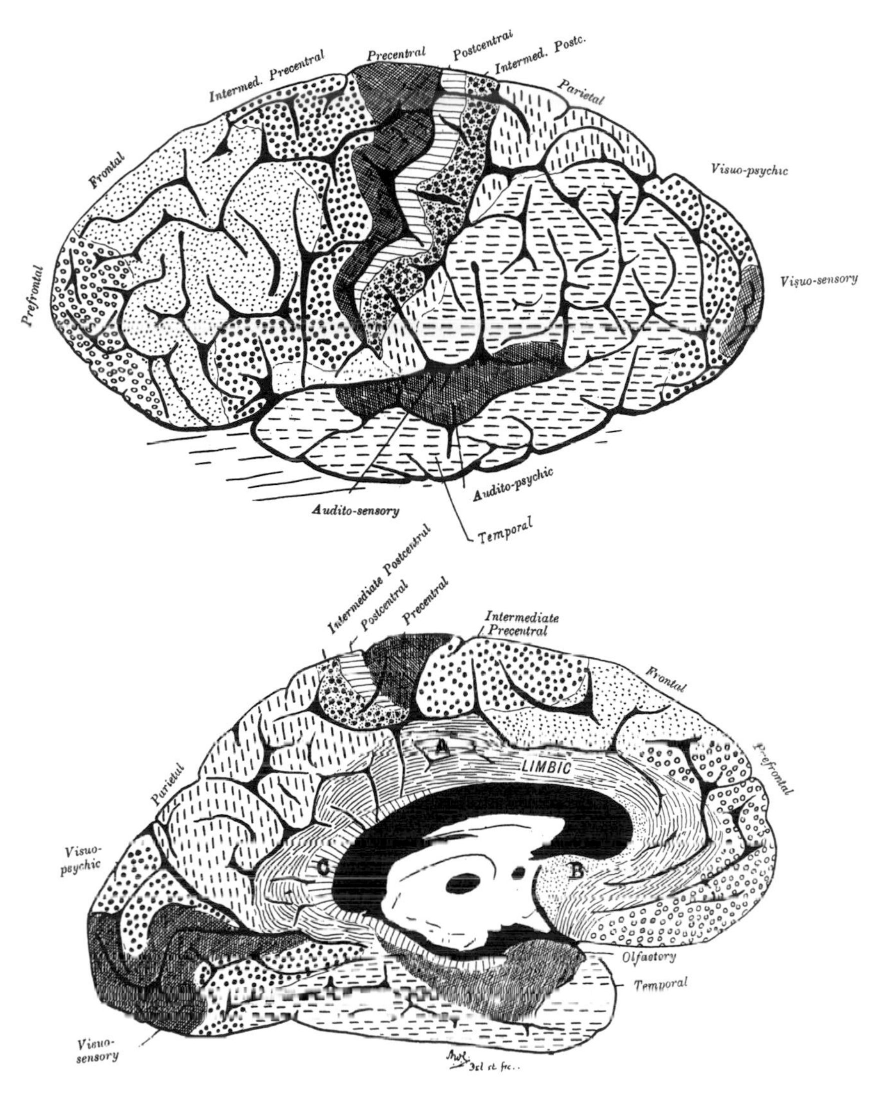

Fig. 7.15. Schema of left lateral (somewhat tilted) and left mesial surfaces of an adult male human brain with areas defined by histological examination of cortical nerve fibers and nerve cells. (From Campbell, 1905, p. 293, Plate I; ×1.)

ing nets (Shepherd, 1991, p. 8). In 1898, Golgi clearly described the dense network within the cytoplasm of nerve cells which was afterward named for him (the "Golgi apparatus," *see* Fig. 7.17, p. 141, left) but actually had been noticed in snail sexual cells 30 years earlier (*see* S. H. Douglas, 1935) and had been drawn by Leeuwenhoek.

In his earlier years, Golgi worked in isolation from academic contacts and without space and equipment beyond his microscope. Later, at the University of Pavia, conditions were not much better, but his fame had spread and investigators from abroad commenced to visit his laboratory. The most notable was the 70-year-old von Kölliker in 1887 who immediately appreciated and supported Golgi's achievement with the silver impregnation method that could make the entire nerve cell visible in stark contrast to its background. Golgi, the "persevering, headstrong, reserved" (Ferraro, 1970, p. 38) clinician–histologist, was too conservative to relinquish his belief that neurons constitute a network which reacts as a whole, thus precluding the possibility of localization of function. Perhaps the severest indictment of his work is that he never clearly depicted or described the details of a network. As Shepherd (1991, p. 266) commented, Golgi's global view of the nervous system was more significant than his erroneous speculation about a protoplasmic reticulum of nerve fibers.

Fig. 7.16. Camillo Golgi's silver impregnation method provided the first clear image of the complete, nonmyelinated neuron but was less satisfactory for myelinated fibers. (From Haymaker and Schiller, 1970, p. 36.) Some of Golgi's drawings are shown in the following figure.

Evidence of the neuron's structural integrity burst forth from the microscopic fields of the beautiful preparations of the great Spanish physician–histologist, Santiago Ramón y Cajal (1852–1934; *see* Fig. 7.18, p. 142). He made Golgi's novel silver chromate method less capricious, selected tissues from newborn or fetal specimens in which the cell processes had little or no myelin sheath—which resists staining—and when it became available adapted methylene blue, which does stain myelin. A man of "intense genius" (Courville, 1970), his photographs and writings reveal a philosophical side to Cajal's character. In a lecture delivered at the Barcelona Academy of Medical Sciences, he branded as flawed the previous work of others and called for "a . . . calm and impartial study of the cerebral cortex," because: "The supreme dignity which surrounds the brain and the awesome complexity of its workings presuppose the existence of an extremely complicated warp, sure to ensnare those who imagine that nature unfolds multifarious exalted phenomena according to schematic formulae. . . ." (Ramón y Cajal, 1892; translation in Rottenberg and Hoffberg, 1977, p. 9). Cajal credited Golgi with having demonstrated that pyramidal axons give off collateral branches and possess arborizations and with identifying two types of cells in the cortex. Throughout his writings, Cajal was consistent in his appreciation of Golgi's work and also in correcting it when appropriate. Golgi, in his speech when they shared the Nobel prize in 1906 (Golgi, 1908, p. 26) lauded "the classic results obtained by Cajal with his reduced silver technique [which] have an eminent place" in research on cellular fine structure, but was not moved to admit the error in his own concept of a neuronal reticulum.

The audacity of Cajal's concepts of pathways and connections was emphasized by Palay (1977, p. 5), noting that the immense scope of the original work is minimally indicated by Cajal's own listing of his major discoveries:

> I have already established that all pyramidal cells project protoplasmic tufts to the molecular [outer] layer, where an infinite number of terminal nerve fibers converge. By virtue of this

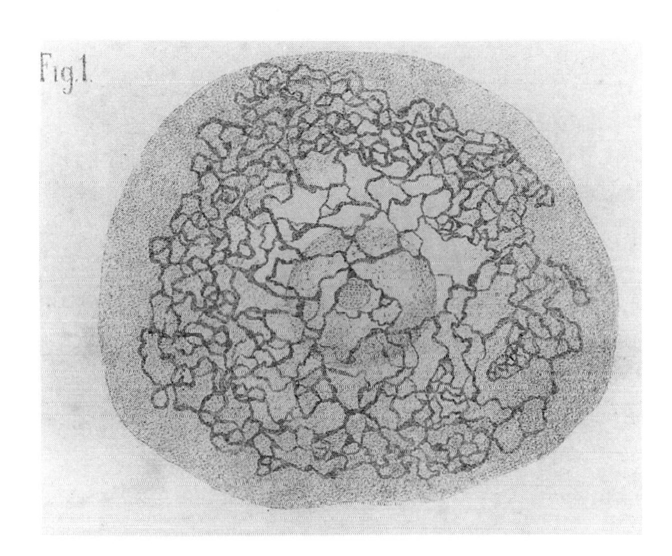

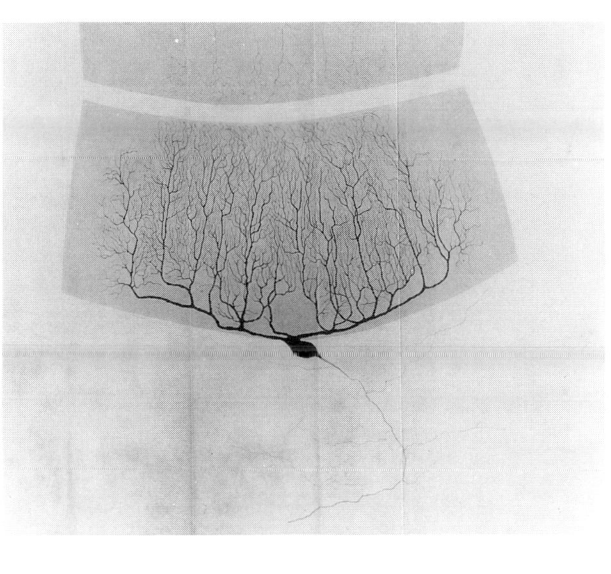

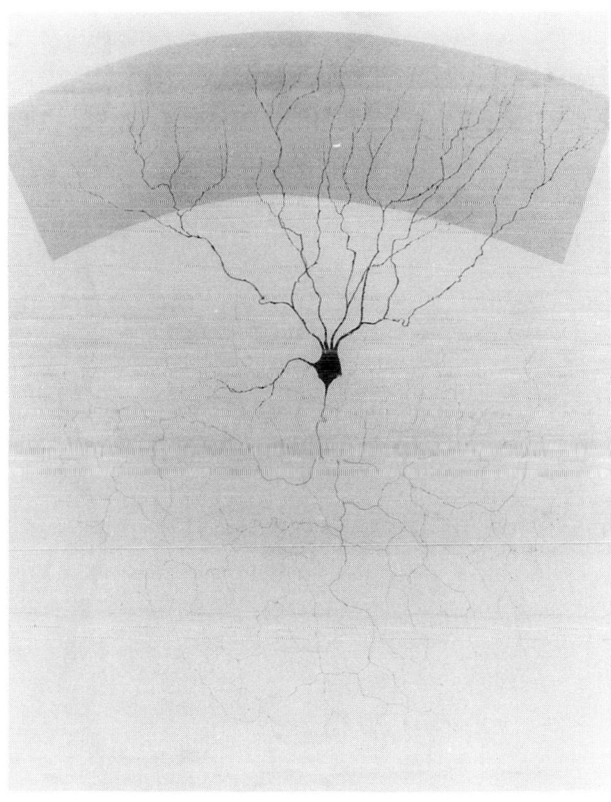

Fig. 7.17. Left—Golgi's drawing of impregnated network ("Golgi apparatus") in a Purkyně cell of an owl. The reality of this network of organelles with secretory activity in the living cell has been debated since its first description in 1867 by La Valette St. George (*see* Hughes, 1959, p. 125 passim). (Drawing from Golgi, 1898, Plate I, Fig. I.) Right—Type I human Purkyně cell (above) and type II kitten cerebellar cell (below); *see* text. (From Golgi, 1883, Plates IV and III, respectively.)

assemblage of axons and dendrites—which must be of great importance as it is present in all vertebrates—each pyramidal cell can be excited by a multitude of other cells. . . . [I]n my opinion these connections are effected by contacts between terminal arborizations and axon collaterals, on the one hand, and cell bodies and protoplasmic expansions on the other. The direction of the impulse traffic is *cellulofugal* in the axis cylinder [axon] and *cellulopetal* in

Fig. 7.18. Santiago Ramón y Cajal did "[m]ore than any other neuroanatomist [to develop] the study of synaptic organization in the central nervous system" (Grundfest, 1957, p. 327). In 1878, Cajal photographed himself in his first laboratory while he was living at Valencia. (From Lain Entralgo, 1978, p. 115.)

> the cell body and protoplasmic expansion [dendrites]. . . . (Ramón y Cajal, 1892; translation in Rottenberg and Hoffberg, 1977, pp. 22–23; *see* Fig. 7.19, p. 143, left).

In this single passage, Cajal expressed his concepts of convergence of impulses, neural connectivity by contact, and the law of dynamic polarity. He showed that the form of the pyramidal cell body is constant throughout the mammalian series, but its size and pattern of dendritic branching are variable. (Recall Fig. 1.5 for Cajal's depiction of the pyramidal cell in phylogenetic and ontogenetic contexts.) Because of the constancy of soma form, Cajal pointed out, "each nerve cell, whatever its functional category, appears to be constructed according to the same model and to possess the same texture and chemical composition" (ibid., p. 26; Fig. 7.19, right). Cajal's surprising attention to cellular chemistry, a foreshadow of modern neurochemistry and his sense of levels of activity are repeated in the same paragraph:

> Science . . . must assume (1) that the internal structure and chemical composition of the cortical cell, rather than its external form, distin-

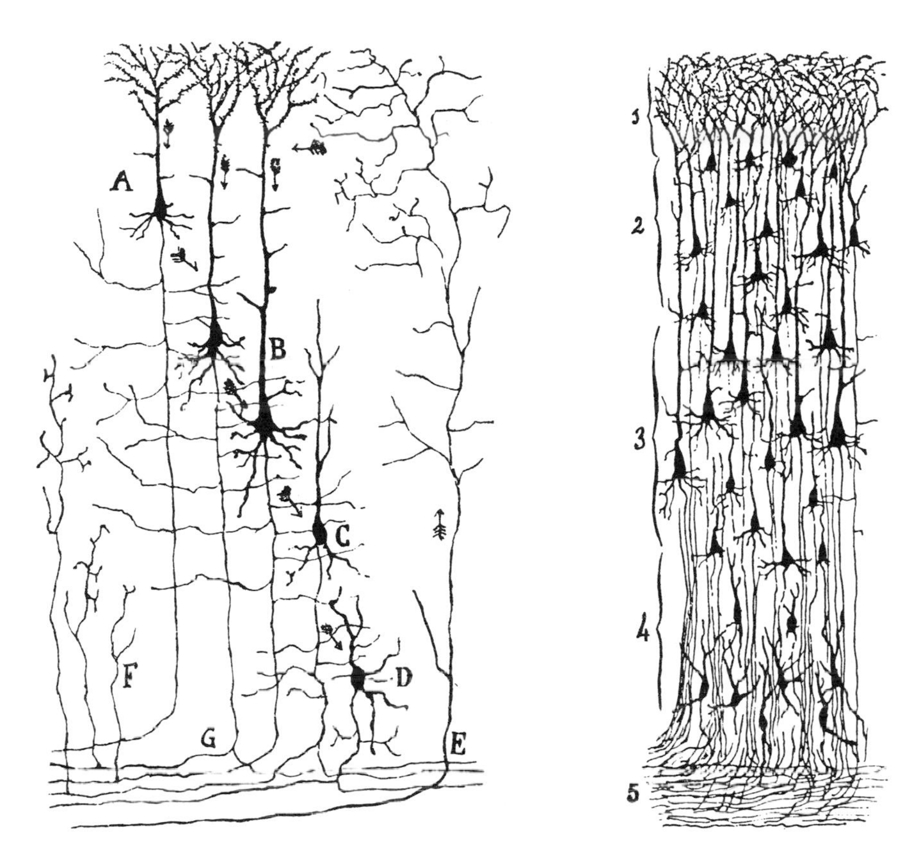

Fig. 7.19. Left—An early diagram by Cajal of dendritic and axonal branchings and the probable direction of the impulse traffic (*see* text). Right—Cajal's drawing of cells in a vertical section of a cerebral convolution, which he held to be representative of all mammalian species. 1—Molecular layer, 2—layer of small pyramidal cells, 3—large pyramidal cell layer, 4—polymorphic cell layer, 5—white matter. Compare the detail of this drawing with Meynert's of 1885, Fig. 7.12. (From Ramón y Cajal, 1892, p. 471, Fig. 13, ×1; 460, Fig. 17, ×0.89.)

> guish it from the medullary or ganglionic cell; and (2) that the mechanical phenomena which occur within the protoplasmic matrix of the psychic cell are not of the same order . . . as those . . . within the protoplasm of cells of a lesser category (ibid., p. 26).

The sustained high esteem for Cajal's achievements rests not only on his exquisite montages of microscopic marvels, but also on his tendency to seek explanations (an ability shared by C. Judson Herrick). Another example was in his Croonian lecture (1894), in which Cajal admitted to a hypothesis regarding cerebral plasticity based not on proliferation of nerve cells, but on their ability to grow new axonal collaterals and dendrites.

Ramón y Cajal was determined to have his work recognized so he demonstrated his microscopic preparations to a critical group at the 1889 Congress of Anatomists in Berlin and convinced some important figures of that time (among them Retzius, Waldeyer, and Kölliker), of his views. Within a few years, he saw his ideas incorporated in two authoritative publications: Waldeyer's[3] definitive journal article of 1891 (as noted earlier) and Kölliker's sixth edition of his *Handbuch* (1852), which "is as much a classic as any of Cajal's great works" (von Bonin, 1960, p. 52).

The institute at the University of Madrid where Cajal and his school established the neuron doctrine as fact and Spain as a leader in neurohistology, throughout the remainder of his life was a small beehive of activity. After his death and the Spanish

[3]Waldeyer is credited with having named the neurone, spelled with the final "e" until Americanized in 1884 by Burt Wilder, a physiologist at Cornell University famous for his collection of vertebrate brains (Hughes, 1959; L. H. Marshall, 1983a).

Civil War in the 1930s, the Cajal Institute "languished during the following decades while the light of excellence was kept alive by a few, albeit courageous, Spanish neurobiologists, such as de Castro, Molina, and Valverde" (Ferrús, 1989, p. 1). A new building now houses the expanded institute (Fig. 7.20) and proclaims the renewal in neuroscience excellence anticipated by the support of the Spanish Research Council.

The recognition of neurons as independent anatomic structures left a small space or cleft between them, which could not remain for long without a name. The circumstance of the christening was related by Sherrington in a Christmas letter to John Fulton:

> [For a new edition of Foster's *Textbook of Physiology* (1897)] I felt the need of some name to call the junction between nerve-cell & nerve-cell [because that place of junction now entered physiology as carrying functional importance]. . . . I suggested using syndesm. . . . & Verrall suggested 'synapse' . . . as that yields a better adjectival form, it was adopted for the book.
>
> The concept at root of the need for a specific term was that, as was becoming clear, 'conduction' which transmitted the 'impulse' *along* the nerve-fibre could not—as such—obtain at the *junction*, [because] a 'membrane' *there* lay across the path, & 'conduction" per se was not competent to negotiate a 'cross-wise' membrane. At least so it seemed to me then; perhaps A. V. Hill & Gasser & Bishop could tell us differently today! (Sherrington, 1937).

In his chapter for the textbook, Sherrington wrote:

> If . . . the axon continues to run and finally ends in the central nervous system, its mode of termination . . . is in the form of an arborescent tuft, which is applied to the body or dendrites of some other cell. So far as our present knowledge goes we are led to think that the tip of a twig of the arborescence is not continuous with but merely in contact with the substance of the dendrite or cell-body on which it impinges. Such a special connexion of one nerve-cell with another might be called a *synapse* (Foster and Sherrington, 1897, pp. 928–929).

The irrefutable proof of a cleft without protoplasmic continuity between neurons was produced by the electron microscope. Although applied to the nervous system in the 1950s (Latta and Hartmann, 1950), Edward George Gray, of University College, London, is credited with the first significant work in that area (Jones, 1984b, p. 23). In 1959, Gray published electron micrographs (*see* Fig. 7.21, p. 146) of the cleft and identified two types of synapses: the asymmetric (type 1 with a denser presynaptic than postsynaptic membrane), and the symmetric (type 2, with equally dense pre- and postsynaptic membranes). He also established that dendritic spines are real and not artifacts. With the discoveries surrounding the independent neuron stretching more than a century, the physiological story of what transpires within the neuron continues to unfold.

MULTIPLE APPROACHES TO THE SUBSTRATES OF VISION

At the beginning of the twentieth century, after decades of attention to issues of evolution and systematics, the interdisciplinary approach to the study of brain and behavior was gaining ascendancy. The advantage, indeed, the necessity of neuroanatomists, to engage the methodology of other disciplines, such as that of the neurophysiologists, and vice versa, became so obvious that by midcentury an interdisciplinary research protocol was de rigueur for neuroscientists at the forefront of their discipline. An excellent example of the complemental relation of different approaches to a biological question is found in the discoveries of how neural tissues are organized, as layer IV of the cerebral cortex exemplifies. As early as 1912, M. Rose in Leipzig had noted the unusual appearance of layer IV in the hand–face area of the primary somatosensory cortex of small mammals. Lorente de Nó found clusters ("glomeruli") of afferent fibers (*see* Fig. 7.22, p. 147) to be the structural basis of the anomaly, and in 1938 he proposed that intercommunicating neurons of the cortex are arranged in vertical columns. The concept received no notice until it was clearly demonstrated in cat primary somatosensory regions by Vernon Mountcastle (1957) who found independently functioning units organized in narrow vertical columns extending from layers II through VI.

Another example of columnar arrangements in sensory cortex was found by Thomas A. Woolsey and H. Van der Loos (1970), who focused on layer

Fig. 7.20. The old and the new in Spanish neuroscience is manifest by the Instituto de Neurobiologia Santiago Ramón y Cajal in Madrid about 1918 (above, foreground) and in 1989 (below). (Photograph of the new institute, kindness of Vice Director Alberto Ferrús.)

IV of the mouse S I region and identified "barrels" as the cortical correlates of the mystacial vibrissae. They determined that the barrels receive specific thalamic projections, thereby identifying them as the primary receiving station for the mechanical displacement of the vibrissae (*see* Fig. 7.23, p. 148).

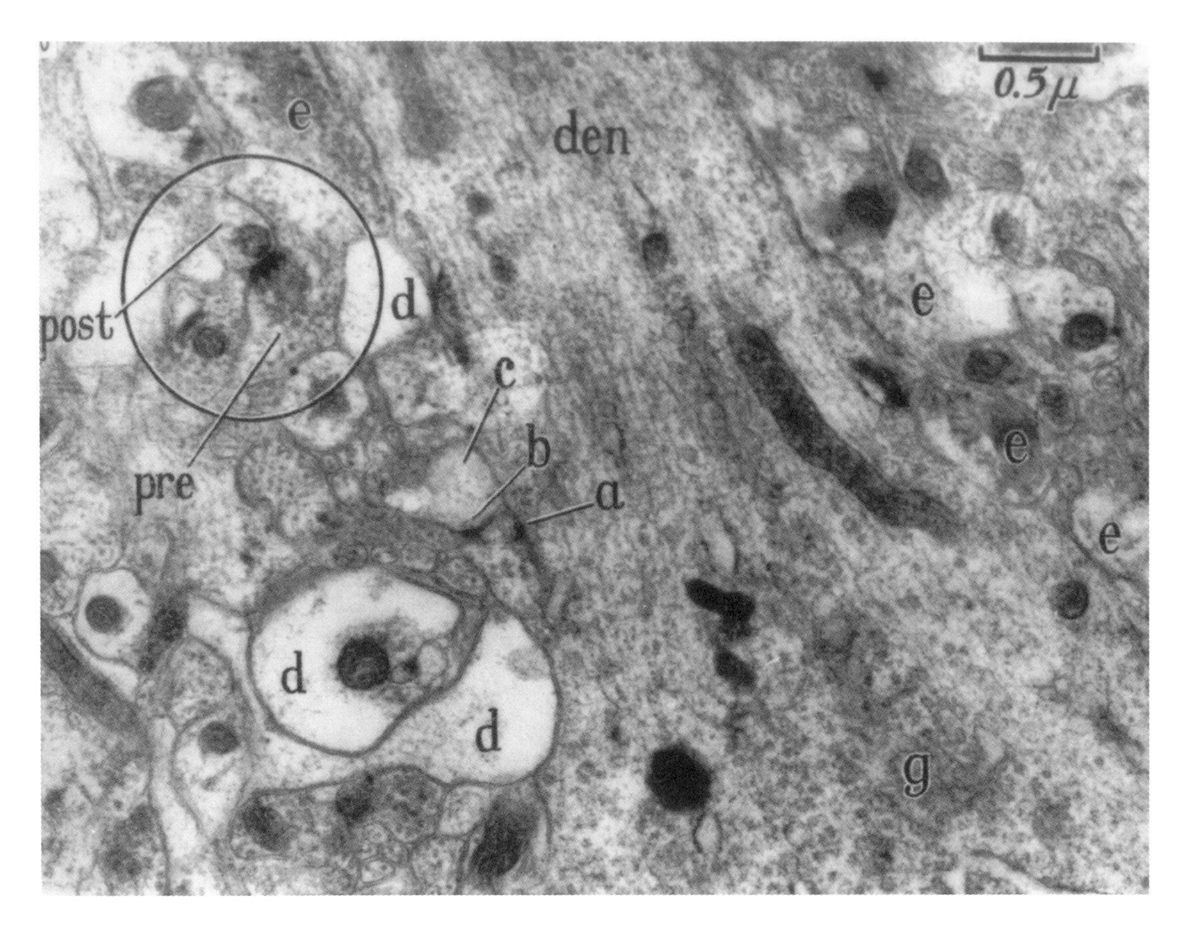

Fig. 7.21. An electron micrograph of the base of an apical dendrite (den) of a pyramidal neuron showing (in circle) a type 1 asymmetrical synapse and(a) a type 2 symmetrical synapse. (From Gray, 1959, Plate 3, Fig. 10, ×1.)

As Hubel and Wiesel pointed out (1972, p. 447), this was the first time that "a correlation of columns was observed anatomically before a physiological identification had been made."

The transduction of barrel anatomy to brain function had a wide appeal to the theorists who were quick to see advantages and lessons to be learned. John Eccles, in a sweeping summary (1979) of what was known of brain function at the time, emphasized the facilitation of impulse spread provided by the barrel concept, which he illustrated in the diagram reproduced in Fig. 7.24, *see* p. 148. A year earlier the subject of barrels had been addressed effectively by Janos Szentágothai (1912–1994), Hungarian neuroanatomist, in his Ferrier lecture of 1978. The "module concept" of units with connections between hemispheres and between cortex and subcortex (*see* Fig. 7.25, p. 148) he believed to be the logical result of viewing neural tissue less as neuronal chains and more as overlapping fields with interlacing spaces of dendritic and axonal arborizations. Szentágothai stressed that the reality of repetitive, vertical columns of about 200-300 μm diameter was established without any doubt. That their discovery was delayed 20 years after their presence was revealed physiologically (Mountcastle, 1957; Hubel and Wiesel, 1959), he attributed to the lack of techniques for tracing complete arborizations that would permit reconstruction of serial sections to show overall patterns of structure. Indeed, the introduction of several novel techniques—the capillary electrode, labeling with radioisotopes or enzymes—spurred studies of nerve tissue organization, and perhaps most notably in the mammalian visual system.

The large proportion of research on the visual system compared with studies of the other senses was noted in Chapter 5. In that context, a series of physiological and anatomical investigations of the

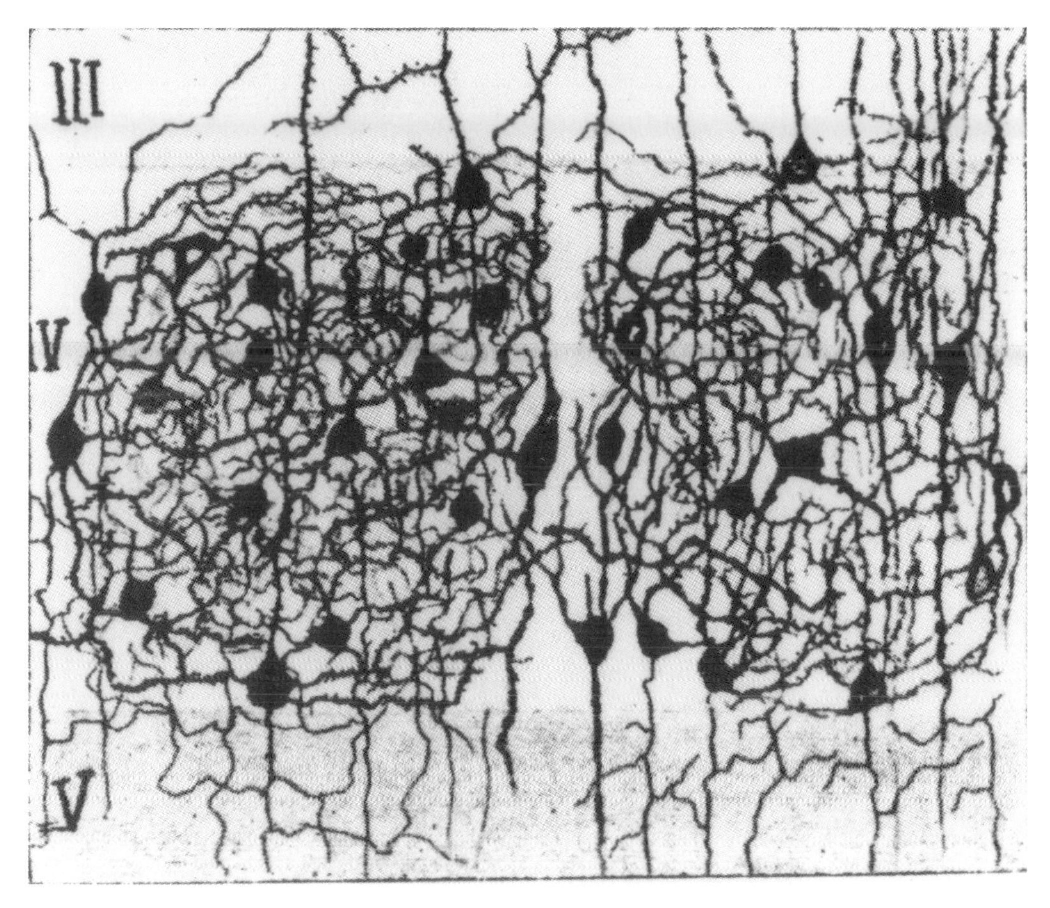

Fig. 7.22. The glomeruli of pyramidal granular layer IV in the rat. III—Supragranular layer, V—infragranular layer. (From Lorente de Nó, 1922, facing p. 78, Plate VII, ×2.)

cellular substrate of the reticulogeniculostriate relationship claimed the attention, briefly at Johns Hopkins University Medical School in Baltimore and later at Harvard Medical School, of an extraordinarily productive team. Canadian-born David Hunter Hubel and Swedish-born Thorsen Nils Wiesel, neurologist and psychiatrist respectively, were recruited in 1958 by an inspired and technically skilled Hungarian-American neurophysiologist, Stephen William Kuffler (1913–1980), and joined the ongoing research program at the Wilmer Institute of Ophthalmology. During the ensuing decades, Hubel and Wiesel investigated primary visual cortex in at least four slightly overlapping problem areas: the responses (receptive fields) of retinal ganglion and lateral geniculate nucleus single cells to visual stimuli; the structural arrangement of striatal (occipital, area 17 of Brodmann) cortex into columns and layers; the application of new anatomical methods to tracing physiological pathways after lesions in the lateral geniculate nucleus; and the postnatal development of vision and the influence of the environment on it. The following brief descriptions of their major findings are derived from the Nobel prize lectures of the respective laureates (Hubel, 1982; Wiesel, 1982).

What Hubel and Wiesel did was to extend into the central nervous system Kuffler's identification (1953) of retinal ganglion cells as ON-center or OFF-center with their receptive fields made up of excitatory cells surrounded by inhibitory cells, respectively. That work had been extended to show that the ganglion cells are more sensitive to a spot of light than to diffuse light (Barlow, Fitzhugh, and Kuffler, 1957). Repeating with the capillary electrode and single-cell recordings the systematic mapping of visual receptive fields that had been carried out with coarse electrodes by Talbot and Marshall (1941), serendipity led to the discovery of the selectivity of striate cortical cells to stimulus

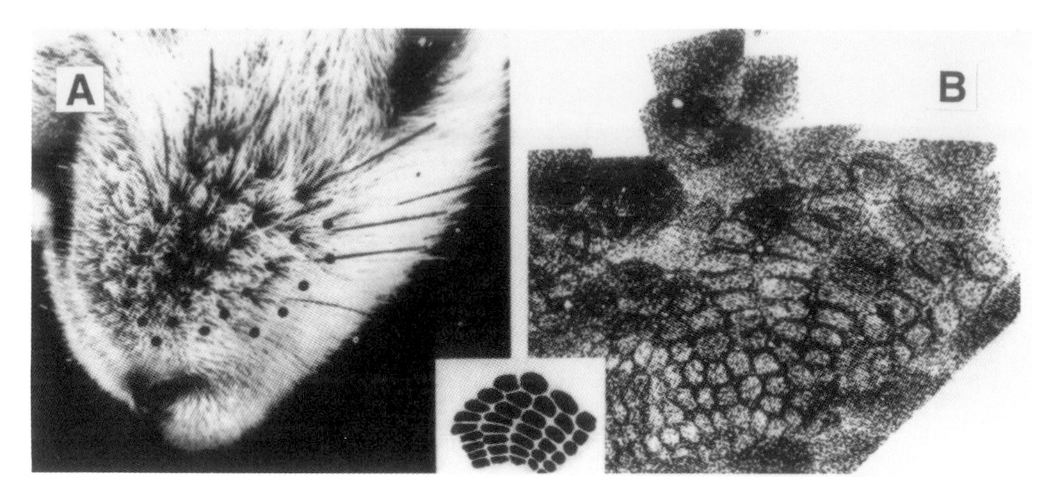

Fig. 7.23. Figure summarizing the one barrel-one mystacial vibrissa hypothesis. **A**—Group of right vibrissae of mouse with dots painted on photo where vibrissae implanted. **B**—Collage of a left (contralateral) postmedial barrel subfield. Insert shows definitely identified barrels in **(B)**. (From Woolsey and Van der Loos, 1970, p. 233, Fig. 15, ×0.8.)

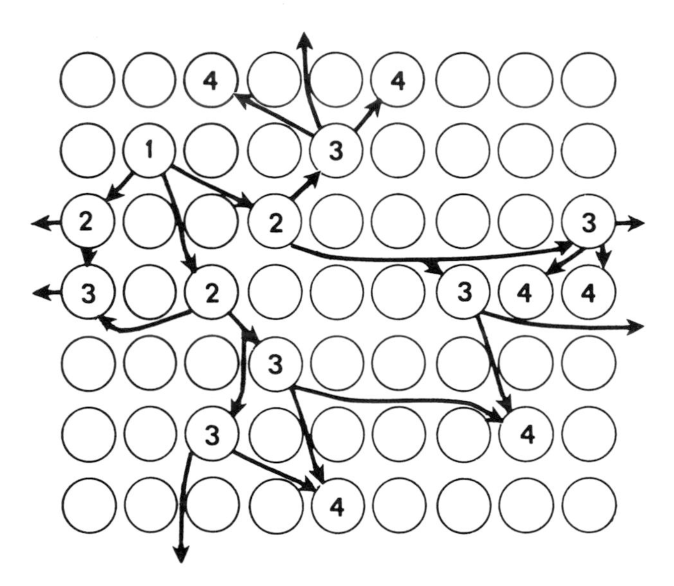

Fig. 7.24. Diagrammatic representation of cortical modules from above. "Strong activation of module 1 . . . leads to transmission effectively exciting modules labelled 2," and so on. (From Eccles, 1979, p. 203, Fig. 9.10.)

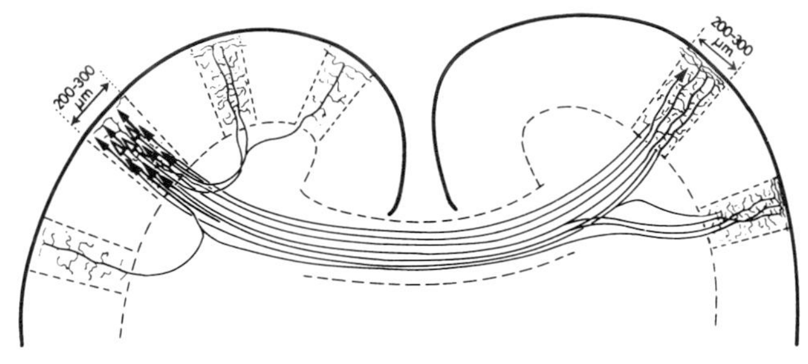

Fig. 7.25. Diagram of corticocortical connectivity showing the highly specific pattern of connections between vertical columns as endorsed by Eccles (1979, p. 152; *see* previous figure). (From Szentágothai, 1978, p. 224, Fig. 3.)

orientation, movement, and speed of movement (Hubel and Wiesel, 1959). Further work showed that single-cell behavior can be linked to brain cytoarchitecture, in this case, orientation columns, which was a new concept:

> It is suggested that columns containing cells with common receptive-field axis orientations are functional units, in which cells with simple fields represent an early stage in organization, possibly receiving their afferents directly from lateral geniculate cells, and cells with complex fields are of higher order, receiving projections from a number of cells with simple fields within the same columns (Hubel and Wiesel, 1962, p. 152).

The interconnections of orientation columns produces, in Hubel's words, "a little machine that takes care of contours in certain orientation in a certain part of the visual field." And he described the function of the columns as the "transformation of information from circularly symmetrical form to orientation-specific form, and the stepwise increase in complexity" (Hubel, 1982, p. 522). The author suggested the genetic implications of the usefulness of a uniform "little machine" in repetitive contexts, so that nature might order numbers of one design rather than one of each of a number of designs.

A third major contribution, announced a decade later, was the result of fine degeneration studies in monkey in which extremely small electrolytic lesions confined to single layers of the lateral geniculate nucleus resulted in degenerative changes in striate cortex, revealed by the Fink-Heimer modification of the Nauta stain. The laminar and columnar distribution of the thalamocortical fibers correlated with the functional differences previously found, and the authors stated: "It is concluded that the long narrow stripes of alternating left-eye and right-eye input to layer IV are an anatomical counterpart of the physiologically observed ocular-dominance columns. Because of this segregation of inputs, cells of layer IV are almost invariably influenced by one eye only" (Hubel and Wiesel, 1972, p. 421). A major discovery, answering the question of where visual inputs fuse, was the high proportion of binocular cells in the striate cortex and the dominance of one eye over the other.

The validation of radioisotopes as pathway tracers (*see* p. 151) allowed Wiesel, Hubel, and Lam (1974) to demonstrate dramatically the normal distribution of ocular dominance columns not only in the lateral geniculate nucleus but also in layer IV of monkey striate cortex. A suitable injection of tritiated proline and fucose into one eye and subsequent exposure of the brain tissue slices to developer, reveal alternate dense versus less dense distribution of precipitated grains to produce a striated appearance in binocular regions of cortex (Fig. 7.26, left), whereas in monocular segments contralateral to the injected eye there are no gaps in the deposited radioactivity (right).

The field of Hubel and Wiesel's research that is most closely related to the human condition and therefore of broadest interest to developmental neuroscientists (and to the popular media) was carried out on neonatal kittens and monkeys (Hubel, Wiesel, and LeVay, 1977). The role of visual experience in normal development has been of interest since Descartes raised the question in the seventeenth century, as Wiesel (1982, p. 553) indicated. In 1963, he and Hubel identified in neonatal animals a "critical period" during which deprivation of visual stimulation may induce permanent blindness. The changes in "functional architecture" were shown to occur in layer IVc of the striate cortex without obvious alterations in the visual media, retina, or lateral geniculate nucleus, except for size of cells in the latter. The earlier study (above) had revealed the normal pattern of merging inputs from the two eyes in approximately equal numbers in layer IVc as regularly spaced bands. After neonatal deprivation, that pattern is thrown out of balance: "We have learned that competition and synchronization of inputs are important factors in forming and maintaining this balance. If the processes are disturbed early in life, the system may be permanently altered" (Wiesel, 1982, p. 589).

The dependence on radioisotopes and autoradiography for the successful conclusion of the studies just described is apparent. In turn, the success of tracers in elucidation of the cytological organization of cerebral cortex is contingent on the identification of axoplasmic flow within the neuron. The movement of some entity along the nerve is an ancient idea (Ochs, 1979), the various forms of the transported entity conforming to the contemporaneous concept of the function of nerves: Descartes described a fiery spirit, Boerhaave a watery fluid, Malpighi saw a fluid like egg albumin, Van Leeuwenhoek (1719) noted the loss of fluid and collapse of the axons when a nerve fiber is cut (*see* Fig. 7.2). And Galvani (1791), although absorbed in promoting "animal electricity," had firm ideas about something flowing through the nerves, and in another context declared:

> Therefore, we believe it most likely that the electrical fluid is prepared by the force of the brain, is extracted from the blood, that it enters the nerves, and that it runs through them internally whether they are hollow and empty, whether, as seems more probable, they carry a very tenuous lymph or another similar, special, very tenuous fluid, as it seems to many, secreted from the cortical substance of the brain. If this is so, the obscure and for so long vainly sought nature of the animal spirits will perhaps finally be explicable (translated in Clarke and O'Malley, 1968, p. 183).

The discovery of Wallerian degeneration (Waller, 1850), based on observation of the results of severing nerve fibers from their cell bodies, clearly indicated that there exists a flow of some

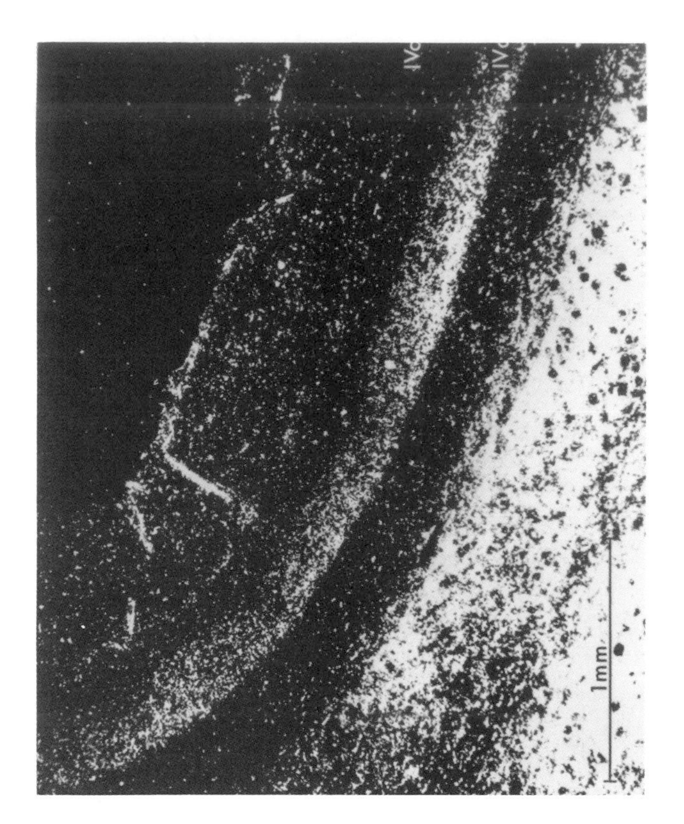

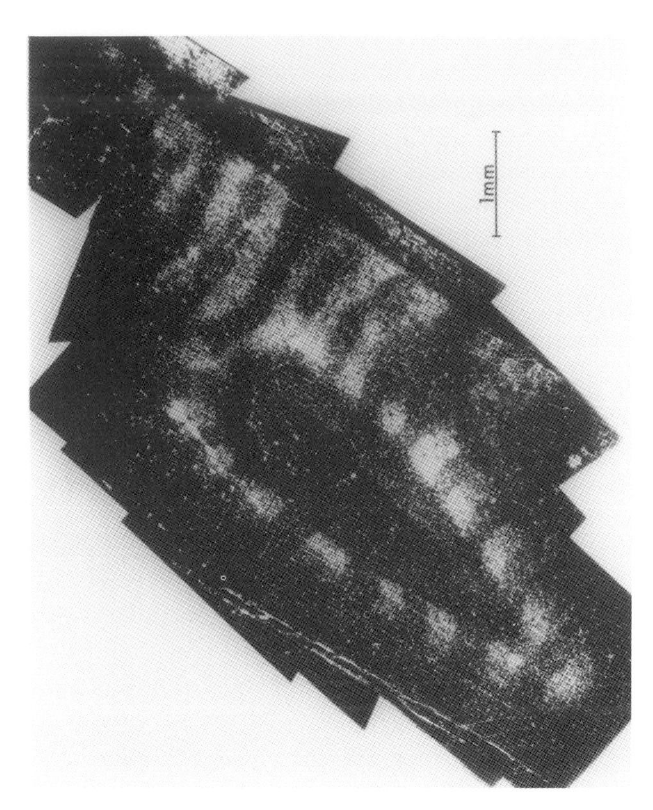

Fig. 7.26. Left—Montage of dark-field photomicrographs from a section of calcarine cortex of monkey, reconstructing the pattern of dense and less dense distribution of silver grains in the terminal projections (layer IVc) from the contralateral eye injected with tracer. Right—Section from stem of calcarine cortex showing continuous band of grains in monocular cortex. (Digitalized reconstructions from Wiesel, Hubel, and Lam, 1974, pp. 277, 278, Figs. 3 and 4 ×1/2.)

metabolic factor(s) along nerves. By the end of the nineteenth century, the idea of transport throughout the axon of a nutrient material appeared in the first textbook devoted solely to the nervous system written by an American, from which Fig. 7.27 is reproduced. The author, Lewellys Franklin Barker (1867–1943), stressed the importance of nutrition throughout the nerve cell: "Exactly the part played by the dendrites, by the cell body, and by the axone in the nutritive processes it is as yet impossible to say, but that each has an important function is certain. . . ." (Barker, 1899, p. 247). Twenty-five years into the next century, the nerve impulse, in the guise of a "neurofibril hypothesis," had been separated from the course taken by the "metabolic influence" but there still was speculation about the character of the latter: "What the nature of the metabolic influences are, it is impossible at present to say. It seems hardly reasonable to think of them as streams of material in the nature of a hormone, emanating from the region of the nucleus and percolating throughout the neurone" (Parker, 1929, p. 170; Ochs, 1979, p. 18).

Both Cajal and the Viennese-American biologist, Paul Weiss (1898–1989), in their regener-ation studies, used compression of the nerve to interrupt the flow of axoplasm, the former concluding that growth cones move along the growing nerve process (1928), whereas the latter believed that it is viscous axoplasm that moves (Weiss and Hiscoe, 1948). In 1964, L. Lubínska, at the Nencki Institute of Experimental Biology at Warsaw, showed that axonal transport is not a simple flow of axoplasm. He described in nerve fibers the retrograde and anterograde movements of radioactive acetylcholinesterase, the enzyme that modulates acetylcholine concentrations and is known to be present in all conducting tissues throughout the animal kingdom (*see* Chapter 8).

Confirmation of the validity of tracing pathways with substances carried by flow of axoplasm was

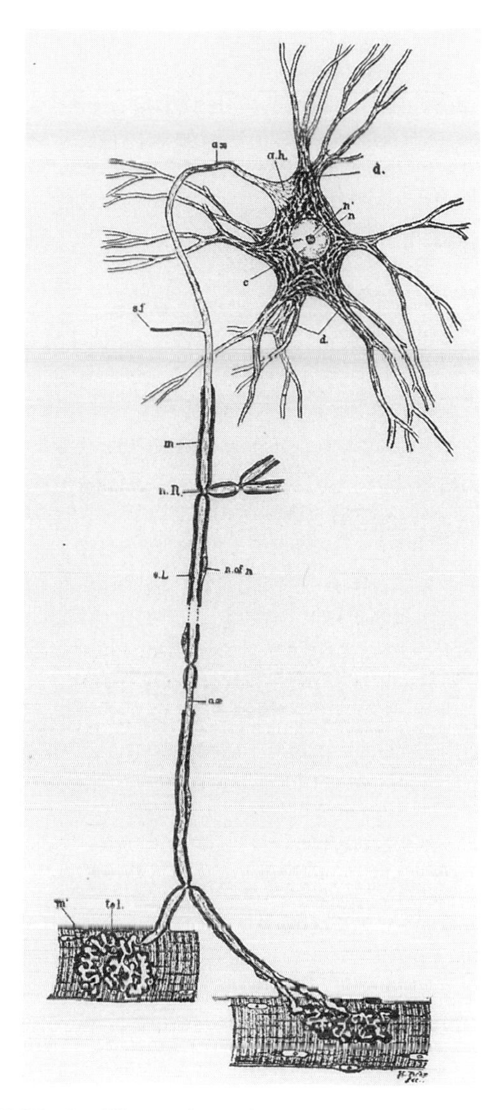

Fig. 7.27. An illustration of what was known about the lower motor neuron from a textbook popular at the end of the nineteenth century. a.h.—Axon hillock, a.x.—axis cylinder, m.—cellular sheath or neurilemma, c.—cytoplasm showing Nissl bodies, d.—dendrites, n.—nucleus, n.'—nucleolus, n.R.—node of Ranvier, s.f.—side fibril, n. of n.—nucleus of neurilemma, tel.—motor end plate, m.'—stripped muscle fiber, s.L.—segmentation of Lantermann. (From Barker, 1899, p. 41, Fig. 17; also p. 898, Fig. 577.)

rapidly forthcoming. Identification of the synaptic terminal with its cell of origin was confirmed autoradiographically by Anita Hendrickson (1969), who found the tracer (intraocularly injected tritiated leucine) in only one type of terminal in the lateral geniculate nucleus. Another technique (Graham and Karnovsky, 1966), using an enzyme, horseradish peroxidase (HRP), served as a tracer of retrograde axonal transport in studies of the central nervous system by LaVail and LaVail (1972). They found that after injection of HRP into chick ocular chamber, the brown reaction product can be detected by light and electron microscopy in the optic nucleus. The very fundamental question of whether or not tracers could cross a synapse and so reveal a postsynaptic pathway beyond the terminal of the receptor neuron was answered by Bernice Grafstein at Cornell University Medical College and a former student of Paul Weiss's. She demonstrated (1971) that radioactive amino acid injected into a posterior eye chamber of anesthetized mice was distributed in significant amounts to striatal cortex contralateral to the origin, compared with little radioactivity deposited in the reference frontal cortex. This result was important on two counts: It constituted direct evidence that the label, presumed to be attached to a protein, had crossed synapses in the central nervous system, and it inaugurated a new wave of cortical and subcortical exploration of receptive fields that had commenced in 1936 (Gerard, Marshall, and Saul) but had yet to be anatomically mapped more precisely than was possible with the cumbersome degeneration methods. Grafstein also found (Specht and Grafstein, 1973) that the radioactivity is concentrated in layer IV of striate cortex, confirming the laborious work of Hubel and Wiesel (1972) using degenerative techniques, as discussed above.

Hubel and Wiesel's comprehensive series of beautiful experiments in one set of functional processes—the visual—disclosed innate mechanisms that are fully developed and maintained by experience. The length and timing of critical periods differ among species, which "suggests that throughout the brain each functional unit has a unique programme of development" (Wiesel, 1982, p. 587). Correlation of that suggestion with the idea expressed by Hubel (*see* p. 148, this volume) regarding "little machines" provides some hint of the broad sweep of the team's conceptualizations.

The enthusiasm for the possibilities of new tracer methodologies and their wide use in various combinations was expressed by a modern investigator: "[W]e are currently moving forward on the crest of a magnificent wave of neuroscientific exploration. The most recent advances focus on tracing successive links in neuronal chains. . . ." (Heimer, 1987, p. 766). That is only part of the story, for radioisotopes have served in deciphering the movement of ions across neuronal membranes, as described in the next chapter.

NONNEURONAL CELLS OF THE CEREBRUM

Even more numerous than the billion neurons in the human cerebrum, the brain's nonneuronal cells have had a more confused, yet shorter history of study. The confusion came in assigning a role to elements that became visible only when the Golgi technique came into use and relating them to their embryologic origins and specific pathology. Rudolf Virchow (1821–1902), the German founder of pathological anatomy and a participant in the revolt against Bismarck, theorized that inflammation of the cerebral ventricles is indicative of the presence of connective tissue, and in 1846 wrote: "This connective substance forms in the brain, in the spinal cord, and in the higher sensory nerves as sort of putty (*neuroglia*), in which the nervous elements are imbedded" (quoted from Somjen, 1988, p. 2). Although Virchow regarded the "putty" as a passive support of nerve cells, he also described (1846) the phagocytic action of neuroglia (Slumberger, 1970).

By 1865, the physiological role of the axon and its processes was sufficiently accepted in Germany that Otto Friedrich Karl Deiters (1834–1863) could state (published posthumously) that cerebral cells without an axon are not nerve cells. Golgi upheld this distinction as the only reliable criterion of glial cells. The demonstrations by Golgi, Nansen, and even Virchow that neuroglia are embryologically derived from the same primitive layer—the ectoderm—as are neurons and hence are not a connective tissue, was one manifestation of the burgeoning dependence on embryology to solve puzzles of classification of cells and tissues.

A forgotten episode in the history of neuroglial function and recently brought to notice (Dieng, 1994) was the suggestion of Carl Ludwig Schleich (1859–1922) that neuroglia moderate neuronal activity through an inhibitory action. A Berlin surgeon who had spent time in Virchow's laboratory, Schleich (1894) reasoned that the "multiplicity of mental events" requires both excitatory and inhibitory neural mechanisms, which, from laws of electrical processes, could not be carried out by the same cell. In concert with the new notion of a synaptic space between neurons, he assigned an inhibitory function to the glia that presumably were inserted as insulators between the excitatory elements; the more "swollen" the glia, the greater the inhibition. This picture of glia as modulators of neuronal activity placed them on equal footing with neurons in the central nervous system, and appeared so ludicrous and without evidence that it could not survive.

Wilhelm His (1831–1904), born into wealth (Rasmussen, 1970, p. 48) in Switzerland and completing his medical education internationally, devised (1893) a classification of tissues according to their histogenesis. His's main interest was in the nervous system, and he conclusively proved that the origins of neural parts of the nervous system are ectodermal and vascular parts mesodermal. He further suggested, in 1889, that embryonal neuroglial fibers may guide the migration of early neuronal cells. Almost two decades later, this idea was taken up by the imaginative Lugaro, whose "intriguing original proposals" (Somjen, 1988) assigned chemotaxis as the guiding mechanism and suggested that the astrocytes, with their perivascular end feet, keep the interstitial fluid "habitable" for neurons and serve a detoxifying function. Astrocyte involvement was invoked 60 years later (Davson and Oldendorf, 1967) and Lugaro's guess (1907) that glial cell processes metabolize excitatory substances foreshadowed work such as that of Stephen Kuffler (1952) on regulation of the neuronal microenvironment by glial potassium ion flux and of Pasco Rakič (1971) on the support provided by glial processes to migrating neurons. The vital roles of both those activities reinforce the equality of neurons and glia as parts of the central nervous system and make Schleich's plea for unity seem not as wild as it was judged by his contemporaries. A view on brain function, inferred from old and modern anatomical data, was promogulated by Robert Galambos (1961, p. 136): Neurons and glia "mutually collaborate to produce behavior. . . Glia [are] genetically charged to organize and program neuron activity."

In 1932, a two-volume treatise titled *Cytology and Cellular Pathology of the Nervous System*, under Wilder Penfield's editorship, indicated that by then the neuroglia had been described and classified satisfactorily. An important article therein was written by Pio del Rio-Hortega (1882–1945), whose "name was to glia what the names of many others . . . were to the nerve cell" (Haymaker, 1970,

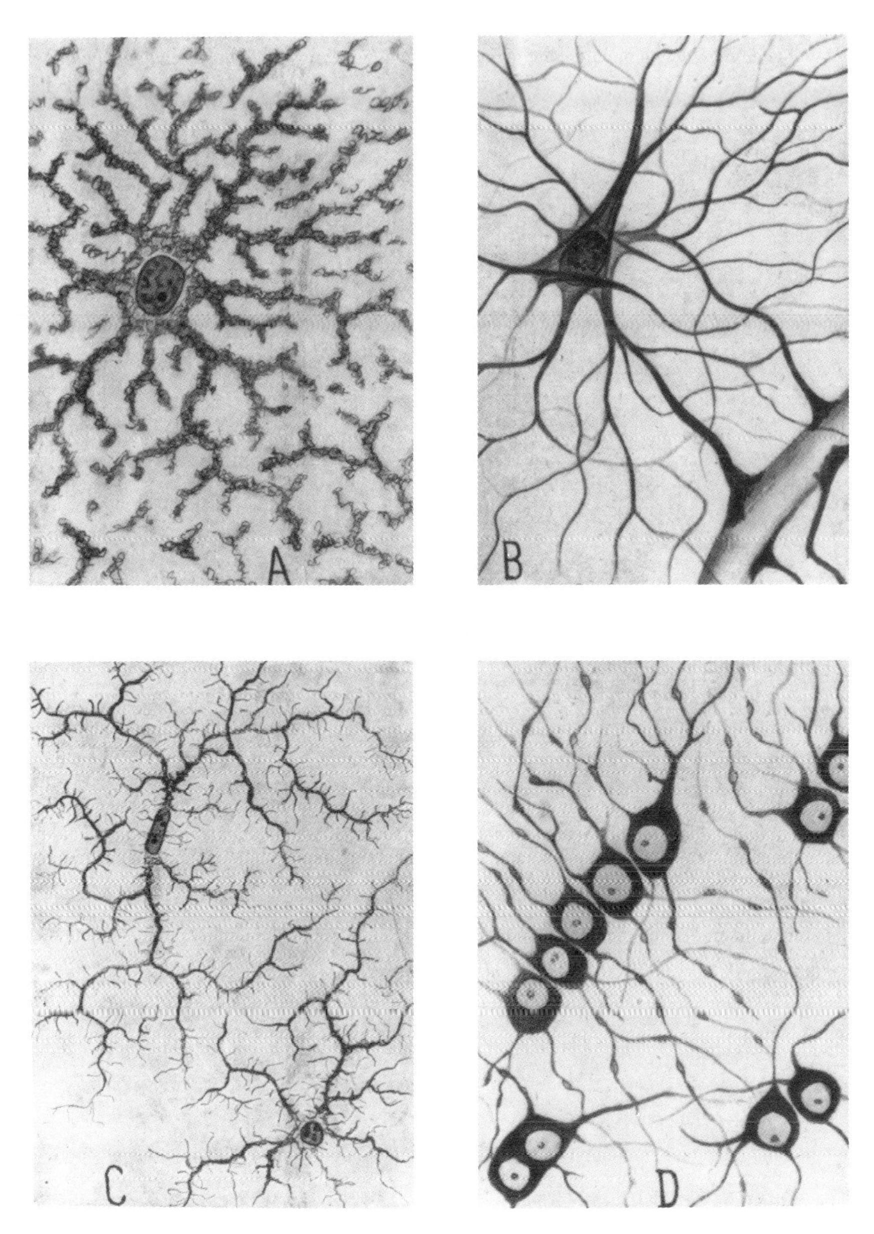

Fig. 7.28. Pio del Rio-Hortega published his illustrations of the four types of glia in 1920. **A**—Protoplasmic neuroglia from gray matter, **B**—fibrous neuroglia from white matter, **C**—microglia, **D**—oligodendrocytes from white matter. (From Rio-Hortega, 1920, p. 48, Fig. 2; also in Somjen, 1988, p. 6, Fig. 4.)

p. 158). While working in Cajal's laboratory, Rio-Hortega introduced (1919) an ammoniacal silver carbonate method of staining tissue and with great skill demonstrated spidery cells—the microglia and oligodendroglia (Fig. 7.28). On publication of the results, Cajal dismissed him, believing that the "Hortega cell" had been described by Robertson (1900) two decades earlier. Penfield, too, had spent some time in Cajal's laboratory, and his lucid paper in *Cytology and Cellular Pathology* dealing with astrocytes and oligodendroglia immediately precedes that of Rio-Hortega (1932) on microglia; those two papers constitute classics on the subject.

Among the many functions proposed for the nonneuronal cells in the central nervous system,

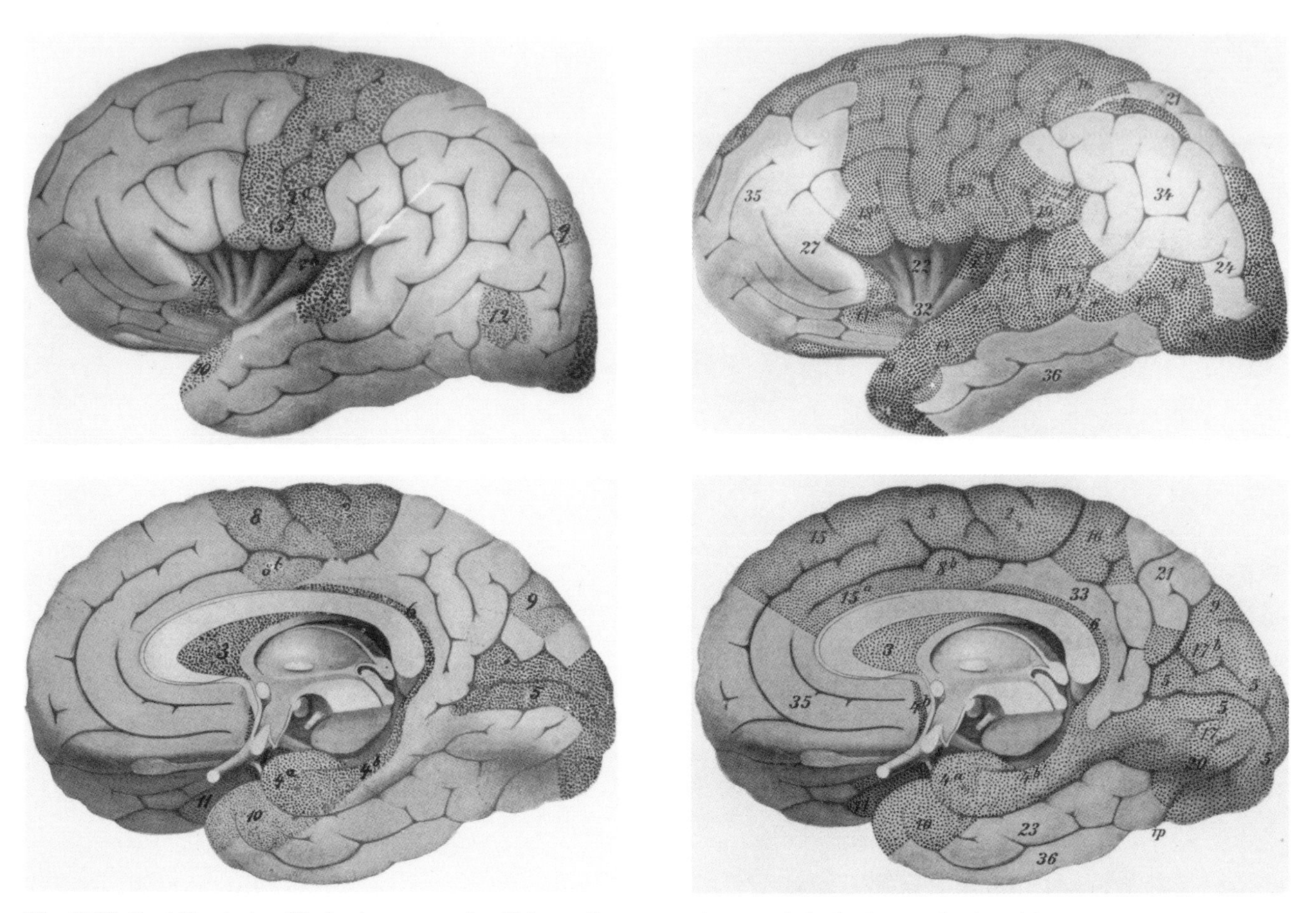

Fig. 7.29. Paul Flechsig of Leipzig summarized his studies on myelogenesis in the human brain with maps of the projection (left) and associated areas (right), numbering the areas in the order in which they become myelinated. Note that the areas with the earliest myelinated fibers include the precentral and postcentral gyri and the visual areas, whereas the last to be myelinated include the prefrontal and lower temporal gyri. (From Flechsig, 1904, p. 1027.)

speculative and real, the production of myelin is one of the latter. From the early drawings of neuronal elements we know that it was difficult to distinguish myelin from axoplasm, for example in Ehrenberg's figure of 1833, Plate IV, Fig 11, showing "myelin" squeezed out of a nerve fiber that is more reasonably axoplasm than myelin (Clarke and O'Malley, 1968, p. 39). The presence of myelin in the central nervous system was first described in 1884, when Thudichum published his "most important single contribution" to the field (Tower, 1970, p. 297): the identification of myelin's lipoid nature. A series of investigators in Thudicum's native Germany recognized and expanded the import of his work (*see* Chapter 8). The structural nature of myelin was visualized by Waldeyer's student, Carl Weigert (1845–1904), who noticed that in brain tissue mordanted in chromic salts, the myelin stained selectively with hematoxylin or acid fuchsin (1882). This technique "opened new avenues to many [demyelinating] diseases of the spinal cord and brain" (Neuburger, 1970, p. 390).

Paul Emil Flechsig (1847–1929) used the presence of myelin to devise a novel map of cerebral architectonics based on the time during pre- and postnatal development when it appears around the nerve fiber. He stipulated that cortical areas containing myelinated fibers at full term are in the "primordial zone," those myelinated during the first month after birth constitute the "intermediary zone," and the "terminal zone" consists of areas with late-myelinated fibers. From those studies, Flechsig formulated his Fundamental Law of Myelogenesis: "[E]qually important nerve-fibers are developed simultaneously, but those of dissimilar importance are developed one after another in a

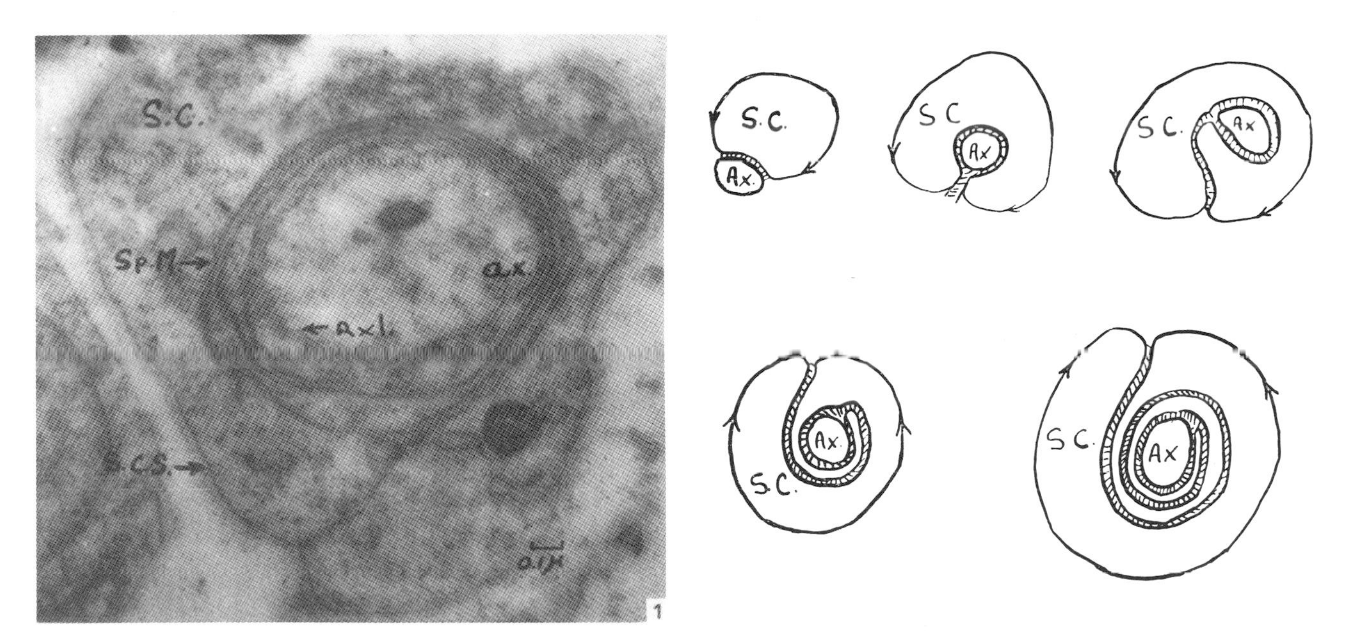

Fig. 7.30. Left—An early electron micrograph of the formation of the myelin sheath, from chick embryo sciatic nerve. Ax—Axon, axl—axolemma, Sp.M.—spiraled membrane, S.C.—Schwann cytoplasm. Right—Diagram of early stages of myelin formation. (From Geren, 1954, pp. 559, 560.)

succession. . . ." (Flechsig, 1901, p. 1027). From comparative myeloarchitectonics, it was apparent to Flechsig (1904) that myelination increases phylogenically with elaboration of the neocortex. Flechsig's map (Fig. 7.29) shows 36 areas numbered in the general order in which they become myelinated. By noting when myelination appears in the fetus and newborn, Flechsig was able to identify the course of intracerebral and spinal cord fiber tracts with greater success than had been possible with degenerative methods, as also had been the case with the evoked potential technique (*see* Chapter 5).

With the demonstration of myelinated nerve fibers in the central and peripheral nervous systems, two new questions arose—where does myelin come from and what is its nature? The first puzzle was partially solved by the publication in 1954 of a classic paper by B. B. Geren. Her inspired interpretation of early electron micrographs of peripheral embryonic nerve fibers showed that the sheath is formed from continuous layers of membrane wrapped spirally around the axon (Fig. 7.30). Herbert Gasser (1952, p. 32) wrote: "The sheath is well-defined; and the fibers are attached to it by ligaments of gossamer fineness, which I have decided to call mesaxons." Evidence for the source of the mesaxon came likewise from electron microscopy. In the developing spinal cord of kittens, M. B. Bunge and her colleagues (1962) photographed for the first time the connections between the myelin sheath and the oligodendroglia.

The physical and chemical nature of myelin has been the domain of the biochemists and biophysicists, whose studies revealed, among other facts, that there are basic differences between central myelin and peripheral sheaths represented by Schwann cells. To cite one example, it was proposed that the higher ratio of phospholipid to cholesterol found in peripheral-nerve myelin than in central myelin (Evans and Finean, 1965) may be related to the occurrence of wider layers in peripheral myelin (Karlsson, 1966) as seen in rat sciatic compared with optic nerves. The interrelatedness of those discoveries is but one example of the overlap of disciplines which constitutes the great strength of neuroscience in solving biomedical problems.

OVERVIEW OF NEURAL FINE STRUCTURE

Knowledge of the fine structure of the cerebrum became available only after discoveries in the peripheral nervous system, for the practical reason that the cerebrum was more inaccessible and complex. The transition from the often artifactual "globule" to their recognition as nerve-cell bodies with attached nerve-fibers took about 40 years, propelled by a consuming curiosity among the early biologists about the revelations of the achromatic microscope. In contrast, acceptance and proof of the complete neuron as an entity encompassed a longer period, from Waldeyer's 1891 enunciation of the neuron as a cell unit until the electron-enhanced images of cerebral synapses (Gray, 1959; 1961) revealing contact without connection. During the period of uncertainty, however, the work of anatomists such as Kölliker, Meynert, and Gennari was establishing the stratified character of the neocortex, and degenerative studies such as those of Waller and von Gudden were endeavoring to identify connections and functions of specific cerebral regions.

The roles of Golgi and Ramón y Cajal in advancing knowledge of the brain are based primarily on their contributions to neural fine structure. Golgi's silver impregnation technique allowed him to see distinctions between axons and dendrites and to recognize association and effector neurons. A systematic description of cortical cells, layer by layer, in human infant and embryo brains as carried out by Cajal, enabled him to construct possible pathways from incoming fibers to the dendrites of effector neurons.

Commencing shortly before the midcentury and after two world wars, investigations of sensory processes at the fine structural level brought to light new types of neural organization, such as barrels, and studies of visual information processing not only demonstrated what sophisticated application of different methodologies can accomplish, but also revealed the exquisite detail and precision of the sense of vision. The extreme specificity of some layer IV cells to directional motion, for example, shown by the work of Hubel and Wiesel, would not have surprised Ramón y Cajal, who expected no less than ultimate perfection of design.

The last cells in the cerebrum to be recognized were the nonneuronal elements, the neuroglia. Reasoning that where there is inflammation there must be connective tissue, Virchow (1846) named them and assigned them a structural role. They were definitively sorted and described by del Rio-Hortega (microglia, 1921) and Penfield (astrocytes and oligodendroglia, 1921). The latter were established (Geren, 1954) as the source of the protective myelin wrapped around cerebral axons, serving the buffering function of the Schwann cells around peripheral axons and later work distinguished central myelin from peripheral myelin, in regard to their source and chemical and physical characteristics. Kuffler and associates (1966) proposed a role of neuroglia in preventing accumulation of potassium in the interstitial space. Another activity of neuroglia (astrocytes, specifically) suggested by Rakič (1971) was as guides to migrating immature cerebellar granule cells.

The sheer weight of reliable knowledge of the substrates of the nervous system has provided an enormous opportunity for the development of concepts of function unprecedented in any other era. A fresh perspective on Nature's bounty was provided by the "father of the electronic information age," Claude Shannon, who said: "That the brain has ten billion neurons probably means that it was cheaper for biology to make more components than to work out sophisticated circuits" (Shannon, 1987, p. 65).

8 Landmarks in Cerebral Neurochemistry

CONTENTS

One thing about the cerebral circulation that has been agreed upon from the start is that it is unique in all essential respects, beginning with the morphological.

(Schmidt, 1950, p. 4)

The chemistry of the brain is one of the newer subdisciplines of neuroscience, having become formalized with the organization of national and international groups such as the International Society of Neurochemistry in 1965. Derived from the surge of interest in the histochemistry and cytochemistry of neurons, neurochemistry is closely related to neuropharmacology and psychoneuropharmacology. Its rapid coming of age marked the emergence of an immensely promising field of research and knowledge in areas as diverse as the biological basis of mental health and the nerve impulse. One channel of basic research on chemical substances, later found to have psychotropic properties (and high interest to the pharmaceutical industry), brought about a therapeutic "revolution" that decimated the mental hospital populations of the midtwentieth century (*see* Swazey,1974). The recognition that the chemistry of the brain, seen as a separate organ relatively isolated from the other parts of the body, is the biological basis of mental health was the key to a vast research domain that extends from human behavior to the level of very specific ionic processes. And because biologic viability depends on energy supply and demand, the dynamics of cerebral vascularity have been of utmost importance in understanding those processes.

THE EARLY CHEMISTS

From the point of view of direct evidence, neurochemistry's early eighteenth century origins were in Hessian Germany where Johann Thomas Hensing (1683–1745; Fig. 8.1), scholar, physician, and chemist at the University of Giessen, was motivated by his reading to systematically search for phosphorus in beef brain. His 42-page academic dissertation (*see* Fig. 8.2, p. 159) establishing him as the first modern neurochemist, was unrecognized until publication of an informative biography accompanied by translation of the Latin text and annotations by Donald B. Tower, a prominent American neurochemist. The title of Hensing's dissertation translates as *The Chemical Examination of the Brain and the Unique Phosphorus from It [which] Ignites all Combustibles* (Tower, 1983, p. 225). From the text, we learn that: "[A] new phenomenon has emerged from the work—truly, light and fire from the brain—which deserves to be submitted, without deception, to the inspection and judgment of the curious" (ibid., p. 249). As Tower and others have pointed out, Hensing's writings are especially interesting for their disclosure of the alchemists' methods in use at the time.

Fig. 8.1. The portrait of Johann Thomas Hensing shows him at about 40 years of age as Professor Ordinarius of Natural and Chemical Philosophy and Professor Extraordinarius of Medicine at the University of Giessen. (From Tower, 1983, frontispiece.)

A century-and-a-half after Hensing, another Hessian, John Ludwig Wilhelm Thudichum (1829–1901; *see* Fig. 8.3, p. 160) made the next important contribution to brain chemistry.[1] Like Hensing, Thudichum received his medical degree from the University of Giessen, but was passed over for a coveted appointment and emigrated to England. In *A Treatise on the Chemical Constitution of the Brain* (1884), he identified the colloidal nature of the brain "bioplasm" and related it to the phosphatides which he knew occurred in plant as well as animal bioplasm. Thudichum's analysis was on one human brain, and his findings, summarized in the table reproduced in Fig. 8.4 (p. 161), are surprisingly modern. Thudichum's work was ridiculed by powerful members of the British scientific establishment as "dilettanteisms . . . an epidemic plague for physiological chemistry" (quoted by Rafaelsen, 1982, p. 297), but when his writings on brain chemistry were translated into his native German 17 years later, just before his death, they commenced to receive the serious attention they merited. Perhaps some of his difficulties in recognition sprang from the fact that he was also the author of books on cookery and viticulture, hardly beneficial to his stature as a serious chemist. As

[1]For a discussion of progress between the times of Hensing and Thudichum, *see* the authoritative account by the British neurochemist H. McIlwain (1958).

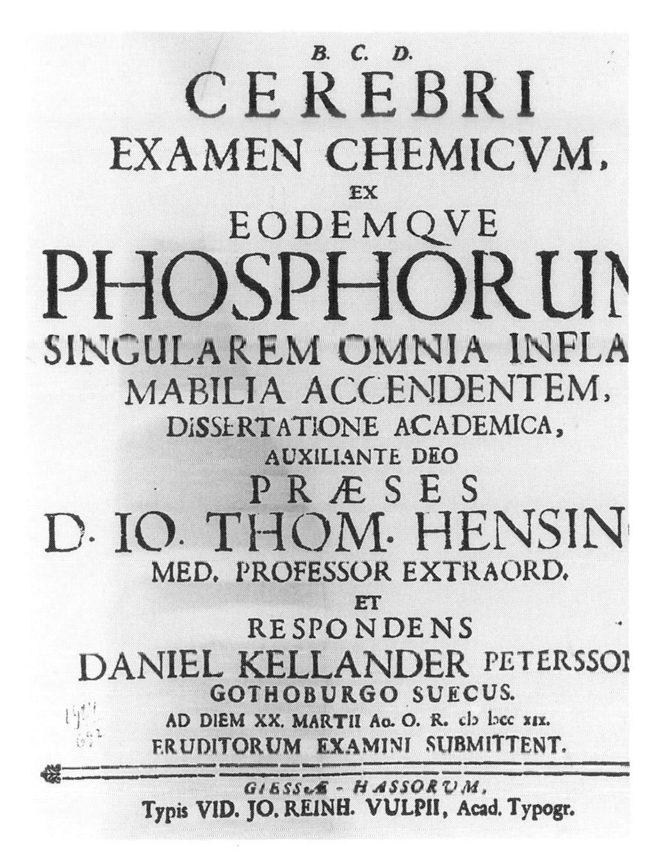

B. C. D.
CEREBRI
EXAMEN CHEMICVM,
EX
EODEMQVE
PHOSPHORU
SINGULAREM OMNIA INFLA
MABILIA ACCENDENTEM,
DISSERTATIONE ACADEMICA,
AUXILIANTE DEO
PRÆSES
D. IO. THOM. HENSIN
MED. PROFESSOR EXTRAORD.
ET
RESPONDENS
DANIEL KELLANDER PETERSSO
GOTHOBURGO SUECUS.
AD DIEM XX. MARTII Ao. O. R. cIↃ Iↄcc xix.
ERUDITORUM EXAMINI SUBMITTENT.
GIESSÆ - HASSORUM,
Typis VID. JO. REINH. VULPII, Acad. Typogr.

Fig. 8.2. The title page of Hensing's treatise on phosphorus in the brain, published in 1719. For translation, *see* text. (From Tower, 1983, p. 188.)

with Hensing, interest in Thudichum's life and work was rekindled by attention from a modern biochemist, David L. Drabkin at the University of Pennsylvania, who wrote extensively about him (1958).

Many of the substances Thudichum isolated from that single brain—lecithins, cephalins, and myelins—he also fractionated, classified, and named; some of his samples exist today in Philadelphia and London in unusually pure form. This talented chemist may be credited with anticipating psychopharmacology by his suggestion that: "Many forms of insanity are unquestionably the external manifestations of the effects upon the brain-substance of poisons fermented within the body" (1884, p. xiii). Unfortunately, in his treatise Thudichum did not discuss specific functional aspects of the compounds he discovered.

The following salute to the complexities of brain chemistry appeared in a textbook of the nervous system at the end of the last century:

> Chemical compounds come into existence . . . of a degree of complexity scarcely approached elsewhere on this planet, and before the nature of which the most advanced organic chemist stands utterly powerless and despondent. It is in the nervous system of all parts of the human body that the delicacy and complexity of the chemistry of the metabolism are most in evidence (Barker, 1899, p. 217).

The author went on to acknowledge the role of Claude Bernard's milieu intérieur: "While emphasizing the maintenance of a certain constancy of function, and consequently of structure, despite the unending chemical alterations going on, we must admit that the metabolism in no individual is perfectly constant" (ibid., p. 219).

The foundation of modern theories concerning chemical transmission of signals from one neuron to another was built on evidence not from the central nervous system, but from the periphery with studies of the nerve–muscle preparation. A vague concept of chemical transmission may be read into the ideas of Giovanni Alfonso Borelli (1608–1679) published toward the end of the seventeenth century. Italian mathematician and physicist, he explained (1681) how agitation of the "nerve juice" in tubes in the brain caused a few drops to be squeezed out and spread to the muscle, thereby causing contraction. One hundred years after Borelli, interest in the chemistry of living tissues had become a major concern of German biophysicists. Among the enthusiasts was Frederick Wilhelm Heinrich Alexander von Humboldt (1769–1859), who wrote in 1796 that "the nerve is a chemical laboratory, and it's remarkable to be able to extinguish and resuscitate [it] at will" (quoted by Dittelbach, 1993).[2] Humboldt discovered the enhancing effect of oxygenated muriatic acid and dilute alkali on the excitability of tissues and opened a field he called "vital chemistry," by which he attempted to explain the "galvanic force," a concept which had spread from Italy a few years earlier.

[2]We are indebted to Michael Dittelbach for drawing our attention to Humboldt's contribution in this field and for graciously permitting citation of his own unpublished work.

Fig. 8.3. Johann Ludwig Wilhelm Thudichum (shown at about 50 years of age) was born near Giessen, received his MD from the university there in 1851, and emigrated to England two years later.

ANIMAL ELECTRICITY

The generation of electrical current by living organisms was called "animal electricity" by its most avid promoter, Luigi Galvani (1737–1798) in Bologna, although he did not discover it (Clarke and O'Malley, 1868, p. 177 passim). Galvani devised ingenious experiments to show that in a suitable closed circuit, an electrical impulse passes through a nerve and produces muscular contraction. As he explained in "Conjectures and Conclusions" of Part 4 of *Commentary*, after describing many types of experiments:

> From what is known and explored thus far, I think it is sufficiently established that there is electricity in animals, which . . . we may be permitted to call by the general name of animal electricity. This, if not in all, yet is contained in most parts of animals; but manifests itself most conspicuously in muscles and nerves. The peculiar and not previously recognized nature of this seems to be that it flows from muscles to nerves, or rather from the latter to the former, and that it traverses there either an arc or a series of men or any other conducting bodies. . . . (Galvani, 1791, transl. by R. M. Green, 1953, p. 60).

The plate from Galvani's publication reproduced here (*see* Fig. 8.5, p. 161) portrays some of the tests that were eagerly carried out by physicists and

TABLE SHOWING RESULTS OF THE QUANTITATIVE ANALYSIS OF A HUMAN BRAIN.

Weights in grammes; third decimals below 5 omitted, above 5 added as 1 to second decimals. Membranes weighed 58 g.

—	1. Weights with Membranes.	2. Weights without Membranes.	3. Quantities analysed.	4. Albumen.	5. White Matter.	6. W. M. insoluble in Ether.	7. W. M. soluble in Ether.	8. Buttery Matter.	9. Kephalin Total.	10. Myelin Total.	11. Lecithin Total.	12. Cholesterin Total.	13. Inosite.	14. Lactic Acid.	15. Hypoxanthin and Alkaloids.	16. Indefinite Extractives.	17. Potassium.	18. Sodium.
Right hemisphere	589	564	465	35·68	18·60	11·63	6·97	21·66	3·34	5·22	5·90	8·93	0·43	0·64	1·30	2·63	0·25	0·40
Left hemisphere	596	570	463	35·06	21·93	12·28	9·65	21·65	—	5·20	—	6·99	0·38	1·00	1·06	2·78	0·39	0·39
Cerebellum	135	129	124	10·94	1·88	1·66	0·22	3·26	1·29	3·76	2·97	1·95	0·05	0·13	0·67	1·49	0·01	0·02
Mesenkephalon and medulla oblongata	34	33	33	2·48	0·64	0·56	0·03	—	0·45	0·23	0·41	1·01	0·06	0·07	—	0·73	0·01	0·03
White tissue	—	—	66	5·70	6·98	4·56	2·42	5·31	0·06	0·16	0·48	2·15	0·14	0·05	—	0·44	0·03	0·04
White tissue for quantation of water	—	—	9·2	0·80	0·97	0·64	0·33	0·74	0·002	0·02	0·07	0·30	0·02	0·006	—	0·06	0·004	0·006
Grey tissue	—	—	46	3·50	0·30	—	0·30	—	0·15	0·02	0·73	0·90	0·69	0·04	0·10	0·31	0·01	0·09
Grey tissue for quantation of water	—	—	6·2	0·46	0·04	—	0·04	—	0 02	0·003	0·09	0·12	0·09	0·006	0·01	0·04	0·001	0·01
Loss in operations	—	—	86·6	6·58	4·12	2·24	1·88	4·14	0·63	0·69	1·10	1·86	0·08	0·20	0·21	0·52	0·07	0·07
Totals	1354	1296	1296	101·20	55·46	33·57	21·84	56·76	5·94	15·30	11·75	24·21	1·94	2·14	3·35	9·00	0·78	1·06

Fig. 8.4. The detail and care of Thudichum's chemical work is attested by an analysis of one human brain, in which 18 measurements were made on six brain regions. (From Thudichum, 1884, p. 257.)

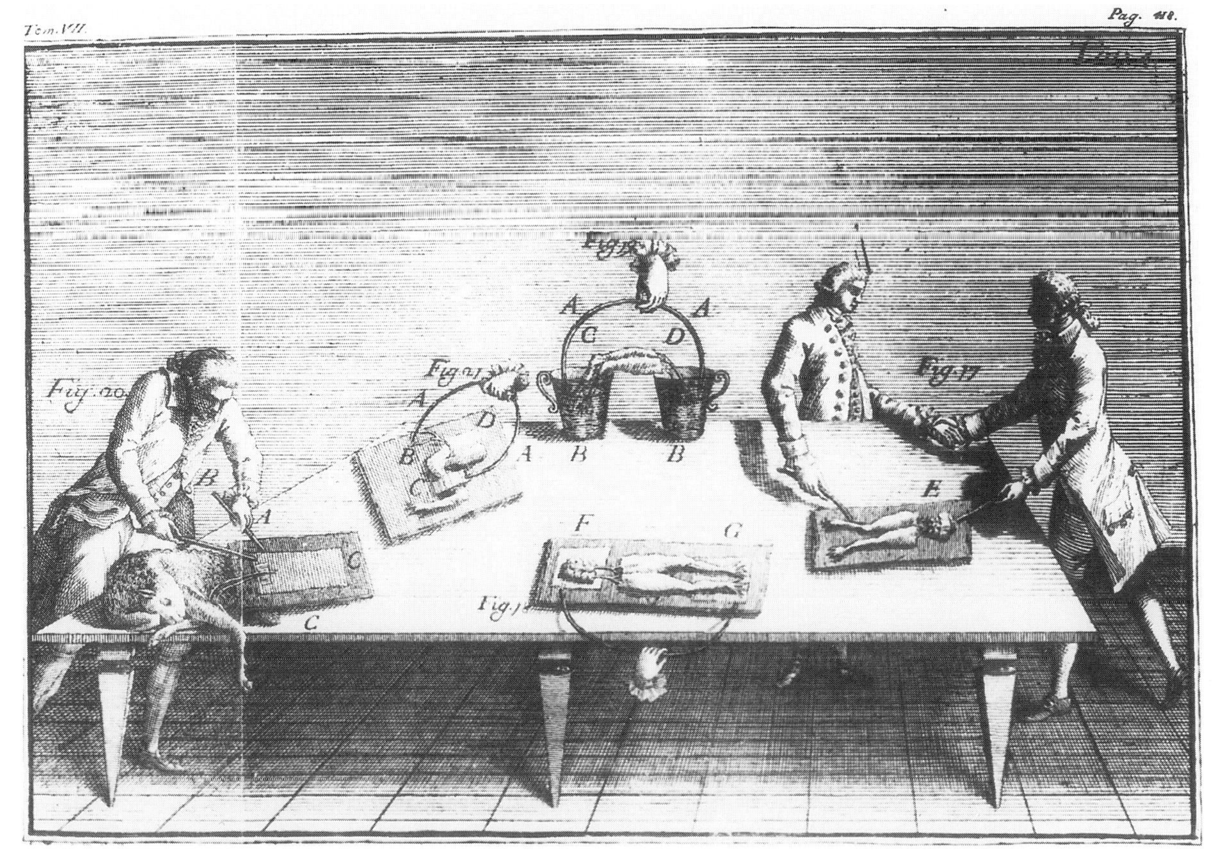

Fig. 8.5. One of Galvani's famous plates, published in 1791, shows some of the methods of making an arc for passage of an electrical current along a nerve to produce muscular contraction. In each figure, a bimetallic source is involved. (From Galvani, 1791, Plate IV, Figs. 17–21.)

Fig. 8.6. Émil du Bois-Reymond, Swiss-German physiologist, brought physics to bear on biological problems and with Helmholtz, Liebig, and others in the midnineteenth century laid the foundation for modern biophysics.

physiologists alike and shows the use of bimetals to create "galvanic fluids" for stimulation of plant and animal tissues. That the effect was a demonstration of "true" animal electricity and not due to a "metallic current" as Volta (1792) claimed, was proven later (Aldini, 1803): the muscle of a pithed or decapitated frog could be made to contract by forming a closed circuit between nerve and muscle surface.

News of the sensational demonstrations had great impact at the time and stimulated many new investigations in Italy and abroad. As Du Bois-Reymond wrote in 1848: "The storm which began with the publication of [Galvani's] *Commentary* in the world of physics, physiology and medicine, can only be compared to that which arose on Europe's political horizon during the same period in the final years of the eighteenth century" (quoted by G. C. Pupilli in R. M. Green, 1953, pp. ix–x). The storm that compared to the upheaval caused by Napoleon's conquests was created by the flurry of speculative explanations of the nature of the new force and was not resolved until the midtwentieth century.

The excited interest in animal electricity flagged somewhat in the early nineteenth century (Clarke and O'Malley, 1968, p. 186), but was revived when another gifted Italian investigator, Carlo Matteucci (1811–1868), built a primitive galvanometer and began to report a series of confirmations and new and significant discoveries. Among the latter, he found that the same muscle current occurs in warm-blooded animals as in frogs, that this action current in one muscle can be led through an attached nerve and induce a contraction in a second muscle, that a current exists between normal and injured muscle (the demarcation current), and most importantly, Matteucci (1838) noted the decrease in current during tetanus of the muscle, the "negative variation" of the action current.

The experiments and ideas propounded by Matteucci were severely scrutinized by the great German physiologist, Émil du Bois-Reymond (1818–1896; Fig. 8.6), one of the "organic physicists" dedicated to a reductionist concept of a unified biology based on physical and chemical principles (Clarke and Jacyna, 1987, p. 209). In 1848, he published the first of two volumes, *Untersuchungen über Thierische Electricität*, from the title pages of which two beautiful and gripping woodcuts are reproduced in Fig. 8.7. Attesting to his career-long commitment to nerve–muscle physiology, the last part of the second volume was not completed until 1884. Du Bois-Reymond (1843) believed the source of Matteucci's "muscle current" to be the potential difference between the negative inner and positive outer surfaces of the muscle at rest. Building a more sensitive galvanometer, he demonstrated convincingly the action current in nerve, and propounded a universal law of electrical stimulation: excitation is not due to the current flow itself, but to the changes that occur at the make or break of the circuit, a positive fluctuation from zero and negative return to zero, respectively. Du Bois-Reymond stated that contractile tissue is stimulated at its boundary by "a thin layer of ammonia, lactic acid, or some other powerful stimulating substance. Or the phenomenon must be electrical" (1877, quoted by Dale, 1937–1938, p. 4P). Dale continued: "His meaning cannot, I think,

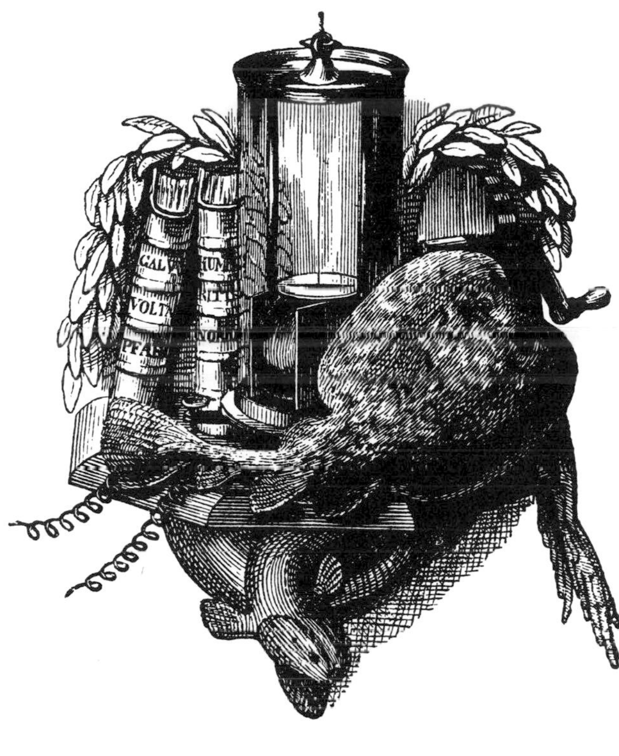

Fig. 8.7. Almost 30 years before du Bois-Reymond considered the possibility of chemical transmission, he published the first of two volumes on animal electricity. The exquisite engravings (×2) adorning the title pages illustrate the contemporary interest in the discharge of electric organs. The names of Galvani, Volta, Pfaff, Humboldt, Ritter, and Nobili are inscribed on the two books in the upper reproduction. The "battle between eel and horse" Humboldt said he had witnessed. (From 1848, vol. I [above] and 1884, vol. II [below].)

be mistaken." Du Bois-Reymond's student, Ludimar Hermann (1838–1914), saw the negative variation as a wave propagated along the nerve by progressive excitability changes from resting to active regions; it did not resemble the passage of current in a wire as was widely thought at the time (Hermann, 1863).

TISSUE EXCITABILITY[3]

At the end of the nineteenth century, although the electrical explanation of conductance along peripheral nerves was acceptable to most electrophysiologists, the more fundamental problem was the excitability of tissues. Again in Germany, modification of tissue excitability by different ion species was systematically tested by F. S. Locke (1895), who recognized that calcium ions are necessary for effective neuromuscular transmission. Ernest Overton (1865–1933; *see* Fig. 8.8, p. 164, left), confirmed those findings and showed (1902) that in addition to calcium, extracellular sodium ions are required for nerve excitability, and suggested that a sodium-potassium exchange is involved in signal conductance. The momentum of German scientific tradition continued with yet another physiologist, Julius Bernstein (1838–1917; Fig. 8.8, right), whose mentor, du Bois-Reymond, had suggested that the measure the conduction velocity of the negative variation in the signal that stimulates the muscle. Bernstein (1868) built an analyzer (for details of the "rheotome," *see* Schuetze, 1983) and produced the first plot of an action potential, with the wave moving in both directions from the point of stimulation and with an overshoot. Combining his physiological measurements with chemical experiments during the next three decades, in 1902 Bernstein identified the source of the electromotive force and proposed a theory based on the selective permeability of the nerve membrane to potassium ions which diffused from inside to outside the membrane and generated the resting potential. Unaccountably, in that paper he did not mention the role of extracellular sodium ions in muscle contraction found by Overton, nor

[3] The relative merits of the terms "excitability" and "irritability" have been of more than passing concern in neurophysiology. They were the subject of a letter from Erlanger to Gasser (10 October 1933) about a question that arose over lunch in Erlanger's department at Washington University, St. Louis. "After a random discussion, the entire group agreed that any differences in the meaning of the words as employed in physiology must be academic." But a few years later they were again corresponding about the problem while preparing their lectures for the Johnson Foundation in Philadelphia (Erlanger and Gasser, 1937).

Fig. 8.8. Two physiologists who continued the strong tradition of German science in the nineteenth century were Ernest Overton (left) and Julius Bernstein (right). Publishing in the same year, 1902, Overton reported that extracellular sodium is necessary to maintain muscle excitability, and Bernstein proposed a theory of the action potential based on selective permeability of the nerve membrane to potassium ions.

his own discovery of the action potential overshoot (Schuetze, 1983). Electrophysiology is enormously indebted to him, however, for suggesting that the inertia-free Braun tube would be of use in the cathode-ray oscilloscope to characterize nerve action potentials (O'Leary and Goldring, 1976, p. 76).

The chemical concept of nerve activity was advanced by an observation of William H. Howell, an American name usually associated with physiology of the circulation. Howell noted the similarity between the effects of potassium depletion and stimulation of the vagus nerve in slowing the heart. He theorized that the nerve fibers "end in what might be designated as an inhibitory substance, which, under the influence of its impulses, is dissociated with the liberation of potassium compounds to which the phenomenon of inhibition is directly due" (1906, p. 293).

The next significant progress was made between the world wars, a period when many technological advances fostered the detailed study of activities along the axon. Among the future "axonologists" in the United States, Alexander Forbes (*see* Chapter 12, in this volume) drew on his wartime experience and built "the first vacuum tube amplifier applied to biology" (Grass, 1980) and used it for recording with the string galvanometer (Forbes and Thacher, 1920).[4] Gasser and Newcomer (1921) introduced the valve amplifier and with Erlanger, Gasser instigated a series of studies of nerve action current that commenced with the discovery of its compound nature (Erlanger and Gasser, with G. H.

[4]Details of the technological developments during this period were authoritatively described by Albert M. Grass in 1980 and again in 1984.

Bishop, 1924). In Britain, Adrian and Matthews developed a capacity-coupled amplifier and inkwriter for recording from the capillary electrometer (Grass, 1984, p. 12) and investigated the Berger rhythm and the all-or-none law of nerve conduction (1934a,b). And in Japan, G. Kato (1924) disproved the accepted theory of decremental conduction in narcotized nerve.

Yet the preceding work dealt with characteristics of the action potential without shedding light on the mechanism, as Andrew Huxley (1995) recently pointed out. Bernstein's theory of membrane permeability to potassium (*see* above) seemed adequate for the moment. Before the Second World War, however, evidence accumulated that subthreshold shocks to certain nerves produced effects (e.g., slow contractions) that marred the all-or-none picture (*see* Rushton, 1977). Just as it became clear that transmembrane measurements were needed to understand the resting and action potentials of nerve, a quantum leap occurred with introduction of the squid giant axon by J. Z. Young (1936). That beautiful animal enabled biophysicist Alan L. Hodgkin and physiologist Andrew F. Huxley, working at Plymouth, England, to insert a capillary filled with saline into the axon and take measurements across the axonal membrane that were "important for two reasons. . . . [T]hey prove that the action potential arises at the surface . . . and they give the absolute magnitude of the action potential as about 90 mV at 20°C." (Hodgkin and Huxley, 1939, p. 711; the complete report did not appear until 1945). Simultaneously, Kenneth C. Cole and Howard J. Curtis (1939) at Woods Hole, Massachusetts, were impaling the squid axon with a metallic microelectrode and making impedance measurements across the membrane; a year later (Curtis and Cole, 1940), they repeated the experiments with a saline-filled electrode, obtaining about the same values for the action potential as had Hodgkin and Huxley. Both teams found the overshoot in the action potential and stated that quantification of the absolute resting potential was not possible, but for different reasons: the English lacking knowledge of the anions inside the nerve fibers, and the Americans having electrode and amplifier limitations. With hindsight, Huxley (1995) judged that Cole's results were better technically, but that he and Hodgkin had taken theirs further by giving mathematical expression to the current's components.

Reminiscing about the postwar research, Huxley (ibid.) explained the inability to recognize the selective permeability of the axonal membrane by evoking the omnipresence of sodium ions, their relatively larger size compared with potassium, and ignorance of Overton's work (which Bernard Katz came across in 1947 and translated into English five years later). Those events, plus the war-time diversion, were all contributory to the delay in their appreciation of the role of sodium flux in the nerve action potential. Two additional technological advances appeared during that period, the reliable glass capillary electrode (Ling and Gerard, 1949) and the "voltage clamp" (Cole, 1949) or feedback circuit to stabilize the membrane potential while measuring the membrane current. The papers that appeared in 1952 (Hodgkin, Huxley, and Katz, 1952; Hodgkin and Huxley, 1952a and b) defined the almost instantaneous, highly selective permeability changes to potassium and sodium, and later, calcium, that account for the nerve impulse. Understanding the mechanism of those changes came much later, beginning with the report of Neher and Sakmann (1976) of the first recordings of currents through single biological channels. The 1952 papers quantitatively analyzing the membrane current were "revolutionary in their impact" and inspired an enormous influx of investigators from other disciplines to neurophysiology (*see* Gardner, 1992). Biomathematical modeling was in the wind and the "cyberneticians" were already taking the nerve impulse into the more complex territory of nerve nets (*see* Chapter 12).

CHEMICAL TRANSMISSION

Leaving the ionic aspects of muscle–nerve conductivity and taking a different approach to neurotransmission, British laboratories at the turn of the century were probing the nature of nerve activity from a pharmacological point of view. In Langley's Cambridge laboratory, Thomas Renton Elliott (1877–1961) was studying the effect of adrenal medulla secretion on nerve and made the first clear statement that transmission of a nerve impulse could be chemical in nature: "Adrenalin might then be the chemical stimulant liberated on each occasion when the impulse arrives at the periphery" (1904, p. xxi). John Newport Langley (1853–1925), in turn, introduced the idea of a "receptor substance," or "synaptic substance," probably not in the nerves, but "in the cells in which they end" (1905, p. 411). As agents acting on those receptor substances, he named not only Elliott's adrenalin but also secretin, thyroidin, nicotine, curare, atropine, and other compounds. Relatedly, the Italian anatomist, Ernesto Lugaro (1870–1940), whose work is often overlooked today, reasoned that because glial processes invest synapses, they must have some relation with neurotransmission, and guessed that they "chemically split or take up" excitatory substances, thus terminating the latter's action. This speculation, with others concerning neuroglial function (*see* Chapter 7, this volume), was made in 1907.

Throughout the first two decades of the twentieth century, chemical neurotransmission in the autonomic nervous system had been only vaguely conceptualized, awaiting a clear demonstration. That materialized with the well known experiment of Otto Loewi (1873–1961), "born during sleep" (Loewi, 1953, p. 35). Loewi had been thinking about neurotransmission years earlier, as the curious story was recalled by J. S. L. Browne (1970, p. 295), and had visited Elliott in 1902, two years before the latter proposed that adrenalin liberated by the nerve impulse acted as a stimulus of muscle contraction. In Loewi's experiment, stimulation of the cardiac branch of the vagus nerve to an amphibian's isolated heart produced *Vagusstoff* in the perfusion fluid which caused slowing when introduced into a second isolated heart (1921; Fig. 8.9).[5] Because Loewi used animals in which the cardiac branch is a mixed nerve containing parasympathetic and sympathetic fibers, he also found that vagus stimulation on occasion produced an accelerating substance. Loewi's professional career was international. Born and educated in Germany, for practical reasons he abandoned his preferred studies in history of art and became a physician, an experience similar to that of his contemporary in Italy, Giuseppe Moruzzi. Turning from clinical work, for nearly 30 years Loewi was professor of pharmacology at Graz, was forced out of Austria by the Nazis, and eventually completed his career in the United States.

The success of the two-toad experiments, providing direct evidence of chemical neurotransmission, intensified the efforts to find pharmacologic and physiologic evidence of neuroactive substances in the entire spectrum of animal species. Perhaps the two busiest laboratories were those of Henry Jallett Dale (1875–1968; *see* Fig. 8.10, p. 168) at the National Institute for Medical Research, Hempstead, England and Walter B. Cannon (*see* Fig. 11.21) at Harvard University, Boston. Both laboratories focused their work on the autonomic nervous system, in which clearcut results were more easily obtained than in the central nervous system. Dale showed in 1914 the obvious relation of acetylcholine's action on the parasympathetic system to that of epinephrine (adrenalin) on the "true sympathetic system," thus reviving interest in "ideas which, by then, had been almost forgotten" (Dale, 1937, p. 230). In spite of Loewi's hesitation to accept the identity of *Vagusstoff* with acetylcholine, this became inevitable when a new method was published by A. W. Kibjakow (1933) from Kazan, "just when it was required" (Dale, 1937, p. 236). Fluid collected after stimulating the cat superior cervical ganglion, on reinjection into the artery of another ganglion (*see* Fig. 8.11, p. 168), produced contraction of the nictitating membrane of the eye.

In Dale's laboratory, Feldberg and Gaddum (1934) confirmed Kibjakow's results and showed acetylcholine to be the responsible substance by five different tests: cat arterial pressure, leech muscle, frog abdominal muscle, frog heart, and rabbit auricle. The pervasive presence of acetylcholine was demonstrated by a series of assays of the animal kingdom by Annette Marnay and David Nachmansohn (1938). They estimated that an acetylcholine splitting enzyme is present in all conducting tissues: nerve and muscle, vertebrate and invertebrate; sympathetic and parasympathetic—the adrenergic and cholinergic systems, so named by Dale.

It is appropriate to quote again from this British pharmacologist's *Harvey Lecture* of 1937 in which Dale stated that the implications of the events and facts so long studied at the periphery were at last being extended to the central nervous system. After noting that John Eccles had used the sympathetic ganglion as "an accessible model of

[5]A good description of the details of Loewi's experiment and the preceding evidence for humoral transmission is to be found in Goodman and Gilman (1970, pp. 408–414).

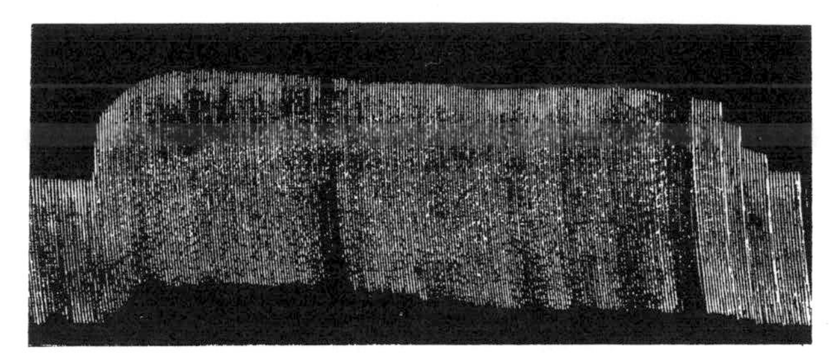

1. 2.
Abb. 3a. Kröte I. 1. Ringer. 2. Vagusreizung.

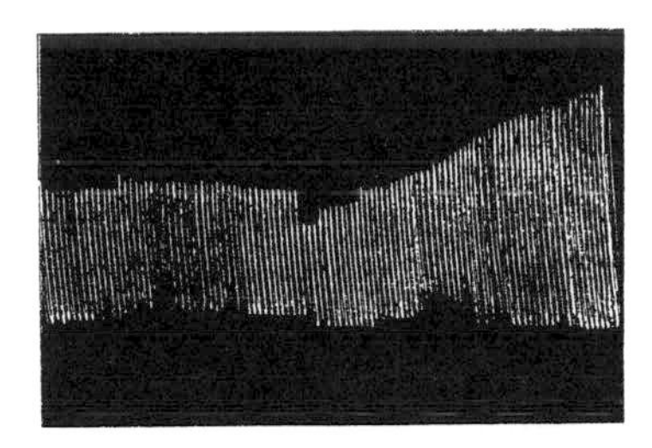

1. 2. 3.
Abb. 3b. 1. Ringer. 2. Ringer aus 25' Normalperiode.
3. Ringer aus 25' Vagusreizperiode.

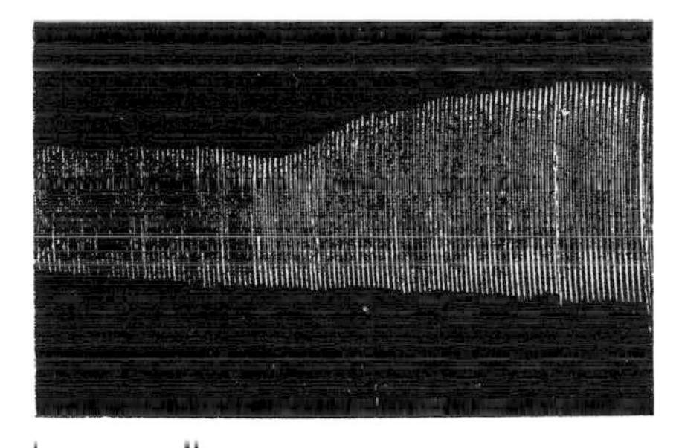

1. 2.
Abb. 3c. Kröte II. 1. Ringer aus 25' Normalperiode von Kröte I.
2. Ringer aus 25' Vagusreizperiode von Kröte I.

Fig. 8.9. Smoked-drum tracings of the movement of a lever attached to an isolated amphibian heart demonstrating neurohumoral transmission. Top—Excitatory effect of vagus stimulation in toad I (2) after baseline perfusion with Ringer's solution. Middle—Also in toad I, perfusion with Ringer's (1,2) and with Ringer's after 25 minutes of vagus stimulation (3). Bottom—In toad II, perfusion with Ringer's from normal period of toad I (1) and with Ringer's from toad 1 after 25 minutes of stimulation (2). (From Loewi, 1921, p. 241, Fig. 3A, B, C.)

the synapses of the central grey matter," Dale wrote that Sherrington "looks upon the transmission of excitation from a motor nerve ending to a voluntary muscle fibre as probably furnishing a pattern, or paradigm, of what happens at a central synapse" (Dale, 1937–1938, p. 243).

The long trail between Galvani's fluid (1791) in the peripheral nerve to the chemical transmission at the synapse (Loewi, 1921) is strewn with bitter arguments; from the denunciation of Galvani's assertion of animal electricity by Volta, who claimed the effects were due only to a bimetallic current (*see* Chapter 7, this volume), to the denial of chemical transmission by Eccles. In his comprehensive analysis of the evidence for the two hypotheses of neurotransmission—electrical versus chemical—Eccles stated: "[T]he presumed chemical nature of the synaptic transmitter in the central nervous system . . . is almost entirely based on an extrapolation from the ACh hypothesis for

Fig. 8.10. Henry Dale, a Londoner, spent time with Langley at Cambridge and Starling at University College, London; it is no surprise that with his training and honors, the brilliant Dale should have been described as "probably the most influential physiologist-pharmacologist of the twentieth century" (Leake, 1970, p. 282). (Portrait (ca. 1918) from Fulton, 1966, between pp. 438 and 439, Fig. 88.)

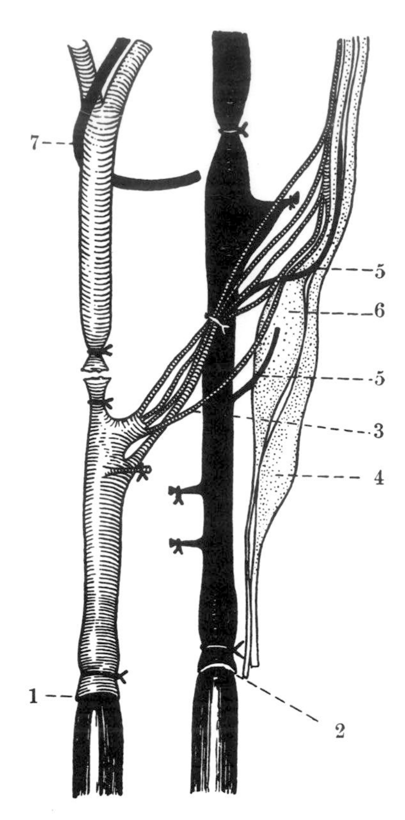

Fig. 8.11. The Russian physiologist, A. W. Kibjakow, invented a method of perfusing the cat superior cervical ganglion, and found evidence for chemical transmission. *See* text for explanation. (From Kibjakow, 1933, p. 435, Fig. 1, ×1.2.)

sympathetic ganglia, for the relevant evidence with the central nervous system is almost negligible" (1936, p. 397). Eccles believed the two ideas were conflicting and "on probation" and that only one hypothesis must cover all neurotransmission.

The "negligible" evidence was provided by Rafael Lorente de Nó, a student of Cajal's and "chief aficionado of the Golgi technique" (Jones, 1984b, p. 20) whose Latin temperament enlivened every discussion in which he took part. Lorente's series of studies on oculomotor (III cranial nerve) stimulation (e.g., 1935) brought questions of neurotransmission squarely into the central nervous system. The final resolution of "spark versus soup at the synapse" came from the gradual accumulation of direct and indirect evidence from many laboratories. An example of the latter type of evidence was reported by Stephen Kuffler (*see* p. 147, this volume). From the results (1946) of intracellular recordings of isolated single nerve-muscle junctions, he could distinguish two systems of transmission at the end plate: slow, tonic contractions with small local end-plate potentials and fast-twitch muscular responses with the typical propagated spike potential. Later Kuffler found

> an irreducible delay between the moment of excitation of the prejunctional nerve twigs and the start of the e.p.p. amounting to about 1 millisecond in the frog at 20°C. This was indeed difficult to reconcile with the basic postulate of electric transmission, namely that the currents produced by the nerve impulse could spread directly to the muscle fiber and excite it. It therefore gave indirect support to the acetylcholine hypothesis (Katz, 1982, p. 242).

The physiology laboratory at Harvard was not as fortunate as Dale's group in the outcome of its careful and detailed work on what was thought to be a hormone "derived from structures under sympathetic control which we suggest be called 'sympathin'," (Cannon and Bacq, 1931, p. 411). Loewi's discovery in 1921 of *Acceleransstoff* in amphibian isolated hearts (*see* above) was extended by Cannon's group to unanesthetized mammals. Cannon was a physiologist and not prepared to suspect that sympathin was related to adrenaline (more precisely, to noradrenaline identified in Sweden by von Euler, 1933).

The resolution of the question of a single versus double mechanism for nerve excitation turned the tide of the "hot" topics in neuroscience. "As soon as it was generally accepted that the effects of nervous excitation were caused by the release of chemical substances, the physiologists found themselves deprived of a vast field which passed into the hands of biochemists and pharmacologists" (Bacq, 1975a, p. 51). Thereafter additional neurotransmitters were identified rapidly in the brain and their excitatory and inhibitory effects were characterized. Ulf von Euler's laboratory found substance P in the brain and smooth muscle (1931) and opened seemingly inexhaustible possibilities into the field of neuropeptides, now another subdiscipline of neuroscience. Serotonin (5-hydroxytryptamine) was isolated from serum in 1947 and seven years later was assigned roles in the brain by two laboratories independently—by Amin, Crawford, and Gaddum (1954), and by Twarog and Page (1954). The ubiquitous inhibitor, gamma-aminobutyric acid (GABA), was found in brain (it was known already in plant and animal tissues) by Eugene Roberts and S. Frankel (1950), who applied the new chromatography technique from Sweden to its isolation and identification.

Additional neurotransmitters and modulators of nerve activity were rapidly identified and their distribution and changes in concentration with behavior and mood in the human brain, as well as comparative studies among species, continue to constitute an enormous reservoir for future research. Among the pioneers in the merging of microneuroanatomy with microneurochemistry was the extremely productive laboratory of Alfred Pope at Harvard's McLean Hospital outside Boston. That group perfected miniaturized methods to quantitatively assay enzyme activities and cellular components of neocortical samples obtained by microdissection and constructed profiles of the distribution of 22 constituents across 20 or more evenly spaced cortical depths (Pope, Caveness, and Livingston, 1952). Prototype profiles are shown in Fig. 8.12, p. 170; the data from normal subjects are useful for comparison with those from diseased or experimental states.

The interdependence of cerebral tissue viability and metabolic substrates has always been an enormous impetus to efforts to understand the cerebral circulation of blood and extracellular and cerebrospinal fluids. Some of the investigations of the intricacies of the blood–brain barrier were described in Chapter 3 on the cerebral ventricles. Here are presented some highlights concerning the more palpable fluid, the blood.

VASCULARITY OF THE BRAIN

The invention of noninvasive methodologies capable of disclosing brain physiology is based on the essential role of blood flow within the cranial cavity. Many centuries before that occurred, however, a relationship between the heart and brain was recognized. Pulsation of the brain was noted in the earliest records of that organ (*see* p. 27, this volume) and was identified with cardiac action: The dura mater "has a movement in time with the pulse" (Rufus of Ephesus, translated in Clarke and O'Malley, 1968, p. 14). Four hundred years previous to Rufus, a structure had been named by Herophilus that would impede the straightforward progress of discoveries about brain circulation—the "rete mirabile." Galen referred to Herophilus and described a dense network of fine blood vessels surrounding the pituitary gland at the base of the ox brain and assumed that this rete mirabile found in brutes also existed in the human. Not until the sixteenth century were doubts raised, notably by Berengario da Carpi, who apologized for not being able to find it in the more than one hundred human brains he diligently dissected: "Thus I believe that Galen imagined the *rete mirabile* but never saw it: and I believe that all others after Galen that spoke of the *rete mirabile* did so on the strength of his opinion rather than their own perception of it" (1521, pp. 459r–v, translated in Clarke and O'Malley, 1968, p. 765). Berengario's 1523 depiction (*see* Fig. 11.1, p. 226, this volume) of the base of the human brain shows no rete mirabile, yet Vesalius published and lectured on it in 1540 (O'Malley, 1964, p. 179). By the time *Fabrica* appeared, however (1543; *see* Fig. 11.2, p. 227, this volume) Vesalius had had second thoughts:

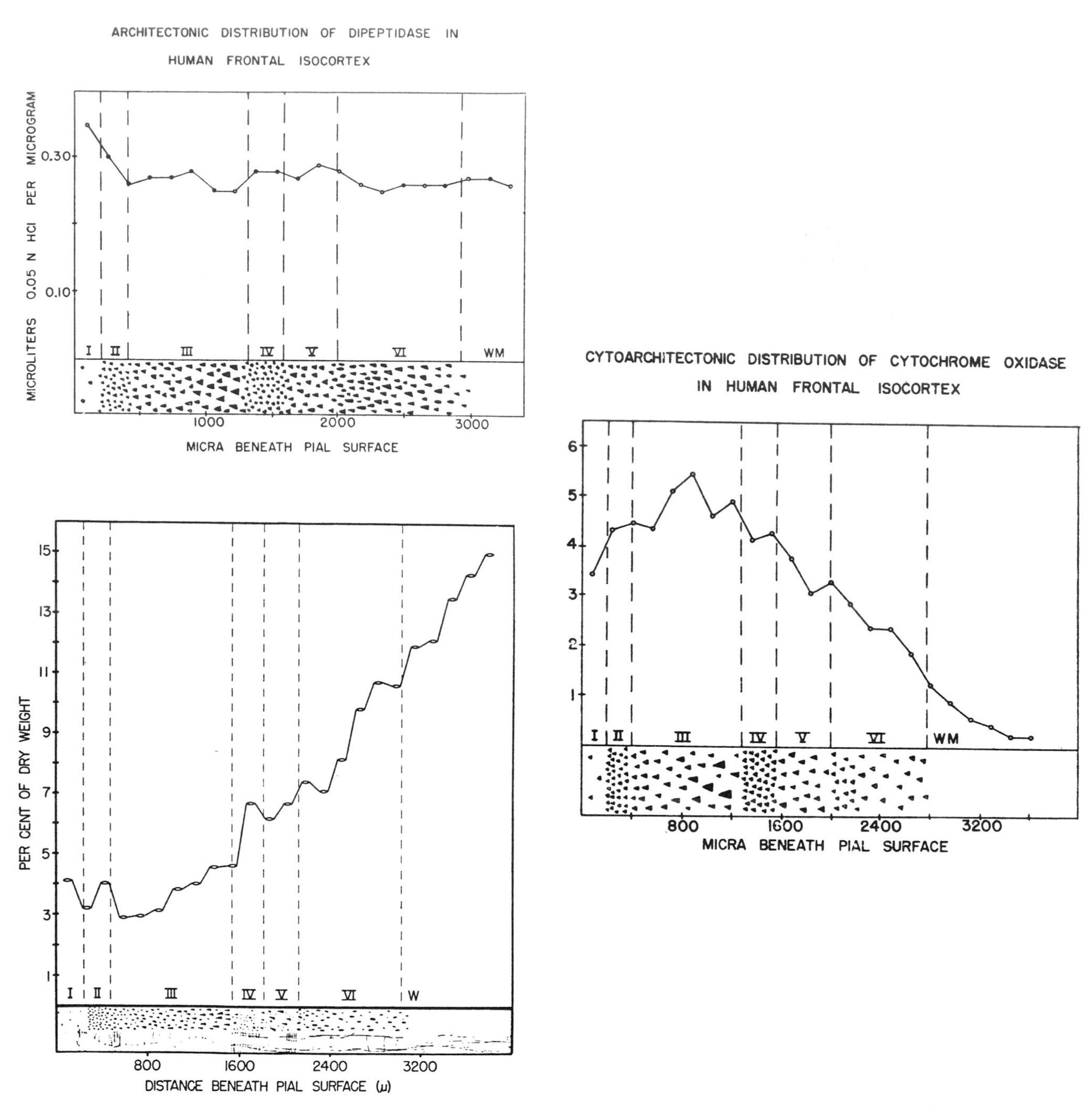

Fig. 8.12. Intralaminar distribution of (top left ×1) dipeptidase, (middle right ×1 1/4) cytochrome oxidase, and (bottom left ×2/3) total cerebrosides in normal human prefrontal cortex. Enzyme activities expressed as units and cerebroside concentration as percentages, both per dry weight. (From, respectively, Pope, 1959, p. 35; Hess and Pope, 1960, p. 211; Lewis and Hess, 1965, p. 215.)

"Indeed, I am now completely astonished at my [former] stupidity and too great trust in the writings of Galen and of other anatomists" (ibid., p. 768). Perhaps the final interment of the alleged human structure was by a Swiss anatomist, Johann Jakob Wepfer (1620–1695). He described the internal carotid arteries as they enter the skull, and asked how could the "little branches" of the rete mirabile join together when the dura mater receives only arteries? (Wepfer, 1658)

Turning briefly from comparative anatomy to the even more complex cerebral physiology, a

TAB. X. LIB X.

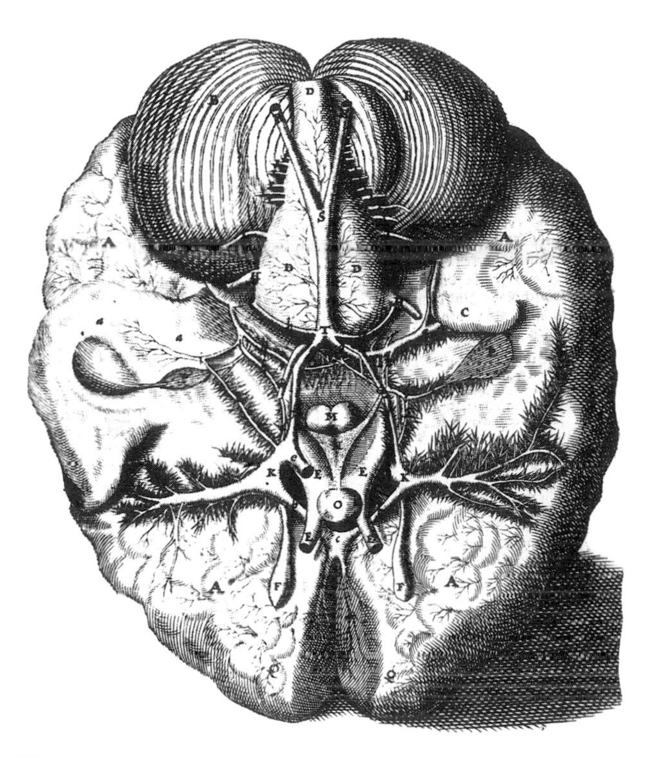

Fig. 8.13. The base of the human brain illustrated by Guilio Casserio, said by Choulant (1945, p. 228) to be the first to draw the great circle of arteries. Note the fuzzy network around the vessels, a remnant of the idea of the rete mirabile in the human brain. (From Casserio, 1627, Tabula X, ×3/4.)

"very interesting scientific speculation" (Kety, 1982, p. 716) was made by an Englishman, John Mayow (1640–1679): "The blood returning to the heart is for the greater part deprived of [oxygen] which it has left in the brain for the production of animal spirit" (Mayow, 1668). Even though the young Mayow has been shown to be unscruplous in his writings (T. S. Patterson, 1931), his statement was indeed far advanced for his time (Cournand, 1982).

The circle of arteries at the base of the brain was first depicted (Fig. 8.13) on copperplates drawn by an Italian, Guilio Casserio (1561–1616) and published posthumously in 1627. "Casserio's plates mark a new epoch . . . owing to the correctness of their anatomic drawing, their tasteful arrangement, and the beauty of their technical execution" (Choulant, 1945, p. 228). When Thomas Willis began investigations of those same blood vessels, he was one of the first to utilize the technique of injecting a substance, dye or India ink, into a major artery and tracing its distribution over cut surfaces. In this manner he explored the fine circulation and worked out the anastomoses of the arteries at the base of the brain that form the "circle of Willis." The focus of the early anatomists on vascularity is evident in Wren's drawings of the dissections shown in Willis's *Cerebri anatome* (1664), in which elegant depictions of the circle are surrounded by stylized gyri (*see* Figs. 4.4 and 9.1, p. 47 and 178, respectively). Clarke and O'Malley (1968, pp. 777–778) note that Willis, thinking about function as well as structure, suggested the circle was a safety device to keep a constant flow of blood to the brain in the event a carotid artery should be occluded.

The persistent interest in the microstructure of the cerebral circulation commenced when the microscope was first used and Leeuwenhoek (1675) described minute vessels going into the pia mater like "roots in earth." Similar studies were carried into the early twentieth century by Richard Arwed Pfeifer (1877–1957), a German experimental neurologist. Using an endothelial vital stain, his work "was technically perfect—there are no finer plates on capillary distribution in the brain than in his *Die Angioarchitektonik der Grosshirundie* (1928)—but he refused to publish his method. 'Work with me for six months . . . and then you will know all about it' " (Haymaker, 1953, p. 25). The essential part the cerebral capillaries play in the blood–brain barrier was described in Chapter 3.

Parallel with the anatomical discoveries related to the cerebral circulation, the physiologists were wrestling with its dynamics. Among them was Alexander Monro Secundus (1733–1817), of the Monro dynasty at the University of Edinburgh, who deduced that the cerebral blood volume within the rigid cranium must be constant, a deduction proven experimentally to be correct by his student, George Kellie (1824). As Kety (1982, p. 718) pointed out, the Monro-Kellie doctrine does not "apply to [regional] blood flow, which may still vary widely within the confines of a constant total blood volume." With direct observation through a glass window, a Dutch ophthalmologist, Frans Cornelius Donders (1818–1889), was the first to report (1850) measurements that evidenced the ability of pial vessels to change their caliber in response to various kinds of stimulation (Clarke and O'Malley, 1968, p. 794).

Based on the Monro-Kellie doctrine, the cranial–vertebral cavity could be used as an oncometer, the basis of a method applied by Charles Smart Roy (1834–1897) and Sherrington (Fig. 8.14) to measure vertical changes in brain volume, which were assumed to reflect altered cerebral circulation. They concluded presciently:

> [T]he chemical products of cerebral metabolism contained in the lymph which bathes the walls of the arterioles of the brain can cause variations of the calibre of the cerebral vessels; . . . in this reaction the brain possesses an intrinsic mechanism by which its vascular supply can be varied locally in correspondence with local variations of functional activity (Roy and Sherrington, 1890, p. 105).

They saw no evidence of intrinsic vasomotor nerves, yet they recorded volume changes that occurred independently of variations in arterial blood pressure. Unfortunately, their elegant experiments and brilliant deductions were overshadowed by the forcefully enunciated conclusions of another British physiologist, Leonard Erskine Hill (1866–1952). He proclaimed (1896) that the cerebral vessels passively follow the systemic blood pressure, behaving "as if they were lead pipes" quoted by Kety (1982, p. 722).

History has blamed Hill for retarding the knowledge of cerebral vascular dynamics for several decades, but with ever-improving methodology, the conclusions of Roy and Sherrington were substantiated and expanded by subsequent investigators. The consensus seemed to be: "This [extrinsic] factor appears to dominate all others. The arterial pressure, in turn, depends on complex reflexes from the cardiovascular centers of the central nervous system, the carotid sinuses, and other reflex mechanisms" (Forbes and Cobb, 1938, p. 202). Regarding intrinsic vasomotor control, their data from direct observation led them to state "that an intracerebral regulation does exist; that in this regulation chemical agents (especially CO_2) play a major part; that cerebral vasoconstrictor nerves . . . are present; the vasoconstrictor nerves are distributed unequally in different parts of the brain" (ibid., p. 213). The important question of oxygen utilization and brain metabolism had become of great interest to those mechanists who hoped to determine the "energy utilization of thought" from measurements of variables such as temperature changes in the brain during different mental tasks.

The experimental trials with animals, in spite of confounding factors such as the presence of the rete mirabile, had set the stage for a reliable method of measuring cerebral blood flow in the human. This was forthcoming from the laboratory of pharmacologist Carl Frederick Schmidt (1893–1988) at the University of Pennsylvania, in association with Seymour S. Kety. Based on the Fick principle, they estimated blood flow by measuring the quantity of an inhaled inert gas, nitrous oxide, removed from the blood as it passed through the brain in a given time. Combined with data on blood oxygen and carbon dioxide concentrations, they calculated the overall metabolic activity of the brain (Kety and Schmidt, 1945, 1948).[6] The method was soon extended to studies of the effects of certain drugs

[6]For a summary of events surrounding the announcement, *see* Kety (1982).

Fig. 8.14. Charles Roy (left) and Charles Sherrington in 1893 at Cambridge University. Their collaboration produced evidence of intrinsic vasomotor activity in cerebral blood vessels. (From Granit, 1966, p. 53, Plate 3.)

and disease states, sleep, and psychic activity in man, and refined, with an inert radioactive gas substituting for nitrous oxide, for autographic estimates of regional blood flow (Kety et al., 1955).

More detailed mapping of regional blood flow changes during various mental and motor activities was achieved by a Swedish group using a method again predicated on the idea that those changes are regulated by the metabolic state of the neural populations involved. From detectors on the exposed brain surface of anesthetized animals the counts of arterially injected krypton-85 indicate changes of blood flow in the underlying cortex (Lassen and Ingvar, 1961). Figure 8.15 (p. 174) illustrates the results in human subjects using Xenon-133 and shows at rest a "hyperfrontal" rCBF and responses to sensory (cutaneous) stimulation and during motor and mental tasks. Like its predecessors, this method yields circumstantial evidence of a correlation of brain function and metabolic activity.

A methodology developed by the Laboratory of Cerebral Metabolism at the National Institutes of Health, Bethesda, Maryland directly discloses cerebral utilization of an energy-rich substrate, 2-deoxyglucose tagged with carbon-14 (Sokoloff et al., 1977). The method is based on the continuously changing use of energy in cerebral cells (and most other tissues as well) coupled closely with the relative rates of uptake which are determined (by autoradiography of serial sections) simultaneously in all parts of the brain. In normal, conscious rats and monkeys, the number of micromoles per 100 g tissue per minute were found to be highest in structures of the auditory system, and the values found

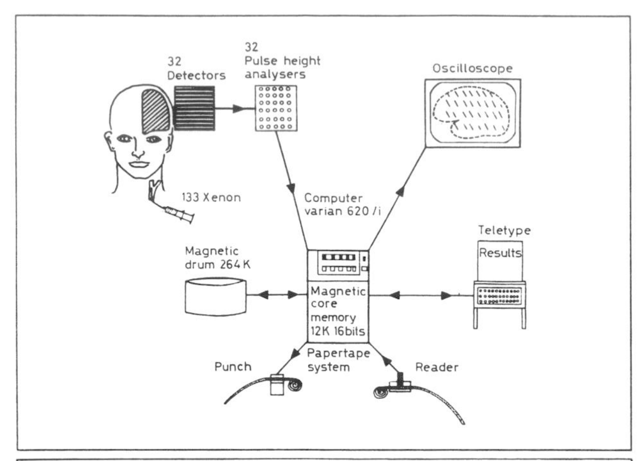

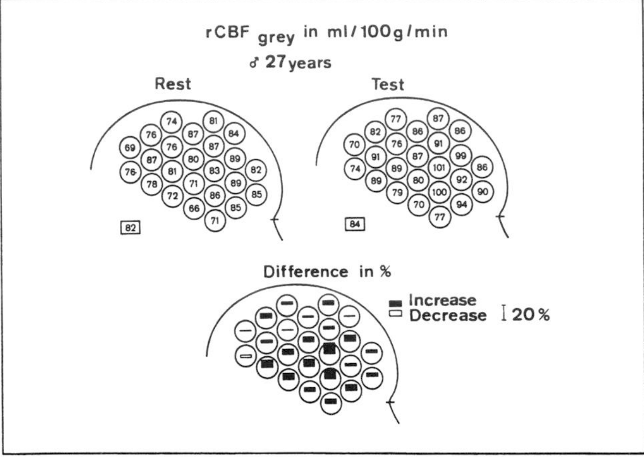

Fig. 8.15. Regional cerebral blood flow estimated from counts of arterially injected xenon-133 in a normal male adult at rest and during a mental task. Areas of large blood flow increase were found in temporal and precentral regions of grey matter. (See text for details. Modified from Lassen and Ingvar, 1972, p. 391, Fig. 6.)

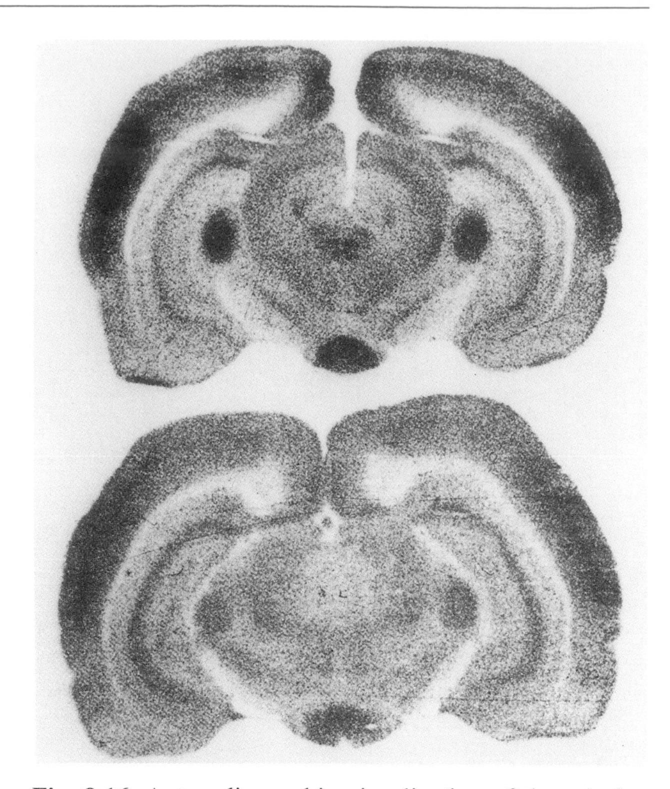

Fig. 8.16. Autoradiographic visualization of the relative rates of glucose utilization in various regions of brain sections from a conscious rat (above) and an anesthetized rat (below). The more intense glucose utilization in the conscious brain is notable. (Digitalized reconstruction from Sokoloff, Reivich, Kennedy, Des Rosiers, Patlak, Pettigrew, Sakurada, and Shimohara, 1977, p. 906, ×3/4.)

in monkeys generally about one-half to one-third those in rats' corresponding structures, probably in accord with differences in the cell density in the two species. Using a derivative of 2-deoxyglucose and positron emission tomography, metabolic mapping of functional activity in clusters of neuronal cell bodies in the central nervous system of man became possible (Fig. 8.16), with vast implications for clinical research and diagnosis.

OVERVIEW OF NEUROCHEMICAL DISCOVERIES

Although the earliest discoveries in neurochemistry were on cerebral tissue—in the eighteenth century, phosphorus and a century later, lipids—during the first half of the twentieth century, ideas on chemical transmission came from work on peripheral neural tissue (the neuromuscular junction and the autonomic nervous system) in subhuman species. The role of ion fluxes in propagation of the nerve impulse was elucidated by midcentury when new technology and a suitable giant nerve cell were introduced. After Loewi proved "slow" chemical transmission in an amphibian, it remained for Dale and his school working on mammalian autonomic nervous systems to find "quick" transmission. Finally, relinquishing his conviction that neural transmission is solely electrical, Eccles, with his associates, demonstrated that these chemical processes occur also in the central nervous system, and identification of a host of new excitatory and inhibitory neurotransmitters ensued. In the words of a Harvey Lecturer:

> The combination of histology with biochemistry is a prototype of a felicitous approach to

> unravelling natural mysteries. The boundary between [them] is not a sharp one but . . . the greatest progress in histochemistry has occurred in studies on distribution of chemical compounds, the histochemistry above all being a topochemistry relating histological structure to chemical composition (Linderstrom-Lang, 1939, p. 214).

Knowledge of the cerebral circulation, on which the neurochemistry of the brain depends, was not forthcoming in breakthroughs, but from bits and pieces of the puzzle. First came the anatomical knowledge, with identification of large and small vessels (Casserio, Willis), and second, the question of how did they function, or what regulated the cerebral blood flow. Again, direct observations, but of the brain surfaces of laboratory animals (Roy and Sherrington, Forbes and Cobb), indicated that regional blood flow is dynamic and responsive to intrinsic and extrinsic factors. Similar responses in human brains shown by noninvasive methods (Kety, Ingvar, Sokoloff) have enabled ever more refined functional mapping and insights into brain metabolism.

9 The Cerebellum

CONTENTS

[T]he Cerebel is a peculiar Fountain of animal Spirits designed for some works, and wholly distinct from the Brain.

(Willis, transl. S. Pordage, 1681, p. 111)

ANCIENT VIEWS AND EARLY EXPERIMENTS

The physical reality of the "little brain" was apparent to the ancients; Aristotle in the fourth century BC wrote: "Behind, right at the back, comes what is termed the 'cerebellum,' differing in form from the brain as we may both feel and see" (transl. D'A. W. Thompson, 1908, p. 494b). Two millennia later, its firm consistency and neat foliate layering were known to occur in all vertebrates in contrast to the dissimilarities of the cerebrum among species: "But the cerebel it self . . . is found almost in all Animals in the same figure and proportion, also made up of the same kind of labels or lappets. . . ." (Willis, 1681, p. 67). During the centuries between those two characterizations of the mammalian cerebellum, many pronouncements regarding its function were made based only on circumstantial evidence. The fatal outcome of occipital wounds was related to the contemporary belief that the cerebral ventricles were the dwelling of the animal spirits. The fourth ventricle was thought to be the most essential to life (*see* Chapter 3), and because of its position between the cerebellum and the medulla oblongata, this ventricle was considered a subordinate part of the cerebellum.

In the second century AD, Galen believed the cerebellum to be the origin of motor nerves and spinal cord and the vermis was seen as a valve to regulate the flow of animal spirits within the ventricles. Galen's views persisted unchallenged into the sixteenth century before they became vulnerable to a new method of dissection introduced by the Italian Constanzo Varolio (1543–1575). He removed the brain from the skull and turned it over, thus enabling a better view of its base than was possible using horizontal slices viewed from above. Varolio recognized the pons as a connection between cerebellum and cerebrum and reasoned that because the auditory nerve seems to arise from the pons it must originate in the overlying cerebellum where hearing is therefore localized. He further thought that the cerebellum is associated predominantly with movement and the cerebrum with sensation. By the midseventeenth century, Johann Vesling (1598–1649) in Padua and others had assigned an additional function to the cerebellum—they designated it the seat of memory. Far from being a new concept, this was a revival of the Galenic tradition,

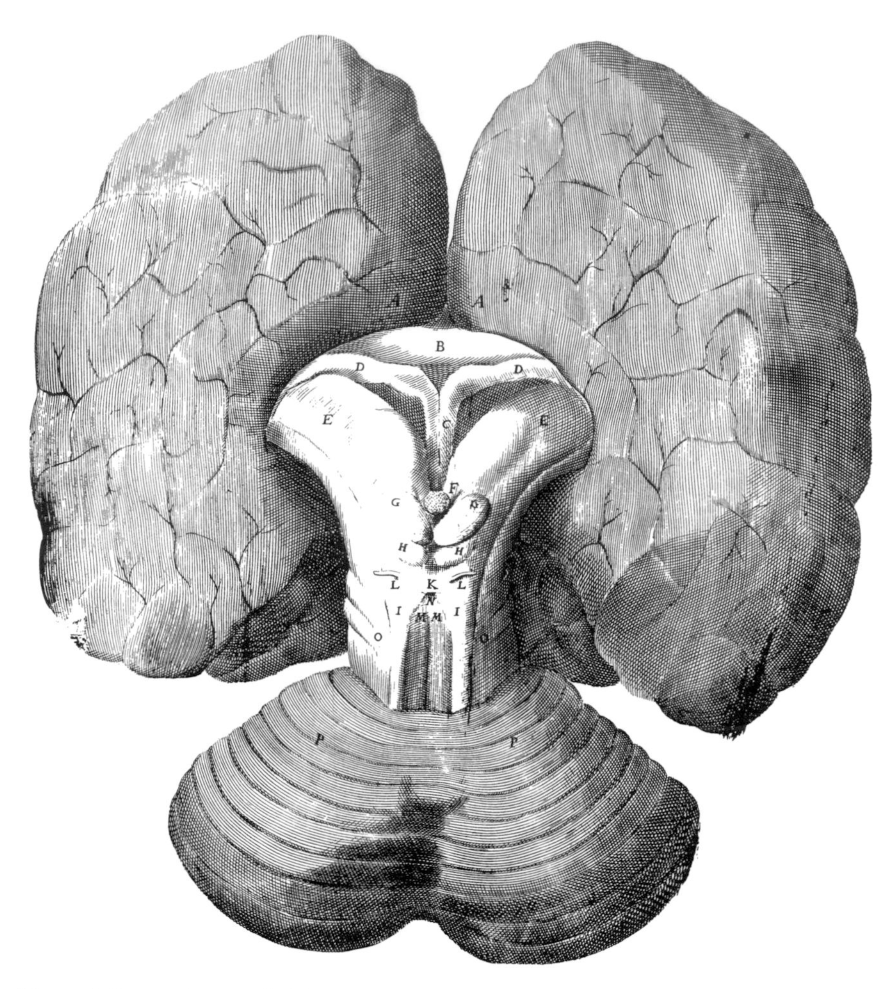

Fig. 9.1. Engraving, probably drawn by Christopher Wren, of Willis's dissection of a human brain from below. In the center is the brain stem, exposed after the cerebellum was separated and turned back. B—Corpus callosum; C—fornix with limbs DD; EE—peduncles; F—pineal body; G,H—quadrigeminal bodies; LL—4th cranial nerves. (From Willis, 1664, facing p. 26, Fig. III.)

based on ideas of the neoplatonists, which had been absorbed and refined by the Arabs and returned to the western world.

The enormous contributions of Thomas Willis to knowledge of the brain made in the mid-seventeenth century included not only the comparative studies already described but also experiments on the cerebellum (Fig. 9.1). Willis assigned the highest control of involuntary movement and vital functions (heartbeat and respiration, for example) to the cerebellum, functions that could be tested physiologically, and "it was the spirit of the times to experiment" (Neuburger 1981, p. 17). In his most famous book, *Cerebri Anatome*, published in 1664, and in English translation in 1681, Willis also assigned a talent for music to the cerebellum.

The following year, Willis's conclusions were challenged by the respected young Danish anatomist, geologist, and later theologist, Nicolaus Steno (1638–1686; Fig. 9.2). In a lecture delivered in Paris before the group of men who became the founders of the Académie Royale des Sciences, Steno included sweeping criticism of both Willis's work and the mechanistic ideas of Descartes. He acknowledged Descartes's introduction of a new method of reasoning based on facts but deplored his and Willis's tendency to blind repetition of unconfirmed antecedent concepts and depictions. The Dane pointed out that Sylvius had noted the discrepancy between some of Descartes's descriptions with what was seen in dissection, and cited the latter's version of the pineal gland as a damning

Fig. 9.2. Nicolaus Steno, although his origins were in Denmark, was painted about 1666 while he was in Florence, showing a young, self-confident man. The portrait is by an unknown artist and hangs in the Uffizi Gallery, Florence.

example of his errors: "The spirits run from all sides of the gland into the cavities of the brain. The gland may perform its functions, though it be inclined sometime to one side, sometimes to the other" (Descartes quoted by Steno, 1669, p. 12).[1] How justified Steno was in his criticism can be judged from the crude, derivative drawing reproduced in Fig. 3.10., *see* p. 37, this volume) In his own dissections, Steno had found that the pineal was not situated in the ventricles nor was it moveable from side to side, instead it was firmly attached to the dorsal surface of the brain. When the lecture was published four years later, and in 1733 translated into English by George Douglas, "[i]ts far-reaching effect was guaranteed . . . by the qualities of the author, who was so knowledgeable in anatomy and physiology . . . and by the clarity of a presentation that united burning enthusiasm with cold irony" (Neuburger, 1981, pp. 22–23). Steno suggested better ways to carry out meaningful scientific investigations that would "tear off the glaucoma from the eyes of those who calmly accept the opinions of past ages" (Steno, quoted by Dewhurst, 1968, p. 46).

Steno was probably one of the first experimentalists to point out that examination of an object with an instrument or a method may inadvertently alter it and so the result may not represent the true object. He rejected the traditional idea that the brain is a gland (*see* Djorup, 1968) and wrote that the cerebellum can be studied best by dissecting the fiber tracts leading to and from it. His drawing of the human brain (Fig. 9.3) shows an artistic quality lacking in most contemporary images.

[1] Translated from the original French by G. Douglas (1733, p. xiii).

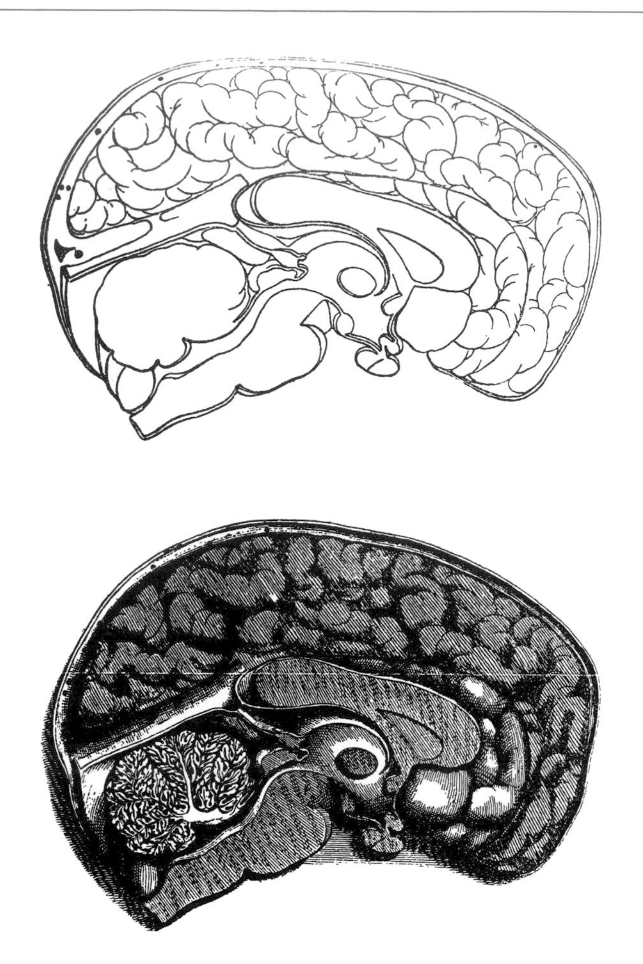

Fig. 9.3. Nicolaus Steno's excellent illustrations warrant special note because they were artistically superior to those of his predecessors as well as more accurate. The four illustrations in his treatise have no labels or captions, but the outline drawings that accompany them suggest that he planned to label the structures and perhaps was rushed into print before doing so. (From Steno, 1669, facing p. 60, ×4/5.)

Although Willis is credited with initiating the experimental study of the cerebellum, perhaps overly so, he can be faulted for concluding from his experiments that the cerebellum is essential to life. The error may be forgiven, however, for a combination of the organ's proximity to the medulla oblongata and technical crudeness compromised survival in animals in which the cerebellum was bilaterally or unilaterally removed. Some historians hold that such poor outcomes discouraged further experimental work on the effects of cerebellectomy (*see*, for example, Neuburger, 1981, p. 259) but more skillful dissection and the drive to determine just which portion of the brain is essential to life seem to have been stimulated by Willis's efforts. Indeed, during the following century, reports of animals surviving, if only for a few hours, after careful removal of the cerebellum, came from France and Italy: Du Verney, 1673 (pigeon, cited by Preston, 1697, p. 461); Pourfour du Petit, 1710 (dog); Kaau-Boerhaave, 1745 (dog); and Morgagni, 1761 (kitten). All found their animals did not immediately expire and their locomotion was difficult, but only Anne-Charles Lorry (1726–1783) noted a disturbance in equilibrium. This astute Parisian experimentalist recorded (1760) his observation of swaying in cerebellum-damaged pigeons without appreciating its significance as a sign of cerebellar dysfunction.

Raymond Vieussens (1641–1716), a French physician educated in Montpellier, was well known in the late seventeenth century for his *Neuro-*

graphia Universalis (1684), which became a publication landmark in neuroanatomy. He commenced dissection of the brain at its base and gently scraped the tissue aside to follow nerve fibers from the medulla forward into the cerebrum, giving a detailed account of the external appearance of the cerebellum (Fig. 9.4), as well as its internal gross features. A century later, an Italian, Michele Vicenzo Giancinto Malacarne (1744–1816), presented advances in the structural knowledge of the cerebellum in his *Nuova esposizione della struttura del cerveletto umano* (1776), the first publication that dealt solely with that organ, describing its gross morphology in great detail and applying names that survive today. The author was handicapped, however, as were all anatomists at the time, by the use of inadequately fixed tissues (*see* Chapter 7). The introduction by Reil (1809) of alkali to the alcohol solution in which tissue was hardened for examination increasing blunt dissection finally made it possible to explore interior brain structures with confidence.

The old idea that the cerebellum controlled vital functions and life itself, held since Willis's time, was finally laid to rest by the great Swiss medical scientist, Albrecht von Haller (1708–1777; Fig. 9.5). He and his followers, operating with great care on cats, dogs, and frogs, concluded that the cerebrum and cerebellum were interchangeable in function, thus stifling any early thoughts of localization (Neuburger, 1981, p. 43). The beautiful copper engravings of the cerebellum (Fig. 9.6) of Johann Christian Reil (1759–1813), a German *Naturphilosoph*, were published between 1807 and 1809 to illustrate papers written in a "rhapsodic manner" (Schiller, 1970b, p. 64) befitting the popular romanticism of the time. Reil nominated the term *nucleus* for a cluster of cells within brain tissue, and described the vermis in man and other animals together with the peduncles and the medullary and caudate nuclei (1807–1808a,b). As Clarke and O'Malley (1968, p. 647) noted, his correlations of cerebellar complexity with ascent of the phylogenetic scale were the most important such observations to that time, so much so that Reil rendered them in italics: *"For it seems that the mere intensity of brain capacity determines the quality and the difference of its function, and again its intensity grows proportionally with its extension and increase in volume in exactly the same way in which the effects of electricity change according to the different degrees of its strength"* (ibid., p. 649).

The technical improvements in anatomical methods had exacerbated the disparity between advancing knowledge of structure and the lagging physiological discoveries of the late seventeenth and early eighteenth centuries. With the advent of the great experimentalists such as von Haller and Flourens, the gap began to narrow. Their discoveries signaled that at last cerebral physiology was catching up with its anatomy and ushered in the productive excitement in "exact knowledge" of the brain that characterized the sciences in the nineteenth century. An outstanding example was the work of "the most colorful figure in the story of the cerebellum" (Fulton, 1949, p. 107), Luigi Rolando (*see* Fig. 4.21, p. 60, this volume). Having followed his exiled emperor to Sardinia, he was working and publishing (1809) in isolation his observations on the cerebral convolutions as mentioned in Chapter 4. The thin and regular folds (lamination) of the cerebellum, resembling a voltaic pile, coupled with the contemporary interest in biological electricity, led Rolando to the idea that this organ functions as an electric motor, controlling movement, especially of the ipsilateral limbs.

By the time the results of Rolando's relatively crude, acute experiments had been translated into French and become known, acceptance of the ideas of Flourens was in ascendancy. Flourens's superior technique assured his animals' longer survival after cerebellectomy and he confirmed Rolando's demonstration that sensation and intellectual functions are unimpaired and the vital activities unchanged after the procedure. "I have shown," Flourens concluded,

> that all movements persist after ablation of the cerebellum; all that is missing is that they are not regular and coordinated. From this, I am led to conclude that *production* and *coordination* of movements consist of two essentially distinct orders of phenomena as well; namely *coordination* in the cerebellum, *production* in the *spinal cord* and *medulla oblongata* (1824, p. 297, footnote 1; transl. M. V. Anker).

Flourens clearly assigned a modulating or regulatory action of voluntary movement to the cerebellum, in contrast to Rolando's idea that the cerebellum has total control of movement.

Like Rolando and Flourens, François Magendie (1783–1855) was interested in the cerebellum (1824c) as part of a larger program of research on

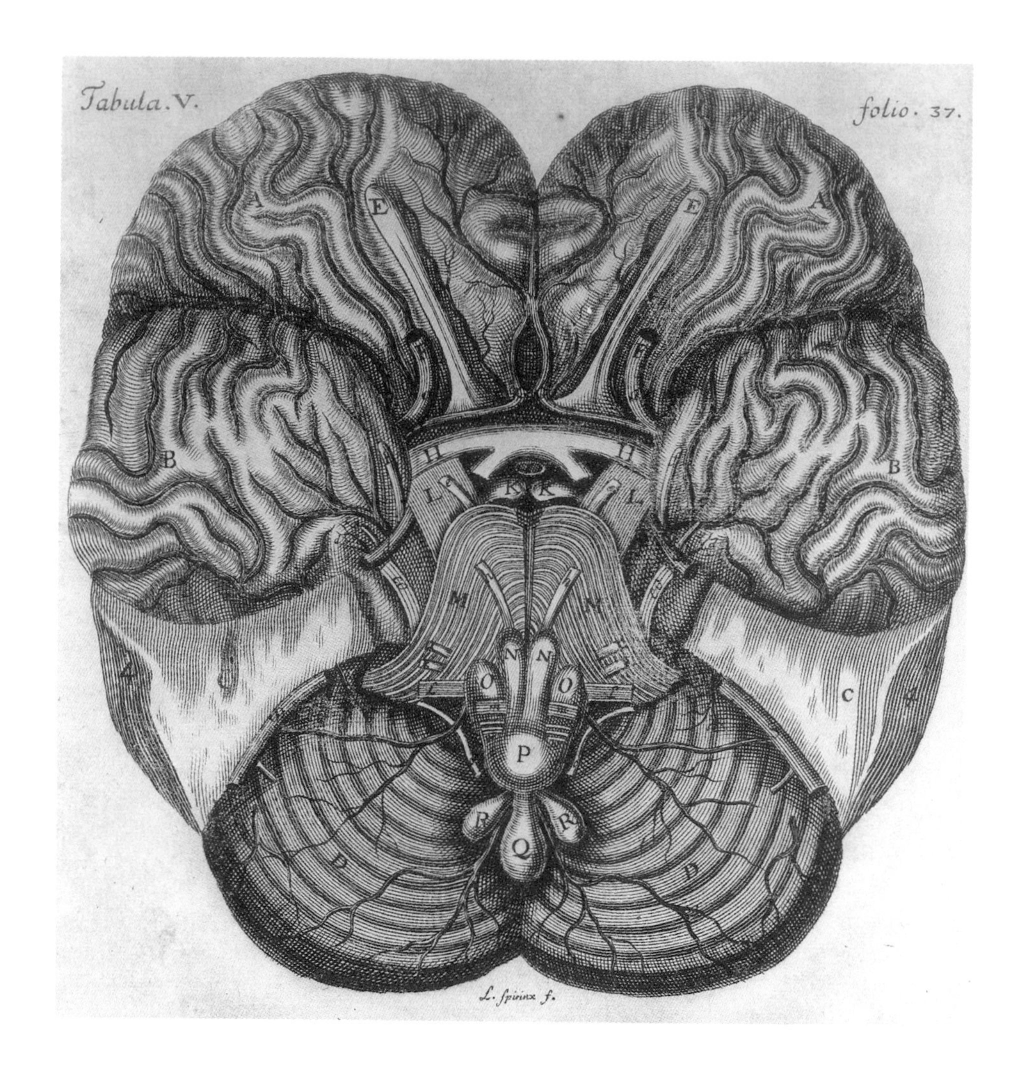

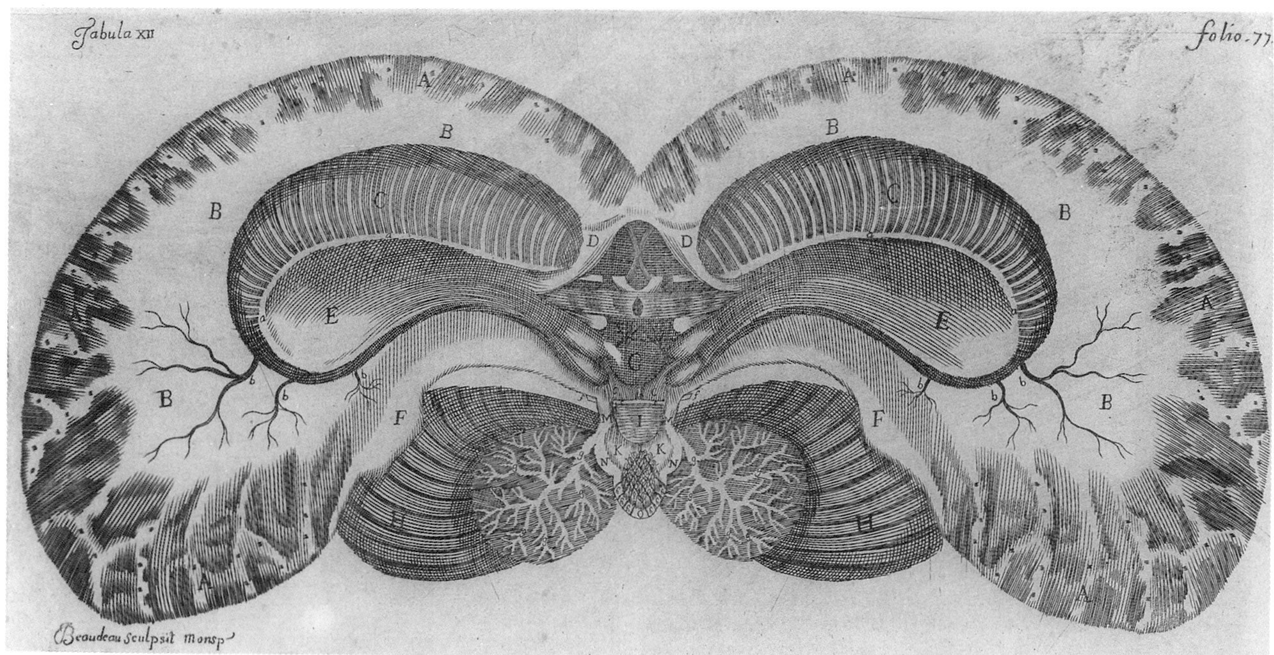

Fig. 9.4. Illustrations of the cerebellum by Vieussens. Above—The underside of the human brain showing the laminated cerebellum (D) to have a surface very different from that of the cerebrum. (From 1685, facing p. 37, plate V, x 4/5). Below—Coronal section of the human cerebellum; g—medullary tracts; K—fourth ventricle with choroid plexus; N—inferior cerebellar peduncle. (From 1685, facing p. 75, Plate XII, ×2/3.)

Fig. 9.5. The many images of Albrecht von Haller invariably show him wearing a wig. In this casual portrait, from an obituary privately published in Bern in 1777, he appears comfortable, alert, and wise, as befits "[p]robably the most outstanding medical scientist . . . of the eighteenth century" (Clarke, in Neuburger, 1981, p. 327.)

the nervous system. In Magendie's Paris laboratory, Michele Foderá, a visiting Italian experimentalist, saw extensor hypertonia after acute cerebellar lesions (1823), an important first observation that has been largely and undeservedly overlooked, (Clarke and O'Malley, 1968, p. 657; Dow and Moruzzi, 1958, p. 5). Magendie himself (1824b), noted the disturbance of equilibrium after ablation of the cerebellum, reminiscent of the report made in 1760 by his compatriot, Lorry (*see* p. 180).

THE RIDICULOUS AND THE SUBLIME

The fourth decade of the nineteenth century, when momentous contributions to cerebellar fine structure were made by Purkyně and Valentin (Chapter 7), was also the decade of what may be deemed Gall's most ludicrous error of localization—the assignment of sex to the cerebellum—a revival of the old idea of the cerebellum as the seat of sensation. In a cartographic scheme of psychological characteristics on the skull ("phrenology"), Gall designated the cerebellum as the source of the generative instinct, and later renamed it the organ of "amativeness." Whether or not from his own generous endowment with this faculty, it was assigned the number one position in his list of 27 faculties (*see* Fig. 4.16, p. 57, this volume) mapped on the skull as indicative of the underlying brain surface. From his flourishing clinical practice, Gall extracted cases which seemed to confirm this relationship and in a monograph, *On the Functions of the Cerebellum*, (written in collaboration with Vimont and Broussais and translated from the French by George Combe in 1838), the section on the "History of the Discovery that the Cerebellum

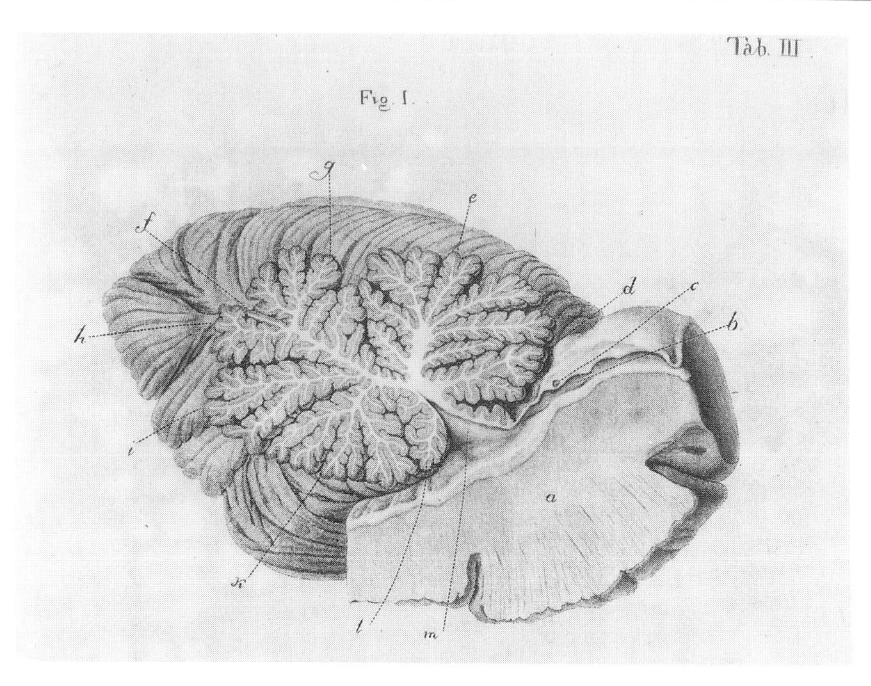

Fig. 9.6. A copper engraving published by Christian Reil of a midline sagittal section through the human vermis. a—Medulla oblongata and pons, b—aqueduct of Sylvius, m—fourth ventricle. (From 1807–1808b between pp. 144 and 145, Plate III, Fig. 1, ×1.)

is the Organ of the Instinct of Reproduction," contains the following passage:

> A young widow, a short time after the death of her husband, was attacked by melancholy and violent convulsions . . . preceded by a very disagreeable tension and heat in the nape of the neck. A few moments afterwards, she fell to the ground in a state of rigidity, . . . [with] the nape of her neck and vertebral column strongly drawn backwards. The crisis . . . was accompanied by a convulsive voluptuousness, and a real ecstasy; after which she continued free from attacks for some time. I frequently supported her during the crisis with the palm of my hand on the nape of her neck, and felt at that spot a great heat. . . . [T]he idea could not escape me, that a connection might exist between the functions of physical love and those parts of the brain situated at the nape of the neck. In a short time, I collected a prodigious number of facts in support of the idea (ibid., pp. 13–14).

Fourteen years after Gall's death, Flourens published the first edition of *Examen de la phrénologie* (1842), which is widely assumed to have dealt the death blow to phrenology. The substance of the treatise was epitomized in the first paragraph:

> Gall's whole doctrine and all phrenology rests on the organs of the brain for, without distinct cerebral organs, there can be no independent faculties and, without independent faculties, there can be no phrenology. But Gall does not say nor has any phrenologist said for him, what is the thing called a cerebral organ.
>
> [H]e did what so many others have done. He commenced with imagining a hypothesis, and then he imagined an anatomy to suit his hypothesis (Flourens, 1845, transl. Meigs, 1846, pp. 74–75).

In the preface to *Examin*, dedicated to the memory of Descartes, Flourens stated: "Each succeeding age has a philosophy of its own. The seventeenth century recovered from the philosophy of Descartes; its eighteenth recovered from Locke and Condillac; is the nineteenth to recover from that of Gall?"

In that treatise, Flourens referred twice to the work of François Leuret, whose elegant publication (1839) on the convolutions of the brain (Chap-

	N	Brain (g)	Cerebellum (g)
Stallions	10	534	61
Geldings	21	520	70
Mares	12	498	61
Total	43		
Average		517	64

Fig. 9.7. Table of average weights of brains and cerebella of horses found by Leuret (1839, pp. 428–429) proving that Gall's placement of sex in the cerebellum was in error.

ter 4) included some unique quantitative data on the cerebellum of the horse that refuted Gall's localization of sex in that organ. More focally, Leuret was testing Gall's statement that castration can cause a notable reduction in the organ's volume. Having chosen the horse for his research, the only source that could supply Leuret with an adequate number of these large animals, without his going into bankruptcy, was a veterinary school. Fortunately, the Royal Veterinary School had been established in 1766 at the town of Alfort near Paris. Moreover, its primary attention was to study the anatomy and diseases of horses: first, because of the preponderance of horses in the environs of Paris in that period and second, because each regiment of the Royal Cavalry could send an officer there for veterinary instruction.

With remarkable enterprise, Leuret persuaded two men at the Alfort school to attend the postmortem examinations of horses, to remove and weigh their brains, and then to dissect out and record the cerebellar weights. When Leuret's monograph was published, he had assembled data on the brains and cerebella of 43 horses. The average weight of the geldings' cerebella, 70 g, was not only above the total average, but it exceeded that of both stallions and mares, each of which was below the average (Fig. 9.7). As Leuret asserted:

> Neither Gall nor his partisans can raise objections to these findings, for none of their assertions concerning the locations of the faculties is based on objective and quantified data such as these. Instead of utilizing the meter and the scale to obtain the facts, Gall preferred simple inspection, using the words: "bigger, smaller, enormously developed, it is easily seen" and others of as little precision which, in reality, have no value at all (Leuret, 1839, p. 430; transl. M. V. Anker).

It does not seem incongruous that the influential Flourens, a precise conservative, should have taken a small detour from his main interests to demolish the ideas of the profligate and free thinking Gall. As for Leuret, he seems to have been interested primarily in good experimental science. Fortunately, his brief excursion into the localization of sex in the cerebellum did not deter him from completing the first volume of *Anatomie comparé.*

The contradictory opinions about cerebellar function were resolved somewhat by the work of Luigi Luciani (*see* Chapter 5), whose expertly decerebellated dogs and apes lived for as long as a year. With such success, Luciani could see the long-term effects and proposed (1891) the accepted theory of cerebellar regulation of tonic and static motor activity. From his experimental results, he assigned a generalized (unitary) refining role ("facilitating effect" in Clarke and Jacyna, 1987, p. 302) to the cerebellum that acts on voluntary movements initiated by higher centers. Luciani dedicated his treatise, *Il cerveletto* (1891), to Rolando, aligning himself with the foes of parcellation of cerebellar function.

INSIGHTS FROM PHYLOGENY AND EMBRYOLOGY

Meanwhile, in Germany the cerebellum was the object of much more factual study of anatomy and embryology. In the English translation of the fifth edition of Edinger's world famous anatomy of the human central nervous system, we read:

> Nowhere else in the animal kingdom does the Vermis cerebelli reach such enormous development as in the great swimmers and the birds. This circumstance, together with the fact that in the same animals there are especially large connections with the tonus nerves of the labyrinth and with the Trigeminus, makes it most probable that in some way or other the cerebellum must be involved in the maintenance of equilibrium. . . . The results of physiological experiments indicate the same thing (Edinger, 1899, p. 102).

Thus did Edinger tentatively confirm a second cerebellar function first noted by Lorry in 1760

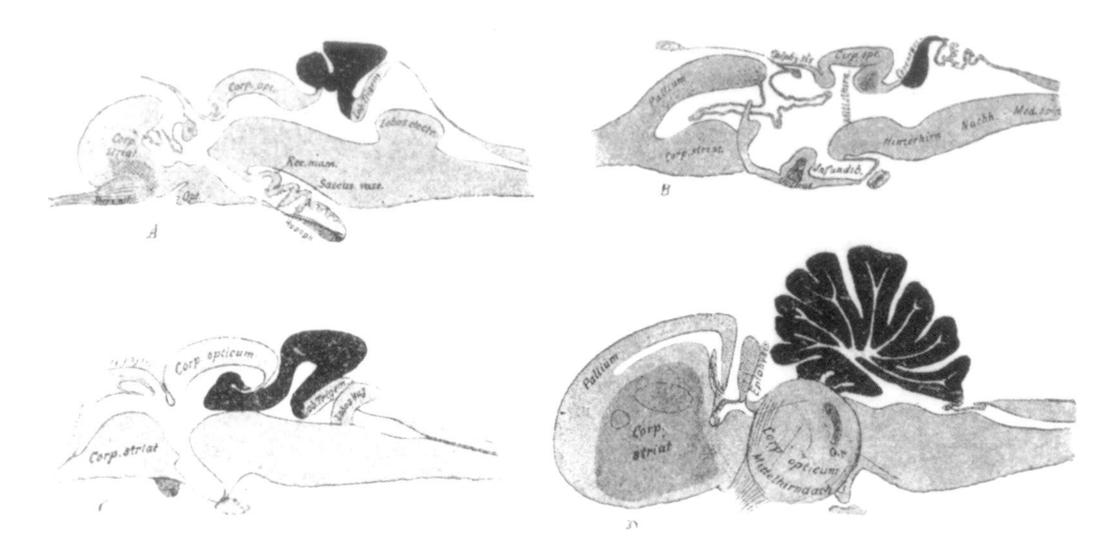

Fig. 9.8. Semidiagrammatic sagittal sections through vertebrate brains, showing the relative size of the cerebellum (in black). A—Ray, B—amphibian, C—embryo of trout, D—bird. Note in C and D the enormous development "in the great swimmers and the bird." (From Edinger, 1899, p. 102, Fig. 55.)

and again by Foderá, Magendie, and Bouillaud in the 1820s.

Edinger identified the cerebellum as phylogenetically one of the oldest parts of the brain (Fig. 9.8). His detailed studies of its anatomy and morphology (*see* Fig. 9.9 and Fig. 9.10, p. 188), combined with the clinical and postmortem observations, led him to make confident statements about function:

> In the cerebellum we have an organ into which nerve-tracts enter from the interbrain, the midbrain, the medulla, and the spinal cord; an organ that in mammals is also indirectly connected with the cerebrum. Into this organ pass bundles from several sensory cranial nerves, especially from the nerves of equilibrium. . . .
>
> It is easily conceivable that in this range of possible connections with tracts from almost every part of the brain is laid the foundation for *coordination of movements* and for the *maintenance of muscle-tonus*: functions which must be ascribed to the cerebellum (ibid., p. 110).

The comparative approach to the anatomy of the enigmatic cerebellum taken by Edinger was also followed in America by two brothers and their associates in the midwestern United States. The elder of the pair (Fig. 9.11, left), Clarence Luther Herrick (1858–1904) in 1891 set the course of their studies into comparative neurology with two publications: "The evolution of the cerebellum" in *Science* and "Illustrations of the architectonic of the cerebellum" in the inaugural volume (Haines, 1991) of the *Journal of Comparative Neurology* of which he was founding editor. His younger brother, Charles Judson Herrick (1868–1960; Fig. 9.11, right) later characterized this published account of the embryonic development of the cerebellum as "the first clear demonstration of its bilateral origin. . . ." (C. J. Herrick, 1924, p. 641, footnote 18). After Clarence's death from tuberculosis at age 46 (Windle, 1979), his brother at the University of Chicago carried on the work so brilliantly initiated and established comparative neurology on solid footing in the United States. The younger Herrick published studies of the cerebellum of *Necturus* (1914a) and the medulla oblongata of *Ambystoma* (1914b) which yielded data for his statement that "in the simplest vertebrates the vestibular centers of the medulla oblongata are the cradle of the cerebellum" (ibid., p. 29). Herrick summarized his work on the tiger salamander and other lower forms with the statement that in mammals the newer part of the cerebellum

> has been elaborated parallel with and in physiologic relationship with the cerebral cortex. The new or central part . . . is very small in the lowest mammals . . . and in higher forms it increases

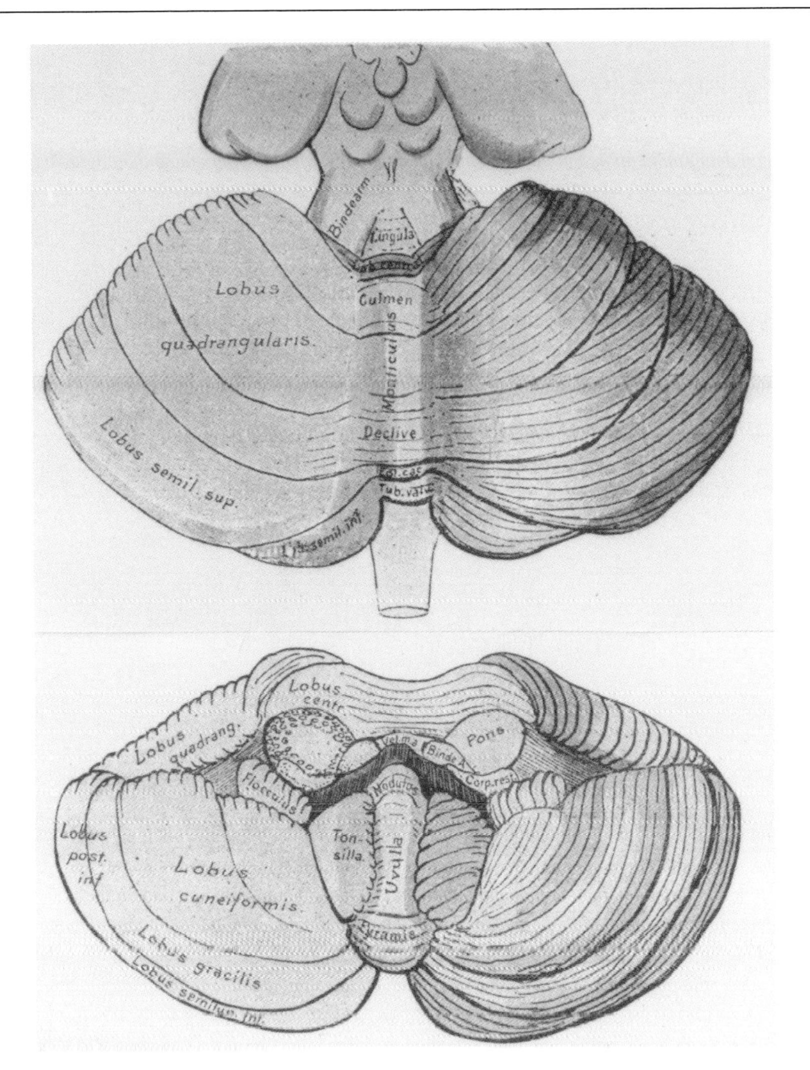

Fig. 9.9. Dorsal (above) and ventral aspects of the human cerebellum with lobes precisely labeled. (From Edinger, 1899, pp. 316, 317, Figs. 203 and 204, ×1.)

> in proportion to the enlargement of the cerebral cortex, until in man it comprises by far the larger part of the organ. . . .
>
> The new part is developed in correlation with the cerebral cortex with which it is in functional relation through the pons and otherwise, probably regulating movements of cortical origin in much the same way that other parts of the cerebellum regulate the simpler reflexes arising from vestibular excitation, muscle sense and other proprioceptive receptors" (Herrick, 1924, pp. 642,643).

Herrick pointed out that the human cerebellum (Fig. 9.12) is highly qualified to be "the great adjustor:" it has a wide diffusion of afferent fibers, exhibits summation and reinforcement of impulses, can deliver rapid discharge of "large batteries of Purkyně cells," and histologically contains no mosaic of cells and, hence, no localization pattern. In fact, Herrick believed that the cerebellar cortex is a representation of the efferent fiber-tracts that happen to lie below it, in contrast to Hughlings Jackson's speculation that the cerebellar cortex is arranged similarly to, but in reverse of, the areas on the cerebral cortex.

When Clarence Luther Herrick went to Leipzig to join the group around Wilhelm Wundt, the most famous philosopher–psychologist of the era, he filled his spare time by translating a popular treatise by the philosopher, Hermann Lotze, and added to the English published edition a long description of the human nervous system (C. L. Herrick, 1885). That slim volume has been said (Windle, 1979) to

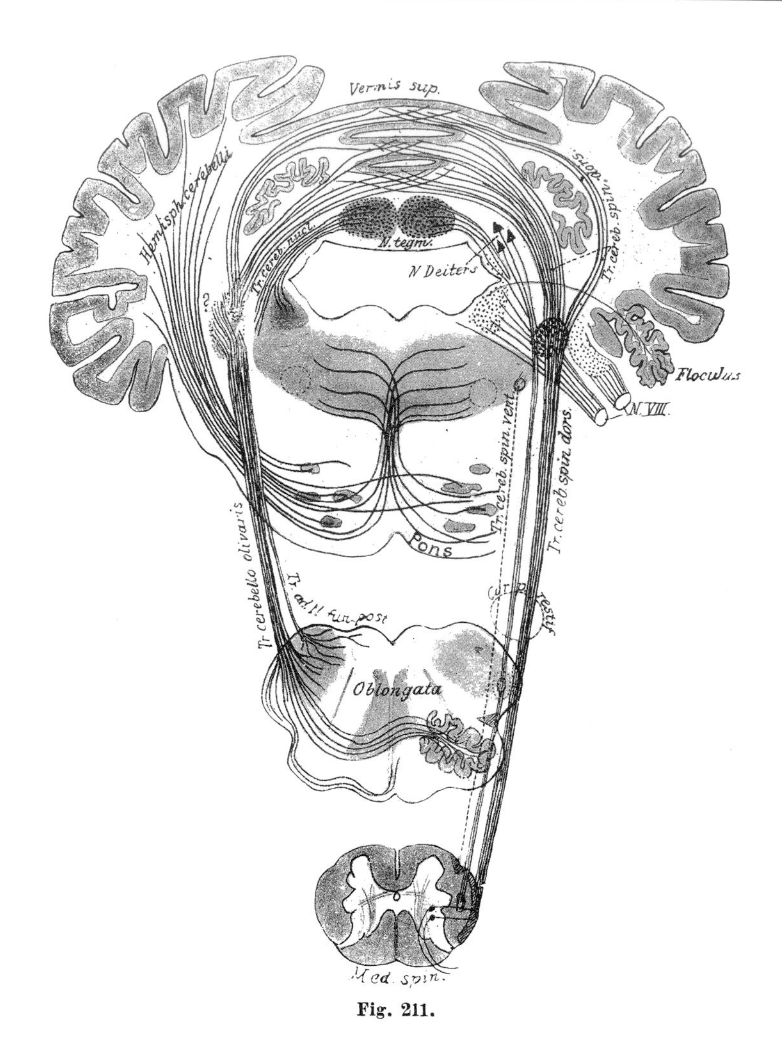

Fig. 9.10. A schema of the origin and course of fibers of the peduncles of the cerebellum, showing "the most important facts at present known regarding the connections of the cerebellum" (Edinger, 1899, p. 324). Especially useful was work directed to tracing degeneration after total or partial extirpation of the cerebellum. (From ibid., p. 325, Fig. 211, x1.)

mark the introduction of physiological psychology as an attempt to understand the close two-way association of brain and behavior. Although many subsequent researchers continued to look at neural and psychological concepts as separate domains, a few strong pioneers dared to acknowledge a multidisciplinary relationship, among them Sperry, Eccles, and Pribram (Lacey, 1985).

CEREBELLAR MAPPING

During the last decade of the nineteenth century, London's National Hospital at Queen Square was graced by the association of many of the most eminent British neurologists of the time such as Hughlings Jackson, David Ferrier, and Victor Horsley. Beyond diagnosis and possible treatment of the neurological disorders they encountered in their practices, they were keen to identify the anatomic basis of the functional changes they observed. Some, like Hughlings Jackson, theorized from detailed physical examination of his patients and others drew conclusions from experimental results as well. Their thinking was in terms of cerebellar representation of the body similar to what the stimulation experiments had shown for the cerebral cortex. Hughlings Jackson, from inference alone, wrote in his verbose style:

Fig. 9.11. Clarence L. Herrick (left) of Denison College in Ohio was one of the first American comparative anatomists to focus on the nervous system. (Impression from Windle, 1979, Frontispiece.) His younger brother, C. Judson Herrick (right), carried on his work and wrote a textbook on the human nervous system used by medical students for many years.

> I submit that the cerebrum represents all parts of the body, and that the cerebellum also represents all parts of the body. But the two representations are in reverse order; . . . the cerebral order is, arm, leg, trunk . . . the cerebellar order of representation is the opposite—trunk, leg, arm. . . .
>
> In manipulating, the cerebrum will be only *chiefly* engaged; in walking, the cerebellum will be only chiefly engaged (1888, p. 317).

The unitary concept of cerebellar action had persisted since Flourens proclaimed (1824) that the organ reacted as a whole, with no localization. Sherrington considered the cerebellum the "head ganglion of the proprioceptive system" (1906, p. 346). Even Rudolf Magnus, who worked with Sherrington at Liverpool during the latter's early career, could be classified as a "universaliser," the term used by Hughlings Jackson (1884b, p. 707; also in Taylor, 1958, vol. 2, p. 74). The unit idea was not accepted by everyone, however, as evidenced by Jackson's championing the idea of cortical representation, and by the midtwentieth century, a complete map of cerebellar localizations had been constructed. The data came from studies that combined the techniques of peripheral nerve electrophysiology with the sensory effects in the central nervous system of the evoked potential.

The report of cortical localizations from Gerard's laboratory at the University of Chicago (*see* p. 90, this volume) mentions three incidents of potentials picked up in the anterior cerebellum in response to auditory stimuli (clicks) or when light was flashed into the eyes of anesthetized animals (Gerard, Marshall, and Saul, 1936, pp. 680, 687). Although no further comment was made then, Gerard may have been referring to those cerebellar potentials when he wrote many years later that "we followed evoked potentials—which we named—. . . into all sorts of regions where sensory impulses were not supposed to go" (Gerard, 1975, p. 468). More focused attention was paid to cerebellar evoked potentials by Robert Stone Dow (1908–1995) at the Rockefeller Institute for Medical Research

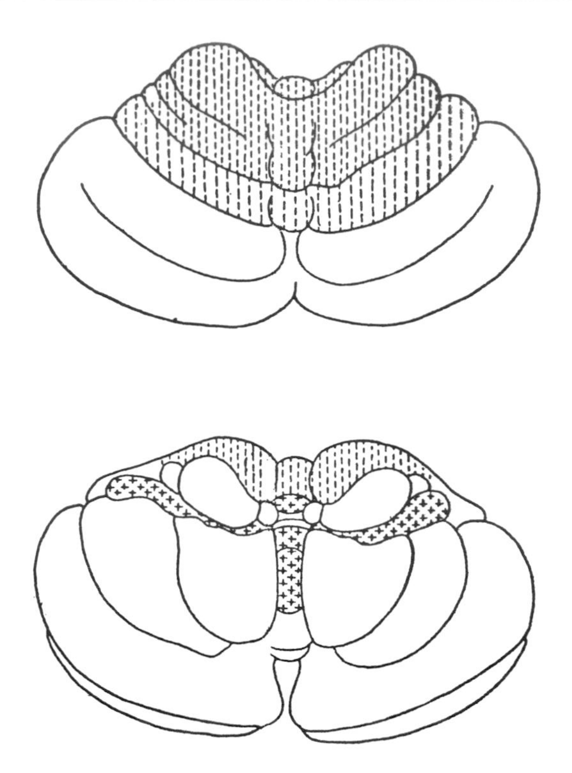

Fig. 9.12. Sketches of the upper and lower aspects of the human cerebellum. The two shaded areas comprise the old paleocerebellum; the unshaded areas are of later phylogenetic origin (the neocerebellum), developed largely in association with fibers from the cerebral hemispheres. (From C. J. Herrick, 1924, p. 642, Fig. 17.)

(now Rockefeller University). He confirmed Adrian's 1935 characterization of the relatively rapid rate of spontaneous cerebellar discharge and reaffirmed the importance of identification of afferent fiber tracts in localization of function (Dow, 1939), verifying with the oscilloscope the connections that had been found earlier by ablation and comparative anatomy studies. Dow tackled the difficult task of correlating the cerebellar points of response to afferent impulses with the sites on the cerebral cortex found by Bailey and von Bonin using strychnine (1942; *see* Fig. 5.20, p. 83, this volume); in cat he found no point-to-point relationships, but in monkey some cerebral areas prominently represented in the cerebellum.

At that time and independently, the team of Raymond S. Snider and A. Stowell at Johns Hopkins Medical School was exploring the same ground and described the receiving areas of the tactile, auditory, and visual systems of the cerebellum in anesthetized cats. They wrote dejectedly that "Unfortunately the experimental facts . . . do not carry with them any hint concerning their significance" (1944, p. 350). Undeterred, a map was published in 1950 (Fig. 9.13) which conferred substance to Hughlings Jackson's speculations made a half-century earlier.

INHIBITORY INFLUENCE OF THE CEREBELLUM

Meanwhile, among the eminent group of men at London's National Hospital, Queen Square, Sir Victor Horsley, whose work on cerebral localization has been described (*see* p. 76, this volume), was among the most active of the experimentalists. In addition to being well connected in British aristocracy, Horsley had personal qualities that assured his success in his chosen field of neurosurgery. "He learned without effort, he had an exceptional memory, he was skillful with his ambidextrous hands, he had daring, he had energy, he had integrity—in short he had from early days the qualities of leadership" (G. Jefferson, 1957, p. 904). Horsley

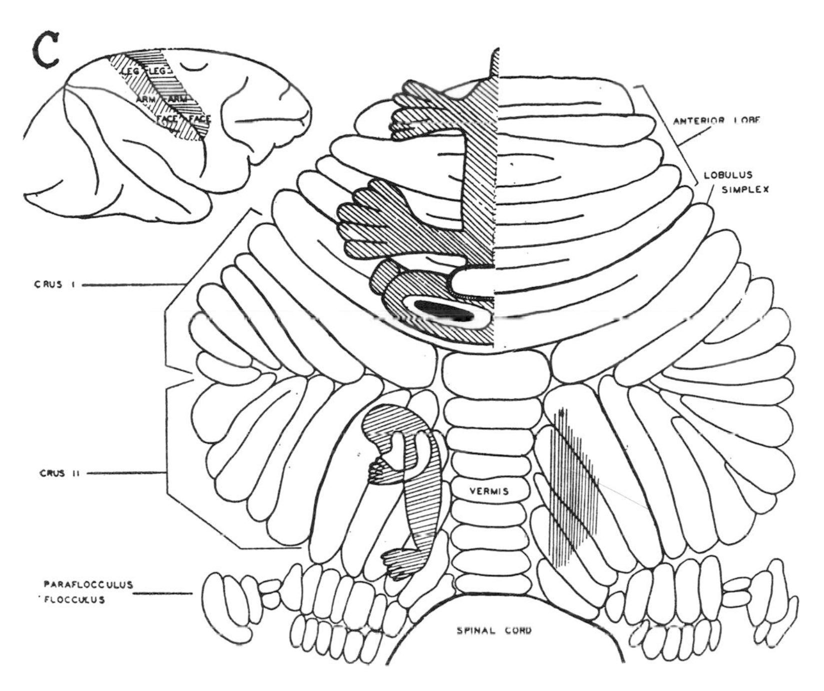

Fig. 9.13. Summary of projections from sensory area I and motor area (shown on upper left insert) of anesthetized monkey to cerebellar areas. Note that the anterior (top) humunculus corresponds to the topography of the cerebral areas (*see* Fig. 2.19) whereas the posterior homunculus is inverted. (From Snider, 1950, p. 204, Fig. 5.)

conducted many of his early experiments at the University of London's Brown Institution, "the only place where in those days at which [sic] recovery experiments could be done on animals of any size" (ibid., p. 904). His explorations of the cerebral cortex led to studies on the excitability of the cerebellar cortex and the possibility of its direct efferent connection with higher or lower centers. An unanticipated finding first seen in 1895, the inhibition of decerebrate rigidity by localized stimulation of cerebellar cortex, was confirmed and extended and a preliminary note sent to the Royal Society (Löwenthal and Horsley, 1897). A week later, Sherrington (1897) sent the report of his own independent observation of the same phenomenon. That discovery "was probably of no less significance for cerebellar physiology than was the famous Hitzig and Fritsch experiment for the physiology of the cerebral cortex" (Moruzzi, 1950, p. 7). The major problem, however, concerned the cerebellar nuclei, and their unreliable accessibility without damage to the overlying cortex.

Horsley had many collaborators and seems to have been generous in acknowledging them. One was Robert Henry Clarke (1850–1926; Fig. 9.14), physician–physiologist, with whom he worked on the cerebellum (Clarke and Horsley, 1905). To overcome their difficulty in identifying the location of their experimental stimulus, Clarke applied geometric principles to the problem of precise and trauma-free electrode penetration into the brain and conceived, designed, and supervised the construction of the first stereotaxic instrument. The preeminent place of "Clarke's instrument," as Horsley always referred to it, in the furtherance of knowledge of the brain is difficult to match; in fact the invention has been compared in importance to that of Galileo's telescope and the microscope of Leeuwenhoek (*see* Tepperman, 1970, p. 304). After a brief description of the stereotaxic instrument delivered at a meeting of the British Medical Association in Toronto in 1906 (Clarke and Horsley), the details of its construction and use were published in *Brain* in 1908 (Fig. 9.15) with a note that experimental data on the cerebellum would follow. Unfortunately, there was no subsequent publication—the close friendship cooled and the collaborators drifted apart.

Fig. 9.14. Shown in riding attire, Robert H. Clarke was a superb athlete. His scientific fame lies in his design of a stereotaxic instrument and its application to accurate localization of precise points within brain tissue.

The principle on which the stereotaxic instrument was constructed lay in determining a reliable baseline. This was accomplished in the rhesus monkey cadaver by inserting a fine ivory knitting needle from one auditory meatus (canal) across the head to the opposite meatus; the same was done for the orbital ridges. When one- or two-millimeter sagittal sections of the frozen head were cut, the transections of the needles served as baseline guides for lining up the superimposed millimeter grid (*see* Fig. 9.16, p. 194, left). The meticulous hand sawing (right) of frozen sections, their staining, and subsequent photographing were often carried out in Horsley's home and a sufficient number of brains were so prepared to make the measurements statistically reliable. It is difficult to determine how much each collaborator contributed. Again quoting Geoffrey Jefferson, who worked with Horsley from 1904 to 1909: "Although it is only too clear that Horsley's part in the instrument's actual design was negligible, it was he who had conceived its necessity. It is not too much to say that Clarke would never have thought of stereotaxis had it not been for Victor Horsley's curiosity about the roof nuclei of the cerebellum. . . ." (1957, p. 908). Only one substantive study using the instrument was carried out contemporaneously in Great Britain and none in the United States until a copy was built in Chicago in 1928 and "kept humming in a range of projects" (Magoun, 1985, p. 250) at Ranson's Institute of Neurology at Northwestern University Medical School, as described in Chapter 11.

While the Horsley–Clarke stereotaxic instrument was being developed and its tremendous potential for exploration of brain function waited in the wings, conjectures regarding the inhibitory influence of the cerebellum were being tested and its role in the complexities of posture and equilibrium probed. Rudolf Magnus (1873–1927), born

Fig. 9.15. Clarke's stereotaxic instrument (left—from above; right—with suspension attached), "one of the most beautiful tools ever introduced into a physiological laboratory" (G. Jefferson, 1957, p. 907), but also deemed "probably the most complex of all the mathematical instruments of physiology" (Paget, 1919, p. 189). (From Horsley and Clarke, 1908, pp. 65, 67, Figs. 6 and 7.)

and educated in Germany, after postdoctoral experience at Edinburgh (Schäffer) and Liverpool (Sherrington), was appointed professor of pharmacology at the University of Utrecht. With a group of productive and happy collaborators (Olninck, 1970), he directed a long and significant series of studies aimed at (among other objectives) defining the cerebellum's participation in maintenance of equilibrium. Kleijn and Magnus (1920, p. 173) reported negative results that proved (italics theirs) "*the pathways serving the labyrinthine reflexes do not run through the cerebellum. . . . [T]herefore the idea that is still very popular, that the cerebellum must be the central apparatus for the labyrinths, must finally be dismissed* (translation from Clarke and O'Malley, 1968, p. 688).

Although the conclusion of Magnus had a temporary negative impact on the association of the cerebellum with maintenance of equilibrium, other facets of its function received full attention. By the early 1940s, Papez had carefully traced the fiber connections of the basal ganglia and with characteristic thoroughness he described the four "progressively overlapping stages" of motor control in the phylogenetic scale and concluded:

> Corresponding to these stages there are roughly speaking four categories of motor pathways connecting directly or indirectly with the motor cranial and spinal nerves which supply the skeletal musculature. . . .
>
> All four of these pathways have accessory cerebellar connections. . . . It appears that the

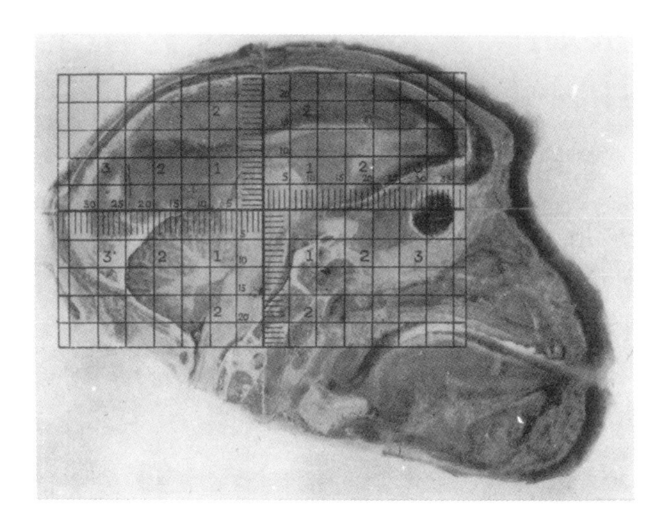

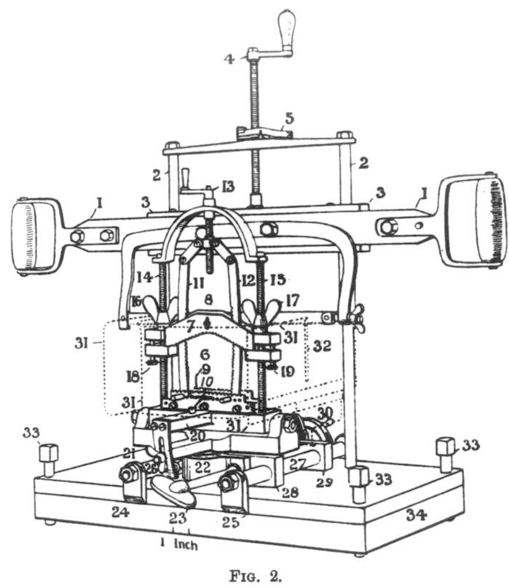

FIG. 2.

Clarke's Saw for cutting sections of frozen heads 1 mm. or 2 mm. thick.

1 Hack saw working on horizontal and vertical guides.
2 Vertical guides for saw.
3 Horizontal guide for saw.
4 Screw for raising and lowering saw.
5 Grip with spring catch, in which screw 4 works.
6 Placed in centre of head vice.
7 Posterior blade of upper jaws of head vice.
8 Anterior blade of upper jaws of head vice.
9 10 Anterior and posterior lower jaws of head vice.
11 12 Vertical jaws of head vice.
13 Screw for approximating vertical jaws of head vice.
14 15 Screw guides for upper jaws of head vice.
16 17 Fly nuts for depressing upper jaws of head vice on screw guides (14, 15).
18 19 Screw for adjusting upper jaws of head vice.
20 Rocking adjustment of head vice.
21 Fly nut for clamping rocking adjustment (20).
22 Rotatory adjustment of head vice.
23 Screw for clamping rotatory adjustment (22).
24 25 Fixed brackets supporting guides of travelling stage of vice.
27 Right guide of travelling stage of vice.
28 29 Travelling stage of head vice.
30 Graduated wheel for moving travelling stage of vice on guides—One complete turn = 2 mm. movement of stage.
31 Zinc tray (shown by dotted lines) to confine CO^2 snow.
32 Dotted lines show gap in sliding side of tray for saw.
33 Levelling screws of foot-plate.
34 Foot-plate.

Fig. 9.16. Left—Sagittal section of frozen head of a monkey with millimeter-scaled glass plate superimposed. Right—Clarke's saw for cutting sections of frozen heads at one- or two-millimeter thickness. The tissue was frozen by encasing the head in halves of a coconut shell filled with chipped ice. (From Horsley and Clarke, 1908, pp. 54, 56, Figs. 1 and 2.)

> cerebellum is concerned with the energizing as well as inhibiting of muscular movements. . . . The synergic effects of the cerebellum are probably due to the dual capacity for excitation and for inhibition (Papez, 1942, pp. 21, 62).

By the end of the decade, the cerebellum was included among the circuits operating to and from the brain-stem reticular core. The concept of a cerebellar-bulbar-reticular pathway was introduced (Snider, McCulloch, and Magoun, 1947) two years before publication of a study in cats of the effects of lesions in cerebellum, cortex, or both, on the antigravity muscles. In the later study, "the gradual diminution of initial spasticity . . . suggest that in suppression, . . . as in other functions . . . the deep nuclei of the cerebrum and cerebellum are not simply relays for efferent cortical influences but, in addition, possess the capacity for intrinsic or independent activity" (Lindsley, Schreiner, and Magoun, 1949, pp. 203–204). Again, the study of muscles had opened the way into the central nervous system.

FROM OLD TO NEW TECHNIQUES

The modern neurophysiologist with the most sustained interest in the cerebellum was Giuseppe Moruzzi (1910–1986; Fig. 9.17) in Italy. As a second-year medical student in Parma, he received an assistantship and was assigned to investigate the cerebellum with Golgi and Cajal techniques. His first paper, in 1930, described the granular layer and was strictly anatomical. Using only the nine-

Fig. 9.17. Giuseppe Moruzzi, in 1961, 10 years after he became director of the Institute of Physiology of the University of Pisa, where "[g]enerosity with the young has been one of the great attractions" (Granit, 1981, p. 461).

teenth-century methods available to him, by 1936 there were five more papers on the cerebellum, the last one of which concerned its inhibitory action. Fully aware of the handicap of his ignorance of modern electrophysiological and surgical techniques (*see* L. H. Marshall,1987), Moruzzi traveled with Rockefeller funds first to Brussels, where he was a guest worker in the laboratory of Frédéric Gaston Bremer (1892–1988). Bremer had confirmed and extended the earlier studies of the inhibitory effect of stimulation of the paleocerebellum on decerebrate rigidity and reported that the cerebellum made no contribution to the original extensor rigidity—in fact he found part of the cerebellar cortex to be nonexcitable (1922). Using cats, he introduced new brain transections, the waking *encéphale isolé* and the sleeping *cerveau isolé* (Fig. 9.18, bottom left), which became classic preparations in neurophysiologic research. While in Brussels, Moruzzi established the association of paleocerebellar with vasomotor activity (1938).

Moruzzi went next to Cambridge to work with Adrian, who, as noted above, had observed the very high rate of spontaneous discharge of the cerebellum compared to that of the cerebrum. In Adrian's

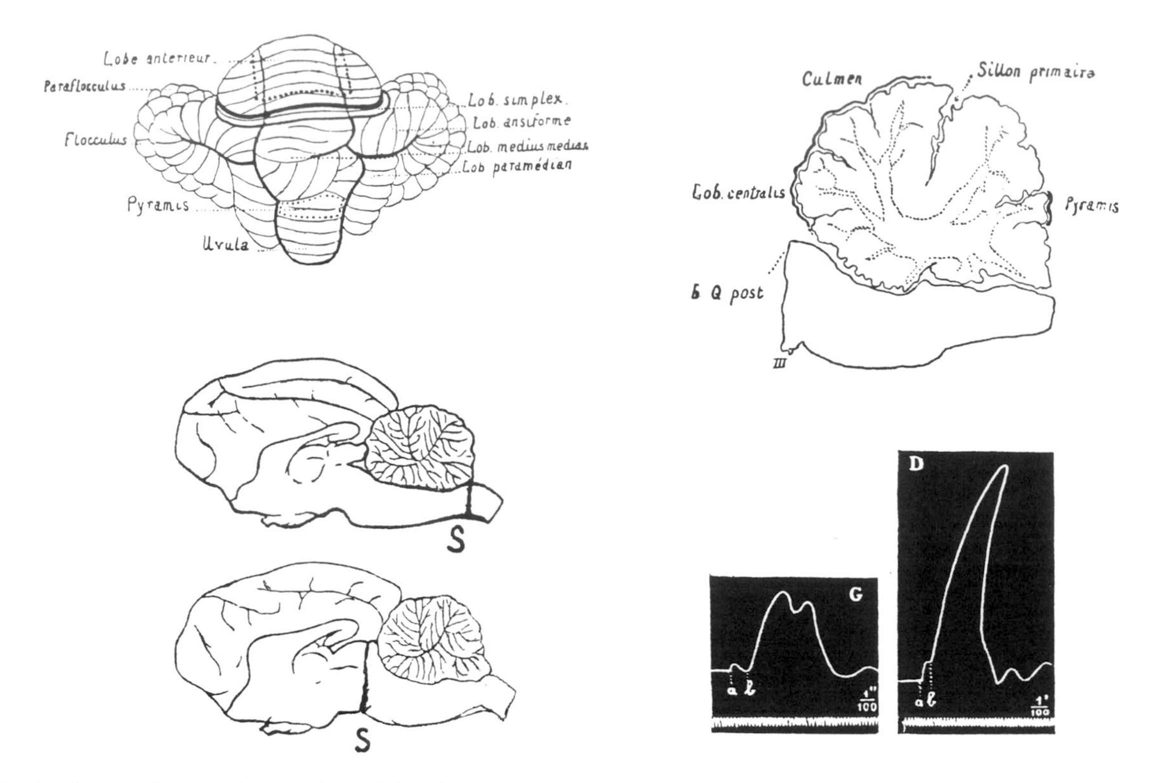

Fig. 9.18. Figures from F. Bremer's publications. Top left—The area (dotted line) on the superior aspect of the cat cerebellum where stimulation inhibits decerebrate rigidity; Bremer discovered that the pyramis should be added. (From 1922, p. 197, Fig. 1.) Top right—The excitable areas (heavy line) of the cat cerebellar cortex. (From ibid., p. 198, Fig. 2B.) Bottom left—Outline of medial cat brain showing the waking *encéphale isolé* (above) and sleeping *cerveau isolé* (below) preparations. (From 1936, p. 461.) Bottom right—Tracings of the right and left knee jerks of a patient with a right cerebellar wound. Note that the right reflex (D) has a shorter latency a b, a larger amplitude of muscle contraction, and the absence of the tonic wave of contraction compared with the left reflex (G). (From 1922, Fig. 4.)

words, "The fact that the normal activity of the cerebellar neurones occurs at such high rates suggests that the cerebellum may perhaps exert an inhibitory influence . . . on some of the structures innervated by it" (1935, p. 33P). Later, it was realized that the high rate of discharge served as a marker of impulses that had passed through the cerebellum after originating elsewhere in the nervous system.

After the Second World War, Moruzzi was invited to Northwestern University Medical School in Chicago, where, by a combination of default and generosity, he joined Horace Magoun's group and participated in the discovery of the brain-stem reticular formation (Chapter 12). Moruzzi later described his "wonderful years abroad" (L. H. Marshall, 1987); at Northwestern, where the Horsley–Clarke stereotaxic instrument had been revived in the late 1920s, Moruzzi again became active in research on the cerebellum, collaborating in single-unit experiments (Brookhart, Moruzzi, and Snider, 1950, 1951) that extended the work on cerebellar suppression of spasticity.

The culmination of Moruzzi's career-long interest in the cerebellum was a thick volume, *The Physiology and Pathology of the Cerebellum* (1958), written in collaboration with the American neurologist, Robert Dow (*see* above), whom he had met in Bremer's laboratory. The authors perceived a need "to relate this [avalanche of] new information from the development of electrical methods of recording to the facts derived from older methods of investigation" (Dow and Moruzzi, 1958, p. 5). Their careers had spanned both domains of experimentation—ablation with subsequent anatomic degeneration and electrophysiological probing of function—and for many years their analysis constituted the only authoritative attempt to relate cerebellar pathology with its function.

The inhibitory influence of the cerebellum noted first by Horsley and by Sherrington in decerebrate

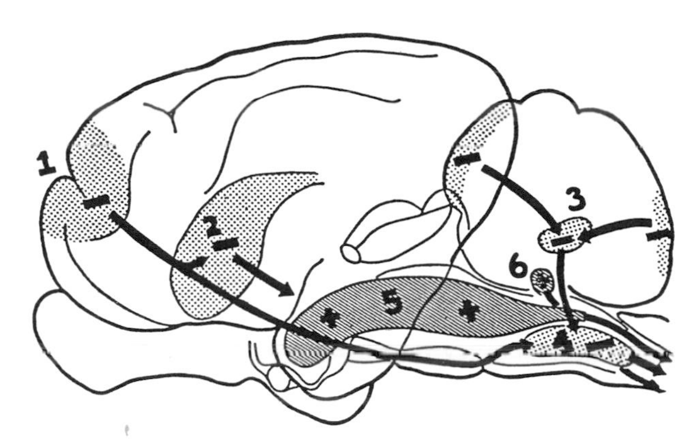

Fig. 9.19. An outline of the cat brain showing the possible pathways of cerebellum inhibition (3) to the brain stem reticular system (4) and connections from the cerebral cortex (1) and thalamus (2). (From Lindsley, Schreiner, and Magoun, 1949, p. 198, Fig. 1.)

Fig. 9.20. The very British Jack-in-the-box analogy of release versus influx theories of spasticity was "prepared especially for this paper by Mr. T. Bryce Spruill." (From Magoun and Rhines, 1947, p. 29.)

rigidity, and suggested by Adrian was irrefutably demonstrated when cerebellar connections with the reticular core were identified. Figure 9.19 diagrams a generalization of the pathways involved, which had assumed global proportions with the demonstration (*see* p. 190, this volume) by Snider and Stowell that tactile, auditory, and visual afferent impulses pass through the cerebellum.

Again, we turn to Hughlings Jackson for the historical development of a theory to explain the inhibition of decerebrate rigidity by cerebellar stimulation. He had, in fact, alternative theories, that of release which he and most contemporaries favored, and that of "influx" which was initially treated with benign neglect. Based on his familiar evolutionary hierarchy of organization of the nervous system into three levels, Hughlings Jackson theorized that disorder of the nervous system produced a negative effect because of loss of the part destroyed and in addition a positive element due to activity of the liberated remainder:

> The higher nervous arrangements evolved out of the lower keep down those lower, just as in government evolved out of a nation controls as well as directs that nation. . . . [T]he reverse process of Dissolution [disease] is not only "a taking off" of the higher, but is at the very same time a "letting go" of the lower. If the governing body of this country [England!] were destroyed suddenly, we should have two causes for lamentation: (1) the loss of services of eminent men; and (2) the anarchy of the now uncontrolled people (1884a, p. 662; also Taylor, 1958, vol. 2, p. 58).

That analogy was transmogrified to a Jack-in-the-box and amusingly depicted (Fig. 9.20) by Magoun and Rhines (1947), from whose treatise on spasticity much of this discussion is taken.

Perhaps not fully satisfied with the release theory, Hughlings Jackson later put forth the idea of influx based on a series of questionable propositions, including the assumption that the spinal centers receive impulses from both the cerebrum and the cerebellum, which normally oppose each other. He wrote: "In accordance with this hypothesis, the rigidity . . . results because cerebral influence being taken off the lowest motor centres . . . cerebellar influence upon those . . . centres is no longer antagonised; there is then unimpeded, and therefore greater, cerebellar "influx" into the lowest motor centres which the cerebrum *has abandoned*" (1898, p. 80; also Taylor, 1958, vol. 2, pp. 426–427). Magoun and Rhines favored the nonantagonized influx to lower centers and stated that "a considerable body of evidence can be marshalled for its support; much more, in our opinion, than can be marshalled for the concept of release. To return to our earlier analogy, according to the *influx* theory Jack jumps out of the box . . . both because the lid holding him down is removed and because there is also a spring inside, which, being unopposed, is then able to push him up" (1947, p. 32).

OVERVIEW OF THE "LITTLE BRAIN"

We leave the cerebellum at this point with its functions that moved like a game of musical chairs, an analogy made somewhere by Francis Schiller, neurologist–historian par excellence. In the early nineteenth century, the uncertainties were symptomatic of the period of social change that prevailed at that time as described in interesting detail by Clarke and Jacyna (1987, p. 297, passim). The trail was highly peripatetic, like most scientific pathways moving from nation to nation and undergoing confusing directional changes in assignment of functions. The multiplicity of observations reported—sometimes of the same phenomenon a half-century or more apart; for example, both Mehée de la Touche (1773) and Foderá (1823) described opisthotonos and head extension after cerebellar injury—confounded the discovery of cerebellar function more than was the case with other organs. Nonetheless, the relatively simple microstructure of the cerebellum (with only three distinctive strata in its cortex) and the early attention from master investigators (Purkyně and his pupil, Valentin) was conducive to the later analysis of its electrophysiologic and chemical properties as a complete neuronal system.

After Purkyně cells were found to be inhibitory—the first suppressor neurons identified in mammals (Ito and Yoshida, 1966)—they were also the first to be identified as GABAergic (Obata, Ito, Ochi, and Sato, 1967). The relatively simple microstructure throughout the cerebellar cortex retarded progress in knowledge of function for about 100 years, from Flourens to Magnus, as it supported the notion of unitary action—the organ acts as a whole without localization. A functional topography was described only after cerebellar excitability and evoked responsiveness were recognized.

Concepts of balance between inhibition and excitation (synergis), facilitation, reverberating circuits, and feedback were first productively explored in cerebellar territory, again due to its relative structural simplicity. Today we know that the cerebellum is connected directly or indirectly with all principal subdivisions of the central nervous system. The Purkyně cell is the final common pathway in the cerebellar cortex and the molecular layer, containing Purkyně cell dendrites, is "organized in a spectacular reticular lattice" (Chan-Palay and Palay, 1970, vol. 1, p. 197). This richness of connectivities provides the substrate for the cerebellum's global regulatory role.

When C. Judson Herrick published *Brains of Rats and Men* in 1926, it was extolled by no less than H. L. Mencken as "perhaps, the best summary of the [now] known facts about mental processes put into English" (quoted from Roofe, 1963, p. ix). In that praised work, Herrick wrote about the cerebellum:

> The indications are that the cerebellar cortex is concerned more with the convergence and summation of diverse sensory impulses than with their refined analysis and redistribution. Which particular motor centers will receive the nervous impulses discharged from the cerebellum is apparently determined less by what is going on in the cerebellum itself than by what systems are in actual function in the rest of the nervous system. . . .
>
> The cerebellum emerged from the vestibular nuclei. In its more highly elaborated forms its functions are enlarged to include some sort of reflex control of all movements of skeletal muscles and of muscular tone. It is highly developed in all active animals from fishes to men; and, although much larger in higher forms, it is not otherwise greatly modified in internal structure or mode of action. The stimuli by which it is activated arise for the most part within the body, and its activities are wholly unconscious (C. J. Herrick, 1926, pp. 30–31).

10 Thalamocortical Pathways and Consciousness

CONTENTS

O, what a world of unseen visions and heard silences, this insubstantial country of the mind!"

(Jaynes, 1976, p. 1)

The relatively orderly series of discoveries of the functional organization of the cerebral cortex, in which anatomy and physiology often shared a certain simultaneity of discovery (as in the production of movement when precentral gyri are stimulated), did not characterize the exploration of structures beneath the cortex. On the contrary, the subcortical regions of the brain were dissected and anatomic structures described and named in profusion before any correct function was assigned to them. The major reason for the discrepancy between the rapid advance of anatomical knowledge and the slow elucidation of its related physiology was, paradoxically, the lack of a methodology to precisely locate structures deep in the living brain. Even though electrophysiologic techniques applied to peripheral nerves were advancing, their utility for investigation of the central nervous system depended on reliable localization. A second hurdle was the confused nomenclature that muddled any new information about the thalamus. In this and subsequent chapters, the history of the discoveries of some prominent subcortical structures and their functions is explored with emphasis on the novel developments in methodology that were essential for further progress.

About three centimeters long and half as wide, in the adult human the thalamus is a group of nuclei clustered into the shape of a football situated on each side of the third ventricle at the base of the cerebral hemispheres. Thirty-some separate nuclei occupy the 6–7 cm^3 of each thalamus. It is now known that all sensory fibers (except olfactory) converge there and synapse before their signals are passed on to cerebral cortex and other destinations.

IN EARLY HISTORY

The term "thalamus," meaning chamber or anteroom, was applied by Galen in the second century to the organ with which he thought the optic nerves were connected and which provided "vital spirits" for vision. A millennium later, Raimondo Mondino de' Luzzi (1275–1326) described but did not illustrate the thalami in his writings and was clear about their function. In Charles Singer's graceful translation of the text republished in Venice in 1495, Mondino declared:

> Before thou dost proceed to the mid ventricle consider the parts between the fore and the mid ventricle. . . . They are of the substance of the brain and are shaped like buttocks (*anchae*). At

Fig. 10.1. Jiri Procháska, a Czech who worked in Vienna, viewed the thalamus as a vague component of the relay station of sensory and motor impulses that resulted in "reflexion," thus providing the link between afferent and efferent limbs of the reflex arc. (From the painting [ca. 1788] by G. Kneipa; presently at the Medical Faculty of the University of Prague.)

> the side of each *ancha*, between the ventricles already mentioned, is a bloodlike substance made like a long or subterranean worm. . . . This worm can lengthen itself by constriction and block the *anchae* closing the way of passage from the fore to the mid part and contrariwise. When a man doth wish to cease from cogitation and consideration, he doth raise the walls and expand the *anchae so* that the spirit may cross over from one ventricle to the other (Singer, 1925, p. 92).

That curious description of the thalamus and choroid plexus in the ventricle must have presented a certain conceptual neatness to Mondino's contemporaries. The assignment of valve-like mechanisms controlling the flow of vital spirits persisted for more than three centuries: as noted, Descartes (1662) believed the pineal gland served the same gatekeeping function (*see* Fig. 3.10, p. 37, this volume).

After a long period of relative neglect of subcortical structures during exploration of the ventricles and gyri, some attention was accorded the thalamus in the second half of the eighteenth century. The concept of the reflex was emerging and speculation arose about where the "reflection" between afferent impressions and efferent activity occurred. Jiri Procháska (1749–1820; Fig. 10.1), a Czechoslovakian ophthalmologist who trained and worked in Vienna, designated a vague *sensorium commune*, which included the thalamus, as the intermediate agent: "[The sensorium commune] seems not improbably to extend through the medulla obligata, the crura of the cerebrum and cerebellum, also part of the thalami optici, and the whole of the medulla spinalis; in a word it is coextensive with the origin of the nerves" (Procháska, 1784, p. 115; transl. T. Laycock in Unzer, 1851, p. 430). As for function, it "changes sensory impressions into motor ones: so that external impressions that are likely to injure our body are followed by motor impressions that will produce movement. . . ." (Procháska, quoted by Neuburger, 1981, p. 244). Procháska divided the sensorium into a part for the soul and consciousness (consisting of the cerebrum only) and a part for the body.

During the first half of the nineteenth century, lines were drawn between those neurologists, especially the French, who believed the thalamus to be motor in function (Magendie and Foville, among others) and those convinced that its nature was sensory. Étienne Renaud Augustin Serres (1786–1868), in his prize-winning two-volume-plus atlas on the comparative anatomy of the brains of the four classes of vertebrate animals (reptiles, fish, birds, and mammals), was more interested in anatomy than in physiology, although he claimed his publication to be the first in the "uniting of all we know about the anatomy, physiology, and pathology of the nervous system" (1827, Preface). Regarding the thalamus, he pointed out (ibid., p. 434) that in the first three classes of animals, the "couche optique" (thalamus) remains underdeveloped, but suddenly in mammals attains its greatest volume; thus, this relatively obscure French neuroanatomist should be nominated one of the earliest comparative neurologists. Even the great Baron Cuvier (1769–1832) in his lessons on comparative anatomy (1800–1805), limited his ideas about the thalamus to the statement (vol. 2, p. 178) that, among the characteristics common to "all animals with red blood," was the fact that they possessed two each of the cerebral hemispheres, the thalami, and the cerebella.

Greater progress in the early knowledge of sensory function of the thalamus was being made outside France. Karl Frederick Burdach (1776–1847; Fig. 10.2) produced his most important contribution to neuroscience (A. C. Meyer, 1970), three volumes on the brain (1819–1826). With a Teutonic talent for diligent systematization, he took pains to report the work of others, described four nuclei in each of the thalami, and noted that of all the "ganglia" they have the greatest influence on mental expression and unify sensory impressions, "thus they allow their owner to realize that he is thinking. . . ." (vol. 2, p. 115). Burdach extrapolated the idea of the *sensorium commune* to placing the seat of consciousness in the thalami, and later wrote "they are the root of consciousness" (1826, p. 290, footnote). In his admirable *Historical Aspects of Cerebral Anatomy* (1971, p. 19), Albert Meyer makes the point that early-nineteenth-century anatomists such as Burdach, concentrated "on elaborating [the] detail of macroscopic anatomy. This entailed the need for a critical re-examination of findings which had remained controversial. . . ." and for that reason historical perspective abounds in their writings, a compulsion that seems to have persisted (cf. Tafel's 1882 annotation of Swedenborg, p. 27, this volume).

Gradually, however, the role of the thalamus as the recipient of sensory input was recognized. In the chapter on the thalamus of his great *Histologie*, Ramón y Cajal (1909, p. 295), wrote in translation: "The role of the inferior lobe of the medial geniculate body in acoustic conduction is undoubted. The results of our histological research accord perfectly with the conclusions of Monakow to prove it." Monakov's studies on sensory input to thalamus are described on p. 98, this volume.

MODERN VIEWS OF CONSCIOUSNESS

The first serious consideration of the neural substrates of consciousness appeared in two English publications, William B. Carpenter's *Principles of Mental Physiology*, published in 1842 with many successive editions and Thomas Laycock's *Mind and Brain* (1860). Although his lectures in Edinburgh were said to be dull, Laycock's influence was pervasive—Hughlings Jackson was one of his students and Laycock's three-volume treatise became widely read. Laycock developed an idea he had published earlier: "[T]he brain, although the organ of consciousness, [is] subject to the laws of reflex action, and . . . [does] not differ from the other ganglia of the nervous system" (1845, p. 298). Sechenov (*see* p. 3, 4, this volume) independently put forth (1863) the same concept of a general reflex pattern throughout the nervous system, a belief which influenced the thinking of Pavlov, Sherrington, and Freud (Amacher, 1964).

William Benjamin Carpenter (1813–1885), English physiologist, not only reaffirmed the thalamus as the center to which almost all sensations are carried but in addition proposed a close connection with cortex:

Fig. 10.2. Karl Frederick Burdach was reared by his widowed mother to distinguish the family tradition in medicine. He did so in research in neuroanatomy and in the authorship of textbooks and an encyclopedia of medical knowledge. (The impression is from Bast, 1928, p. 35 of a lithograph drawn by the artist, Kriehuben, presumably in 1832.)

> The *Sensory Ganglia* constitute the seat of consciousness not merely for impressions on the Organ of Sense, but also for changes in the cortical substance of the cerebrum so that until the latter have reacted downwards upon the Sensorium, we have no consciousness either of the formation of ideas, or of any intellectual process of which these may be the subjects (Carpenter, 1859, p. 757).

As James O'Leary (1956, p. 189) commented, "Thus a hundred years ago, albeit crudely, Carpenter already had two ideas that prevail today: those of a diencephalic center for consciousness and of a to-and-fro relation between thalamus and cortex." Recognition of the return link between cortex and thalamus is further elaborated in Chapter 12.

Arguments surrounding the involvement of the thalamus in consciousness continued along two lines—what constitutes the state of consciousness and does it have a physical substrate? Some measure of the seriousness with which those questions were regarded, especially by the psychologists, was expressed by William James, the Harvard-based propounder of "the stream of thought." In the last chapter of his introductory textbook, he wrote: "Something definite happens when to a certain brain-state a certain 'sciousness' corresponds. A genuine glimpse into what it is would be *the* scientific achievement, before which all past achievements would pale" (1899, p. 468). For himself, James offered the idea that consciousness is a "breath moving outwards, between the glottis and the nostrils . . . the essence out of which the philosophers have constructed . . . consciousness" (1904, p. 491). A half-century later, this ethereal view was echoed by the Harvard neurologist, Stanley Cobb: "It is the integration itself, the relationship of one functioning part to another, which is mind and which causes the phenomenon of consciousness. There can be no center. There is no one seat of consciousness. It is the streaming of impulses in a complex series of circuits that makes mind feasible" (1952, p. 176).

Fig. 10.3. Margaret Floy Washburn taught psychology at Vassar College for many years and was the second woman to be elected to the U.S. National Academy of Sciences (1931). She is shown in a 1927 snapshot taken at the symposium on "Feelings and Emotion" to dedicate a new psychology laboratory at Wittenberg College in Ohio, when she received an honorary degree.

At the turn of the century, American psychologists were debating the animal mind in full voice. A widely used college text on the subject was published by Margaret Floy Washburn (1871–1939; Fig. 10.3) in 1908 with new editions at approximate decades to 1936. She crisply stated her belief in how consciousness should be studied: "While consciousness exists and is not a form of movement, it has as its indispensable basis certain motor processes, and . . . the only sense in which we can explain conscious processes is by studying the laws governing these underlying motor phenomena" (1928, p. 104).

Washburn's contemporary, Robert Mearns Yerkes (1876–1956), famous for his studies of learning in apes confined in primate colonies, pointed out that Jacques Loeb, eminent biologist and student of learning in paramecia, "accepts associative memory as the criterion of consciousness, and then adds, quite safely, 'The criteria for the existence of associative memory must form the basis of a future comparative psychology' " (Yerkes, 1905, p. 145). As Yerkes was writing of animal psychology, he did not directly confront the question of human consciousness, but instead made a Jacksonian pronouncement on levels of consciousness: "[W]e may safely say that mere ability to learn is common to all animals, and that it is indicative of a low grade consciousness; ability to learn associatively . . . is a sign of a higher grade of consciousness . . . there is no one criterion of the psychic which can be accepted as a sign of all forms and conditions of consciousness" (ibid., p. 147).

The central role of the thalamus in mediation of sensation and affect in the state of consciousness was the subject of extensive discussions in the writings of that formidable team of English neurologists, Henry Head and Gordon Holmes (Fig. 10.4): "[W]e believe that the essential organ of the optic thalamus is the centre of consciousness for certain elements of sensation. It responds to all stimuli capable of evoking either pleasure and discomfort, or consciousness of a change in state. The feeling-tone of somatic or visceral sensation is the product of thalamic activity. . . ." (Head and Holmes, 1911, p. 181).

Twenty years later, another influential pair of Englishmen, George C. Campion and G. Elliot

Fig. 10.4. Left—Henry Head, a Quaker, was editor of *Brain* for 15 years and made outstanding contributions to knowledge of sensory systems. He "brought order out of chaos by his vivid thought and refined clinical method" (Denny-Brown, 1970, p. 451). Right—Gordon Morgan Holmes, an Irish neurologist who had trained with Edinger and Weigert, was a meticulous observer, known especially for his description of cerebellar deficits. His attention to detail complemented beautifully the more conceptual approach of Head, and their joint paper (1911) on the sensory effects of cerebral lesions is a classic.

Smith, went further and proposed that within the thalamus resides a center which influences all thinking: "[T]he thalami which Head and Holmes regarded as the central seat of consciousness for the affective aspect of sensation, act also as central propagators of streams of neural impulse to all the 'engrammic systems' or 'neural schemata' which form the neural bases of our thought-processes. . . ." (Smith in Campion and Smith 1934, p. 107). Smith had already elevated the thalamus to prominence as an agent of evolutionary pressure, calling attention to "the neopallium [whose] expansion provoked the vastest revolution that ever occurred in the cerebral structure. It came into being to form a receptive organ for fibres coming from the thalamus, whereby . . . all the non-olfactory senses—secured representation in the cerebral cortex" (ibid., p. 25, 26).

While firm evidence of the sensory input to the thalamus accumulated, tentative statements about the importance of connections with the nearby hypothalamus came from two careful and highly respected experimentalists. Both Walter B. Cannon and S. Walter Ranson believed the hypothalamic "drive" was essential for maintenance of a conscious state of wakefulness; on the basis of close observation of monkeys with thalamic and hypothalamic lesions, the latter proposed that the "active hypothalamus discharges not only downward . . . but upward into the thalamus and cerebral cortex. The upward discharge may well be associated with emotion as a conscious experience" (Ranson, 1939, p. 18).

By midtwentieth century, observations from the clinic were expanding but not necessarily elucidating the problem. Although some neurosurgeons believed that the cortex was involved in consciousness, the extent of the involvement was unclear, whereas others were equally certain the cortex was in no way concerned (*see* Dandy, 1946). The English neurosurgeon, Hugh Cairns, in his Victor Horsley Memorial Lecture, expressed this dichotomy:

> I have the impression that some . . . overestimate the part which the cortex plays in maintaining consciousness, while others attribute to the brain-stem and thalamus degrees of consciousness which are more correctly assigned to the cortex. . . . The fact that these disturbances arise from lesions in the brain-stem and thalamus [indicates] that there are nervous pathways in these parts of the brain which are essential to the maintenance of crude consciousness. . . . A healthy cerebral cortex cannot by itself maintain the conscious state (1952, pp. 113, 141).

Similar observations and conclusions were made by G. Jefferson and Johnson (1950), French (1952), and Barrett, Merritt, and Wolf (1967).

Important as they were, most of those reports were of passing interest to their authors. There was, however, one neurosurgeon for whom the study of consciousness and the human brain was a career-long preoccupation. A year before his death, Wilder Penfield published a volume on *The Mystery of the Mind* (1975). Unlike his predecessors and critics, Penfield provided a synthesis of his half-century of studies of the exposed unanesthetized brains of more than 1000 conscious human subjects (*see* Chapter 5). In general terms, he differentiated two major systems interrelating the higher brain stem and cerebral cortex. The first was an automatic, computer-like mechanism comprising the sensory-motor regions, and the second an interpretative mind-mechanism involving the newer temporal and prefrontal cortical regions. Penfield postulated that the two are functionally related, but only the latter presents interpretations of their experiences to consciousness. As noted earlier, in the 1930s he "proposed the term 'centrencephalic system' as a protest against the supposition that cortical association or cortico-cortical interplay was sufficient to explain the integrated behavior of a conscious man" (1936, p. 68).

On completion of the manuscript for *The Mystery of the Mind*, Penfield sent copies to a philosopher, a neurosurgeon, and a neurologist. The neurologist, Sir Charles Symonds, began his reply in true British fashion:

> As Hughlings Jackson said, we are differently conscious from one moment to another. It is a function, presumably, of synaptic activity, now here and now there. It seems to me more probable that its representation is in the cortex than in the diencephalon, having regard to the relative numbers of neurons available. . . .
>
> The synaptic activity associated with consciousness is continuously present except during sleep. The explanation for this appears to be that the reticular formation in the brain stem in some way facilitates, or "drives," the higher centres, and that in sleep the activity of the reticular formation is inhibited. Here the relationship of consciousness to the brain-stem seems well-established (1975, pp. 96–97).

Many advances in brain research have been accompanied by offensive contention, but that surrounding the seat of consciousness was unusual in twice involving the same person, but on opposite sides of an issue, first at the beginning and then at the end of an intervening century. The circumstances constitute a prime example of the recycling of an idea. In 1860, Laycock complained that Carpenter had plagiarized the concepts put forth in his *Mind and Brain*, but Carpenter succinctly replied that this was impossible, because he could not understand them. In condemning Penfield's "centrencephalon of the brain stem," the irrascible British neurologist, Francis Martin Rouse Walshe (1885–1973) stated that Carpenter's concept of a dominating role of the thalamus in consciousness appeared to be the prototype of Penfield's views, that Penfield had "produced a speculative hypothesis that is virtually a replica of Carpenter's and that it might have been formulated a century ago. . . . As we read it, we find ourselves back in the intellectual climate of mid-nineteenth century, pre-Jacksonian imaginings" (Walshe, 1957, p. 538).

This persistent interest in the neural substrate of consciousness was exemplified in 1948 by an outstanding exhibit presented at the annual meeting of the American Medical Association in Chicago by

Fig. 10.5. Left—George N. Thompson, a practicing psychiatrist in Los Angeles, collaborated with Nielsen in demonstrating brain pathology associated with loss of consciousness (*see* the following figure). He is shown as president of the Society of Biological Psychiatry, of which he and Nielsen were founding members. Right—Johannes M. Nielsen was the most prominent of the early neurologists in Southern California, and director of the specialized treatment and research center for aphasic disorders established by the U. S. Veterans Administration at Long Beach after the Second World War.

two neurologist/psychiatrists from Los Angeles. Titled "The Area Essential to Consciousness: Cerebral Localization of Consciousness Established by Neuropathological Studies," Johannes M. Nielsen and George N. Thompson (Fig. 10.5) showed photographs of sagittal and coronal sections of human brains illustrating diencephalic damage associated with long-term loss of consciousness and "concluded that the engramme system essential to crude consciousness is located where the mesencephalon, subthalamus, and hypothalamus meet" (Fig. 10.6, right).

During the early twentieth century, even though the central nervous system had been pushed off center stage by an expanding interest in endocrinology and the autonomic nervous system, as historian John Burnham (1977) indicated and Cannon typified, neurologists and neurophysiologists continued to chip away at the mind/body problem. They were well aware that control of visceral processes resides in the central core of the brain—psychosomatic medicine was just around the corner. However, the "visceral" system did not fully replace the intellectual appeal of the mystery of the conscious mind and symposia on the subject continued to be organized at intervals, one of the most influential of which was the study week of the Pontificia Academia Scientiarum convened in 1964 by Sir John Eccles (*see* Fig. 10.7, right, p. 208). He brought together an international assembly of distinguished neuroscientists to talk about *Brain and Conscious Experience*, the title of the ensuing publication (1966) which he edited. Twelve years later, Eccles provided a remarkable survey and synthesis of broad topics related to the human experience of consciousness in his first series of Gifford Lectures (1979) at the University of Edinburgh. His initial concept limited commun-

Area Essential to Consciousness

Cerebral Localization of Consciousness as Established by Neuropathologic Studies

George N. Thompson and J. M. Nielsen

University of Southern California School of Medicine and Los Angeles County General Hospital

LOS ANGELES CALIF.

CONCLUSIONS

1. It is concluded from these case studies that bilateral thalamic and hypothalamic lesions result in impairment of consciousness to the degree of a lethargic stupor from which the patient can be partially aroused.
2. Either bilateral thalamic lesions alone or hypothalamic lesions alone may produce this syndrome.
3. The depth of stupor does not seem to depend upon the extent of involvement of these structures.
4. Destructive lesions of the junction of the hypothalamus and of the subthalamus with the mesencephalon result in deep coma from which no degree of recovery is possible.
5. Both the stupor and the coma are permanent and irreversible.
6. It is concluded that the engramme system essential to crude consciousness is located where the mesencephalon, subthalamus, and hypothalamus meet.
7. Pathological sleep or stupor may result from lesions just above this area, in either the hypothalamus or thalami, if the lesions are bilateral.
8. A specific nuclear mass essential to consciousness is unknown to us. The structure destroyed by the lesion which we describe may be a crossroads, that is an intercommunicating fiber system.
9. A specific nuclear mass may lie adjacent to the periventricular grey matter.

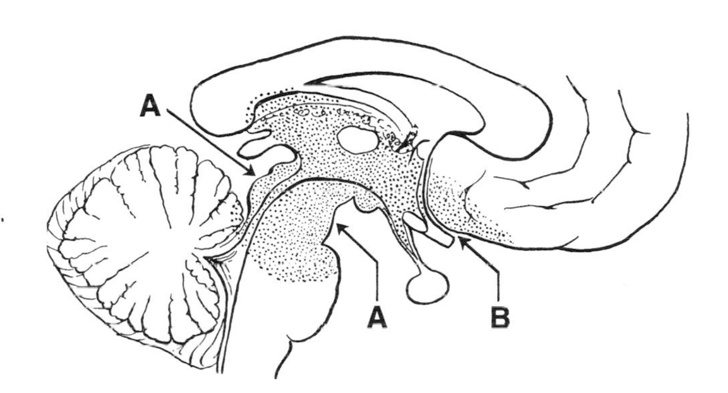

Fig. 10.6. Left—Two panels from a series exhibited at the American Medical Association meeting in 1948, showing brain sections from patients who had had long-term unconsciousness before death. (Courtesy of George N. Thompson.) Right—Diagram of monkey brain on which stippling indicates the boundaries of the region found experimentally to be essential for consciousness. (From G. Jefferson, 1958, p. 734.)

ication between the brain and the environment ("the world") to what was then called the "dominant" hemisphere (*see* Fig. 10.8, left, p. 209). In the light of the revealing work of Roger Sperry and his group on patients with commissurotomy ("split brain") demonstrating laterality of function (*see* Chapter 6, this volume), Eccles moderately revised his views, postulating that the minor hemisphere could communicate with the world, but only indirectly through the commissure to the other side (Fig. 10.8, right).

Sperry's own views on consciousness were crystallized by the thorough testing he and others carried out not only on patients with therapeutic commissurotomies but also on a case of agenesis of the corpus callosum, studies which revealed that the whole brain is more than the sum of its parts. In his words:

> Another thing to come out of all this . . . is a modified concept of the nature of mind and consciousness. . . .

Fig. 10.7. An impromptu photograph of three eminent neuroscientists of the midtwentieth century: the Australian, John C. Eccles (right) with two Frenchmen, Paul Dell (left), and Alfred Fessard (center) taken at a meeting in the 1960s.

> This is a view that postulates the presence in the brain of mental as well as physiological forces, and contends further that the phenomena of conscious awareness play an important active role in shaping and directing the flow pattern of cerebral excitation. . . . [They] interact with and largely govern the physiochemical and physiological aspects of the brain process. It obviously works the other way round as well, and thus a mutual interaction is conceived between the physiological and mental properties (Sperry, 1968, pp. 135–136).

A pioneer investigator of the neurophysiology of consciousness, Benjamin Libet at the University of California, San Francisco, and his associates applied sophisticated electronic techniques to the unanesthetized human brain and found (Libet, Wright, Feinstein, and Pearl, 1979) that electrical stimulation of the exposed sensory cortex must be maintained 500 ms to 1000 ms before subjective sensation is experienced, whereas the neural events of the evoked cortical potential are seldom prolonged beyond 25 ms to 50 ms. Such dissociation in timing of physical and mental events raised difficulties for theories of psychoneural identity, again reinforcing Cuvier's words, "Although the brain is much studied no one has not left something to be discovered by his successors" (quoted by Serres, 1827).

THALAMIC FUNCTIONS AND EFFERENT PROJECTIONS

Returning to the midnineteenth century, the thalamus was then regarded as the true visual center, an idea from Galen's time and the focus of a dissertation (1834; *see* Fig. 10.9, p. 210) by the Dane, Sophus Augustus Wilhelmus Stein (1797–1868). He listed more than 14 synonyms for "thalamus" then in use and, based on careful dissections, he described the structures in man (*see* Fig. 10.10, p. 211) and in many lower animals as the origin of the optic nerves; Stein also thought the thalami connect with the entire cerebral cortex.

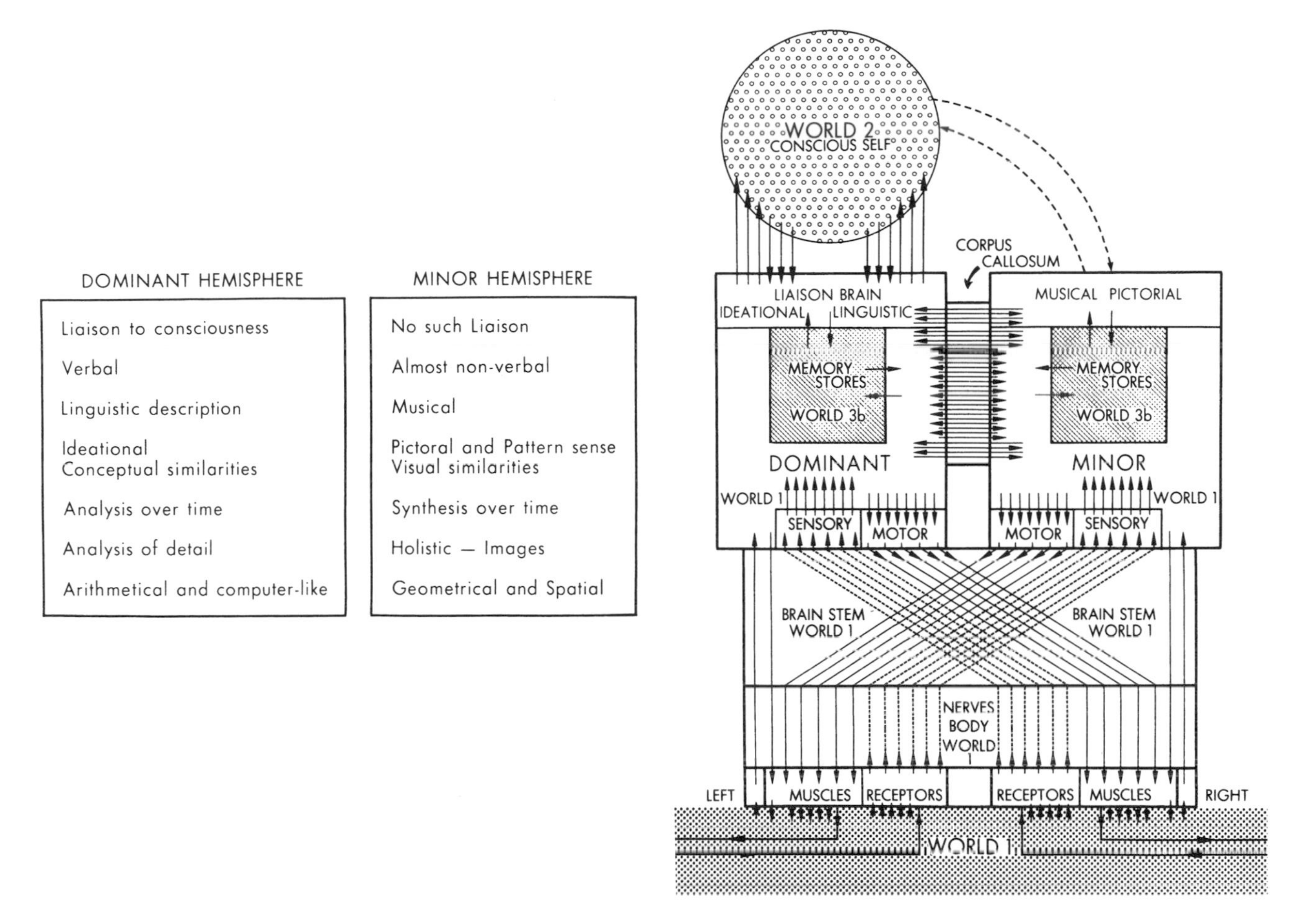

Fig. 10.8. Left—A comparison of the specific activities of the two cerebral hemispheres from the viewpoint of dominance of one hemisphere over the other, as understood during the 1970s. Right—Eccles's geometric diagram of the principal lines of commun-ication from peripheral sensors to the cerebral hemispheres, showing only a tentative participation (broken lines, top) of the minor hemisphere in consciousness. (From Eccles, 1979, pp. 222, 223, Figs. 10–7, 10–8.)

The European concept of sensory thalamic function recorded by Stein and earlier by Burdach was echoed in the English *Cyclopaedia* (1835), edited and extensively authored by Robert Bentley Todd (1809–1860). He described the optic thalami as "the principal foci of sensibility without which the mind could not perceive the physical change resulting from a sensitive impression" (vol. 3, p. 722 M). In a more sweeping context, he envisioned the encephalon as a series of centers serving intellect, volition, sensation, muscular coordination, emotion, and respiration and deglutition (ibid, p. 722 N). Although the ordering of those centers presaged Hughlings Jackson's hierarchic levels, their self-containment without interconnections precluded any idea of levels of control or inhibition.

A more substantively correct treatise on the connections and functions of the thalamic nuclei was published in 1865 by Jules Bernard Luys, a Parisian of imposing presence (*see* Fig. 10.11, p. 212). A biographer (M. B., 1897, p. 141) wrote that Luys's synthesis of what was then known about the cerebrospinal nervous system gave "une impulsion nouvelle et durable" to French thought along those lines that had been abandoned after Gall's theories were discredited (*see* Ritti, 1897). Luys's excellent three-dimensional depictions (*see* Fig. 10.12, p. 213) endeared him to a modern peer in that genre, the American neuroanatomist, Wendell J. S. Krieg, who wrote that "Luy's [sic] book . . . marks the beginning of knowledge of thalamic function" (1970, p. 56). With schemata, three-dimensional drawings, and even crude photographs, Luys presented an organized concept based on his own and others' experimental evidence and his extensive clinical observations. He revived Procháska's

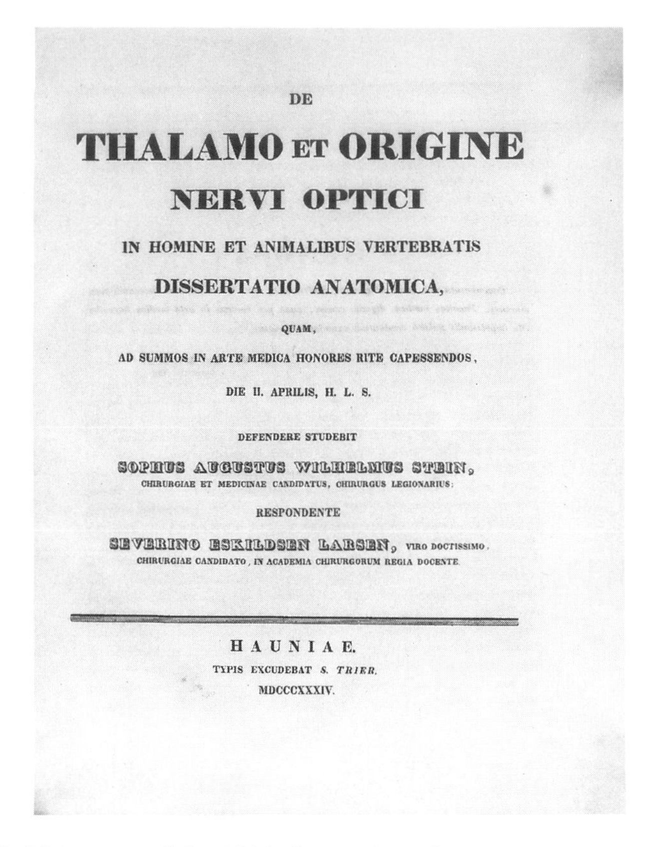

DE

THALAMO ET ORIGINE

NERVI OPTICI

IN HOMINE ET ANIMALIBUS VERTEBRATIS

DISSERTATIO ANATOMICA,

QUAM,

AD SUMMOS IN ARTE MEDICA HONORES RITE CAPESSENDOS,

DIE II. APRILIS, H. L. S.

DEFENDERE STUDEBIT

SOPHUS AUGUSTUS WILHELMUS STEIN,

CHIRURGIAE ET MEDICINAE CANDIDATUS, CHIRURGUS LEGIONARIUS;

RESPONDENTE

SEVERINO ESKILDSEN LARSEN, VIRO DOCTISSIMO,

CHIRURGIAE CANDIDATO, IN ACADEMIA CHIRURGORUM REGIA DOCENTE.

HAUNIAE.

TYPIS EXCUDEBAT *S. TRIER.*

MDCCCXXXIV.

Fig. 10.9. Title page of the 1834 dissertation of S. A. W. Stein in which he compared the brains of many vertebrates including humans and described the thalami as the origin of the optic nerves. That publication was the earliest of several distinguished monographs written solely on the thalamus.

appellation, *sensorium commune*, and suggested that each of the then-known four thalamic nuclei had connections to specific cortical areas and served a specific sensory modality.

The credit for initiating serious work on the thalamus held by Luys is shared by Bernard von Gudden (1824–1886) who substantiated the earlier evidence with new approaches to the problem of connections between thalamus and cortex. Familiar with the latest technical advances in tissue preparation emanating from German laboratories, Gudden's experiments on brain function were also notable in that his animals were operated neonatally and studied as adults. In 1870, his published results show that destruction of specific areas of the rabbit cortex produces atrophy of certain thalamic nuclei.[1]

One of the assistants in von Gudden's Munich laboratory was Constantin von Monakow (1853–1930; Fig. 10.13), the Russian–Swiss son of a wealthy nobleman. In his private Zurich laboratory, later associated with the university, von Monakow extended his mentor's experiments and, again in rabbits, demonstrated atrophy of the contralateral superior colliculus after enucleation at birth, and in other studies complete degeneration of the lateral geniculate body (whereas the remainder of the thalamus was intact) by neonatal removal of the occipital lobes (1882).

Each of the three preceding investigators had a modern partisan who claimed for the man he championed a preeminent role in promoting knowledge of the thalami. In *Founders of Neurology* (Haymaker and Schiller, 1970), Kreig described the discoveries of Luys (p. 56), Papez wrote of von Gudden (p. 45), and Yakovlev eulogized von Monakow (p. 487). Historians have conferred that distinction, however, on von Gudden, recognizing the novelty

[1]For a concise review of those studies, *see* Walker (1938, p. 7).

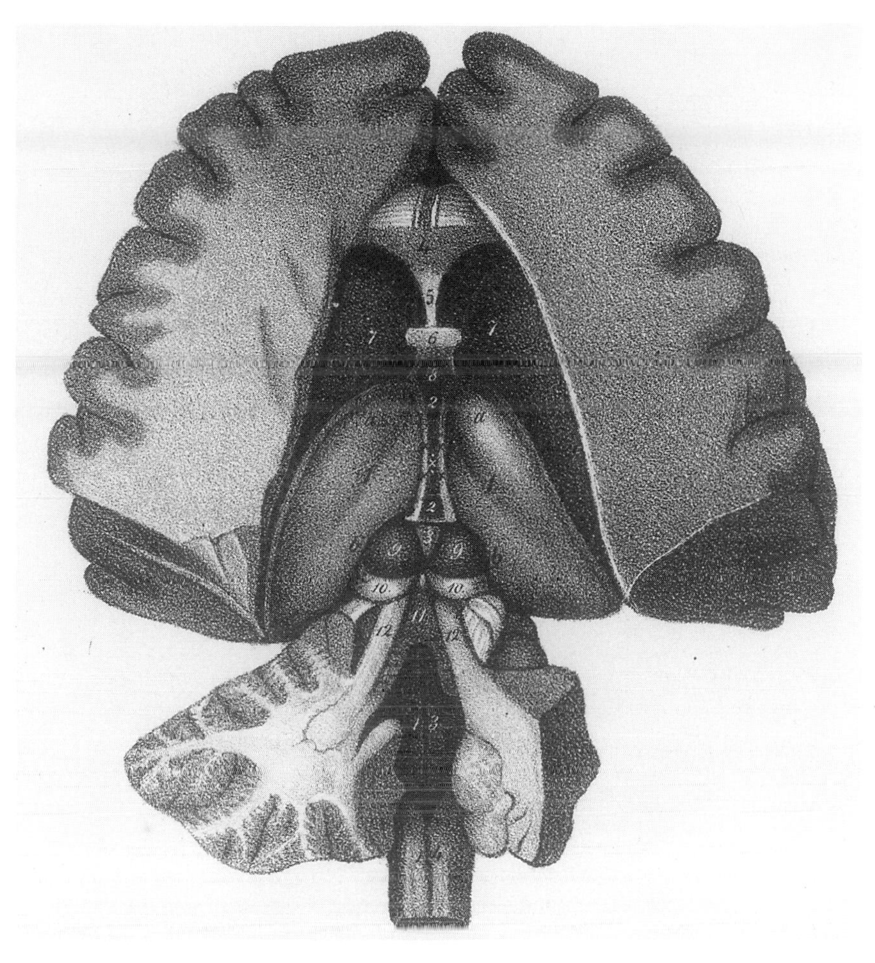

Fig. 10.10. Stein's dissection of a human brain showing it "true to nature" (ad nat. verit.). 1—Thalami nervi optici; 2—third ventricle; 3—pineal gland; 4—anterior corpus callosum; 5—septum pellucidum; 6—anterior fornix; 7—corpus striatum; 8—anterior commissure; 9,10—corpora quadrigemini; 11—valvulae magnae; 12—cerebellar peduncles; 13—fourth ventricle; 14—inferior medulla oblongata. (From Stein, 1834, Fig. 1.)

and significance of his chronic preparations (*see* Clarke and O'Malley, 1968, p. 606, passim).

The work of August Henri Forel (1848–1931; *see* Fig. 10.14, p. 215) contributed to the knowledge that the thalamic nuclei are separate in function as well as in cellular morphology. This Swiss psychiatrist-neuroanatomist made a clear statement (1887) of what was later called the neuron doctrine, based on pathological and functional evidence, within two months of that of his countryman, Wilhelm His. Additional support for the idea of thalamic nuclei as distinct units emerged when Flechsig claimed (1896) that differentiation between certain thalamopetal cells could be based on the progressive myelination of their projections, which constituted "gateways" to the brain and its functions when they continued to the cortex.

Many years later, a crucial point was raised by the Hungarian neuroanatomist, Stefan Polyak, then at the University of Chicago. He asked if those cortical projection areas were "to be regarded exclusively as 'gateways' of the cerebral cortex for the incoming impulses . . . or, do they participate likewise in higher integrative processes, . . . thus depriving the intercalated regions of the exclusive monopoly of these higher processes?" (Polyak, 1932, p. 213). Polyak's ablation experiments on monkeys had shown the unexpectedly wide extent of the somatic sensory cortex to include the precentral "motor" region.

The early microscopical studies of Luys and Forel, followed by those of Nissl (1889), demonstrating that the thalamic nuclei are distinct units, received physiological support when it became

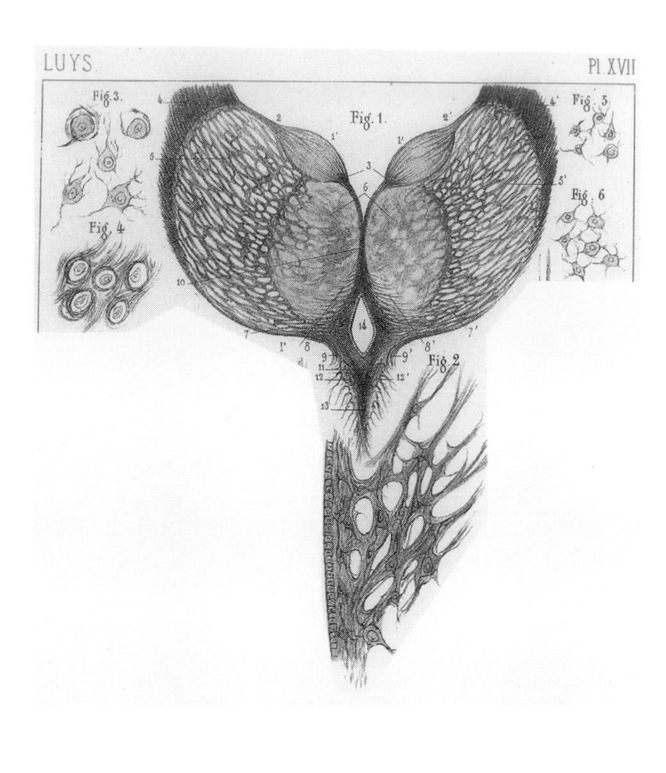

Fig. 10.11. Jules Bernard Luys's talent for draughtmanship created orderly depictions of the connections of the thalamus. 1—Anterior nuc.; 2—afferent fibers to ant. nuc.; 3—efferent fibers; 4—converging fibers from cortex to external thalamus; 5—secondary converging gray fibers which lose themselves in center (6) of thalamus; *see* original for additional legends. (Portrait from obit. signed M. B., 1897, p. 141; drawing from Luys, 1865, facing p. 39, Plate XVII, Fig. 1.)

possible to confine a lesion to a single nucleus so that the resulting behavioral deficit was relatively unaffected by damage to surrounding parts. Two French experimentalists, Sellier and Verger, used bipolar electrolysis "to study carefully with methods not used by our forerunners" (1898, p. 706) the nuclei of the couche optique. They applied (usually) a current of six milliamperes and varied the duration according to the desired size of the lesion,[2] then eight to ten days later tested the animals' perception of body position, heat, and touch. The modern nature of their experiments is striking; they also verified histologically the size of the lesions and illustrated them in the published report. The authors concluded that the "central ganglions" serve if not the same at least very similar functions as the cerebral convolutions, in other words, a specificity of function. That principle was elaborated about 30 years later from a laboratory where Dusser de Barenne had instituted the use of strychnine (neuronography) to explore thalamocortical connections. A systematic mosaic of connections constructed with von Gudden's retrograde atrophy method applied to newborn rabbits demonstrated that projections from anteroposterior sites on two medial thalamic nuclei are represented on cortex in inverse order (Stoffels, 1939; Fig. 10.15).

Coincident with the revelations about the thalami from the experimental laboratory, a scattering of clinical observations denoting thalamic involvement was appearing in the literature. The first description of a case of thalamic pathology with specific symptoms was published in excruciating detail in England in 1825 (Hunter) and a simi-

[2]This was the first application of the method for lesion-making selected by Clarke and Horsley to use with their novel stereotaxic instrument (*see* Chapter 9).

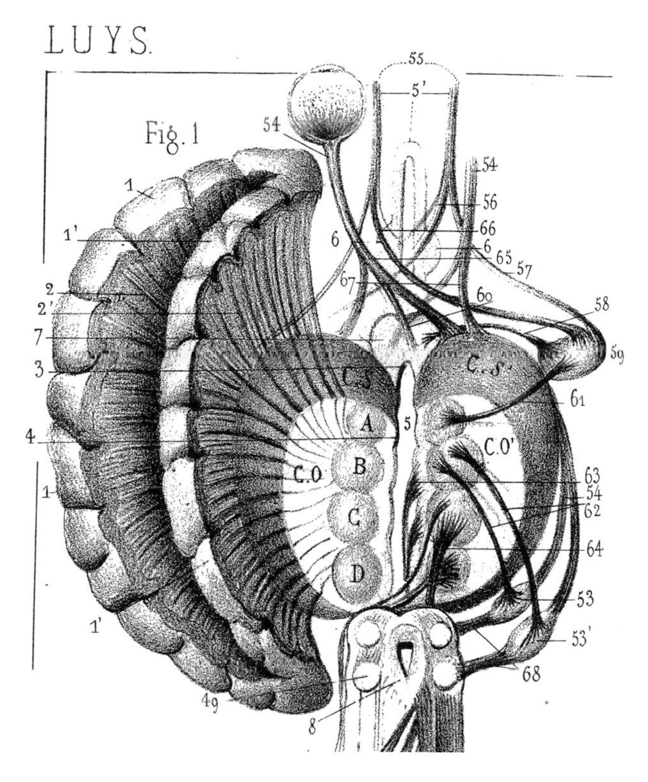

Fig. 10.12. A theoretical figure by Luys showing fibers converging on the thalamus. 1—Cerebral cortex; CS—corpus striatum; CO—couche optique (thalamus); A,B,C,D—anterior to posterior nuclei; 2,3,4—superior fibers from the cortex converging on the thalamus after passing through the corpus striatum; 5—walls of third ventricle; see original for additional legends. (From Luys, 1865, facing p. 4, Plate I, Fig. 1, ×3/4.)

lar patient was followed by Richard Bright (1837) at Guy's Hospital. Among the later neurologists with such cases, James Crichton Browne (1840–1938) stands out (Viets, 1938). This energetic and intelligent director of the West Riding Pauper Lunatic Asylum noted that the sensory cortex and the thalami share the same arterial blood supply. From observations on seven patients, he concluded that the thalami were "probably" sensory ganglia, but more interestingly he proposed a mechanism to explain the impairment of reflexes on the side opposite the lesion. Browne wrote (1875, p. 252 passim): "The sensorial ganglia . . . almost invariably consult the cerebrum before dealing with the impressions which they receive. . . ." He envisioned an "encephalic loop current" as part of a reflex pathway to explain the experimental observations made on himself: he felt pain after, not before, withdrawing his toe from a noxious stimulus, there having been a "sacrifice of time" in transmission of the loop current. It seemed to him that "undulations of nerve force are being perpetually diffused centrifugally from or through the optic thalami" and it was not difficult to understand how a lesion might create an "enfeeblement" of reflex activity.

Eight decades after Browne, George Bishop at Washington University championed the spinothalamic tracts' primitive function in pain transmission. In an unpublished talk to the Washington University Medical Center early in 1958, Bishop elaborated on his finding that the ventral lateral thalamic nuclei do not carry pain to the somesthetic cortex; in addition to experimental evidence, he invoked phylogeny, pointing out that the thalamus is the pain center in primitive brains and therefore the pain tracts are the classical spinothalamic projections situated in the more medial thalamic nuclei.

During Browne's enlightened directorship at West Riding, David Ferrier conducted experiments there on cats and dogs that disproved Meynert's (1872a) theory, concurred in by the French, that the thalamus has motor function. Ferrier's contemporary and associate at University Hospital, Queen

Fig. 10.13. Constantin von Monakow, photographed during the late nineteenth century, in his early career as a clinical and experimental neurologist, demonstrated the close relationship between cortex and thalamus in auditory and visual functions.

Square, Victor Horsley, like his colleague was caught up in the experimental approach to the many neurologic questions confronting them in their practices. When Horsley suggested a research topic to an American postdoctoral guest in his laboratory, it was the optic thalamus, "about which little was known" (Sachs, 1909, p. 95). Ernest Sachs, Sr. spent the greater part of two years in London, carrying out carefully executed experiments on cats and macaques with the novel stereotaxic instrument (*see* Chapter 9). His relatively discrete electrolytic lesions of the thalamus confirmed von Monakow's 1895 work showing connections between thalamus and prefrontal cortex. Although he contributed little new information, because he used the capricious Marchi stain for his histological preparations (Walker, 1938, p. 9), Sachs was among the first to electrically stimulate the thalamus. In the clinic, the thalamic syndrome, a condition also called thalamic hyperesthetic anesthesia, is the best known of the several neurologic disorders first described by Dejerine from his wide experience at the Salpêtrière and Bicêtre hospitals in Paris. He related thalamic disease or damage to a combination of specific symptoms, dissociated those due to motor deficits from sensory loss, and ascribed both to derangement of the vascular supply (Dejerine and Roussy, 1906).

The comparative and embryological studies of Edinger in Europe and the Herricks in America (*see* Chapters 1 and 9) contributed to sorting out the morphologic complexities of the many thalamic nuclei. An interest in function, however, relatively flagging during the early twentieth century, was refocused by Le Gros Clark in two lectures to the Royal College of Surgeons in 1932. Their publication in *Brain* represented the second historically significant treatise devoted solely to the thalami.

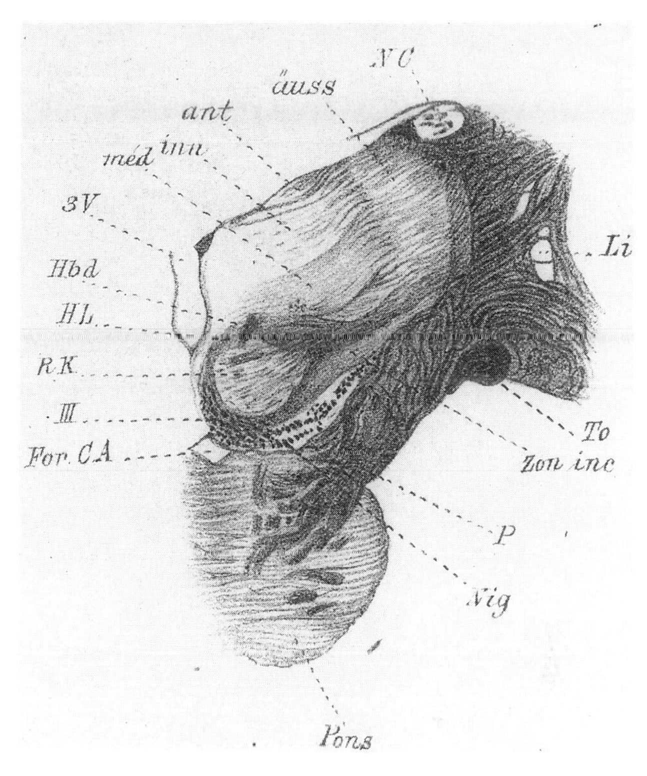

Fig. 10.14. August Forel's portrait was painted by Oskar Kokoshka and hangs in the Kunsthalle, Mannheim, Germany Forel's drawing shows the thalamic nuclei. NC—Caudate nuc.; 3V—third ventricle; ant, inn, anss—superior, inner, outer thalamic nuc.; med—centre median; Hbd—habenula; HL—long bundle of the crest of Meynert; RC—red nuc.; III—oculomotor nerve; For CA—ant. foramen; Nig—substantia nigra; P—pes pedunculi; Zon inc—zona inserta; To—optic tract, Li—lenticular nuc. (From Forel, 1877, Plate VII, Fig. 7.)

Clark set out to "clear the terminological atmosphere," then evaluated the early work showing that thalamic fibers constitute the most direct route to the cortex, and presented evidence that the thalamus is "the anatomical equivalent of the very threshold of consciousness" (Clark, 1932, p. 407).

By 1938, thalamic cytoarchitecture and connections with the cortex were fairly well known, the former in part through Cajal's work and the latter assisted by neuronography studies (Dusser de Barenne and McCulloch, 1938). In addition, Papez had concluded, from a detailed degeneration study of serial sections of the brain of a dog from which one hemisphere had been removed six months previously, that "the dependence of the neothalamic nuclei on the cortex is evident" (1938, p. 118). The year 1938 also saw publication of the third important monograph on the subject, *The Primate Thalamus*, by the Canadian-born neurologist and neurosurgeon, A. Earl Walker. Based in part on extensive experimental studies (e.g., Walker, 1936) this milestone treatise provides an example of phylogeny being called on to "prove" an idea. Explaining that the group of midline thalamic nuclei is "relatively constant throughout the mammalian scale," Walker wrote: "Because it is present in the thalami of most primitive animals not possessing appendages, it must be related to the axial portion of the body" (Walker, 1938, p. 239). Relatedly, the author invoked phylogenetic correlations in suggesting the functions of other, nonmidline nuclei, but although the intrinsic connection between some of them with intralaminar nuclei was known, their functions were "unclear" or "even less clear" (ibid., p. 243). Among the details of the specific representation of efferent thalamic projections to cortex

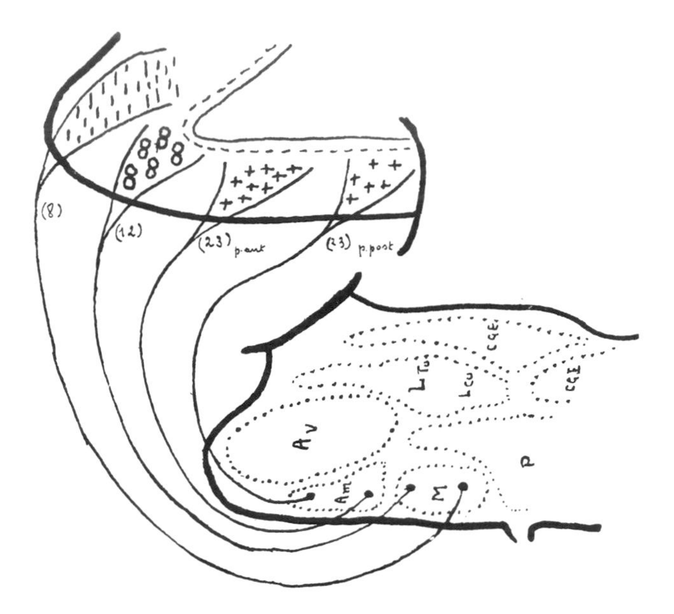

Fig. 10.15. A systematic mosaic of inverse relationships between cortical fields (top) and thalamic nuclei (bottom) was found in the rabbit after cortical ablations. P—Pretectum; Av—nuc. ant. ventralis; M—nuc. medialis; Am—nuc. ant. medialis; Ltu—nuc. lateralis, partio tuberculosa; Lcu—nuc. lateralis cuneiformis; CGI, CGE—corpus geniculatum int., ext. (From Stoffels, 1939, p. 822, Fig. 69.)

advanced by Walker's experiments on subhuman primates, an illuminating finding was that "the intensity of the thalamic projections to the different cortical areas is not at all uniform" (ibid., p. 190), as shown in Fig. 10.16. He pointed out that functionally distinct groups of thalamic nuclei revealed physiologically in the laboratory, in many clinical cases can be correlated with specific pathological syndromes, thus merging theoretical mechanisms based on anatomical relationships with clinical applicability. Walker's volume concluded with this paragraph:

> This discussion of its pathology has emphasized the complex function of the thalamus. It is the mediator to which all stimuli from the outside world congregate and become modified and distributed to subcortical or cortical centers so that the individual may make adequate adjustments to the constantly changing environment. The thalamus thus holds the secret of much that goes on within the cerebral cortex (ibid., p. 277).

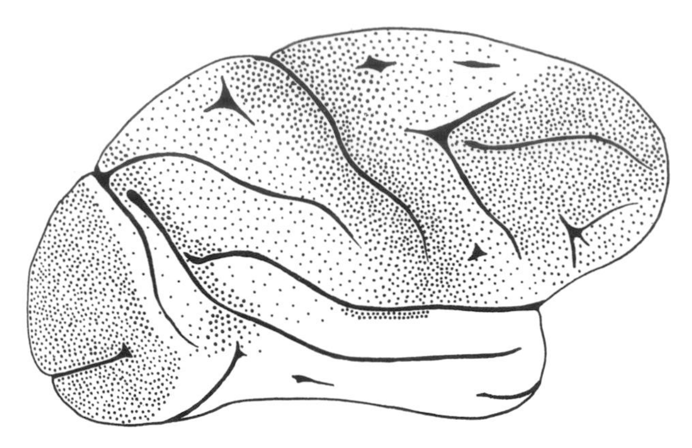

Fig. 10.16. Stylized primate right cortical hemisphere on which the relative densities of the thalamic projections to various cortical areas are depicted. Projections are most numerous in the prefrontal, pre- and postcentral, and occipital areas. (From Walker, 1936, p. 191, Fig. 68, ×4/5.)

The intimacy of the association between the environment and thalamic mediation was revealed only after suitable histological methods were at hand that could bring into focus the complete picture of information transfer from the outer world to cortex. In brief mention (*see* Chapter 7 for details), the introduction of radioactive markers to trace normal pathways from retina to thalamus to specific layers in primary visual cortex showed not only the strict distribution of the projection terminals but also their correlation with the physiologically demonstrated orientation-dominant and hemispheric-dominant behavioral responses. This rare example of knowledge of function preceding the discovery of form concerns the work of Grafstein, Hubel, and Wiesel, and their associates from the late 1950s to the early 1970s.

THE DIFFUSE PROJECTIONS

Concomitant with the massive technological effort associated with the Second World War, electronic amplification and recording methods were advanced to where they could provide meaningful information in many areas of endeavor. After the horrendous, worldwide disruption of basic biomedical research, progress in electronic technology achieved during the war effort was particularly (but not exclusively) applicable to furtherance of neurophysiologic investigations. This was especially notable in studies of the central nervous system, into which the evoked potential technique

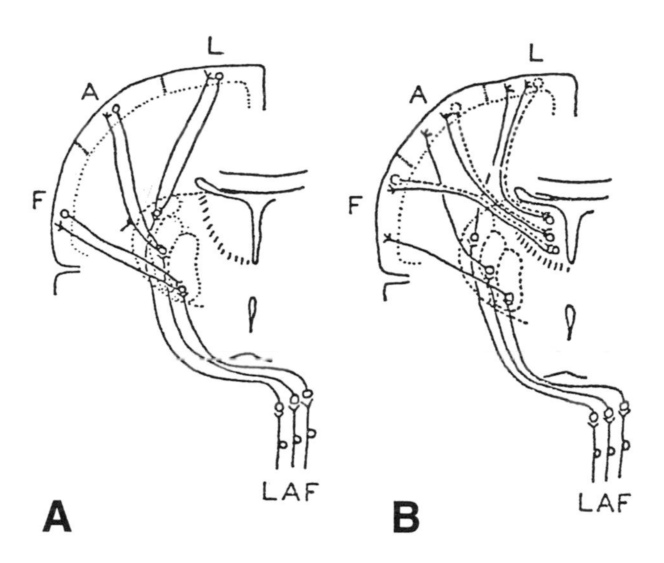

Fig. 10.17. By comparing a diagram (A) from Dusser de Barenne and McCulloch (1938) with their own schema (B), Dempsey and Morison illustrated their conclusion that data from recruiting and spontaneous cortical potentials suggest a second and diffuse thalamocortical system involving the dorsal medial nucleus and sensory cortex (dotted lines), in contrast to the well known specific circuits between the ventral lateral thalamic nuclei and cortex. L, A, F—leg, arm, face. (From Dempsey and Morison, 1942b, p. 306, Fig. 5.)

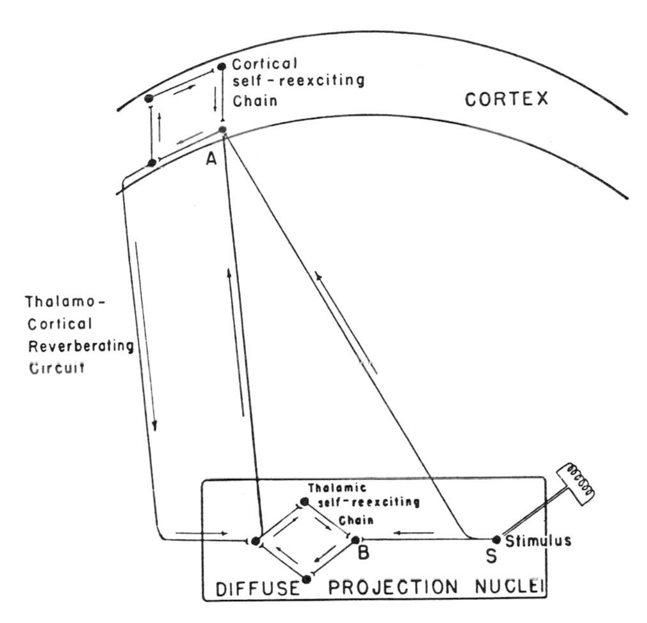

Fig. 10.18. Diagram of the mechanism proposed for development of the recruiting response, based on physiologic similarities of cortical and thalamic responses, as reverberating circuits A and B. (From Verzeano, Lindsley, and Magoun, 1953, p. 194, Fig. 15.)

moved cephalad from the peripheral nervous system, where the axonologists had used it to great advantage. At this time also electroencephalography, a technology of proven clinical value, became a research tool essential to the demonstration of a nonspecific relationship between thalamus and cerebrum.

The groundwork for identification of the widespread influence of thalamus on cortical activity was carried out by Robert S. Morison and Edward W. Dempsey at the Harvard laboratory of physiology in the early 1940s. They found the physiologic relation between the two regions while stimulating the smaller intralaminar nuclei of the cat thalamus: a single shock produced over most of the cortical surface a train of negative potentials of gradually increasing amplitude which crested, then faded away. These spindle-shape "recruiting" responses were also obtained by repetitive stimulation, waxing and waning during five-per-second stimuli to the nuclei. Later, Morison justified the diffuse image: "I think we were the first to use the term 'diffuse.' We did it because there are places in this system . . . from which the entire system can be activated" (1954, p. 110). Because of its diffuseness, the response was considered to be nonspecific, in contrast to the already known specifically localized cortical response to stimulation of the larger thalamic nuclei, the augmenting response. Dempsey and Morison continued their initial observations (as did many other neurophysiologists) and soon the recruiting response was recognized as similar to the spontaneous cortical rhythm and resembled it in detail (Dempsey and Morison, 1942a,b; Fig. 10.17). Later they wrote, "Evidence is presented that a type of spontaneous effect . . . results from activity of the thalamic relay nucleus and its cortical projection" (1943, p. 296). That same year, it was proposed (Rose and Woolsey, 1943) that the nuclei reticularis thalami might serve as the distribution system for the diffuse projection (*see* below). The pursuit of the mechanism of the recruiting response became widespread among neurophysiologists; among them the active group at Northwestern University Medical School in Chicago (later at the University of California, Los Angeles), in the course of examination of the diffuse projections to cortex, diagrammed their concept of reverberating circuits (Verzeano, Lindsley, and Magoun, 1953; Fig. 10.18). Eventually, intracellular recording demonstrated that neurons participating in the recruiting response are also involved in EEG desynchronization (arousal) (Purpura and Shofer, 1963). Yet in spite of intense investigation, the recruiting response "remains as one of the largely unexplained phenomena in the electrophysiology of the cortex and thalamus" (Jones, 1985, p. 632).

A sharp focus on the diffuse projections was brought about at the midtwentieth century by the discovery in Magoun's Northwestern laboratory of the ascending reticular activating system (*see* Chapter 12). The finding that the pattern of cortical phasic activity (shown by the EEG) is profoundly altered by stimulation of the brain-stem reticular core was quickly substantiated (Jasper, 1949), but the import of an ascending arousal system was obscured by the attention accorded the other diffuse projection to cortex, the thalamic system. That situation was evident at a meeting a few months after the Northwestern work was published.

The symposium, held at Atlantic City, was presided over by Earl Walker and signaled unprecedented progress in understanding the physiology of thalamocortical relationships. The keynote paper by two prominent figures in the field who collaborated extensively—first at Johns Hopkins and later at the University of Wisconsin, Madison—the neuroanatomist Jerzy E. Rose and the neurophysiologist Clinton N. Woolsey summarized the state of knowledge of the intricate organization of the thalamic nuclei: "The available data concordantly suggest that the mammalian thalamus consists of three divisions different from each other in their phylogenetic and ontogenetic development, and in their relations to the cortex" (Rose and Woolsey, 1949, p. 402). Based on a fruitful coupling of cortical mapping using evoked potentials with histological studies of embryologic development, they characterized the epithalamus as independent of the "endbrain" and reduced in size in higher forms, the dorsal thalamus as projecting to primary sensory cortical areas which become progressively constricted in higher mammals, and the ventral division consisting of the independent ventral lateral geniculate body and the nucleus reticularis thalami as projecting "upon a large number of cortical fields" and "capable, presumably, of evoking generalized cortical activity." Reminiscent of C. J. Herrick's thesis that the increased size of the cortical association areas relate to the higher nervous functions, Rose and Woolsey stated: "[I]t appears that a very prominent feature of the phyletic development of the mammalian neocortex consists in the growth and differentiation of those sectors which are intercalated between the primary projection fields. . . . [T]hose which are so separated, drift, so to speak, farther apart the more highly the cortex is developed" (ibid., p. 401). The configuration on the cortex of primary and secondary somatosensory areas illustrated by Rose and Woolsey (Fig. 10.19) shows that in lower mammals the concentration of thalamocortical axon terminals favors the primary areas (White, 1979, p. 281). In humans, however, "the total extent of the primary visual, auditory and somatic sensory fields is relatively small and the total cortical surface of these fields together with the primary limbic fields and the primary motor area is probably less than 15 percent of the total cortical surface" (Rose and Woolsey, 1949, pp. 400–401).

Another seminal paper at the Atlantic City symposium was presented by Herbert Jasper from Montreal. He described some characteristics of the diffuse thalamocortical system which "seems well established" to exist "with independent projection to the cortex overlapping that of the better known specific . . . systems" (Jasper, 1949, p. 405). And he concluded, ". . . there exists a separate regulatory system involving thalamus and other brain stem structures which acts upon the cortex, controlling the form and rhythm of the background upon which afferent impulses must act. . . ." (ibid., p. 418). In discussing the paper, Magoun said, perhaps with a tinge of sarcasm: ". . . today, this diffuse system appears to compete seriously with all the rest of the thalamus together in its functional significance" (Magoun, 1949, p. 420). Thus did the two Titans of diffuse systems duel politely in subtle competition.

To close the symposium, Walker lamented the isolationism of thought about the thalamus brought about by nonstandard nomenclature, and offered some conciliatory remarks regarding the nice correlation between what was known about "Jasper's thalamic reticular system" and Magoun's "lower reticular centers," summarizing the proceedings thus: "There appear to be at least three mechanisms of thalamortical [sic] integration. The first is that of the well known thalamic relay systems, the second that of the thalamic elaborative or associative system and the third that of the thalamic reticu-

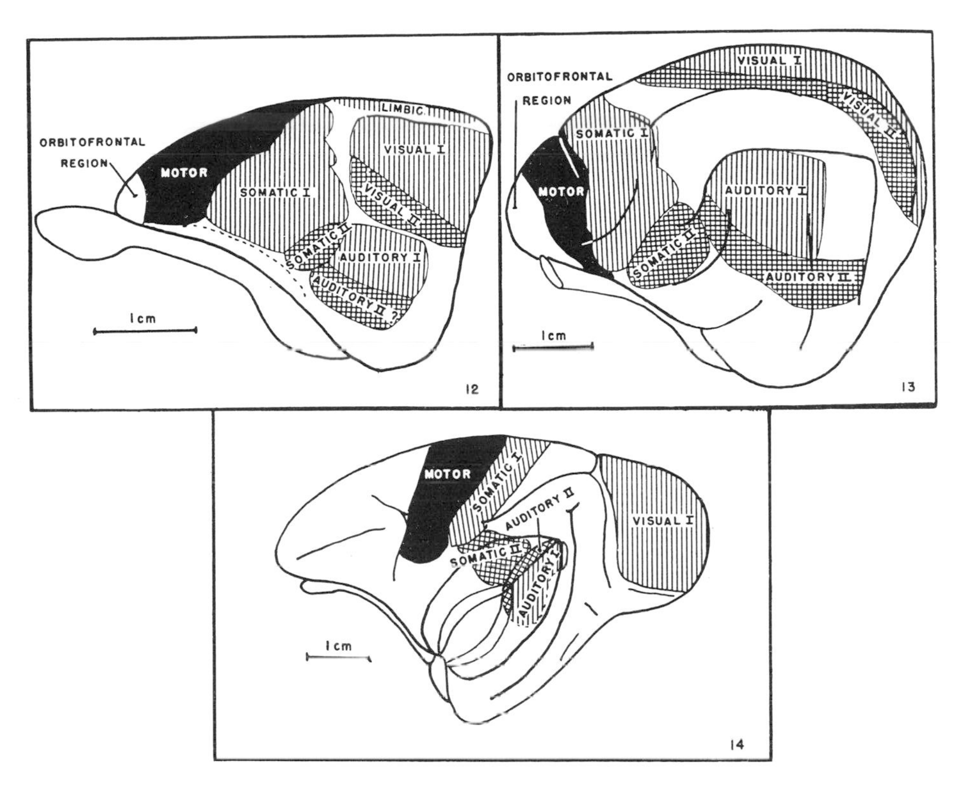

Fig. 10.19. Diagrams of the cortical fields of rabbit (left), cat (right), and monkey (below) to show primary (hatched) and secondary (cross-hatched) projection areas of visual, acoustic, and somatic afferent impulses. Note that higher position on the phylogenetic scale coincides with smaller primary sensory areas and larger secondary areas. *See* text for comparison with human brain; images not in scale. (From J. Rose and C. N. Woolsey, 1949, p. 400, Figs. 12–14.)

lar system" (Walker, 1949, p. 453). The chairman also noted the advantage conferred by the semi-independence of the thalamus and cortex manifested by thalamic spiking without concomitant cortical spikes, "since it allows the thalamus to react and adjust to extraneous stimuli without affecting cortical activity or consciousness" (ibid., p. 453).

The differentiation of thalamocortical pathways and their nuclei as specific and nonspecific in function opened at least two questions that demanded further research. The first was at what level do the linkages between specific and nonspecific systems occur? Jasper and Ajmone-Marsan (1952) found an interconnection at the cortical level for visual and intralaminar responses. Regarding a more caudal interconnection, the report (Nauta and Whitlock, 1954) of anatomic connections of intralaminar nuclei with ventralis posterior and other specific relay nuclei was welcomed as implying the possibility of a diencephalic interaction between specific and nonspecific systems: "The basal ganglia are the traditional relay points in the extrapyramidal descending paths in the motor system. The above observations suggest in addition the basal ganglia serve in much more general cerebral processes" (Magoun, 1954b, p. 114).

The second question was the reintroduction of the problem of structure that had been addressed 75 years earlier (notably by Luys and Forel), but with a new twist: Are there morphologic differences between specific and nonspecific neurons? Scheibel and Scheibel (1966) solved that problem with their Golgi-stained comparative studies of the ventrobasal nucleus of the thalamus, a specific system, and the diffuse centre median-parafascicular complex. The ventrobasal pattern of organization, which includes bushy terminal arborizations accommodating repetitive firing, implies a correlation of organization with specificity and physiological differentiation. In contrast,

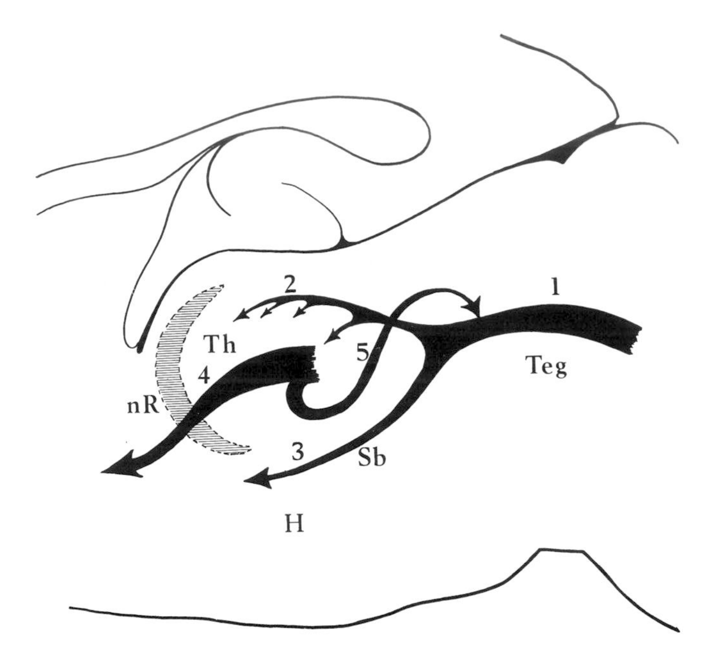

Fig. 10.20. A graceful schematic delineates the possible rostral courses of axons from brain-stem reticular core (1) and thalamic nonspecific (4) systems. Teg—Tegmentum; 2—branch passing dorsally to thalamic intralaminar (Th) and dorsomedial fields; 3—branch passing ventrally through subthalamus (Sb) and hypothalamus (H), to nuc. reticularis (nR). The thalamic system sends a caudal branch back to the tegmental level, perforates the nuc. reticularis, and continues rostrally. (From Scheibel and Scheibel, 1967, p. 84, Fig. 15.)

neurons of the nonspecific fields are homogeneous, with the afferents seeming to disappear smoothly into the neuropil matrix. This symbiosis of structure and function affirmed the individuality not only of the thalamic nuclei but also of the neurons within them.

The existence of an inhibitory action of thalamocortical neurons, as suggested by Jasper (*see* below), was supported by the Scheibels' extension (1967) of their studies on the structural organization of a specific and a nonspecific thalamic nucleus to the corticopetal projections of nonspecific nuclei. They examined the relation of intralaminar axonal and dendritic neuropils to adjacent specific and associational nuclei, as well as projections to the cortex and drew a circuit model (Fig. 10.20) that compared the pathways to cortex of the brain stem with those of the thalamic system and showed the dendrites of nucleus reticularis traversed by fibers coursing between thalamus and cortex. Their findings plus what was known of reticularis neurons led to the proposal that reticularis neurons may be inhibitory to their specific and nonspecific targets alike. Later, intracellular recordings (Schlag and Waszak, 1970) revealed that reticularis neurons discharge profusely when other thalamic neurons are silent, in striking agreement with the Scheibels' hypothesis.

The idea of possible inhibitory mechanisms operating on thalamic projections was suggested by Jasper in experiments using intracellular recording predominantly from the thalamic reticular system: Cats and monkeys habituated to a tone can be aroused by a novel tone after ablation of the auditory cortex (Li, McClennan, and Jasper, 1952).[3]

> These observations suggest the possibility that the function of the reticular system in normal adaptive . . . behavior may be more in the nature

[3]This work was deemed a milestone contribution because it introduced cellular (unitary) studies (Steriade, 1981, p. 330).

Fig. 10.21. This rare photograph of Elizabeth Crosby (center) was snapped at Northwestern University Medical School during a visit by a group of prominent neuroanatomists, about 1940–1941. From left: A. T. Rasmussen, C. Judson Herrick, Crosby, O. Larsell, and the host, S. Walter Ranson.

of a prevention of a general arousal reaction to all stimuli, with a control of selective responsiveness to significant stimuli. . . . This implies that inhibitory rather than excitatory functions may be most important. . . . (Jasper, 1958, pp. 321–322).

The Montreal group believed that activation is characterized by varying degrees of excitation and inhibition "in a matrix" of patterns "held in the dendritic meshwork of the cortex" (ibid., pp. 330–331).

At the conclusion of the second edition of *The Waking Brain*, Magoun attempted to assimilate the focal inhibitory system into the larger concept of an integrated mechanism:

The consequences of the action of this mechanism are the opposite of those of the ascending reticular activating system . . . for internal excitation. The principle of reciprocal innervation . . . would appear relevant to the manner in which these two higher antagonistic neural mechanisms determine the alternating patterns of brain activity manifest as wakefulness and light [nonREM] sleep (Magoun, 1963, p. 174).

A comment made by Elizabeth Caroline Crosby (1888–1983; Fig. 10.21), one of the great comparative neuroanatomists of the twentieth century, raised a relevant and thought-provoking semantic question. In her invited discussion of thalamic connections, Crosby (1972, p. 89) asked: "Would inhibition of the activity of a neuron be regarded as a supportive function? Or should such a term be applied only when there are possible excitatory effects?" The answer applies to all situations in which inhibition is a passive rather than an active event, and relates to Hughlings Jackson's dictum that the outcome of neurological dysfunction may be due to the mere absence of the removed process or it may be the effect of a new process (*see* Chapter 9).

Elizabeth Crosby's reputation as a dedicated and knowledgable authoritory on nervous systems throughout the animal kingdom is legendary. Her doctoral dissertation under C. J. Herrick at the University of Chicago, was "The forebrain of alligator mississippiensis," and became a classic publication (1917), opening doors to her internationally. She showed the reptilian representation of cortex to be the precursor of more sophisticated cortex in higher animals, a comparative approach that underwrote her life-long interest in brain evolution. Her extensive knowledge of the human brain was inspirational in her clinical work with neurologists, neurosurgeons, and psychiatrists. Perhaps more influential than personal contacts were the reference volumes, *The Comparative Anatomy of the Nervous System of Vertebrates, Including Man* (Kappers, Huber, and Crosby 1936), and a dissection guide, *A Laboratory Outline of Neurology* (Herrick and Crosby, 1918). Both works were published collaboratively, but Crosby's hand fashioned them during her seven decades of research, interpretation, and teaching.

After the two groups working on cortipetal inputs from brain stem and thalamus had reported their research in 1949, a decade of intensive activity in Chicago, Montreal, and elsewhere worldwide continued to add new interpretations and findings. Jasper's so-called thalamic reticular system did not escape scrutiny by Magoun's group. In cats and monkeys, their studies yielded "results suggesting that the . . . system is organized for mass thalamic influence upon associational cortex" (Starzl and Magoun, 1951, p. 146) and serves as a "mechanism [by which] the electrical activity of a large portion of the cerebral mantle can be brought under control by stimulation of a very tiny area of the thalamus" (Starzl and Whitlock, 1952, p. 464). Combined with the "zone of collateralization" of the brain stem reticular activating system, there was provided a "uniquely complete liaison with the external environment" (ibid., p. 464).

A definitive account of thalamic anatomy and physiology was brought to the modern period by a fourth publication focused solely on that organ. Outwriting in scope and depth Stein's dissertation, the lectures of Le Gros Clark, and Earl Walker's monograph, Edward G. Jones, a New Zealand-born neuroanatomist, contributed *The Thalamus* (1985). In more than 800 pages and with a bibliography of about 2300 references, the volume signaled the core position of research in this region of the brain. In sorting out the thalamic nuclei, Jones stated: "The hallmark of the classical system is specificity" (ibid., p. 117), in which inputs to the thalamus seek specific nuclei, are systematically distributed according to the receptor sheet they represent, and may also end on cells of a particular physiologic class.

A definition of the anatomy of the diffuse thalamocortical system is not yet possible, but there are as many versions of its function as there are investigators of what it does. An early definition spoke in generalizations: "Thus, the diffuse thalamic projection system seems to represent a mechanism for widespread simultaneous distribution of incoming impulses to large subcortical and cortical areas—a system organized, essentially, for associative and integrative functions (Verzeano, Lindsley, and Magoun, 1953, p. 183). A quarter-century later, a necessarily more sophisticated point of view regarding the nonspecific midline thalamic nuclei was that they consist of "networks of interneurons [that] form the substrate for multiple interactions that lose the labels of their specific origins and produce the convergent activities of excitation and inhibition" (Brazier, 1977, p. 205). This perspective benefited from the information that in addition to excitatory arousal impulses to cortex, the brain stem reticular formation can also forward inhibitory signals, as already noted.

OVERVIEW OF THALAMOCORTICAL PATHWAYS AND CONSCIOUSNESS

The thalamus was known in antiquity, was recognized as having something to do with vision, and therefore was named the optic thalamus. By the late eighteenth century, Procháska (1784) viewed the dual thalamus as a relay station for reflexes, the *sensorium commune* in which sensations are transformed into movement. Early in the next century, the philosopher/psychologists were in serious debate over the mind–brain question and the relation to it of sensation and consciousness and Burdach wrote that the thalami are "the root of consciousness." An argument that lasted three decades (Lacey, 1985) centered on the emotions and whether they are initiated in motor and visceral activities which feed back to the neocortex before being felt, as the James–Lange theory postulated. Or, following Cannon, are affective states translated in thalamus into central and peripheral signals for "fight or flight" (Fig. 10.22).

In the meantime, the neurologist–anatomists, represented by Luys, von Gudden, and Monakov, established the identity and independence of the thalamic nuclei, thus laying the foundation on which rests the specificity of thalamic projections to cortex. When it was realized that all sensory inputs except olfactory form synapses in thalamus, the correlation of longstanding coma with thalamic

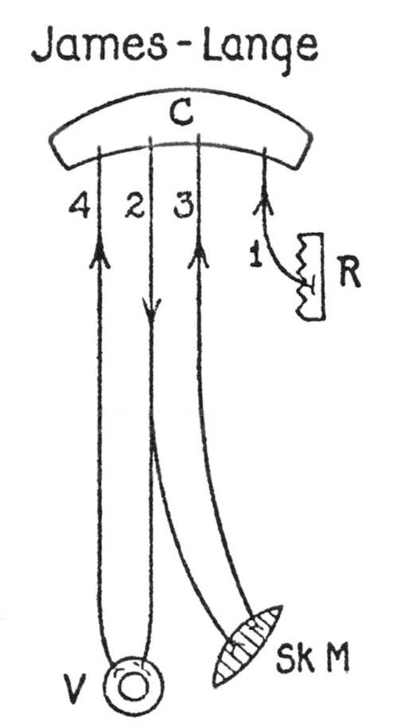

Fig. 1. Diagram of nerve connections in the James-Lange theory.

Fig. 2. Diagram of the connections in the thalamic theory.

R, receptor. *C*, cerebral cortex. *V*, viscus. *Sk M*, skeletal muscle. *Th*, thalamus. *P*, pattern. The connecting lines represent nerve paths, with the direction of impulses indicated in each instance. Cortico-thalamic path 3, Fig. 2, is inhibitory in function.

Fig. 10.22. Two diagrams of "that whale among the fishes, the theory of emotions" (M. F. Meyer, 1933), outlining the James–Lange theory and that of Cannon and Bard; *see* text. (From Cannon, 1931, p. 282, Figs. 1 and 2.)

lesions demonstrated postmortem in patients settled the question of the seat of consciousness.

In addition to the classic specific thalamocortical pathways, at midtwentieth century the existence of a diffuse system was implied by the electrophysiological studies in experimental animals (Dempsey and Morison). Clear evidence of a diffuse arousal effect on cortex of impulses from brain stem reticular core (Moruzzi and Magoun) was quickly followed by Jasper's reports of diffuse projections from thalamus to cortex and their inhibitory action. The subsequent flurry of activity produced a wealth of confirmatory and new findings and a later evaluation (Steriade) that the original observations had stood the test of time. Regarding the gaps in knowledge of thalamocortical interrelations, "only the application of a wide variety of approaches can provide the information necessary to understand how the cortex receives and processes its thalamic input" (White, 1979, p. 301). This chapter presents only a partial story; the key to the pervasive influence of the thalamus lies in the fact that the cortex projects back to it in profusion and thus establishes an effective means of control by higher centers. The discoveries of corticothalamic pathways and functions is addressed in Chapter 12 and the rewards of reading these sections in sequence is obvious.

11 The Pituitary–Hypothalamic Axis

CONTENTS

"*Actually, there are few, if any, endocrine glands in the body that are not affected by the nervous system.*"

(Ganong, 1978, p. 197)

Tucked beneath the thalamus of the adult human brain is the tiny hypothalamus representing 0.5% of the total brain volume. The diminutive size is deceptive, however:

> [T]he hypothalamus is without peer in its authority over body adjustments to our external and internal environments. . . . [It] regulates body temperature, hunger, thirst, sexual activity, goal seeking behavior, endocrine functions, affective (emotional) behavior, and the activity of the visceral nervous system (Diamond, Scheibel, and Elson, 1985).

With so many known functions, it is not surprising that the separate hypothalamic nuclei and the specificity of their actions were identified piecemeal, with contradictions and erroneous conclusions along the way. The pervasive influence of that bit of brain tissue could not be established until its intimate relation with a nearby small and inaccessible glandular body, the pituitary (hypophysis), was untangled. Their axial, or interactive relationship was essential knowledge before the clinical as well as experimental observations involving those two regions of the brain would make sense. Further complicating the details of the pituitary–hypothalamic axis (and the history of that knowledge), is the fact that the human pituitary consists of two lobes with independent embryological origins, innervations, and vascular circulations.

Those early anatomists who carried out human dissections found the pituitary lying conspicuously at the base of the brain. Figure 11.1 reproduces a woodcut published in 1523 by Jacapo Berengario da Carpi in a short version of his enormous *Commentary* on an anatomical treatise written two centuries earlier by Mondino de' Luzzi of Bologna. Berengario's role as the link between medieval Galenism and Renaissance observational anatomy was substantial (Clarke and O'Malley, 1968; 1996) and accelerated the acceptance of anti-Galenic ideas proposed two decades later by Vesalius. In the text surrounding the illustrations, Berengario wrote:

> Through this foramen the spirit and some humidities . . . pass out to a certain vacuity stretching toward the basilar bone near the place where there is a certain glandulous flesh under the crossing of the optic nerves.
>
> This vacuity is called lacuna by Mundinus, head of the rose by Avicenna, and embotem by others because it is broad above, narrow below, and surrounded on all sides by a thin panniculus as far as the basilar bone [third ventricle] (Berengario, 1535, p. 143; transl. by Lind, 1959).

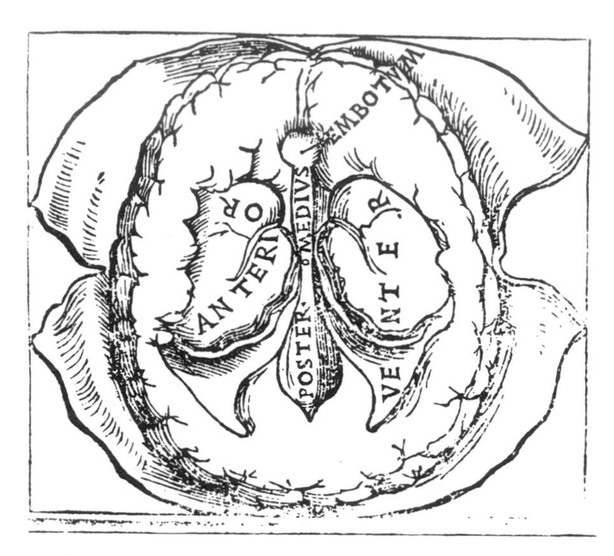

Fig. 11.1. The lower half of a wood engraving from page 56 of the 1523 edition of *Isagogae brevis* by Jacapo Berengario da Carpi, published in Bologna and showing the lateral ventricles (ANTERIOR, VENTER) and the "embotum" or pituitary gland through which the ventricular wastes supposedly drained. "Berengario's contribution . . . was to integrate text and illustration" (O'Malley, 1964, p. 20, ×4/5).

Berengario's description was again visualized in unusual detail (Fig. 11.2) in the woodcuts prepared for Andreas Vesalius to illustrate *De humani corporis fabrica* (1543). This large folio volume (*see* also Chapter 3) "is, without question, one of the great achievements of the human spirit... When all is said, the most permanant achievement of Vesalius was his triumpant enlistment of graphic art in the service of biological structure in general and of anatomy in particular" (Singer, 1952, pp. xxi and xxvi). Vaesalius expanded on Berengario's text but did not follow his predecessor's method of organization: Whereas Berengario adhered to the arrangement established by Mondino (O'Malley, 1964) and presented the complete anatomy (muscles, nerves, blood vessels) region by region, Vesalius chose to describe an entire physiologic system throughout the body. In the chapter "On the Infundibulum and on the Gland that receives the phlegm of the brain, and on the other passages that purge it," Vesalius had this to say about the pituitary:

> In form it is exactly like the top of a funnel . . . being wide and round above but gradually contracting into a long narrow tube . . . ending with a point in the pituitary gland.
>
> It is of glandular substance but harder and more compact than other glands. It is covered everywhere by tenuis [membrane] which arises from the lower end of the funnel [=infundibulum] or from the membrane [=endocranium] that lines the skull from which the aura here retracts.
>
> All these things minister to the evacuation of the phlegm from the brain (Vesalius, Chapter XI, translated in Singer, 1952, pp. 52–53).

Vesalius did not subscribe to the Galenic idea that the phlegm is purged through the cribiform plate before passing to the pituitary, rather, he "hypothecates vaguely passages to take phlegm from gland to nose." For comparison, Fig. 11.3 reproduces a modern depiction of the human pituitary gland by the acknowledged founder of comparative neurology, Ludwig Edinger.

The centuries-old notion that pituita was formed in the brain and appeared as a nasal excretion was dispelled in two stages. First, in Germany, the nasal mucus was shown to be the product of mucous glands lining the nasal cavity by Conrad Victor Schneider (1660). Next, the cribiform plate was proven to be impervious to pituita or any other fluid by the Englishman, Richard Lower (1672), who has been characterized as a typical seventeenth-

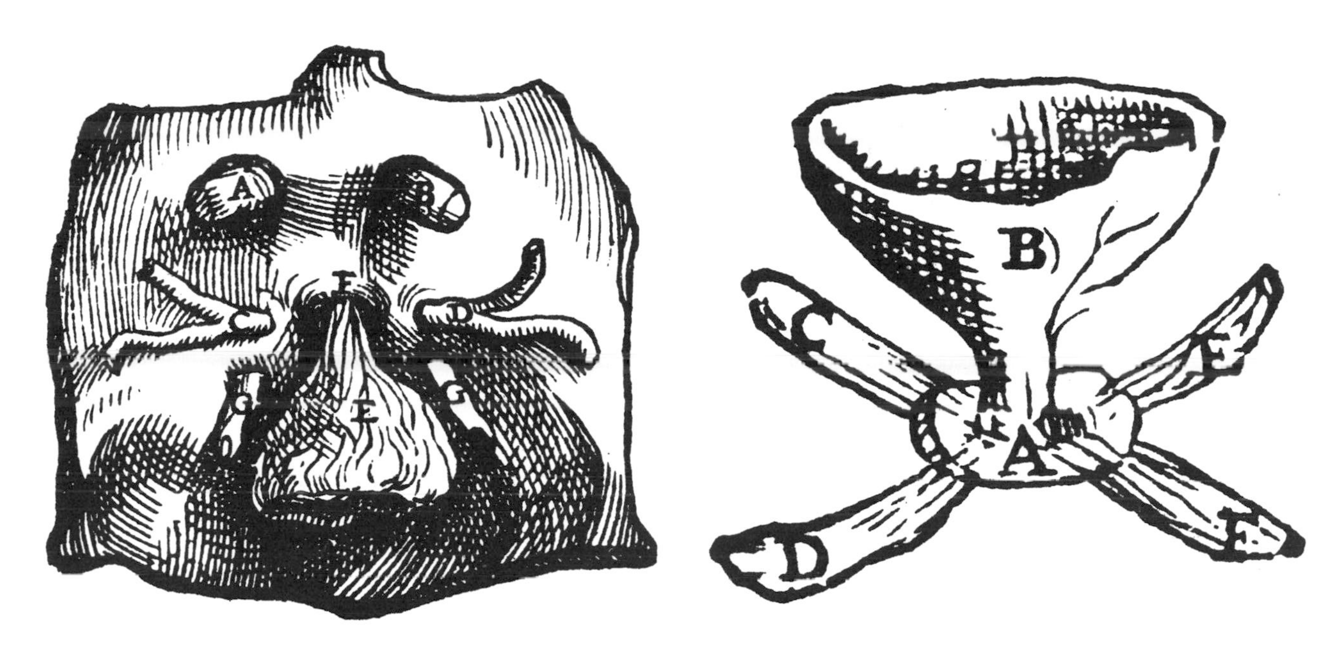

Fig. 11.2. Two of six woodcuts showing the pituitary gland from the great *Fabrica* of Vesalius. Left—that part of the skull cavity lined by dura attached to the sphenoid and ethnoid bones. A, B—Optic nerves; C, D—arteries; E— "here lies hanging the funnel which receives the phlegm descending from the third ventricle; F—the aperture through which the end of this funnel-like basin reaches the gland that receives the phlegm of the brain" (Vesalius, 1543, p. 610, Fig. 15; translated in Singer, 1952, p. 114). Right—the erect basin (B) by which the phlegm trickles down into the basin (A) below it; C, D, E, F, imaginary outlets. (From Vesalius, 1543, p. 621, Fig. 18; Singer, 1952, p. 116, ×2/3.)

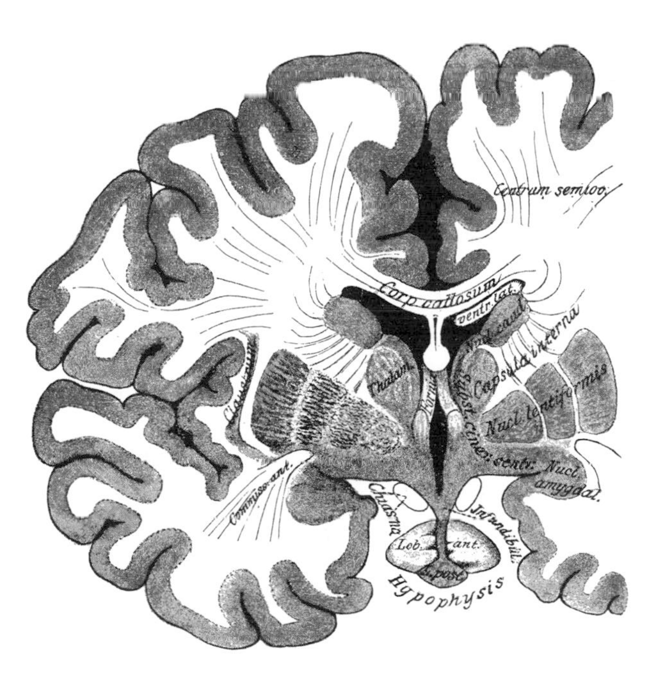

Fig. 11.3. Ludwig Edinger's drawing of the frontal section of an adult human brain, showing the pituitary gland (hypophysis) at lower center. (From Edinger, 1899, Fig. 127, ×3/4.)

century experimental physiologist (E. Clarke, in Neuburger, 1981, p. 339). Lower applied fluid under pressure to the cephalad side of the plate and found that none passed through to the nasal side.

Further knowledge of the anatomy of the pituitary gland or interest in its function remained at a standstill for 200 years until case reports from the clinic began to call attention to that region of the cerebrum. While the interest of the physiologists was held by the electrical activity of the nerves, it became increasingly apparent to practicing physicians and pathologists that the base of the brain was involved in several physiological functions. In 1840, a Dr. Mohr, lecturer at the University of Wurzburg, described a case of obesity which he believed to be caused by pressure from a pituitary tumor. Shortly afterward, the Viennese pathologist, Carl Freiherr von Rokitansky (1804–1878; Fig. 11.4), made two statements in his *Handbuch* that presaged a connection between the brain and digestive system: "Gelatinous softening of the

Fig. 11.4. Carl Freiherr von Rokitansky was among the group of continental pathologists whose medical competence and keen observation in the autopsy room elevated their discipline to an art. The engraving is dated 1846.

stomach commonly runs a subacute course. . . . It is frequently founded upon a demonstrable affection of the brain . . . and this fact renders it probable that there is a similar causative nexus. . . . Perhaps the proximate cause may be a morbid condition of the vagus, and to extreme acidification of the gastric juice" (1841, vol. 2, p. 36 of the translation). This observant pathologist also noted that infections in the vicinity of the third ventricle may be associated with gastric hemorrhage.

Rokitansky's observations were a brilliant example of the ascendancy of pathological anatomy in Europe in the midnineteenth century (*see* pp. 63 and 152, this volume). The reputation of the Vienna school in this connection was furthered by Ludwig Mauthner, who reported (1890) that somnolent patients with epidemic encephalitis (the "Nona") at autopsy showed involvement of the base of the brain posterior to the sella turcica, the bony housing of the pituitary gland. From yet another direction, acromegaly was recognized by Pierre Marie in Paris (1886) and Oskar Minkowski in Prussia (1887) as a pituitary disorder. Relatedly, a Scottish pathologist, Byrom Bramwell (1847–1931; Fig. 11.5), in his text, *Intracranial Tumors* (1888), made important correlations:

> Tumors of the pituitary body are in many instances attended with an excessive development of the subcutaneous fat: and in some cases with the presence of sugar in the urine, or with simple polyuria (diabetes insipidus). . . . Whether these symptoms are due to the fact that the pituitary body itself is diseased, or whether, as seems more likely, to the secondary results which tumors in this situation produce in the surrounding cerebral tissue, has not yet been decided" (ibid., pp. 164,165).

What had been decided, clearly, was that brain influences a great variety of body processes.

PITUITARY POSTERIOR LOBE: THE NEUROHYPOPHYSIS

Toward the end of the nineteenth century, the attention of physiologists and pharmacologists was at last drawn to the pituitary, not by interest in the gland *per se*, but by curiosity about the actions of

Fig. 11.5. Sir Byrom Bramwell, a Scot, was among the early pathologists who recognized a connection between pituitary tumors and body fat and polyuria.

its injected extracts. In an era when glandular replacement therapy was an exploited medical fad (Greep, 1974), there was curiosity in the experimental laboratory about the changes in blood pressure after administration of whole-gland extracts from different organs. Oliver and Schäfer at University College, London, reported (1895) a presser response in anesthetized laboratory animals after injection of whole-pituitary extract. The separate embryologic origins of the pituitary lobes had been known since 1838 (Rathke, cited by Herring, 1908) and the active principle was localized in the posterior (neurohypophyseal) lobe by the American cardiovascular physiologist, William H. Howell (1898), at Johns Hopkins. A nerve-fiber bundle between the hypothalamus and pituitary gland was traced in the rat by Ramón y Cajal (1894) a few years before Howell's colleague, Harry J. Berkley (1898), found "[p]eculiar structures resembling nerve end-organs" and colloid-containing vesicles that were fully described some dozen years later by an Englishman, Percy Theodore Herring (1908), after whom the vesicles are named. The scope of Herring's work in Schäfer's Edinburgh laboratory has not been appreciated (*see* Heller, 1974, p. 111): Herring not only confirmed the accumulation of colloid-containing material in mammalian posterior pituitary tissue, but looked for and found it in nonmammalian species, thus substantiating its phylogenetic status in evolution. A half-century later, studies carried out in France on the antidiuretic effect in rats of cockroach-derived material

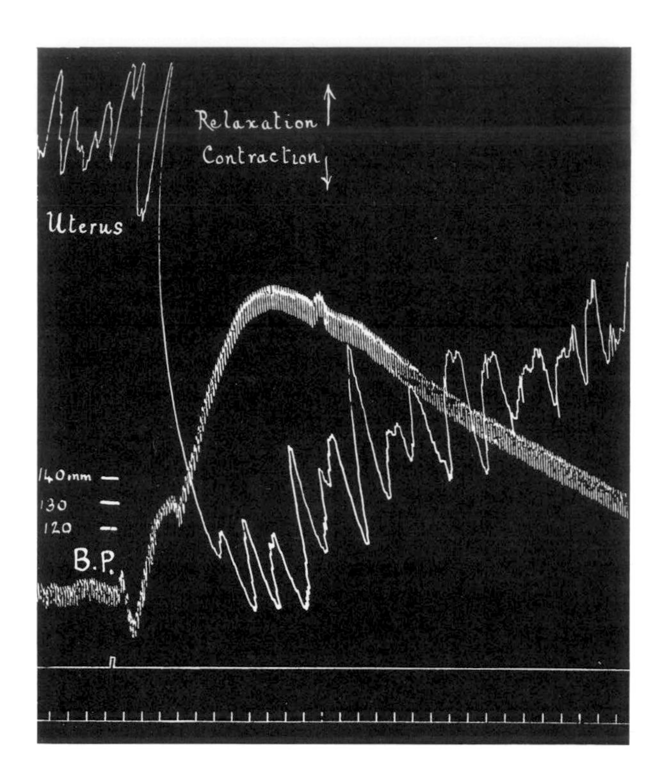

Fig. 11.6. The first published recording of the effect of oxytocin (top trace) on the pregnant uterus. At the signal (third trace), ox-pituitary extract was injected into a vein of an anesthetized cat; blood pressure (B.P.) was also recorded. (From Dale, 1906, p. 195, Fig. 26, ×4/5.)

showed that the neurosecretory substance is not class-specific (Stutinsky, 1953).

The oxytocic effect of posterior pituitary extracts was discovered "almost by accident" by the British pharmacologist Henry Dale in 1906. Sir Henry related the circumstances 50 years later at the first published symposium on the neurohypophysis (*see* Heller, 1974, p. 108). He had been comparing the anomalous depressant action of intravenously administered adrenaline under certain experimental conditions in pregnant cats and expected the extract from the posterior pituitary to act similarly; on the contrary, the effect was the reverse—blood pressure increased and the uterus contracted (Fig. 11.6). A second activity of posterior lobe extract was as a galactogogue, discovered and satisfactorily demonstrated by two American physiologists, Isaac Ott and J. C. Scott (1911), using pregnant goats.

The diuretic effect of posterior pituitary extracts was recognized in Schäfer's laboratory after he moved to London and continued his two-decades' interest in the gland (Magnus and Schäfer, 1901). This action, claimed to be independent of the vascular effects already known, became the focal point of many studies, but the experimentalists with their tunnel vision, "missed discovering the antidiuretic action in the unanesthetized mammal, now generally agreed to be one of the main physiological functions of the neurohypophysis" (Heller, 1974, p. 105).

The independent and simultaneous discoverers of the antidiuretic action were two physicians, R. von den Velden (1913) at Düsseldorf and F. Farini at Venice (1913). Each had patients with disease or damage to the pituitary accompanied by polyuria and, interpreting this to be due to the gland's impaired function, they easily controlled the excessive water loss by administration of extracts of the posterior lobe. The classic experiments in pituitary physiology were carried out by Starling and Verney in England, who included posterior lobe extracts in their investigations of water and chloride excretion by the isolated dog kidney perfused with blood from a second dog pumped through the heart–lung preparation of yet a third dog. They concluded that the normal kidney is under the control of a "regulating mechanism . . . which has presumably been produced during phylogenetic development in response to the request for an increasing control over salt and water metabolism" (1925, pp. 360–361); as shown by the isolated kidney, without that blood-borne "harmonic influence," the mammalian kidney reverted to a urine characteristic of fish and amphibian classes. By reversibly poisoning the renal tubules with hydrocyanic acid, Starling and Verney found that the hormone's action is on the tubules.

The concept of economy of mammalian water balance was reinforced by Hans Heller, working at University College and Oxford. He discovered that when heated, the antidiuretic fraction of commercial posterior pituitary preparations remains active longer than does the pressor substance. Heller's studies on rabbit kidneys *in situ* (1940) showed that in small doses, the heated extracts inhibit diuresis in anesthetized mammals and that the relatively large amounts of unheated extract used in the early experiments produce diuresis through their vasoconstrictor action. Additional pieces of the water-balance puzzle were furnished by Chambers, Melvillo, Hare, and Hare (1945) at the University of Pennsylvania. Using conscious dogs, they found that antidiuretic hormone secretion depends on the osmotic pressure of the blood plasma: Experimentally produced hypotonic plasma decreases secretion, whereas a hypertonic condition increases secretion. Those and other data led Verney (1947) to suggest the presence of osmoreceptors in the hypothalamus. Later, Share (1962) proposed the possibility of the posterior pituitary–hypothalamus system being subject to vascular volume receptors in the walls of the left atrium and pulmonary veins. The presence in the posterior pituitary of two constituents, oxytocin and the misnamed vasopressin, became an accepted reality with the determination of their chemical structures by Vincent du Vigneaud and coworkers (1953a, b). By then, further elucidation of neurohypophyseal function had become closely associated with the hypothalamus, as shown below. But what of the other, anterior, lobe of the pituitary?

PITUITARY ANTERIOR LOBE: THE ADENOHYPOPHYSIS

Acromegaly and adiposity with sexual immaturity had been described as accompanying tumors or inflammation of the base of the brain, as already noted. Events at the beginning of the twentieth century accentuated interest in those two conditions. In 1901 Harvey Cushing, at the Johns Hopkins Hospital, operated three times on a fat, sexually immature adolescent who, at autopsy, revealed a tumor which had nearly obliterated the anterior pituitary. At about that same time, a pituitary cyst from a similar patient was successfully diagnosed by Alfred Frölich, a Viennese neurologist and Cushing's long-time friend. Almost 40 years later, when invited to open the discussion of a paper at a meeting of the Association for Research in Nervous and Mental Diseases, Frölich acknowledged: "We were wrong in 1901—it was not the pituitary but the hypothalamus—but in 1901 all we knew of the hypothalamus was its position below the thalamus" (1940, p. 722).[1]

Mindful of his incorrect diagnosis (1906), Cushing, the perfectionist, was motivated to embark on a long course of clinical and experimental neuroendocrine research. In 1912 he published in great detail 40 cases of surgical intervention in pituitary disorders, related the clinical observations to his experimental findings, and even suggested (1909) a possible relation of insufficiency of the gland to epilepsy. A collection of his *Papers Relating to the Pituitary Body, Hypothalamus and Parasympathetic Nervous System* (1932) documented Cushing's persevering interest and authority on the subject. From his early experimental work on dogs, he distinguished between excess function (as in gigantism) and deficiency of function of the anterior lobe (as in dwarfism). The missed diagnosis was redeemed in 1909 by his skillful removal of part of the anterior pituitary in an acromegalic man who lived a normal life for many years thereafter.

During the next decade or so, as interest in endocrinology commenced an ascendancy, three groups of experimentalists contributed significantly to anterior pituitary discoveries. An Austrian neurosurgeon, Bernhard Aschner, used an improved buccal approach in puppies (1912) to remove the anterior pituitary lobe and produced adiposity, genital aplasia, and other signs. The same operation in adult dogs, however, resulted in no fat deposits. As long as the hypothalamus was undamaged, Aschner found that as little as one-third of the anterior lobe supports normal thyroid, gonadal, and adrenal activity.

In France, Camus and Roussy (1920), in experiments not reported until after the First World War, used the same buccal approach and contrasted hypophysectomy (no polyuria) with a "piqueur" injury of the hypothalamus (accompanied by polyuria). They found that the maneuver resulted also in glycosuria, which they believed to be the first time urinary sugar loss had been produced experimentally. The French findings were confirmed and extended by two scientists visiting in Cushing's laboratory, the American neurologist-neurosurgeon, Percival Bailey, and the Belgian neurophysiologist, Frédéric Bremer. They set out to examine reports which seemed to show "that in the base of the brain just over the pituitary is a very important center, intimately connected with the entire visceral nervous system" (Bailey and Bremer, 1921, p. 774). From their evidence and contrary to their host's belief in the dominance of the pituitary gland, they relegated the posterior

[1]For a lively account of that era, *see* Anderson (1969).

Fig. 11.7. Philip E. Smith's skill in the technique of hypophysectomy in very small animals spurred progress in neuroendocrinology as he was able to show the multiple effects of anterior lobe removal and replacement therapy. (Photograph from Sawyer, 1988, p. 26, Fig. 3, left.)

pituitary extract, which had never been shown to be produced physiologically, to the status of a pharmacologic agent with antidiuretic, vasopressor, and oxytocic properties.

By the third decade of the twentieth century, the improved assay and operative techniques and an explosion of research activity in a new field brought about a rush of productivity. In the words of a participant, Roy O. Greep (1974, p. 4):

> No period in the history . . . of the anterior lobe can compare with the 10-year span from 1925 to 1935. This was the era of one spectacular advance after another. . . .
>
> In terms of specific advances . . . the veil of ignorance was lifted from the anterior lobe. . . . The central role of the pituitary in controlling somatic growth and the functions of the thyroid, the adrenal cortex, mammary glands, ovaries, and testes were firmly established. . . . By 1935, the pituitary had come to enjoy such a heady state of supremacy as to be referred to as conductor of the symphony of endocrine glands.

There was much work to be done, however, before that extravagant reputation was attained.

Identification of the growth hormone was the culmination of work in the laboratory of Herbert M. Evans at the University of California, Berkeley. While there, Philip E. Smith (1884–1970; Fig. 11.7) mastered the technique of hypophysectomy in tadpoles and newborn rats and later (1930) characterized "a constant syndrome" resulting from the procedure: cessation of skeletal growth and weight increase; regression of adrenal cortex, thyroid, and gonads; and decrease in size of the viscera. He found that daily homotransplants of anterior pituitary tissue could restore those changes to or near normalcy (Fig. 11.8). Evans and his team, using the same technique of ablation and restorative therapy in young dogs, had similar results and went on to produce the opposite condition, gigantism, by administration of excess anterior pituitary extract (1933; *see* Fig. 11.9, p. 234). That work proved that a single agent produces acromegaly and dwarfism, and validated Cushing's finding, a quarter century earlier, of the opposite effects of too much or too little available growth hormone.

Another breakthrough in identification of anterior pituitary functions was the demonstration by Collip, Anderson, and Thomson (1933) of separate substances regulating the adrenal cortex and the thyroid gland. In completely hypophysectomized young rats, the atrophied adrenal cortex (shown histologically in one adrenal) could be restored by treatment with the adrenotrophic principle (in the second control gland), whereas the thyroid remained that of an untreated hypophysectomized rat. Thyroxin replacement with a highly purified extract of anterior pituitary, lacking both prolactin and growth activity, was also tested successfully (Anderson and Collip, 1933). It should be recalled that earlier Smith (1930) had succeeded in restoring thyroid atrophy in hypophysectomized rats with crude extract or homotransplants of pituitary tissue.

The most organized research attack, probably because of its obviously close relationship to pervasive societal problems, was in the reproductive field. Funded by the Rockefeller-supported Bureau of Social Hygiene through an elite committee of the National Research Council, one of the most

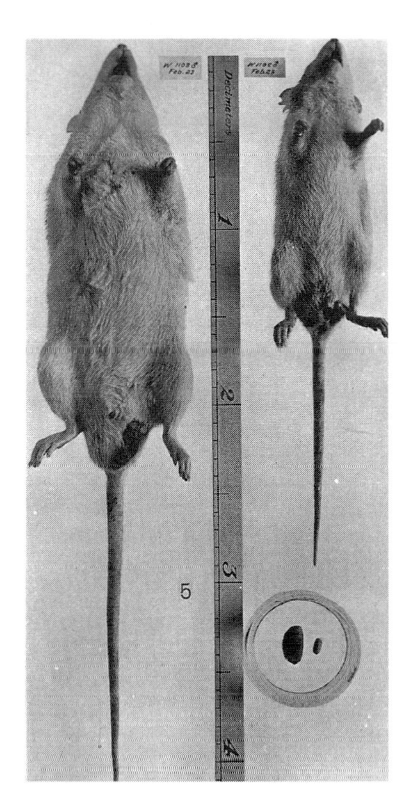

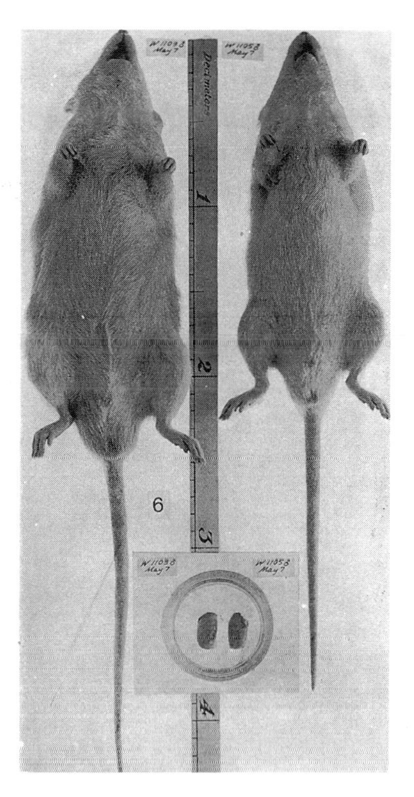

Fig. 11.8. The first conclusive experimental demonstration of pituitary influence on growth and the gonads. Left pair—A normal control rat at 125 days of age (left) and a littermate hypophysectomized at 89 days with a testis from each animal (circle). Right pair—The same rats after an additional 74 days during which daily transplants were made in the hypophysectomized rat (right); below are the remaining testis from each rat. Scale in decimeters. (From P. E. Smith, 1930, pp. 261, 263, Plates 2 and 3.)

active of those programs was at the University of Chicago under Frank R. Lillie, of freemartin fame. On the basis of a series of experiments on hundreds of rats, it could be asserted that: "The gonads function only when they are forcibly stimulated by certain secretions that are normally provided by hypophysial activity. . . . Hypophysis activity is to some extent controlled by gonadal secretions, for when the gonad hormones are present in effective amounts, hypophysial activity is lowered" (Moore and Price, 1932, p. 61). That concept of a "reciprocal influence" is one of the historically earliest examples of a biologic feedback mechanism and led to formulation of the "Moore–Price Theory." Dorothy Price (1899–1980) described vividly her work with Moore and particularly how she arrived at the explanation of their experimental results. After being exhorted by Moore to *think*, ". . . I sat down at my desk and thought! I simplified the problem. . . . Then quite suddenly a plausible explanation occurred to me. . . . If . . . the anterior pituitary controlled the secretion of male hormone and the male hormone, in turn, controlled the secretion of gonadotropin by the anterior pituitary, there would be a splendid scheme for a seesaw balance" (Price, 1975, p. 228).

The question of whether or not the pituitary gland has a significant nerve supply has been debated since Jean-Marie Bourgery in 1845 described what he thought was a "ganglionic plexus" that served a[illegible] an intermediary between the brain mass, spec[illegible] cally that part near the origins of the first fiv[illegible] nial nerves, and the great sympathetic trunk [illegible] believed the plexus was involved in ps[illegible] and instinctive functions. Later see[illegible] not find Bourgery's plexus (was [illegible]

Fig. 11.9. After daily injections for six months of growth hormone (purified anterior pituitary extract) the dachshund on the right had increases in skin, skull, and body size, but not in the length of the extremities compared with the untreated littermate on the left. (From Evans, Simpson, Meyer, and Reichert, 1933, p. 430, Fig. 1.)

the gland's portal plexus, by chance?). During the intervening century when the tangle of endocrine disorders and secretions was undergoing resolution, the role of the central nervous system became increasingly evident. Among the many relevant studies, those carried out in Austria by Hohlweg and Junkmann (1932), using the positive estrogen feedback mechanism, seemed to indicate an intermediary "hypothalamic sex center," a more complex conception of feedback control than Price's "seesaw." More than an additional decade of tentative interpretations of conflicting data ensued before Sawyer, Everett, and Markee (1949, p. 231) at Duke University wrote that the sex center "would be a locus in the nervous system at which sex hormones alter the sensitivity to extrinsic stimuli and thus alter hypophyseal secretion. . . . [T]he hypothalamus would appear to be the most likely site of the gonadotrophin sex center." Later, Hohlweg declared (1975, p. 167) that "[t]he discovery of the cybernetic control of endocrine glands gained clinical significance" and illustrated his statement by the diagram reproduced in Fig. 11.10.

THE HYPOTHALAMUS

In the early 1930s, when "the mystery of the hypothalamus must have been already in the air" (Krieg, 1975, p. 18), the probing of hypothalamic–pituitary interconnections received a tremendous impetus from a new quarter—the American Midwest. At Chicago, the founding director of the Institute of Neurology, Northwestern University Medical School, Stephen Walter Ranson (1880–1942) revived the use of an instrument he had seen briefly in 1911 on a visit to Sir Victor Horsley's laboratory in London where it had been designed by Horsley's associate, Robert Clarke, to further their resarch on the cerebellum (*see* Chapter 9). Ranson had also seen in use (Sachs and Fincher, 1927) at Washington University School of Medicine an early model purchased by Ernest Sachs, an American neurosurgeon who had been a guest in Horsley's laboratory. Realizing the great potential of the instrument for precise exploration and experimentation in the interior of the living brain (but probably not envisioning its application to diagnosis and treatment of certain types of brain disease in humans) he had a unit constructed locally from the detailed drawings published in *Brain* (Horsley and Clarke, 1908) and initiated a broad program of research on several fronts.

The stirring details of its use by Ranson's group were breezily related firsthand by Walter W. Ingram (1975) and by Horace W. Magoun (1985), both of whom were extensively involved in keeping the instrument "humming in a range of projects on the role of the hypothalamus and lower brain stem in visceral integration, emotional expression, and the regulation of feeding, fighting, mating, and other vital behaviors, all of which functions contribute to the well-being of the internal environment, as well as of the individual and the race" (Magoun, 1985, p. 250). The Northwestern studies dealt initially with the control of posture and movement, and soon spread to exploring millimeter by millimeter the effects of electrical stimulation of the hypothalamus and other regions of the dien-

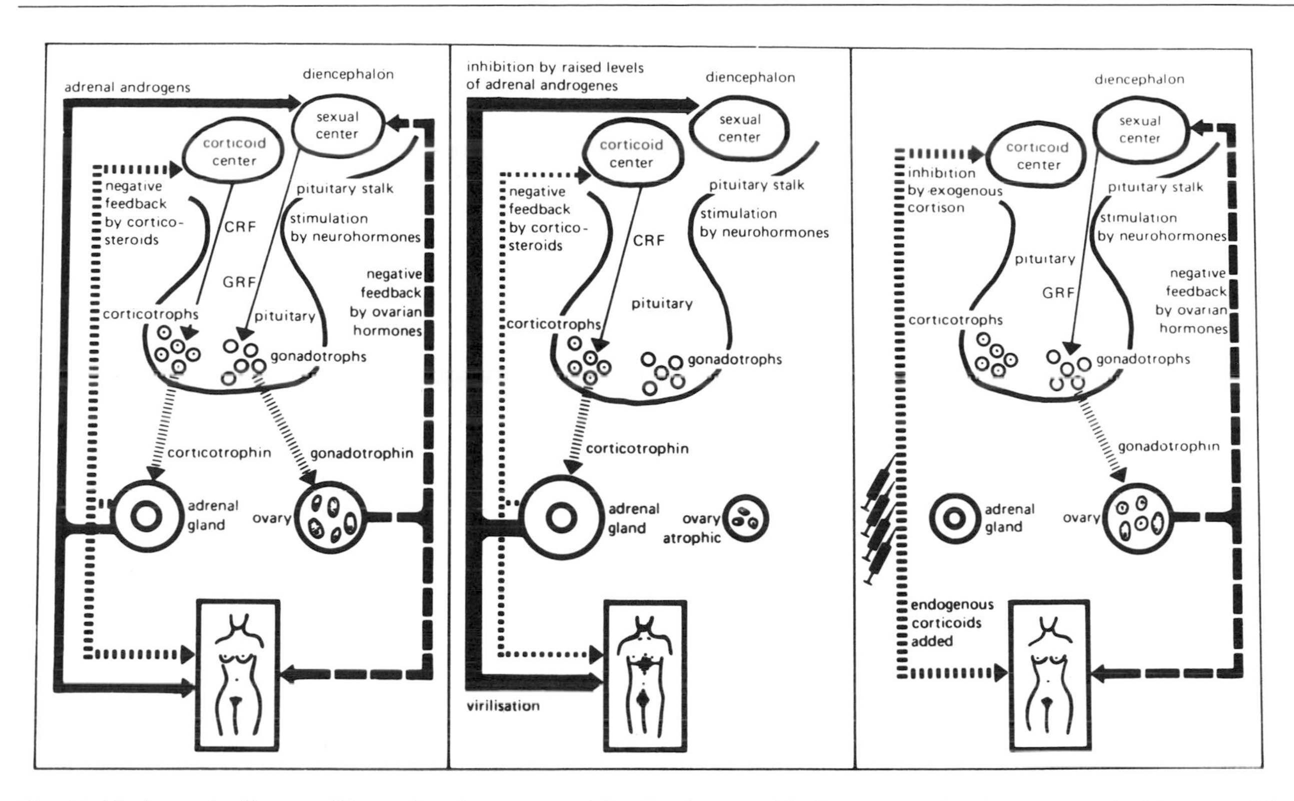

Fig. 11.10. An early diagram illustrating the concept of feedback control in the neuroendocrine system, in this case of the adenogenital syndrome. (From Hohlweg, 1975, p. 168.)

cephalon. A surprising finding was a persistent polyuria in animals that had bilateral lesions placed stereotaxically to interrupt the supraopticopituitary nerve tracts without damage to the posterior pituitary (*see* Fig. 11.11, p. 236). The precision of the lesions and the unequivocal demonstration of diuresis without pituitary damage indicated that a "higher control was in effect" and "[t]he importance of the hypothalamus began to emerge gradually as it became disentangled from the pituitary" (Anderson, 1969, p. 2).

Although the hypothalamus had been known for several centuries, it was not described definitively until 1895 in the *Nomina Anatomica* (His, 1895). The American, C. L. Herrick (1892b), and Edinger (1896) in Germany used the term "preoptic region" when writing of the phylogenetically ancient hypothalamus that can be identified in all vertebrates. "Throughout the whole vertebrate scale there is no part of the forebrain that has maintained so constant a general arrangement during phylogeny and ontogeny" (Fulton, 1932, p. 61).

The presence of scattered globules of colloid material in the posterior lobes, which are essentially neural structures, first by Berkley (1894) and later by Herring (as noted above), made sense only after the droplets were shown to contain antidiuretic hormone and oxytocin. Wolfgang Bargmann, while at Frankfurt in 1933–1934, had seen those strange inclusions in sections of posterior lobe tissue (Fig. 11.12) under Ernst Scharrer's microscope. After the hiatus of the Second World War, Bargmann (1949) published his own findings: Using the Gomori method which selectively stains axons, he saw in dogs' brains the complete structure of hypothalamic projections coursing to the posterior lobe of the gland. He found that the Herring bodies are swellings of those axons whose endings surround the capillaries where they release hormones into the blood stream (see Fig. 11.13, p. 237). By 1953, Bargmann and collaborators had determined that the colloid material contains antidiuretic, vasopressor, and oxytocic activity, and three years later they demonstrated the presence of colloid throughout the distribution of the supraoptic tract.

The winding trail from 1928 and the work of E. Scharrer to universal recognition of a neuro-

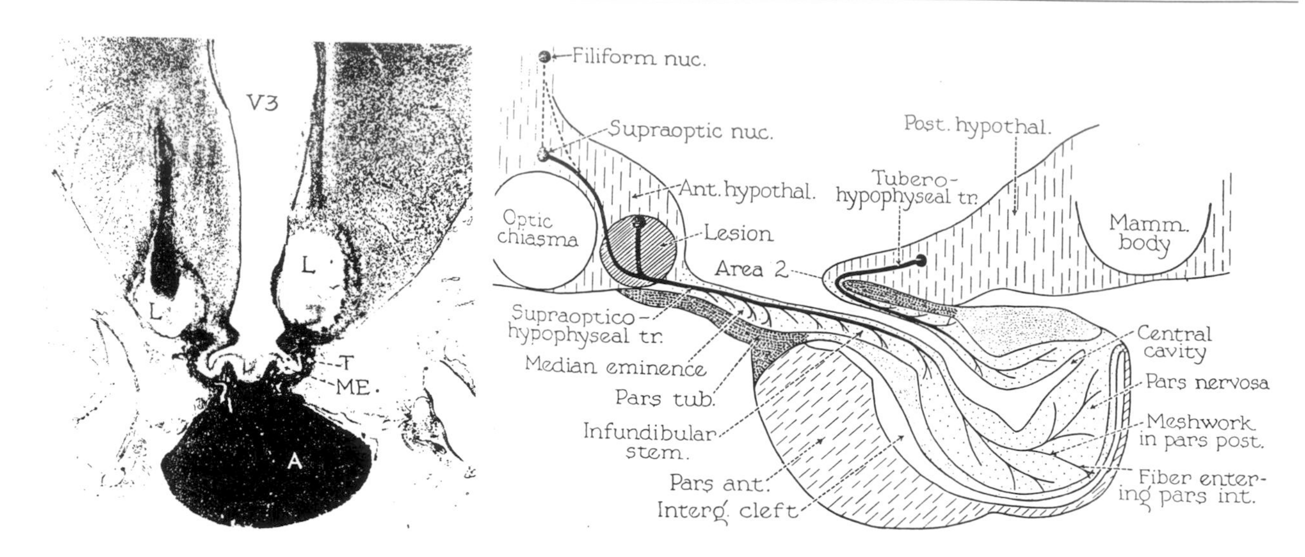

Fig. 11.11. Left—Photomicrograph of cat brain section showing tracts of electrodes positioned stereotaxically in bilateral sites to produce lesions (L) that destroyed nerve fibers between hypothalamus and pituitary (A). Right—Diagram of the region demonstrating the intact pituitary, median eminence (ME), and tuber cinereum. (From Fisher, Ingram, and Ranson, 1938, pp. 2, 3, Figs. 1 and 2.)

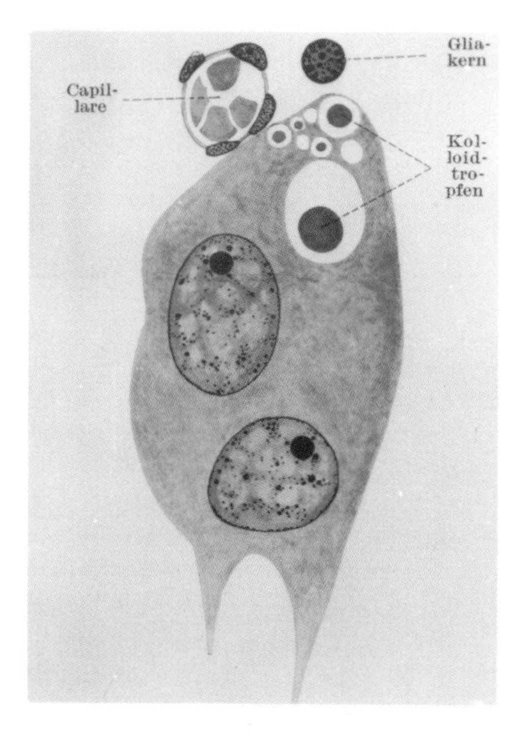

Fig. 11.12. A binucleate cell from the human hypothalamus showing vacuoles containing colloid material. a—Capillary; b—nucleus of glial cell; c—droplets of colloid material. (From E. Scharrer and Gaupp, 1933, p. 770, Fig. 6, ×1.)

secretory apparatus in 1953, when the first conference on neurosecretion was held, "was long and arduous, and for many years lonely. There was much uncertainty, but never any real doubt about the final outcome" (B. Scharrer, 1975, p. 259). Ernst had found hypothalamic neurosecretory cells

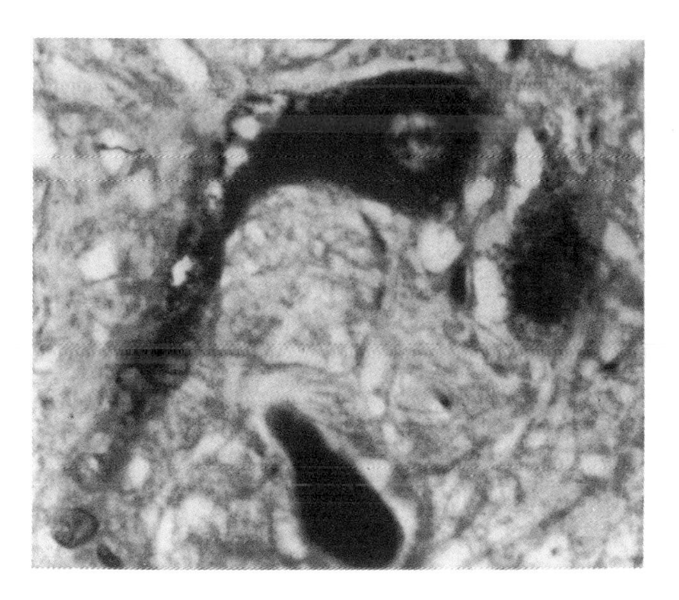
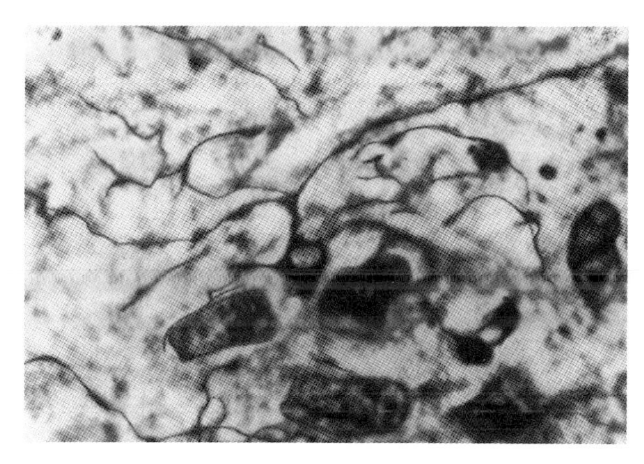

Fig. 11.13. Left—Secretion-filled cells from the supraoptic nucleus of a dog, with isolated vacuoles in the extension; Gomori stain. Right—Gomori-stained network in the posterior lobe of the pituitary. (From Bargmann, 1949, pp. 616, 624, Figs. 5 and 17 ×4.)

related to pituitary function first in a teleost fish, Bertha had identified such cells in invertebrates, and in 1952 they advanced the concept that a neuron with dual properties served the evolutionary purpose of making possible "effective communication between the neural and the endocrine centers, each of which operates in its own way" (ibid., p. 262). The electron microscope yielded conclusive anatomic evidence for such intimate relations between neural and hormonal elements by visualization of colloid-containing vesicles in hypothalamic nerve endings in close juxtaposition to capillaries, as Barbara G. Monroe's beautiful micrographs testified (*see* Fig. 11.14, p. 238). The new field of neuroendocrinology was well on its way, due in no small measure to the painstaking work of morphologists.

The unusual anatomical relationship between the hypothalamus and the pituitary's anterior lobe was clarified by investigations that began with the suggestion of a Bucharest pathologist who had noted in the autopsy room a cluster of blood vessels around the pituitary stalk (*see* J. D. Green, 1956, p. 14). Gregory T. Popa with U. Fielding (1930), while at University College, London, found capillaries in the tuber cinereum passing along the stalk in large trunks and ending in the sinusoids of the anterior lobe, which they called "veins," believing the flow to be away from the pituitary (*see* Fig. 11.15, p. 239). The erroneous idea was corrected by Bernardo A. Houssay (1887–1971), Argentinean endocrinologist whose research was mainly on the metabolic effects of pituitary hormones. From observations in the living toad, the direction of flow was shown to be from hypothalamus to anterior lobe (Houssay, Biasotti, and Saminartino, 1935). The first convincing demonstration in mammals, including humans, appeared in the paper of Wislocki and King (1936). From histological studies, they determined that blood in the portal system must pass from capillaries in the median eminence to the sinusoids in the anterior pituitary, and they wrote: "The arrangements of the vascular circulation of the hypophysis are the most complex in the body and less is known . . . about its blood supply than that of any other organ or tissue" (1936, p. 441).

The accumulation of evidence for the humoral activity of the hypothalamus–pituitary axis generated suggestions for a mechanism from several sources. One of the earliest was that of Joseph Clarence Hinsey (1901–1981; *see* Fig. 11.16, p. 240), who participated in the revival of the stereotaxic instrument in Ranson's laboratory at Northwestern University. With Joseph Markee, Hinsey proposed (1933) that nervous control of the ante-

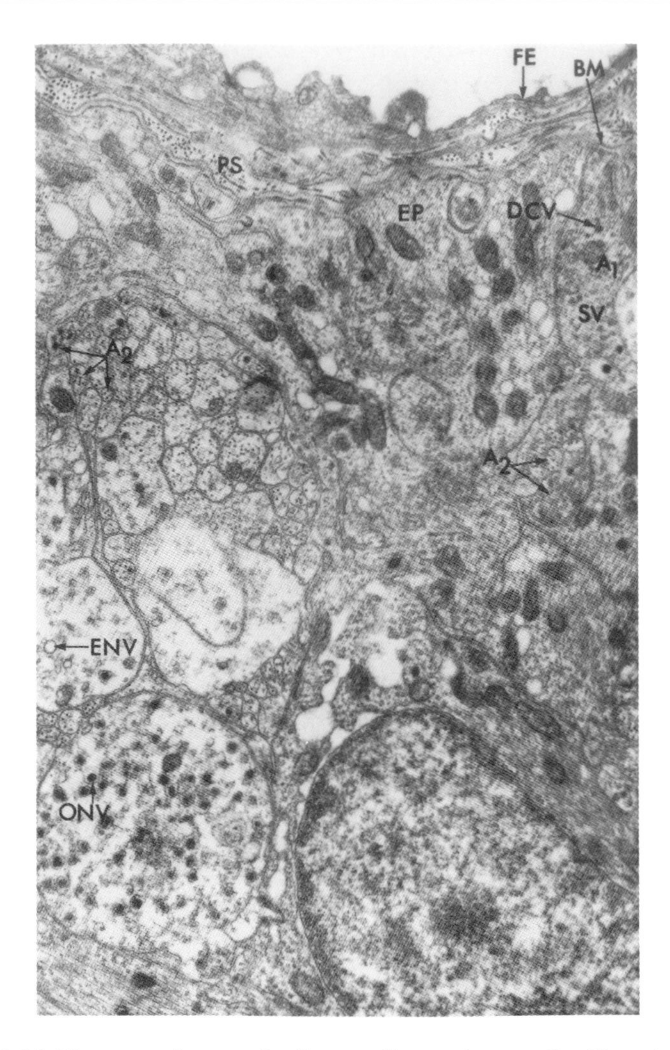

Fig. 11.14. Electron micrograph of rat median eminence. Capillary (top) has a fenestrated epithelium (FE) separated from the neural elements by perivascular space (PS) and basilar membrane (BM). Abutting the capillary are small axon endings (A1, A2) containing synaptic vesicles (SV) and dense core vesicles (DCV). Below are large axons with osmiophilic neurosecretory vesicles (ONV) and a few clear vesicles (ENV). (From Monroe, 1967, p. 426, Fig. 14, ×1.)

rior pituitary might be through humoral transmission of stimuli from the hypothalamus and posterior pituitary; their paper "is often quoted as a landmark in the budding field of neuroendocrinology" (Sawyer, Everett, and Hollingshead, 1971, p. 132). Markee later readdressed the problem and produced significant evidence for such a mechanism, demonstrating with associates that controlled electrical stimulation applied directly to the rabbit's anterior pituitary did not activate the gland, whereas application to the basal hypothalamus was effective (Markee, Sawyer, and Hollingshead, 1946).

Soon, however, acceptance of the concept of humoral transmission of neural stimuli became unavoidable, the result of the detailed reinvestigation of the portal system and innervation of the pituitary carried out in England in 1948 by Geoffrey Wingfield Harris (1913–1971; *see* Fig. 11.17, p. 241), initially a disbeliever, and joined by John D. Green (1917–1964; *see* Harris, 1972 for references). The successful replacement of ablated gland tissue became one criterion of proof that the tissue was, indeed, secreting the hormone in question, and a crucial problem was whether or not regeneration of nerves and blood vessels in the

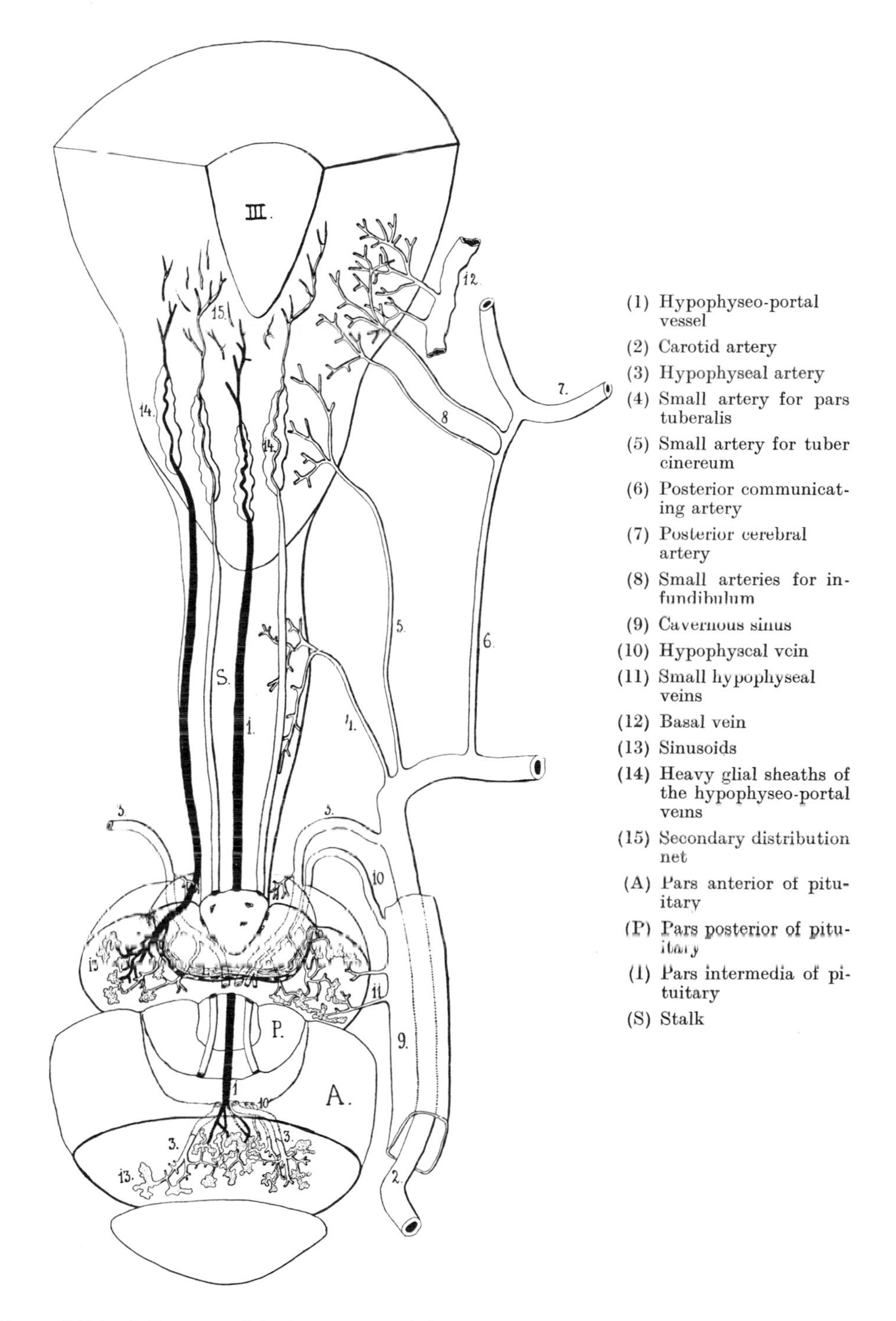

Fig. 11.15. The first published diagram of the human portal circulation between pituitary and hypothalamus, erroneously showing the branches of the carotid artery (2, 3, below) coursing into the pituitary and the portal vessels (1, solid) and passing rostral to the hypothalamus. Later studies found the blood flow to be in the opposite direction. (From Popa and Fielding, 1930, p. 90, Fig. 1.)

transplanted tissue could occur. In collaborative experiments initiated in Cambridge in 1949 (*see* Jacobsohn, 1975) and completed in Lund, Harris and the Swedish investigator, Dora Jacobsohn (1954) showed that return of function was possible when transplants of anterior pituitary tissue were

Fig. 11.16. Joseph Clarence Hinsey (above) with J. E. Markee, based on their own work and that of Cannon, proposed that "pathways from the hypothalamus must activate the posterior lobe of the hypophysis which in turn may exert an influence on the anterior lobe by hormonal transmission" (1933, p. 273). (Photograph from Sawyer, 1988, p. 30.)

made into sites close to the hypothalamus and not elsewhere. Harris (1958) likened the regulatory action of hypothalamus over pituitary to that of the reticular system on cerebral cortex activity. Three examples of hypothalamus–pituitary functional dependence may be mentioned. In 1955, Roger Guillemin reported that the presence of hypothalamic cells is necessary in a pituitary cell culture for hormones to be formed, adenocorticotrophic hormone (ACTH) in that case. A second example is the elegant experiment of Nikitovitch–Winer and Everett (1958) which showed loss of cyclic reproductive activity in rats after the pituitary was transplanted to a site beneath the renal capsule and restoration of function when the gland was reimplanted under the same animal's median eminence. Also, in 1961 Philip Smith found that effective substitution treatment in stunted young, hypophysectionized rats was possible only if the homotransplants were made to sites near the base of the hypothalamus. As Jacobsohn wrote (1975, p. 202): "The role played by the hypophyseal portal vessels in the neural control of the anterior pituitary gland is now common knowledge. Fortunate are those who, like myself, were able to participate in the search of the secret of this . . . beautiful device."

Returning to the 1940s, Harris had been working on the pituitary from his graduate days, when he developed an ingenious method for remote electrical stimulation of the gland in the unanesthetized rabbit, as described in his thesis submitted at Cambridge University in 1944. The method was extended to a series of hypothalamic stimulations to assess their effect on gonadotrophin, adrenocorticotrophin, and thyrotrophin secretion. He confirmed (1948) Markee's earlier findings of absence of secretory response to electrical stimulation of the pituitary gland. In his masterly Dale Lecture, delivered after almost 30 years of research, Harris expressed the current view: "[N]erve terminals of various hypothalamic tracts release neurohumoral agents into the primary plexus of the portal vessels in the median eminence and . . . these agents are carried to the anterior pituitary where they act on the different glandular cells to excite or inhibit their secretory rate" (1972, p. x). His flair for encourag-

Fig. 11.17. Geoffrey W. Harris trained and held appointments in both anatomy and physiology at Cambridge, then moved to the Institute of Psychiatry in London, and finally developed the Neuroendocrinology Research Unit at Oxford. His research contributions from about 1937 covered the entire range of hypothalamic function.

ing and stimulating those with whom he came in contact contributed greatly to his influence in establishing the interrelationship of brain and endocrines.

In spite of evidence of neural pathways for some pituitary–hypothalamic functions, and the pioneering effects of precise faradic stimulation reported by Ranson's group (from 1930), Markee and associates (1946), Harris (1948), and others, real proof that neuronal activity accompanied neurosecretion had to await the advent of even more sophisticated methodology. The early studies utilized grossly unnatural stimuli under nonphysiologic conditions—electrical shocks or other insults to masses of cells in anesthetized subjects—and often depended on subjective observations. The introduction of single-unit recording of impulses evoked by a physiologic stimulus revolutionized all neurophysiology and the neuroendocrinologists were not slow to exploit the new techniques. Some of the earliest to do so looked for changes in firing rates of hypothalamic neurons coinciding with variations in blood osmotic pressure, the same physiologic stimulus used in studies of posterior pituitary control of water balance (*see* earlier this chapter). Cross and Green (1959) at the University of California, Los Angeles, produced the first direct evidence of that relationship in mammals. They injected hypertonic solutions into the common carotid of anesthetized rabbits and recorded a doubling of the firing rate in some supraoptic (hypothalamic) neurons (*see* Fig. 11.18, p. 242), presumedly osmoreceptors. There were no simultaneous assays of an end-product, however—blood or tissue hormone concentrations, output of urine, or hematocrit value—to bolster the assumptions. Recording from another region of the hypothalamus, Brooks, Ishikawa, and Koizumi (1966) found in postpartum, lightly anesthetized cats that increased firing of hypothalamic or stalk single cells after osmotically induced stimulation was accompanied by milk ejection, which constituted evidence of oxytocin release. But in both those cited studies, the subjects were anesthetized. A partial improvement in experimental conditions was reported by Nakayama (1955). Using conscious rabbits fitted with chronically implanted electrodes, he recorded increased nonspecific electri-

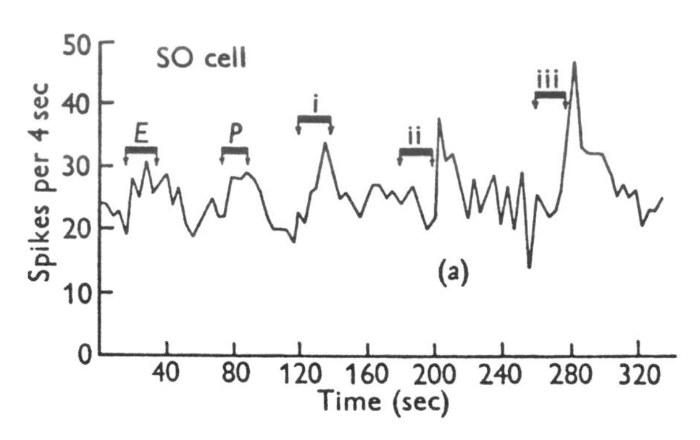

Fig. 11.18. Oscilloscope record of rate of firing of a supraoptic neuron accelerated by osmotic stimuli illustrating the graded effect of increasing the rate of injection of 9% NaCl: i—0.006 mL/s, ii—0.013 mL/s, and iii—0.031 mL/s; E—ear pinches, P—paw pinches. (From Cross and Green, 1959, p. 562, Fig. 5 above.)

cal activity (EEG) from both pituitary lobes when epinephrine was injected. The ultimate experiment—recording single cells' responses to physiologic stimuli in behaving subjects accompanied by a measurable change in end-product—was achieved in 1974 by Dufy, Dufy-Barbe, and Poulain. They found in conscious rabbits in estrus that vaginal stimulation resulted in increased electrical activity of hypothalamic cells (in the premammillary nuclei) followed by a rise in luteinizing hormone in the blood.

The slow progress made in the early years of endocrine research due to the laborious and imprecise bioassays then available was relieved somewhat after introduction of sensitive chemical and tissue-culture techniques, in addition to precise stimulation and lesion methodology. Especially supportive was the attention of the biological chemists; although they did not yet view themselves as neurochemists (organizational activity waited 30 years; *see* Chapter 8), the biochemists and incipient neuroendocrinologists were quick to undertake the isolation and characterization of substances extracted from various endocrine glands. The first successful isolation of a tropic substance was the thyrotropic hormone by Anderson and Collip at McGill University in 1933, mentioned above, and the earliest pure pituitary protein to be crystallized was prolactin in 1937 (White, Catchpole, and Long).

Acceptance of the presence of important releasing factors in the pituitary portal blood shifted the attention of many biochemists to their identification. A preliminary question was the role of the catecholamines in brain function, in addition to their known vasomotor action. Sawyer, Markee, and Everett (1950) demonstrated that norepinephrine (noradrenaline) was involved in the release of gonadotrophins and concluded: "Little doubt can remain that an adrenergic mediator is employed in the mechanism by which the nervous system controls LH-release from the anterior pituitary" (p. 541). The problem became of special interest to Marthe Vogt, émigrée from Germany and working in Edinburgh, who had noted (1943) that adrenal venous blood contains a protective factor against stress. She later found (1954) that blood concentrations of norepinephrine in the brain are too high to be accounted for by its vasomotor nerve endings and identified the hypothalamus as the source, thus confirming her view that the hypothalamic catecholamines, pituitary activity, and stress-related adrenal effects are linked.

The human adrenal glands were depicted first by Bartholomeo Eustachio, in 1552 (Fig. 11.19), and 150 years later were described as a solid organ morphologically divisible into cortex and medulla (Cuvier, 1805, vol. 5, pp. 242,243). Their separate functions were proposed by Rudolph Albert von Kölliker (1817–1905), a Swiss-German comparative anatomist renowned for his studies of fine nerve structure (Fig. 11.20) comparable to those of Ramón y Cajal. Kölliker considered

> the cortical and medullary substances as physiologically distinct. The former may, provisionally, be placed with the so-called

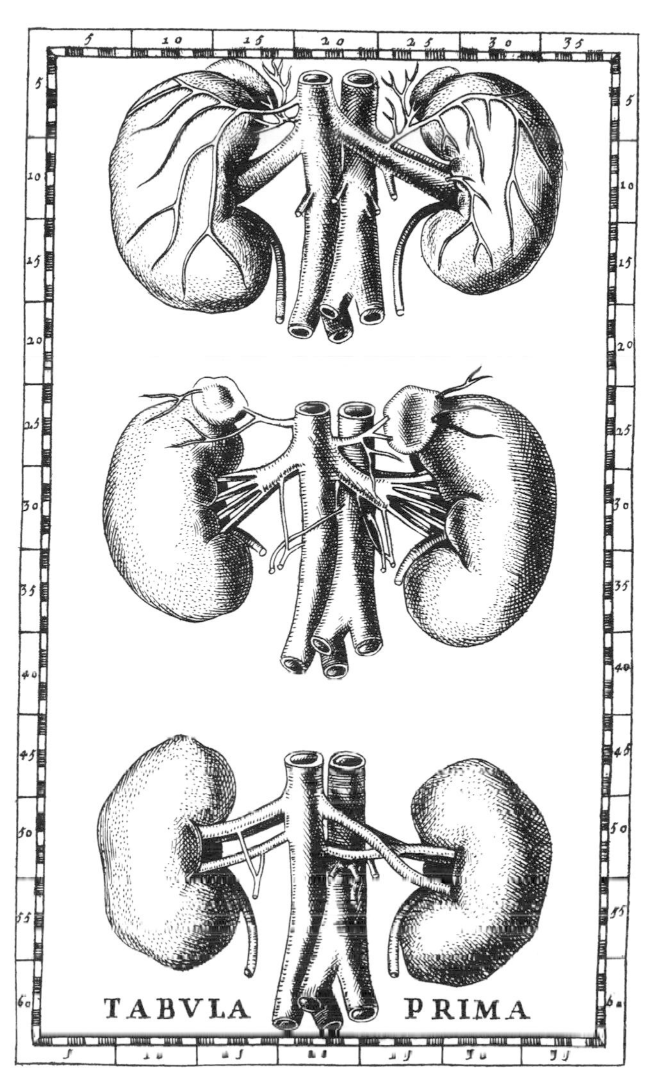

Fig. 11.19. The first published drawing of the human adrenal glands lying above the kidneys was engraved in 1552 by Bartholomeo Eustachio, who also discovered the tube between the middle ear and pharynx that bears his name. Top—Ventral view showing arterial distribution; middle—dorsal view. (From Eustachio, 1714, Plate I facing page p. 12, ×1.)

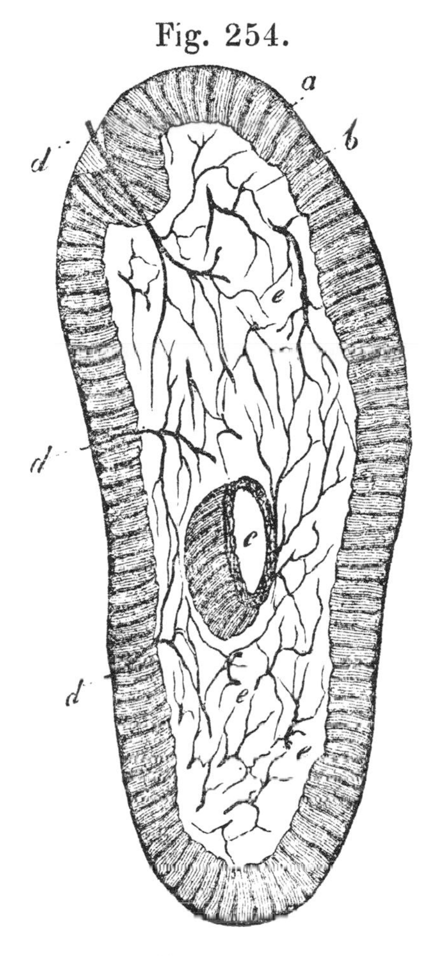

Fig. 11.20. Among his many contributions to the fine structure of the human nervous system in his famous *Handbuch*, which was translated into English two years after publication, Albert von Kölliker described the adrenal gland as having two distinct kinds of cells. a—Bovine cortical cells; b—medullary tissue; c—central vein surrounded by cortical cells; d—nerves traversing cortex and ramifying within the medulla. (From Kölliker, 1852, p. 490, Fig. 254, ×4/5.)

'blood-vascular glands' and a relation to secretion assigned to it; whilst the latter, on account of its extremely abundant supply of nerves, must be regarded as an apparatus appertaining to the nervous system. . . . (Kölliker, 1852; translated and edited by G. Busk and T. Huxley, 1853, p. 127).

The detection of large amounts of epinephrine in the adrenal gland (Takamine, 1901; Abel, 1903) was followed by studies of its conversion from norepinephrine (Holtz, 1938) in the chromaffin cells of the adrenal medulla. The early demonstration of the dependence of catecholamine secretion on calcium ions (Houssay and Molinelli, 1928), was confirmed and in isolated posterior pituitary preparations was found for neurohypophyseal hormones also (W. W. Douglas and A. M. Poisner, 1964) leading to a theory of stimulus-secretion coupling (W. W. Douglas, 1968). The usefulness of this system in neuroscience research lay in its application to other systems: "From the '50s onward, the emphasis shifted from a study of adrenal medullary function *per se* to an outlook where the adrenal chromaffin cell was regarded as a model neuron in general, a model sympathetic neuron in particular" (Carmichael, 1989, p. 93).

The physiologic trigger for release of ACTH was known to be stress, as the work of Hans Selye and Walter Cannon had made clear. The mechanism of its release from the anterior pituitary proved elusive from the beginning, in spite of its examination by two highly competent laboratories. Murray Saffran at McGill University, with two younger collaborators, one of whom was Andrew Schally, embarked in 1953 on in vitro study of the problem. The other laboratory was organized the same year at Baylor University in Houston by Roger Guillemin, who had just moved there from the University of Montreal, where he had been Selye's student. Guillemin's interest was to show the peptide nature of the hypophysiotropic hypothalamic substances. Shortly after publication of the McGill work (Saffran, Schally, and Benfey, 1955) in which the term "corticotropin-releasing factor" (CRF) was coined for the ACTH-releasing activity, Schally left Montreal to join Guillemin for several years of intensive work to isolate and identify CRF. The hormone was partially purified in 1958 (Schally, Saffran,and Zimmerman) and later was claimed to be the first direct demonstration of a hypothalamic factor regulating pituitary function, in contrast to the many indirect functional studies (Schally, 1978).

Incontrovertible proof of a thyrotropin-releasing factor (TRF) was announced in 1962 (Guillemin, Yamazaki, Jutisz, and Sakiz) and in 1969, TRF was isolated and characterized independently by Guillemin's group (Burgus, Dunn, Desiderio, and Guillemin) and by Schally and his associates (Bøler, Enzmann, Folkers, Bowers, and Schally). In an autobiographical account of the antecedent events, Guillemin considered the discovery of TRF "the major event in modern neuroendocrinology, the inflection point that separated confusion and a great deal of doubt from real knowledge" (1978, p. 239). In that same memoir, Guillemin wrote:

> There is no doubt in my mind that Selye's studies on stress and his remarkable observation of the involvement of the pituitary-adrenal axis in response to stress were powerful and persuasive incentives for the early studies. . . . Selye, through his stress concept, had thus a major stimulatory role in orienting the early efforts in neuroendocrinology toward the study of the hypothalamus-pituitary ACTH-adrenal cortex functional relationships. Strangely enough, and unwittingly on Selye's part, this is probably about the worst thing that happened to nascent neuroendocrinology. The search for CRF was to prove so complex and baffling that it is still not satisfied . . . more than 20 years after it started" (ibid., p. 232).

The corticotropin-releasing factor was finally purified in 1981 by Vale, Spiess, Rivier, and Rivier.

Stress research in the first half of the twentieth century was dominated by two productive investigators of contrasting background, character, and investigative approach. The younger, Hans Selye (1907–1982), was born in Vienna, educated in Prague, Paris, and Rome with specialization in biochemistry, and became a Canadian citizen. The object of his career was promotion of the "complex topic of stress as applied to every aspect of daily life or medicine" (Selye, 1979, p. xiv). He had observed that patients with diverse diseases had a common "syndrome of just being sick," a condition which he reproduced experimentally by challenging rats with an array of sufficiently noxious experiences. The adaptive mechanisms for the stereotyped stress response had a common pathway through the pituitary, he believed, and with his associates he showed that stress increased the production of ACTH, the catecholamines, corticoids, and CRF. Although many of Selye's conclusions did not survive, the intellectual ferment he created combined with his magnetic personality did much to bend the course of endocrinology and its application to medicine.

The life and career of Walter Bradford Cannon (1871–1945; Fig. 11.21) were in sharp contrast to Selye's, although their interests in stress overlapped. The "compleat" physiologist, Cannon's approach to stress was embodied in his classic text, *Bodily Changes in Pain, Hunger, Fear and Rage* (1929). The prologue to the presentation edition of Cannon's autobiography states: "Over a period of twenty years (1911–1931) Cannon and his many students obtained much evidence that under conditions of physiological stress . . . the sympathetic system and its constituent part, the adrenal medulla, act to produce visceral adjustments which are

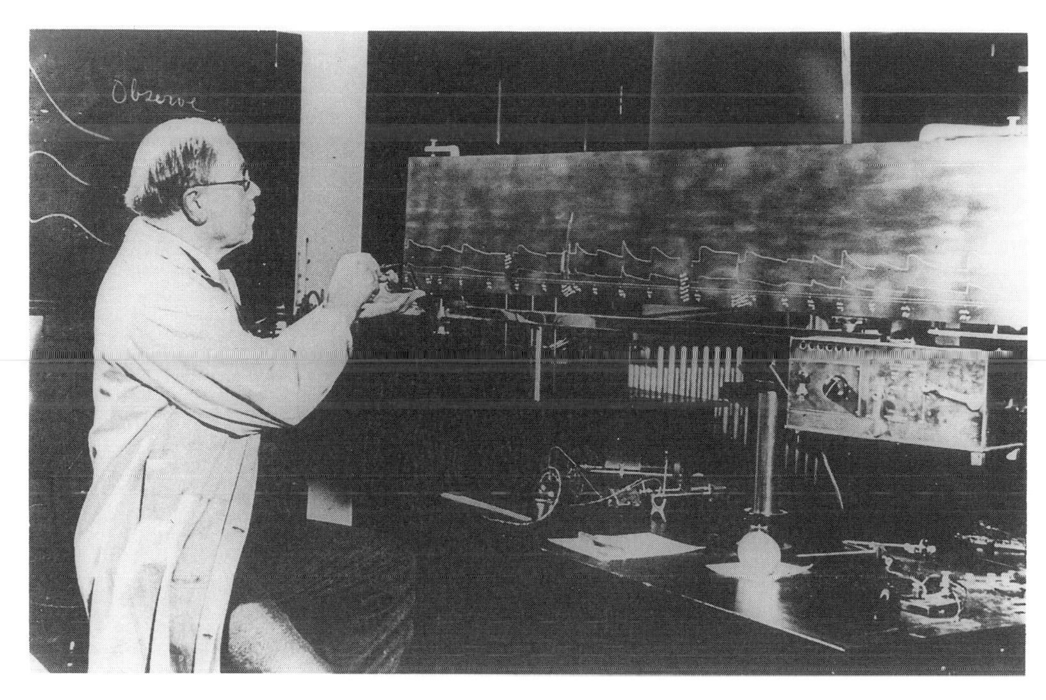

Fig. 11.21. Walter Bradford Cannon's scientific career was confined to Harvard University, where he succeeded to the chairmanship of the first physiological laboratory in the United States and enhanced its prestige at home and abroad. This 1940 photograph shows Cannon with his "long-paper" kymograph and famous admonition to his students.

nicely adapted to the preservation of the individual" (Bard, 1968, p. vi).

Cannon had found that the denervated heart "demonstrated that if the adrenal glands are intact and are normally supplied with their sympathetic nerves, every condition which results in a generalized discharge of sympathetic impulses causes an accelerated heart beat" (Cannon, 1945, p. 104). The tachycardia was caused by what was scoffingly referred to by another towering American physiologist, A. J. Carlson, as the "tail hormone," to explain the so-called fight or flight response (*see* also p. 169, this volume). The hormone was called *sympathin* "[b]ecause the substance is derived from structures under sympathetic control, when they are influenced by sympathetic impulses. . . ." (Cannon and Bacq, 1931, p. 411). The hotly debated question of whether or not increased adrenal medulla secretion accompanies sympathetic activity was settled by Cannon's series of sympathectomized animals rendered unable to cope with stress although they could live contentedly for years in a controlled environment (*see* Bacq, 1975b).

One of Cannon's most distinguished students, Philip Bard (1898–1977), wrote of his own research with Cannon:

> In searching for conditions that induce vigorous sympathoadrenal medullary activity . . . [h]e suggested that I attempt to ascertain the locus of the central mechanisms essential for this ["sham rage"] which so closely resembled the behavior of the infuriated normal animal and is accompanied by widespread sympathetic discharges. . . . It turned out that the region requisite for sham rage . . . is situated in the caudal half of the hypothalamus (1973, p. 10).

That conclusion was the outcome of Bard's experiments published in 1928, undertaken in response to Cannon's suggestion and based partly on the neglected results of the fruitful collaboration of two Viennese physicians, neurologist Johann Paul Karplus (1866–1936) and physiologist Alois Kreidl (1864–1928; *see* Fig. 11.22, p. 246). Their series of eight papers published between 1909 and 1927 described faradic stimulation of the surface of the cat hypothalamus which reproduced almost

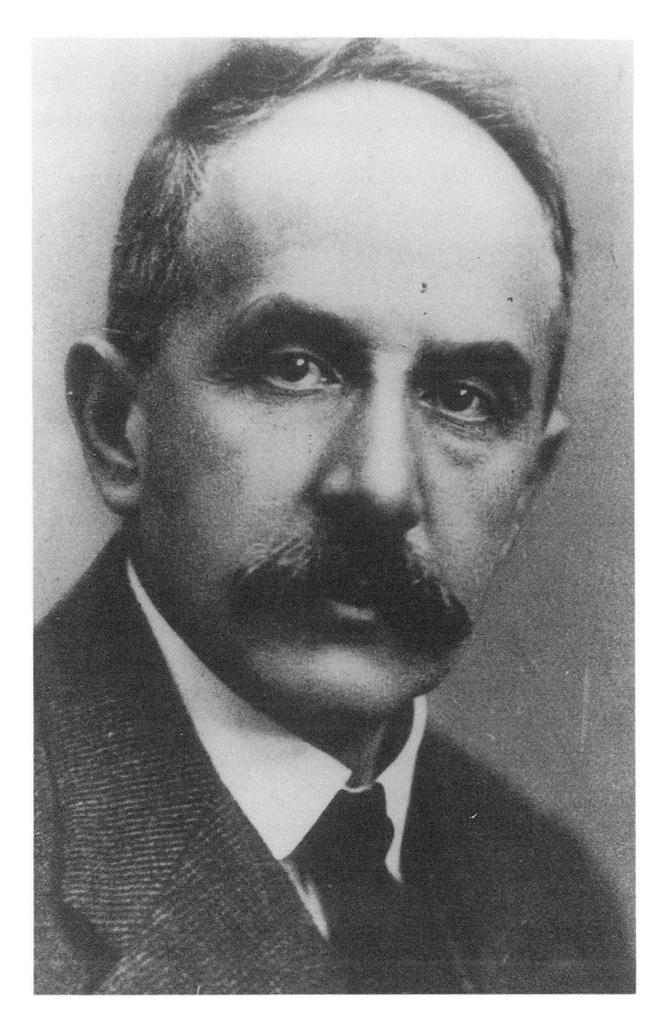

Fig. 11.22. J. P. Karplus (left) and A. Kreidl (bookplate by Emma Lowen Statt) were pioneers in the study of central mechanisms of "vegetative" function, which they discovered could be reproduced by hypothalamic electrical stimulation.

all the various effects of sympathetic nervous discharge. They concluded that there is a hypothalamic vegetative center, a "discovery [that] forms a foundation stone not only of the study of physiological regulations but also of clinical neurology" (Wang, 1965, p. 524). Wang pointed out that Walter Hess at Zurich, using conscious cats, confirmed and extended the findings of Karplus and Kreidl, as did Ranson and his Chicago group using both anesthetized and conscious cats.

On the basis of many careful brain stem transections, Bard concluded that: "the hypothalamus is not an autonomic center; but rather a part of the brain that contains neural mechanisms requisite for complicated patterns of behavior—such as the display of emotion and defense against heat and cold—in which there are autonomic components" (1973, p. 11). An evaluation of that work by Bard's successor as chair of physiology at Johns Hopkins stated: "It provided an objective definition of a neural center, a concept that guided the later studies of many investigators" (Mountcastle, 1977, p. 297).

OVERVIEW OF THE AXIS

The crazy quilt of discoveries—identifications and characterizations—of the hypophysiotropic hormones was superimposed on a matrix of neuroanatomy—nerve fibers, tracts, and blood vessels—before a picture of the pituitary–hypothalamic axis came into focus. As that image emerged, enormous vistas of therapies and controls were opened for investigation such as, but far from limited to, reproductive and growth processes. The initial rejection and final acknowledgment of Otto Loewi's finding in 1921 eased the way to much later acceptance of the notion of humoral transmission and may be said to have pointed the path to neuroendocrinology. Embedded in this history are examples of the importance of histology, such as the Gomori-stained whole neuron with its colloidal inclusions and the electron microscopically visualized axon endings close to the vascular bed. The final proof of neurosecretion was contributed by electrophysiologic recordings of activity in single neuronal units that coincided with evidence of secretion. As a system essential for the survival of the organism and the species, the many components of the pituitary–hypothalamic axis that bridge the endocrines and the brain have various degrees of essentiality. This dynamic system is effective because it is more than an axis, it is a confederation in which the adrenal gland constitutes a vital component.

12 Three Major Integrative Systems

CONTENTS

The amount of information in a system is a measure of its degree of organization.

(Wiener, 1948, p. 18)

Again addressing the oscillating, cause-and-effect relationship that has illuminated the development of knowledge of brain anatomy and physiology, we direct attention to three great systems on which the integrated behavior of higher animals depends: the ancient limbic system, the classical sensory or thalamocortical system, and the nonspecific ascending reticular system. Named in the order in which they were recognized and will be discussed, in each system an anatomic substrate was described before its "use," or function, became apparent. The English-American neurophysiologist well known for her historical writings on the nervous system, M. A. B. Brazier, equated this triad of structures with the functions necessary to maintain the conscious state (1963, pp. 748–749). First, "transmission of the sense-labeled impulse bearing the message from the periphery to the brain" is achieved by "the classical afferent system, ascending laterally through the specific thalamic nuclei to specific cortical sites." Second, "awareness that the message has arrived" occurs because "the ascending sensory systems in the midbrain core and medially placed thalamic nuclei are profoundly implicated." Receiving and storing the message, the final requisite of the conscious state, "is served by the third of the three systems named: the limbic system, and in particular the hippocampal system." This chapter considers the manner by which each of those integrated anatomic circuits was recognized as a functional entity. They cannot be envisioned as neatly demarcated nor are they competitive; rather, they lack agreed-on boundaries and it is clear that they interact profusely, each feeding into and receiving information from the others.

THE LIMBIC SYSTEM AND MEMORY

"Modified by evolution, the structure plays an important part in man; this is the great limbic lobe of Broca." (Schiller, 1979, p. 247.)

Much of the history of the limbic system consisted in demonstrating what it is not—a sensory system entirely engaged in olfactory functions. Its constituent parts were known and named by the early anatomists, but the idea that the elements functioned as a cohesive entity did not emerge until the last quarter of the nineteenth century. Initial recognition that such was the case came from that remarkable French neurologist-anthropologist-politician and superb comparative anatomist, Paul Broca, whose localization of speech in the left frontal lobe was described in Chapter 5. Broca was characterized by a modern biographer as "[b]old but not foolhardy, radical but not extreme, and fanatic only in his quasi-religious belief that facts were infallible" (Schiller, 1979, p. 234). Broca published a lengthy review of "Le grand lobe limbique . . . dans le série des mammifères" (1878a) in which he explained his choice of the term "limbic" for a group of structures surrounding

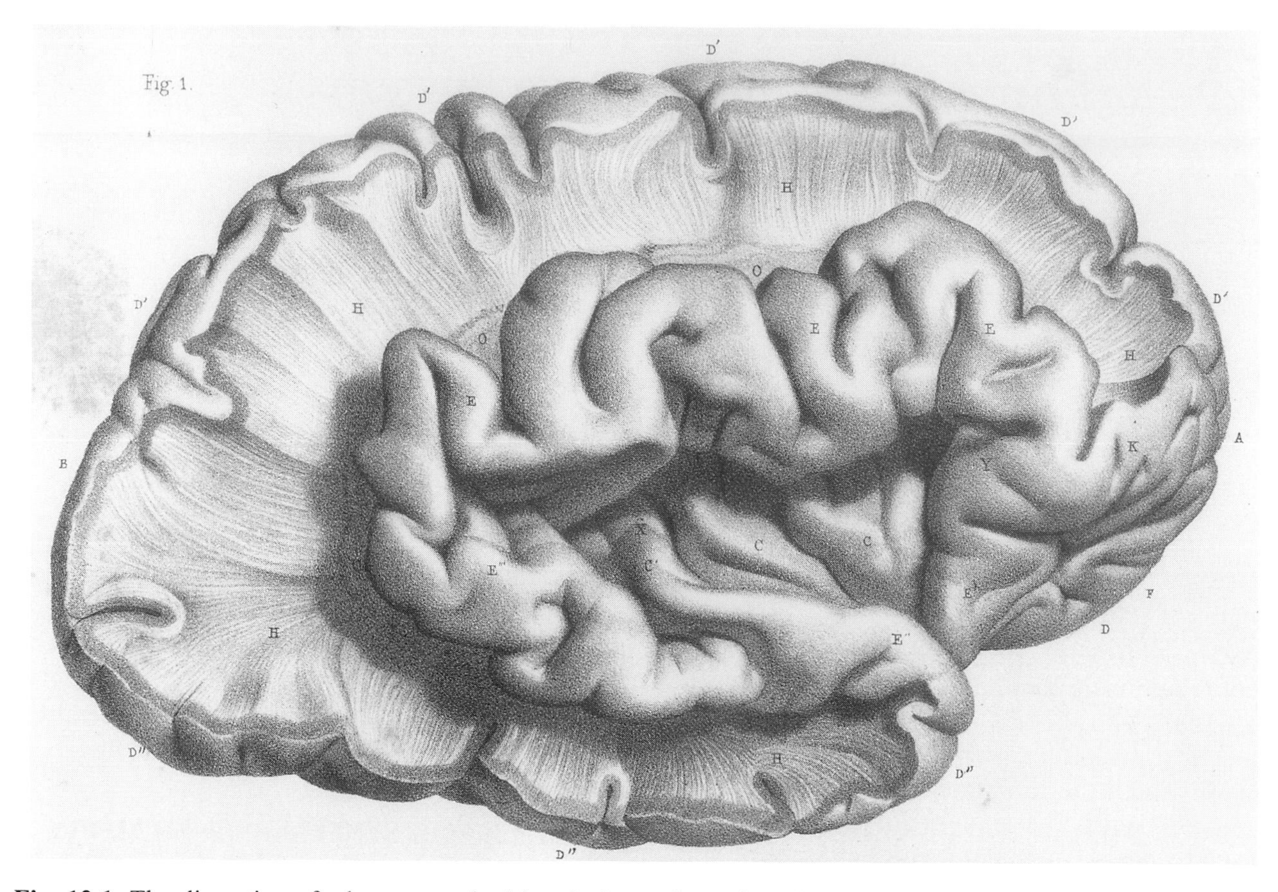

Fig. 12.1. The dissection of a human cerebral hemisphere, shown in a drawing by E. Beau, illustrated Achille-Louis Foville's characterization of the convolutions surrounding the sylvian fissure as a "drawstring." (From Foville, 1844, Plate 7, Fig. 1, ×1/2.)

the brain stem: The meaning is not restrictive, implies no function or theory, and has no definite shape so it can apply to all mammals. Several earlier terms were available—Pierre Gerdy called it the "annular convolution" (1843) and Foville (1844) the " 'drawstring convolution,' because he believed that its base was formed by a bandelet of circular fibers. . . . This bandelet is artificial." (Broca, 1878b, p. 387). The resemblance is seen in Foville's illustration reproduced in Fig. 12.1. Schiller (1979, p. 256) pointed out that Broca did not mention Rolando's description in 1830 of nerve fibers coursing from the olfactory bulb in a hemicircle to the uncus: the omission seems to indicate that the scrupulous Broca had not read it.

Broca dissected the brains of many mammals, and "saw the functional coherence of the parts and so discovered this entity," thereby consolidating his place in history as a preeminent comparative anatomist, "equaled [by C. J. Herrick, Ariens Kappers, and F. Tilney and H. A. Riley] but never surpassed" (ibid., p. 255). Broca concluded that the presence of the limbic lobe is a common denominator among mammals because he found it in all mammalian brains he examined (1878a; Fig. 12.2). His suspicion that the lobe's function is not confined to olfaction in spite of its close association with the primordial olfactory bulb is vividly illustrated by his own observations of a dog out hunting:

> The dog may hesitate, investigate again, the sensory center come[s] into play, intelligence deliberate[s] once more to modify the plan of the chase. . . .
>
> Pursuing or pursued, an animal makes use of the olfactory lobe for striving; of the frontal lobe, for control. . . . [T]he olfactory system always

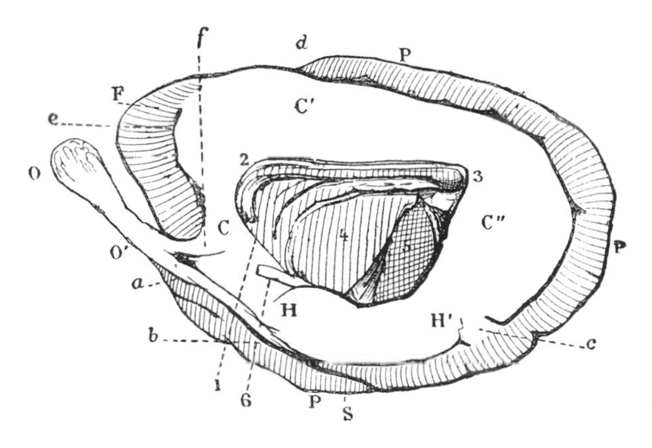

Fig. 12.2. Paul Broca chose "limbic lobe" to name the structures illustrated by Foville because the term "implied no theory" and "is applicable to all mammalian brains." The otter's brain (above) is one of many he dissected to demonstrate the ubiquity of the structure (C, C', C") among mammals. (From Broca, 1878a, p. 399, ×1.)

Fig. 12.3. James Papez and his artist wife, Pearl, collaborated on publications that dealt with the minutiae of the microscopic anatomy of fiber tracts and nuclei, thus building a solid base for his theory of emotion. (From MacLean, 1978, p. 10.)

> functions in two ways: first as a means of *investigation*, second as a means *of action*. . . .
>
> The first function is called *sniffing*; it takes place in the sensory center to which the olfactory lobe transmits its impressions. . . . By contrast, what we might call *olfactory action* . . . has the simplicity of a reflex. . . . It is the direct transformation of sensation into movement without intervention of the will. . . . (ibid., p. 445, passim; translated in Schiller, 1979, p. 266).

Broca's idea of an extended limbic lobe function received little attention during the six decades after he announced it, but his inferences (Broca, 1879) about its olfactory component persisted, and by the 1890s the entire limbic lobe was referred to as the rhinencephalon or "smell brain." That restricted concept came apart under the virtually simultaneous impact of three proposals that were directly or indirectly concerned with the nature of emotion and its neural substrates. The psychologists especially were groping for a firm explanation of emotional responses, as witness the U.S. National Research Council's organization in 1926 of the Committee on Experimental Study of Emotion with Margaret Floy Washburn (1871–1939) as chairman.

On a second front, the microanatomists had been constructing the foundations on which theories of emotion would rest by their diligent pursuit of tracts and nuclei; the contributions of Golgi, Ramón y Cajal, and Lorente de Nó are discussed elsewhere. The first important breakthrough came in 1937 when an American anatomist, James Wenceslas Papez (1883–1958; Fig. 12.3) from his isolated laboratory at Cornell University in Ithaca, New York, published a remarkable paper in which he described the circuit that later carried his name and cautiously suggested that it might constitute the neural substrate for emotion, thus substituting a circuit for a center. In that classic account of imaginatively related facts, Papez wrote that he presented

> anatomic, clinical, and experimental data dealing with the hypothalamus, the gyrus cinguli, the hippocampus, and their interconnections. Taken as a whole, this ensemble of structures is proposed as representing theoretically the anatomic basis of the emotions. . . . The term "emotion" as commonly used implies two conditions: a way of acting and a way of feeling. The former is designated as emotional expression; the latter, as emotional experience or subjective feeling. The experiments of Bard (1928) have demonstrated that emotional expression depends on the integrative action of the hypothalamus. For subjective emotional experience, however, the participation of the cortex is essential (Papez, 1937, pp. 725, 726).

Papez believed that the substrate of emotion had evolved phylogenetically from mechanisms essential to nutrition and reproduction, that is, the taste and smell of food and products of the sex glands had formed the drives for feeding and mating, two essentials for survival of the species.

Immediate interest in Papez's proposal was minor, perhaps due to both his modesty in pushing his own ideas and the appearance the following year of two notable papers, in neither of which his work was cited; in all fairness, however, the timings were very close. In his paper, the experimental psychologist, Karl S. Lashley (1890–1958), later famous for his search for the engram of memory (1950), postulated the thalamus to be an important site of emotion, thus aligning himself with Cannon. The great value of Lashley's penetrating analysis of emotion (1938) was in its identification of two major areas of uncertainty in the then current concepts of the neural substrate of emotion. The first and more inscrutable concerned the affective aspects of emotional experience, in turn a feature of the more general problem of the basis of all consciousness and subjective experience. The second uncertainty had to do with the identity of forebrain mechanisms and processes that normally hold emotional behavior in check, the elimination of which in "decortication" releases diencephalic mechanisms, leading to an excessive expression of emotional behavior. This idea recalls the Jack-in-the-box analogy of Rhines and Magoun (*see* Fig. 9.20, *see* p. 197, this volume). As Lashley commented in his discussion of emotional states, "Although we may assume that the increased excitability of the motor centers is a result of withdrawal of inhibition, a survey of the evidence leaves some doubt as to the source of the inhibition" (ibid., p. 46).

Additional support for the limbic circuit concept of emotion materialized with serendipity from a new quarter in an attempt to isolate the site of action of the psychomimetic drug, mescaline, by experimental psychologist Heinrich Klüver teamed with neurosurgeon Paul Bucy. A few months before Lashley's review appeared, they presented (1938) the results of a two-stage bilateral temporal lobectomy that included significant portions of the amygdaloid nuclei and hippocampi of the basal forebrain, carried out initially in an unusually ferocious monkey.[1] The variety of abnormalities that followed the operation in this and subsequent subjects included a marked taming of this wild animal and its development of hypersexuality (*see* Chapter 5 for details of those studies and the priority of Sanger Brown and E. A. Schäfer).

A dramatic clinical confirmation of the findings of Klüver and Bucy was reported by Terzian and Ore in 1955 in a human subject. After bilateral temporal lobectomy, including most of the uncus and the anterior part of the hippocampus and amygdala, carried out for the relief of intractable psychomotor epilepsy, the postoperative syndrome reproduced all signs reported in monkeys except the oral tendencies and in addition there appeared a serious deficit in memory.

The changes in emotionality in "the temporal lobe syndrome" were supportive of Papez's proposal, replacing the earlier smell-bound thinking and even distracting some behavioral scientists from their preoccupation with the neocortex. A decade after the influential studies just described, a young neurologist, Paul D. MacLean (Fig. 12.4), organized and elaborated Papez's views in two publications: "Psychosomatic Disease and the Visceral Brain. . ." (1949), which enormously extended and enlivened its Papezian forebear and in the follow-up paper, "Some Psychiatric Implications. . ." (1952) revived Broca's term "limbic" and referred to the lobe and its brain stem connections as the "limbic system," the term that is current today. In essence, whereas Papez saw through binoculars, MacLean envisioned a wider scenario.

Based on his clinical research and comparative neurobehavioral studies, MacLean later proposed the "triune brain" (Fig. 12.4, right) to explain normal and pathologic human behavior. His concept of the evolutionary development of the human brain emphasized the key position of the limbic or paleomammalian brain (*see* p. 25, this volume), as evidenced by a relatively superior metabolic activity: the limbic cortex exceeds the neocortex in turnover of protein, a measure of the demand for new RNA in memory formation (Flanigan, Gabrieli, and MacLean, 1957; Hydén, 1969). That demand apparently occurs in spite of the fact that in phylogenetically higher mammals the neocortex, where

[1]Information from interview of Paul Bucy by K. E. Klivington, titled "The Papez Memorabilia, Boston, April 8, 1981," with permission.

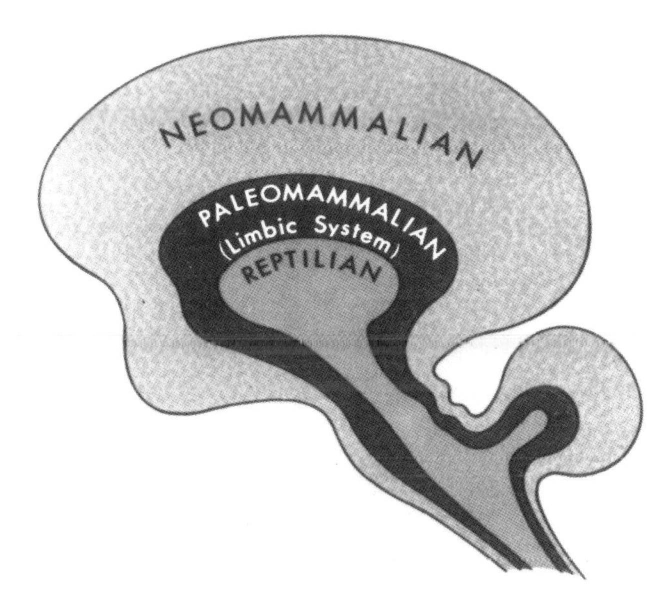

Fig. 12.4. As chief of the Laboratory of Brain, Evolution, and Behavior at the National Institute of Mental Health, Paul D. MacLean (left) directed studies on reptiles and squirrel monkeys that reinforced his concept of the "triune brain." MacLean's symbolic representation (right) of the successive overlay of neural tissue with new functions during evolutionary time was first published in 1967, p. 377, Fig. 2. Photograph ca. 1957.

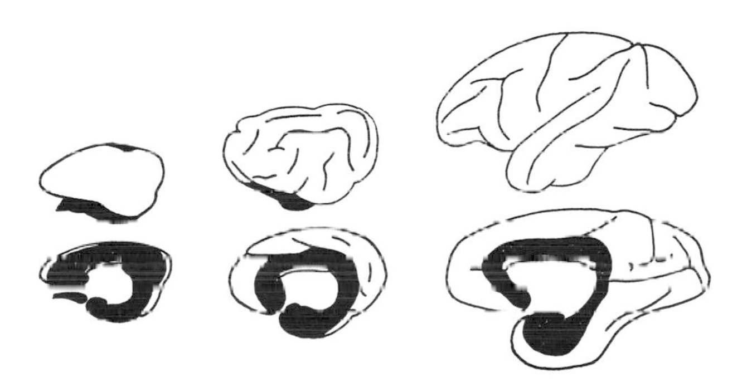

Fig. 12.5. The evolutionary old cortex (the limbic lobe) is a "common denominator" in mammalian brains with the neocortex mushrooming around it as the phylogenetic scale moved forward. (Adapted from MacLean, 1954, p. 106.)

"higher" nervous functions take place, has ballooned around the two older brains, as shown in Fig. 12.5.

MacLean gathered support from his own studies and those of others to conclude that the traits distinguishing mammalian from reptilian forms are three "cardinal" behaviors: caring for the young, audiovocal communication, and play. From the "momentous significance" of the evolutionary acquisition of those behaviors, he extrapolated to write that "the history of the evolution of the limbic system is the history of the evolution of mammals, while the history of the evolution of mammals is the history of the evolution of the family" (1990, p. 247).

The first association of memory with a specific part of the limbic system, the hippocampus, seems to have been made by Vladimir Mikhailovich Bekhterev (1857–1927; *see* Fig. 12.6, p. 254), the Russian neurologist–anatomist whose career was contemporaneous but not parallel with the work of Pavlov. Bekhterev described (1900) the brain of a patient with memory deficit and hippocampal degeneration, another example of a clinician leading the way to new insights, as was the case with the discovery of acromegaly (*see* Chapter 11) and Bright's correlation of sensation with the thalamus (*see* Chapter 10). The hippocampus was targeted in memory deficit also by G. Elliot Smith in his Croonian Lectures (1919), a series presented by distinguished figures representing the contemporaneous accepted wisdom. Attempts at replicating human memory loss in the experimental animal (Orbach, Milner, and Rasmussen, 1960) did not clarify the problem of the exact site of memory deficit, however, nor had Lashley been able to find the site of the engram of memory in the rat brain in his dedicated, almost desperate search (1950). Rather, we turn from the experimental studies to

Fig. 12.6. The elderly Vladimir Bekhterev is shown with two assistants in the unheated Reflexology Laboratory at the State University in St. Petersburg. In his psychological work Bekhterev used muscular responses as an indicator, in contrast to Pavlov's more easily quantified drops of saliva.

the careful and persevering testing of a few human subjects for real insight into limbic functions.

A relationship of limbic structures to memory and learning was made clear by a Canadian neurophysiologist and student of Donald Hebb's, Brenda Milner, in her repeated study of H. M., whom she first saw in 1955 (B. Milner, 1992). Two years previously, H. M. had been operated bilaterally with a medial temporal lobe resection which included the radical removal of limbic structures: the amygdala, hippocampus, and parahippocampal gyrus. Milner found initially that her subject's limits of retention were dependent on the absence of attention-distracting intervening elements. Then in experiments carried out on the same patient in 1960 and with a series of complicated psychological tests, Milner uncovered what she described as a "diencephalic memory system," and supposed that there might be many different memory systems in the brain.

In spite of statements in two influential textbooks, Obersteiner's (1890) and Schäfer's (1898), that some mammals have no sense of smell, yet possess a limbic system, the early notion that the hippocampal part of the system is "an important olfactory 'centre' " was one of those "conceptions [that] survive almost like proverbs. However, some general suspicion of the truth of this credendum seems to be just about to dawn" (Brodal, 1947, p. 179). Opening his review with those prescient words, the Swedish neuroanatomist, Alf Brodal, presented anatomic evidence that olfactory fibers had not been traced to the hippocampus proper and concluded that "[r]ecent physiological experiments have yielded no support for the conception that the hippocampus has important relations to the sense of smell in mammals, nor does clinical evidence seem to favour this view" (ibid., p. 218).

The hippocampus, according to Frederick Tilney (1938), who could see no resemblance to the sea horse, had been named "with too loose a rein" by the Renaissance anatomist, Arantius (1587, pp. 44–45), yet it became the most thoroughly investigated part of the limbic system for good reason. That its major component offered great advantages for electrophysiological studies because of its simple and highly oriented structure compared with other cortical areas was recognized by Renshaw, Forbes, and Morison (1940, p. 75) who stated, "Our results demonstrate that particularly interesting deductions may be made from . . . data obtained in regions of the nervous system where the spatial arrangement of cells is particularly simple, as it is in the hippocampus." Thirty-five years later, and for the same reason—two completely separate cell populations, pyramidal and granular—it was proposed as a model system for research on neuronal plasticity (Lynch and Cotman, 1975). Additional advantages are that the surgical approach to the hippocampus is relatively free of damage to its blood supply and easy to isolate from other structures, thus facilitating excision or investigation in vitro. Those factors were especially conducive to single-unit research undertaken to probe for the source of complex action potentials displayed in the patterns recorded on the electroencephalogram (EEG). The question had been a recurring challenge to the "brain wavers" made by the "axonologists" who were working at the periphery of the nervous system and announcing exciting discoveries about the compound action potential and conductance in the axon. The axonologists, strutting on the boardwalk at annual meetings in Atlantic City (*see* L. H. Marshall, 1983a, p. 631, fn 23), delighted in taunting their peers about a perceived lack of progress in interpretation of the complex brain wave patterns and what they reveal of how the central nervous system works. On their part, the "EEGers" were searching for answers in several directions. Gibbs and Gibbs (1936) had already shown that, of the brain regions they tested electrically in cats, the hippocampus had the lowest threshold for seizure. The findings of J. D. Green and his associates (summarized in Green, 1959; Fig. 12.7), elicited a proposal that the origin of the slow hippocampal (theta) wave in the rat is dual, "generated by the hippocampal pyramids between the distal part of the cell body layer and a level near, but not at, the termination of the apical dendrites" (ibid., p. 270). In Japan, somewhat similar studies in rabbits targeted the sites more precisely: "the somata and apical dendrites of the hippocampal pyramids are activated in a seizure discharge" (Taira, 1961, p. 198; *see* Fig. 12.8, p. 257).

In another direction, B. R. Kaada (1951) at the University of Oslo focused on the behavior-modulating effects of the limbic system. He found that electrical stimulation produced opposite effects depending on the region that was stimulated: from septal sites the effect was inhibitory for motor, reflex, and autonomic responses, whereas from the cingulate gyrus it facilitated those same responses. With that research as a background, the next step was to more closely identify an excitatory dendritic locus on the hippocampal pyramidal cells and to postulate that "a dendritic location of excitatory and a somatic location of inhibitory synapses may well hold true for other cortical pyramidal cells" (Andersen, Blackstad, and Lømo, 1966, p. 247). The Scandinavian group proposed a new model of the phenomenon, reproduced in Fig. 12.9 (p. 257), which suggests the possibility that fibers carrying impulses from thalamus en route to cortex send collaterals to inhibitory neurons. Acknowledged as "somewhat at variance with previously formulated theories," Andersen and Andersson (1968) stressed that there may be many facultative pacemakers governing cortical phasic activity. A decade later Winson and Abzug (1977) reported that the behavioral state of the subject influenced transmission of impulses in the hippocampus.

By the middle of the twentieth century, it was clear that the key to limbic system function must be sought from combined anatomic, physiologic, and behavioral sources. Some of the pioneers in the application of the multidisciplinary approach were concerned that reliable experimentation in integrated research is subject to certain hazards when scientists cross disciplinary lines to work in an unfamiliar field. Robert A. McCleary at the University of Chicago, a physiological psychologist, enunciated a strong caveat in the first issue of a published series, *Progress in Physiological Psychology* (Academic Press, 1967), dedicated to fos-

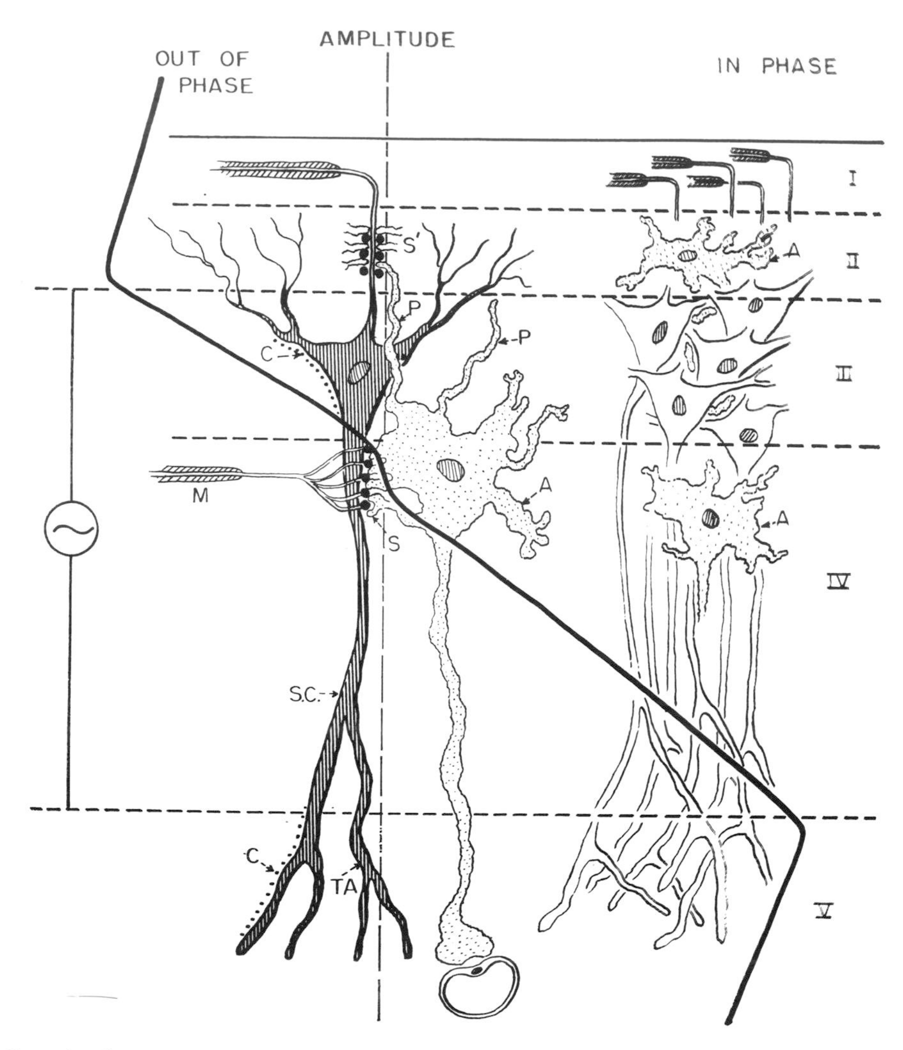

Fig. 12.7. John Green's diagram showing the single-unit potential superimposed on a pyramidal cell with a nearby astrocyte (A), to illustrate his theory of a dual source of the hippocampal theta wave (*see* text). (From Green, 1959, p. 268, ×1.)

tering the momentum which that discipline was then experiencing as it struggled to catch up with neurophysiology and neuroanatomy in studies of brain and behavior. Concluding his review of the response-modulating functions of the limbic system, McCleary wrote:

> [T]he general importance of considering the possibility of multiple deficits resulting from a particular lesion . . . is related to the obvious fact that experimental lesions most commonly are gross compared to the intricacy of the structures ablated. . . . [T]he remedy is the same in any case: the use of batteries of behavioral tests in evaluating the effects of lesions. The 'one lesion, one test' study is certainly no longer justifiable, if indeed it ever was (1967, p. 266).

This statement is one of many examples of the experimental psychologists' endeavor to ensure that neuroscientists from other disciplines adhere to rigorous testing and interpretation of behavioral data.

As MacLean pointed out (1990, p. 412), probably there is no clinical condition that provides more windows for viewing the neural substrate of the human psyche than psychomotor epilepsy. It has produced "crucial evidence" that the limbic system is fundamentally involved in emotion, as in the aura occurring at the beginning of a seizure

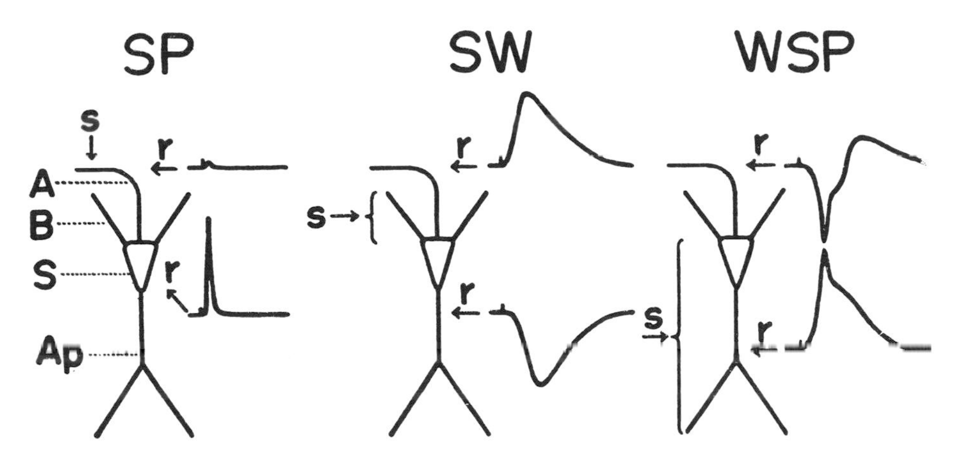

Fig. 12.8. Schema of types of evoked potentials (SP—spike; SW—slow wave; WSP—wave-and-spike) elicited by stimulation of different levels of hippocampal cells in rabbit brain. A—axon; B—basal dendrite; Ap—apical dendrite; S—soma. (From Taira, 1961, p. 193, Fig. 2. ×1.)

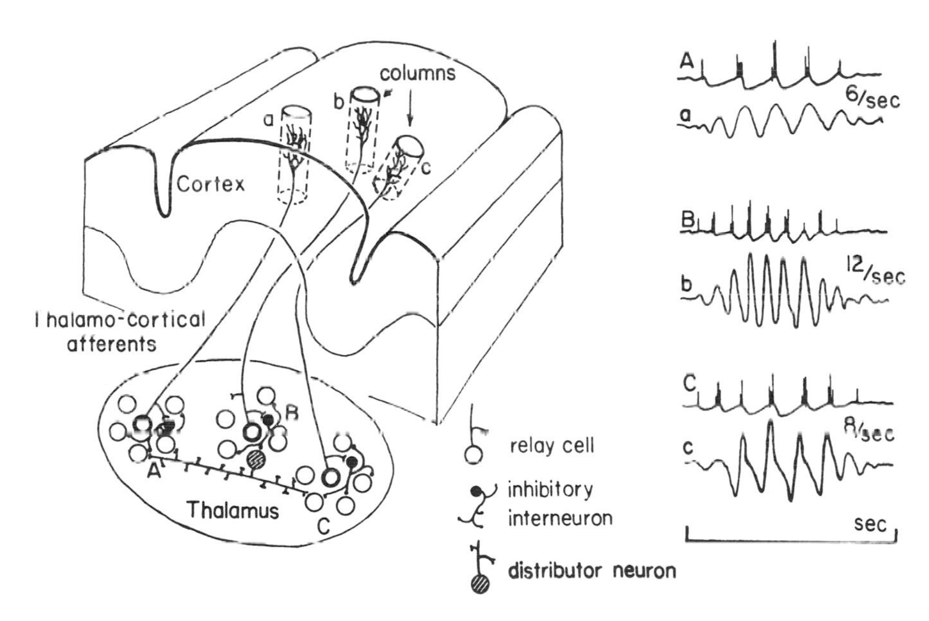

Fig. 12.9. A sophisticated model of cortical phasic activity showing different intraspindle frequencies and times of onset and stopping. Left—A, B, C: Thalamic nuclei send axons to cortex (a, b, c) and collaterals to interneuronal cells some of which are inhibitory (black). Right—imaginary spindles at thalamus (A, B, C), at cortex (a, b, c). (From Andersen and Andersson, 1968, p. 59, Fig. 5.9. ×4/5.)

when patients may experience a variety of subjective states ranging from intense fear to ecstacy. Those emotions were reported by patients of one of the earliest active groups working on the EEG in America, at Massachusetts General Hospital (Gibbs, Gibbs, and Lennox, 1938). Termed "temporal lobe epilepsy" (Lennox, 1951), a more fitting name is "limbic lobe epilepsy" because it involves structures beyond the temporal lobe (Fulton, 1953, p. 77; Glaser, 1967). By means of stimulation through depth electrodes chronically implanted at several sites, Pagni (1963) in Milan produced clinical signs of spontaneous seizures in human subjects and evidence that the hippocampus, Ammon's horn, and the amygdaloid nucleus constitute a functional unit.

Limbic structures other than the hippocampus were also subjected to careful investigation and one of the most interesting was the septal region. Psychiatrists at Tulane University reported some patients who experienced sensations of pleasure when the septal region was stimulated (Monroe and Heath, 1954, pp. 348, 560). The phenomenon was replicated by two Canadian psychologists, working in the laboratory of Donald O. Hebb at McGill University in Montreal. James Olds and Peter Milner taught rats to self-stimulate their septal regions electrically in positive reinforcement experiments. The bar pressing for a pleasurable reinforcement became addictive to the point of exhaustion, and Olds and Milner wrote: "[W]e have perhaps located a system within the brain whose peculiar function is to produce a rewarding effect on behavior. . . . In septal area preparations, the control exercised over the animal's behavior . . . is extreme, possibly exceeding that exercised by any other reward system previously used in animal experimentation" (1954, p. 426).

In the opinion of some members of the distinguished group of neuroscientists assembled by David McKenzie Rioch (1900–1985) American neuropsychiatrist, at the Walter Reed Army Institute of Research during and after the Second World War, the plethora of studies that followed the discovery of the apparent "pleasures" of self-stimulation did not adequately analyze "the interrelationships between the physiological events associated with self-stimulation and the behavior of the self-stimulated animal" (Porter, Conrad, and Brady, 1959, p. 43). Such an analysis revealed that in some chronically implanted monkeys, self-stimulation of the hippocampus produced seizure activity and "the maintenance of high lever-pressing rates appeared to be positively correlated with . . . seizure patterns. In contrast, seizure discharges following self-stimulation of the amygdala produced suppression of the lever-pressing rate" (ibid., pp. 53–54), findings in the behaving animal which reinforced the physiologic events that had been found in single units of the hippocampus mentioned above.

A potentially important insight on limbic system function was provided in work of a group of Canadian neuropsychologists (Goddard, 1967; Goddard, McIntyre, and Leech, 1969). Describing a process which was later felicitously named "kindling," they reported that in experimental animals seizures can be induced by repeated application of subthreshold brain stimulation. They found the amygdaloid bodies are the most easily kindled regions, that no demonstrable histologic damage is sustained, the effect is transsynaptic, and the epileptogenic changes are permanent; Goddard related the permanence to the mechanisms of learning. Later evidence suggested that alterations are localized at the synapses (Racine, Newberry, and Burnham, 1975). Eventually it was suggested that

> we should seriously consider the possibility that the underlying mechanisms of learning and synaptic enhancement are one and the same. . . .
>
> The machinery by which this happens need not be pathological at each synapse. The pathology may lie only in the number of synapses involved, and the number of neurons brought into synchronous activity (Goddard, McNaughton, Douglas, and Barnes, 1976, p. 363).

The report of kindling had been preceded by somewhat similar observations of Watanabe (1936, cited by Alonzo-de-Florida, 1994, p. 206) in Japan on dogs subjected to subthreshold cortical stimuli, and by a series of experiments carried out in the laboratory of José Delgado at Yale. Alonzo-de-Florida and Delgado (1955) used chronically implanted electrodes stereotaxically placed in cat amygdala to produce a state of electrical activity characterized by spontaneous seizures.

As a tribute to the high priority accorded research on the limbic system, the University of Toronto mounted a conference on "The Continuing Evolution of the Limbic System Concept" in 1976. One-third of the presentations concerned kindling, led off by Graham Goddard's review, "From Iconoclasm to Orthodoxy." When the papers were published two years later, Goddard's interesting title had become: "Synaptic Change in the Limbic System," and the monograph simply *Limbic Mechanisms*. In 1975 the editors of a two-volume collection of papers starkly titled *The Hippocampus*, wrote: "Thus the enigma of hippocampal function, although slowly yielding its wrappings, as yet has lost none of its appeal or challenge" (Isaacson and Pribram, 1975, p. 439).

CORTICOTHALAMIC CONNECTIONS AND CYBERNETICS

> "[M]inute anatomy—those deserts of detail without a living functional watercourse, only a mirage from unverified speculation." (Rushton, 1977, p. 85.)

A distinctive feature of another major system of neural circuitry in the human brain was noted as early as 1839 by Carpenter in his textbook. He believed (*see* Chapter 10, this volume) that the impulse traffic between thalamus and cortex was in both directions, a concept not soon proven experimentally but one which generated speculation and search for the anatomic pathways. A century later, the imagery of a return loop was considerably broadened by British investigators: "[T]he conception of a thalamocortical circulation of neural impulse will in the future come to be found as fundamental for the neurology of what we colloquially call 'thought' as the conception of a circulation of the blood is for modern physiology!" (Campion in Campion and Smith, 1934, p. 97). Campion's coauthor, G. Elliot Smith, subscribed to the same idea: "The circulation of the thalamic and cortical currents maintains this constant state of readiness and is a vital and essential part of consciousness and mind" (ibid., p. 24). They were writing about the neural basis of thought, but the state of readiness referred to was muscle tone, again an illustration of the historical fact that entrée to the physiology of the nervous system was by way of the experimental study of muscular action.

Two decades later, however, it was not necessary to call on consciousness, mind, and the neuromuscular junction to explain the nervous system, for the experimental evidence of two-way circuits had become abundant. In Walker's words (1949, p. 250), "It is . . . evident that as well as receiving a spatially well-organized system of fibers from the thalamus, the cerebral cortex sends to that ganglion a system having, if not quite as precise, at least a certain organization." He recalled that Head and Holmes in England had labeled this centrifugal system the basis of inhibition, and the Germans concurred, whereas the French said the concept was not compatible with the clinical findings. To others, (Wallenberg, Brouwer, D'Hollander, for example), the corticothalamic pathways were a means by which the cerebral cortex "can modulate the sensitivity of the primary receptive centers to render them more susceptible to incoming impulses, in other words, it is a mechanism of sensory attention" (ibid., p. 251).

As noted in Chapter 10, Luys knew as early as 1865 that the four then-recognized nuclei in each thalamus had representative cortical connections. He also depicted converging fibers from cortex to the external thalamus (*see* Fig. 10.11, right, p. 212, this volume). Flechsig described a "descending efferent mechanism" as well as his "gateways" to the brain, ideas that were apparently based on the work of his associate, von Tschisch (1886), according to Mettler (1972, pp. 1, 2). Monakow, too, studied corticofugal projections (1895), coupling cortical ablations with degeneration in localized thalamic regions. At the turn of the century, a summary of views on information flow between cortex and thalamus in clinical material had been provided by Dejerine (1901), but at that time the polarity of the pathways was conjectural, especially in human material. With the continuing appearance of ever more reliable anatomic data, however, it became increasingly clear that Carpenter's suggestion of an interactive relation between thalamus and cortex was correct.

The strongest evidence was seen through the microscope lenses of one of the period's master histologists, Ramón y Cajal (*see* Chapter 7, this volume), who found (1903) a corticothalamic tract and asserted that it carried impulses in both directions (*see* Fig. 12.10, p. 260). Mention has already been made (see Chapter 10, this volume) of Sachs's report (1909) of fibers between frontal cortex and thalamus, based on anatomic and physiologic data obtained with the first stereotaxic instrument constructed by Horsley and Clarke. Head and Holmes were believers, too: "The functions of this organ [thalamus] are influenced by the coincident activity of the cortical centres, and this control is effected by means of paths from the cortex to the thalamus which probably end in the lateral nucleus" (1911, p. 151).

Santiago Ramón y Cajal's last and perhaps most eminent student, Rafael Lorente de Nó (1902–1990; *see* Fig. 12.11, p. 260), followed the master's example in expert manipulation of the staining techniques for nerve tissue that had become available and brilliantly continued the investigation of the auditory system. When he emigrated to America he added expertise in electrophysiology to his competence in otolaryngology and during his tenure at Rockefeller Institute for Medical Research, he published a comprehensive study of nerve physiology (1947) that demonstrated his "extraordinary versatility, breadth of interest, and diligence" (Kruger and Woolsey, 1990, p. 2). In his section on nerve activity in Fulton's *Textbook of Neurophysiology* (1938), Lorente reemphasized the physiologic rightness of a two-way communication between thalamus and cortex:

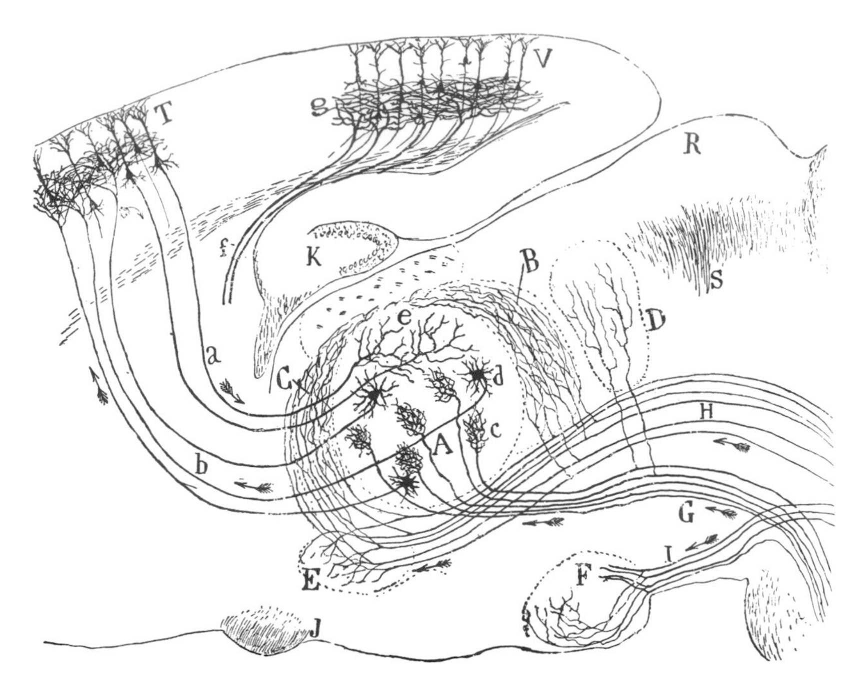

Fig. 12.10. This composite drawing of a thalamic nucleus by Santiago Ramón y Cajal was published in 1903, p. 339, Fig. 3. He repeated it in 1911, first to illustrate sensory paths to thalamus and again with a different caption to show corticofugal thalamic connections. A—Sensory nucleus of thalamus; T—sensory motor center; V—visual cortex; a—corticothalamic fibers; b—thalamocortical fibers; B, C—accessory nuclei of trigeminal nerve; D—post. nucleus of thalamus; E—nucleus zona incerta; F—ext. mammillary nucleus; G—medial ribbon of Reil; H—central path of 5th nucleus; I—pedicle of mammallary body; J—optic chiasa, K—Ammon's horn; f—sup. optic path. (From *Histologie . . .*, vol. 2, 1911, p. 876, Fig. 548; p. 501, Fig. 323, ×1.)

> Since the impulse conducted by a fibre necessarily passes into its collaterals and [since] branches of the descending axons are distributed in the same territories as the cortical afferents, there can be no doubt that the effect of the impulses entering the cortex depends largely upon the impulses at that moment circulating through the descending axons as a result of the existing cortical activity. The intracortical distribution of the axonal branches is as systematic as that of the dendrites (Lorente de Nó, 1938, pp. 307–308).

The earliest experimental study of corticothalamic projections that were both carefully timed

Fig. 12.11. Rafael Lorente de Nó was distinguished by his ebullient red bow tie, gracious Spanish manner, and relentless pursuit of his beliefs: "I want to make it stronger!." (Photograph by Dr. Emilio Decima in early 1980s.)

(staining) and systematically controlled (lesions) were carried out by F. d'Hollander. His work was initiated at Louvain under the tutelage of van Gehuchten, but the records and slides were destroyed by the guns of August, 1914, and only in 1922 were the studies repeated and the results published. The author lamented the complexity of the thalamic structure—"c'est un encéphale miniature"—and divided the corticothalamic paths into two groups, superficial and deep with short and long fibers, respectively. D'Hollander decorticated adult rabbits serially from frontal to occipital poles and 15 days later determined the effect on the thalamus as revealed by careful Marchi staining. Although he admitted this was not yet direct proof, he believed his results showed that the judgment of there being few such fibers needed revision; Lorente confirmed those findings and made the revision.

During the 1930s, voluminous evidence accumulated for what came to be termed "feedback" of information passing between cortex and thalami. In a presentation to the Boston Society of Psychiatry and Neurology, B. Brouwer from Amsterdam embellished his talk with slides, drawings, and glass models, then much in vogue, depicting centrifugal and centripetal brain systems. The neuronography of Dusser de Barenne had been used in his laboratory, and he reported: "A very remarkable fact is this, that many fibers descending from various parts of the cerebral cortex, go back to all their [thalamic] nuclei" (Brouwer, 1933, p. 624).

Dusser de Barenne argued that the bilateral "thalamic syndrome" of acute cutaneous hypersensitivity after application of strychnine to a small area of sensory cortex, in spite of decortication (with novocain) of the surrounding areas, must be due to setting "on fire the cortex of the whole sensory arm area and . . . those [representational] portions of the optic thalamus" (Dusser, 1924, p. 284). In the assumption of a close functional relationship between cortex and thalamus (Fig. 12.12), Dusser joined the company of those clinical investigators, among them Head and Holmes, Monakow, and Dejerine, who entertained the idea of an interactive information flow between cortex and thalamus.

Although this line of inquiry was not immediately pursued during the decade of the 1940s because of competition from the more compelling concepts of diffuse and specific projections of the thalamic nuclei, as described in Chapter 10 and below, nonetheless, the ground was ready for an entirely new approach to understanding neural networks and needed only the cultivation by scientists with different mind-sets to come to fruition. The recognition of neural nets as aggregates of Cajal's discrete structures making nonconnecting contact with each other through Sherrington's synapses opened the door to computational theory and intelligent control technology. Because neural networks are flexible, capable of analyzing problems with many variables, and self-propagating with feedback and feedforward loops (J. R. Zweizig, personal communication, 1989), they lent themselves to the symbiosis of elements from mathematical and physiological sciences from which cybernetics evolved.

The union was nowhere more productive than in the collaboration of a few people around Warren Sturgis McCulloch (1898–1968; *see* Chapter 5, this volume), first at Yale, then the University of Illinois, Chicago, and finally at the Massachusetts Institute of Technology. From an early interest in philosophy, McCulloch trained as a psychiatrist; his attainments eventually gained him memberships in an unusually broad array of professional societies representing the fields of neurology, anatomy, physiology, mathematics, biological psychiatry, and arts and sciences. Such diversity was lodged in a man who was dubbed a "rebel genius" (Gerard, 1970) and whose intense eyes, unfashionable beard, and abrupt manner did not inspire confidence. Nonetheless, he was a magnet to those neuroscientists who could conceptualize beyond impulse conduction and the neuromuscular junction.

The first modeling of the functional possibilities of neural networks was perhaps that of Sigmund Exner (1846–1926), Austrian physiologist, published in 1894 (Fig. 12.13). Crystallization of the neural network concept into cybernetics came much later, but there were clues scattered, albeit sparsely, throughout the work of neurophysiologists, some of whom have been mentioned. Recall, for example, that Gall had cast "a long shadow" on networks (*see* p. 55, this volume); Forbes (1922) suggested feedback in mammalian spinal recruitment responses; Lorente de Nó (1938) was convinced that feed-back loops accounted for afterdischarge; and Zed Young (1938) found in squid giant nerves "self-reexcitatory" circuits that he proposed might be involved in memory. When McCulloch moved to Chicago in 1941, he soon took under his wing a child prodigy, Walter H. Pitts Jr.,

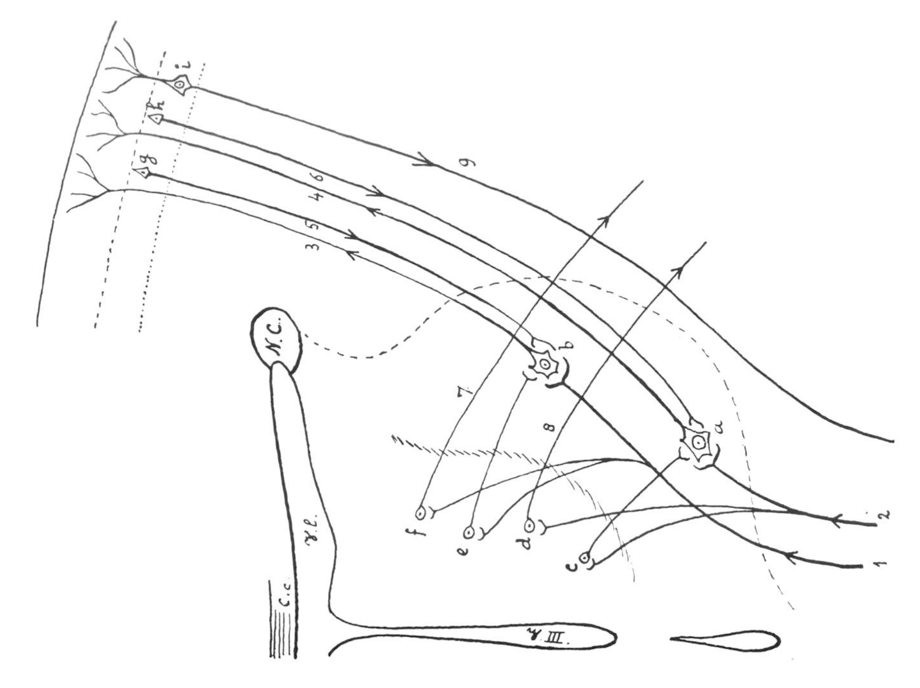

Fig. 12.12. A concept of reciprocal action between cerebral cortex and thalamus as diagrammed by the "father" of neuronography. a, b—Ventrolateral thalamic nuclei; c, d, e, f—medial thalamic nuclei; g, h—corticothalamic neurons; 1, 2—afferents from the periphery; 3,4—corticopetal fibers to sensory cortex; 5,6—corticothalamic fibers; 7, 8—extrapyramidal fibers; i,9—corticofugal, pyramidal fibers. (From Dusser de Barenne, 1935, p. 285, Fig. 79, ×1.)

with whom he collaborated in formulating a theorem for neural nets which was "the apex of scientific achievement for both men" (Heims, 1991, p. 41). Published as "A logical calculus of the ideas immanent in nervous activity" (McCulloch and Pitts, 1943), the theorem is based on the following premise: "If and only if the net excitation of a neuron during the brief period of latent addition exceed the neuron's threshold voltage will transmission take place across the synapse and a pulse be generated and travel from the neuron along its axon toward other neurons" (Heims, 1991, p. 41).

That 1943 article, still discussed and cited at the end of the century, reflected the interchange of ideas among the Chicago group and the Rosenblueth/Wiener nucleus at MIT, plus John von Neumann at the Institute for Advanced Study at Princeton. It generated a great deal of excitement among some biomathematicians and neuroscientists about the possibilities of understanding brain function by pairing logical analysis of the romanticized machine-organism with detailed experimental neurophysiology. McCulloch, however, had larger dreams of drawing humanists into the discussion and tightly shepherded a series of conferences funded by the Josiah Macy Jr. Foundation to which he invited psychologists, social scientists, anthropologists, even a philosopher. In the Macy series (1946–1956) the formal presentations and organized discussions of sensory processing are embedded in publications which can be mined in detail, as Steve Heims (1991) has done so interestingly.[2] The resolution of how visual form (or that of any other modality) is perceived did not ensue after 10 annual meetings in the cybernetics series nor the famous Hixon Symposium on cerebral mechanisms in behavior at Caltech in 1948 at which the discussion of McCulloch's paper, "Why the mind is in the head," was almost four times longer than the paper.

On an individual basis, too, McCulloch endeavored to spread ideas of neural nets among his con-

[2]The Macy series was a sharp contrast to the earlier "axonology" meetings (1930–1942) at which the nature of the nerve impulse dominated the informal agenda and the records are almost nonexistent. Ralph Gerard and Lorente de Nó were the only regular participants in both series, marking them as survivors of the transition from axonologist to cybernetician.

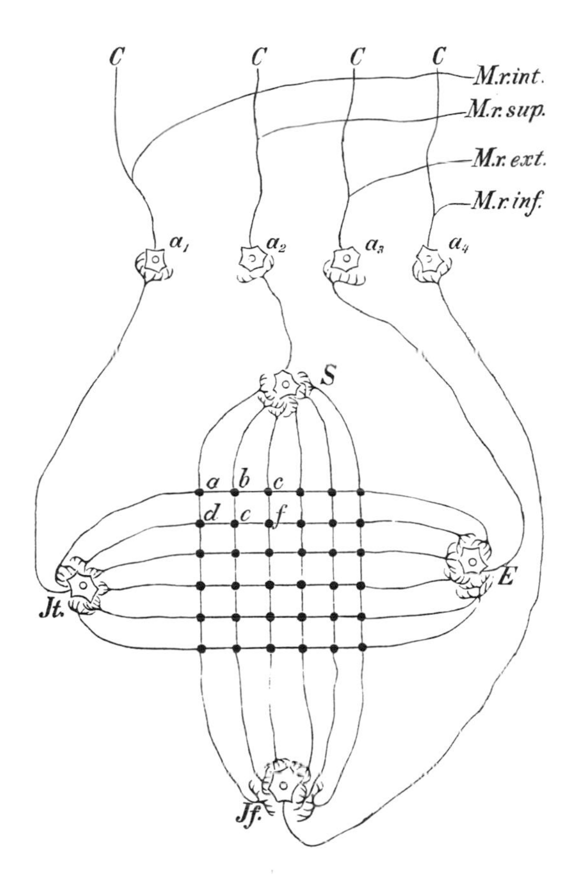

Fig. 12.13. The earliest diagram of a neural network, representing a center for visual perception of movements. a–f—Points where fibers from the retina enter the network. S,E,Jt,Jf—Cells representing terminals from any stimulus point. a_{1-4}—Centers closely associated with the nuclei of four external eye muscles (M. rectus inf., sup., ext., int.). C—Fibers to cortex as the organ of consciousness. (From Exner, 1894, p. 193, Fig. 53, ×4/5.)

temporaries in other disciplines. For example, he sent an early version of Wiener's *Cybernetics* to Wolfgang Köhler, the "Granddaddy of Gestalt psychology" (McCulloch, quoted by Heims, 1991, p. 236) to nourish Köhler's interest in carrying out some neurophysiological measurements of steady currents in the brain's visual apparatus. Probing the ancient problem of perception was in full swing, with the Pitts–McCulloch group endeavoring to "mechanize" perception using the individual neuron as the unit of activity (*see* Lettvin et al., 1959, "What the frog's eye tells the frog's brain"), whereas the Köhler school was promoting perception as an isoelectric phenomenon encompassing the entire cortex.

Looking over the enormous volume of research stimulated by the ideas of feedback mechanisms and loop circuits, it is apparent that the interest in the return pathways from cortex to thalamus and the educated guessing about possible routes was conducive to a resurgence of illustrative artistry in neuroscience. Attractive model diagrams flourished and became so useful a method of depicting ideas that no one decried their abundance as had been the case in the late nineteenth century in trying to explain the aphasias (Fig. 5.7). Perhaps the diagrams were inspired subliminally by the cyberneticists' "loopy" models, and again illuminated presumed pathways for information exchange between brain stem, cortex, and the interposed thalamus. Pitts and McCulloch (1947) updated Exner's prototype, thus documenting the progress in theory and knowledge achieved in the intervening half-century (*see* Fig. 12.14, p. 264). We note two other examples from the proceedings of a major conference convened explicitly to redeem the neglect of the corticothalamic thoroughfare while attention had been largely on the neural traffic in the reverse direction. In 1972, the Parkinson's Disease Research Center at Columbia University's College of Physicians and Surgeons brought together neuroscientists interested in corticothalamic projections and sensorimotor processes. At that conference, the report from Scheibel's laboratory emphasized the key role of nucleus reticularis thalami in the cortical influence on thalamic neurons (*see* Fig. 12.15, p. 265). Another vivacious diagram embodied the ideas from a laboratory at the University of Oslo, Norway, showing three major paths between cortex and thalamus (*see* Fig. 12.16, p. 265). With degeneration studies and silver impregnation of brain tissue, axons from areas S I and S II and M I and M II in cat were traced and a substrate established for the cortical influence on sensory input (Rinvik, 1972).

Rinvik's summarizing diagram above is misleadingly simple compared with the confusing array of discoveries of the organization of systems that are so vital to the quality of life. In John Eccles's many writings he has consistently emphasized the interconnectedness of the systems that maintain the organism's contact with its environment. This legacy from Sherringtonian integration was expressed in a confusingly crowded yet curvy diagram of the interconnections of thalamus and cortex with the limbic system (Fig. 12.17), presented

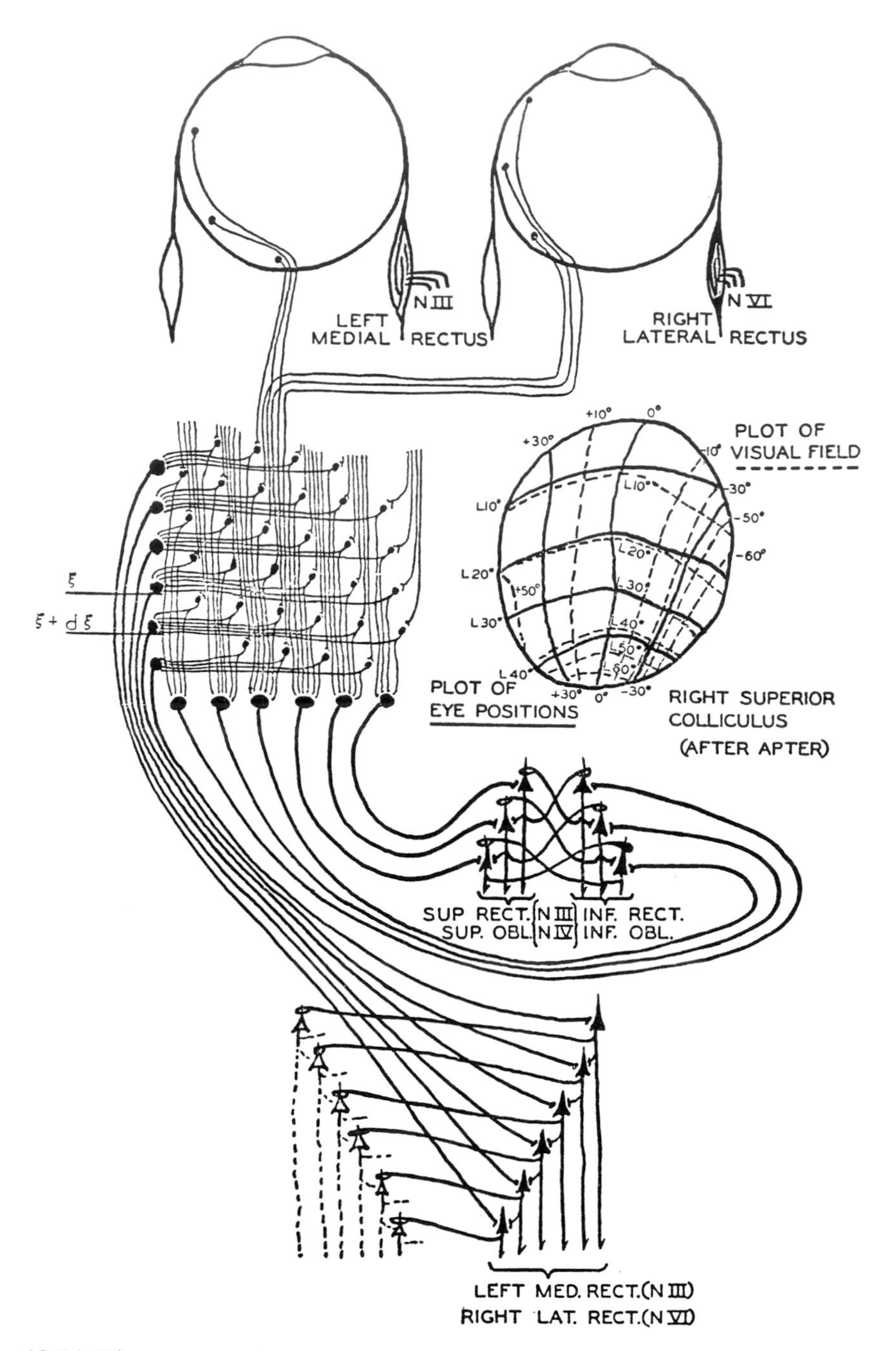

Fig. 12.14. Diagram reproduced for comparison with the previous figure shows ocular afferents to left superior colliculus whence they are relayed to the motor nuclei of the eyes. An inhibiting synapse is indicated as a loop about the apical dendrites. (From Pitts and McCulloch, 1947, p. 141, Fig. 6, ×1.)

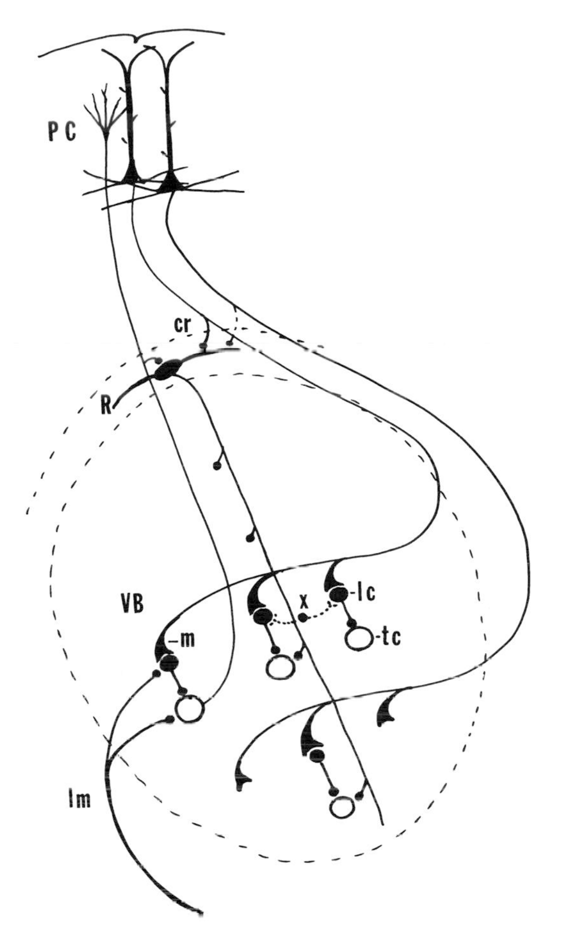

Fig. 12.15. A simplified view of the relation between corticofugal systems and sensory thalamus that shows axonal projections from pericruciate cortex (PC) and nucleus reticularis thalami (R) establishing contacts with linear arrays of thalamocortical circuit (tc) and local circuit (lc) interneurons arranged in rostral-caudal sequence. VB—ventrobasal complex; lm—medial lemniscus fibers; cr—input from cerebral cortex; x—cells without axons. (From Scheibel, Scheibel, and Davis, 1972, p. 152, Fig. 15, ×4/5.)

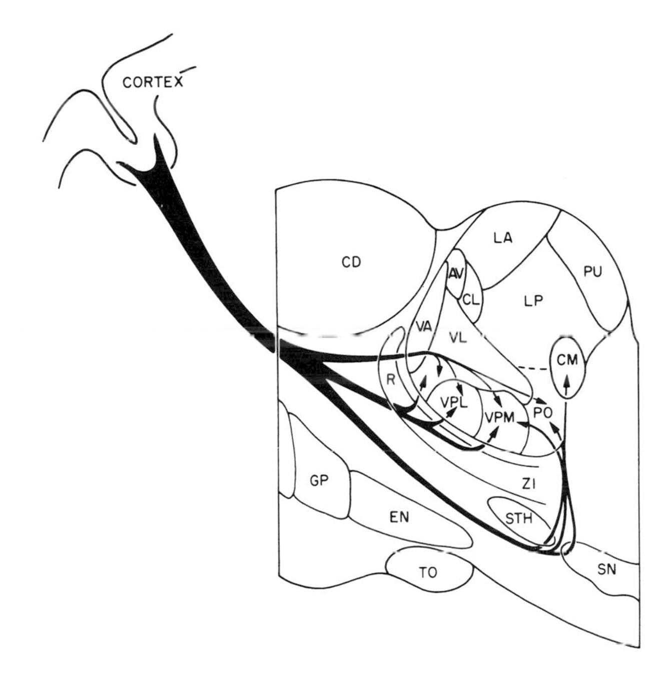

Fig. 12.16. The three major pathways of fibers from sensory and motor areas of the cerebral cortex to thalamus. AV—nuc. ant. ventralis thalamii; CD—caudate; CM—centrum medianum; EN—entopeduncularis; GP—globus pallidus; LA,LP—nuc. ant., post. lateralis thalamii; PO—post. group thalamic nuclei; PV—pulvinar; R—nuc. reticularis thalamii; SN—substantia nigra; STH—nuc. subthalamicus; TO—tractus opticus; VA, VL—nuc. ventralis ant., lat. thalamii; VPL, VPM—nuc. ventralis post. lateralis, medialis thalamii, ZI—zona incerta. (From Rinvik, 1972, p. 71, Fig. 5, ×4/5.)

in one of Sir John's more recent philosophical-physiological series of lectures, *The Human Mystery* (1979).

THE BRAIN-STEM RETICULAR FORMATION AND AROUSAL

"There is no better description than Sherrington's '. . . functions for which the words neither motor nor sensory are fitting'." (G. Jefferson, 1958, p. 731.)

The most ancient of the three major functional systems considered here has roots in a primitive neuropil which is, in C. J. Herrick's words, "the mother tissue from which the specialized central functional systems have been derived in vertebrate phylogeny. . . . [A]n intricate tangle of thin unmyelinated fibers from various sources . . . representing no specific modality. . . ." (1961, pp. 628, 629). The mass of neuronal processes and somata that constitute the major part of the central nervous systems of early vertebrate species evolved into ever more complex structures in parallel with the species' ascents along the phylogenetic scale. The evolutionary growth of the reticular formation of the brain stem pushed the established motor and sensory neural structures laterally, surrounded some thalamic and hypothalamic nuclei, and encompassed clusters of neurons such as the red

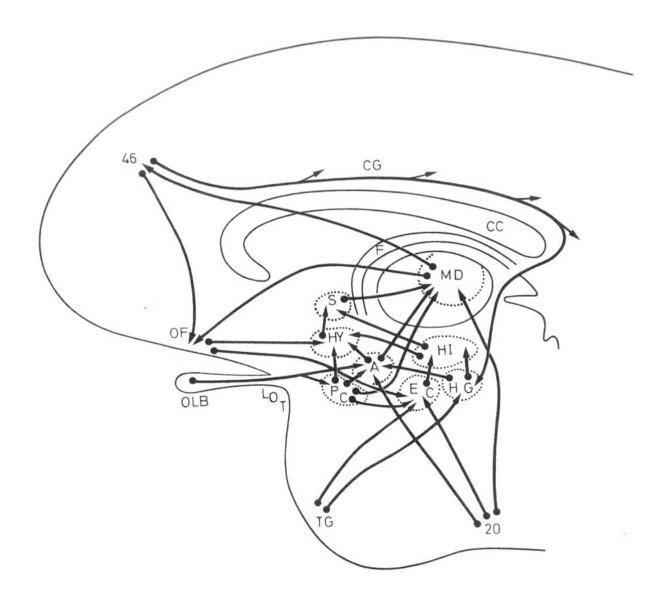

Fig. 12.17. The schematic drawing of interconnections between neocortex, the mediodorsal thalamus (MD), and the limbic system is an attempt on the part of one of neuroscience's great synthesizers to summarize on one schema what was known at the time. OF—Orbital prefrontal cortex; TG—temporal gyrus; HG—hippocampal gyrus; HI—hippocampus; S—septum; F—fornix; CC—corpus callosum; OBL,LOT—olfactory bulb, tract; PC, EC—piriform, entorhinal cortex; A—amygdala; HY—hypothalamus; CG—cingulate gyrus. (From Eccles, 1979, p. 175, Fig. 8–14, ×1.)

nucleus, substantia nigra, the subthalamus, and other brain stem elements, more than 100 in all (Olszewski, 1954). The process of cephalization involved groups of cells that developed

> parallel to the elaboration of sensory inputs which became successively important in the animal's economy. Such a picture of central nervous system major levels linked to one another by ascending and descending fiber paths serves as the primitive apparatus of integration. . . . This way of looking at a reticular system as a main component of the nervous system is supported by . . . currently derived information (Bishop, 1958a, p. 416).

The brain-stem reticular formation in higher mammals extends from the lower medulla oblongata (bulb) forward to the interface of the mesencephalon with the diencephalon, although, like the regions essential to consciousness (*see* Chapter 10, this volume), its boundaries and functional components have never been sharply defined. This deep region of the brain was usually neglected by early anatomists who found it difficult to see or dissect. During the late nineteenth and early twentieth centuries, while anatomic and physiologic investigations focused on spinal reflexes below (Sherrington) and cerebral localization above (Broca, Ferrier), the intervening brain stem was largely unattended (*see* Magoun, 1958a, p. 11). The situation quickly changed, however, in midcentury: As was facetiously commented by a participant in the heyday of reticular system discoveries: "It was not too absurd to say that wherever any really interesting fun was going on in brain research, that part was immediately claimed as part of the reticular system" (G. Jefferson, 1958, p. 729). The inadequacy of the early anatomic definitions was highlighted when neurotransmitters and their related enzymes were found in different nuclei of the reticular formation and were correlated with sleep states and EEG changes (Jouvet and Michel, 1958). Those findings suggested that a neurochemical "map" would more meaningfully collate the reticular formation's anatomy and physiology (*see* A. B. Scheibel, 1987, pp. 1057–1058).

The existence of motor and sensory pathways, centrifugal and centripetal respectively, between the spinal cord and neocortex, only hinted at by the early French ablation studies, was made factual with Hitzig and Fritsch's crude physiological experiments and the detailed cell and fiber tracings of Golgi and of Ramón y Cajal with their new stains. The pyramidal tracts seemed to fulfil the needs of the organism for a prompt reaction to rapid transfer of information from the external environment. The first idea of extrapyramidal connections between spinal cord and brain-stem reticular formation sprang largely from work of O. Kohnstamm and F. Quesnel, during the first decade of the twentieth century and were summarized in a short notice in the *Neurologische Centralblatt* (Quesnel, 1907). Working with puppies and using degenerative techniques, on the basis of Marchi-stained preparations and the work of others, they hypothesized a "centrum receptorium medullae oblongatae" for converging sensory impulses, which were then conveyed by the formatio reticularis to thalamus and thence to cortex. "The notion of a multineuronal pathway of pain and temperature conduction involving the brain-stem reticular formation was

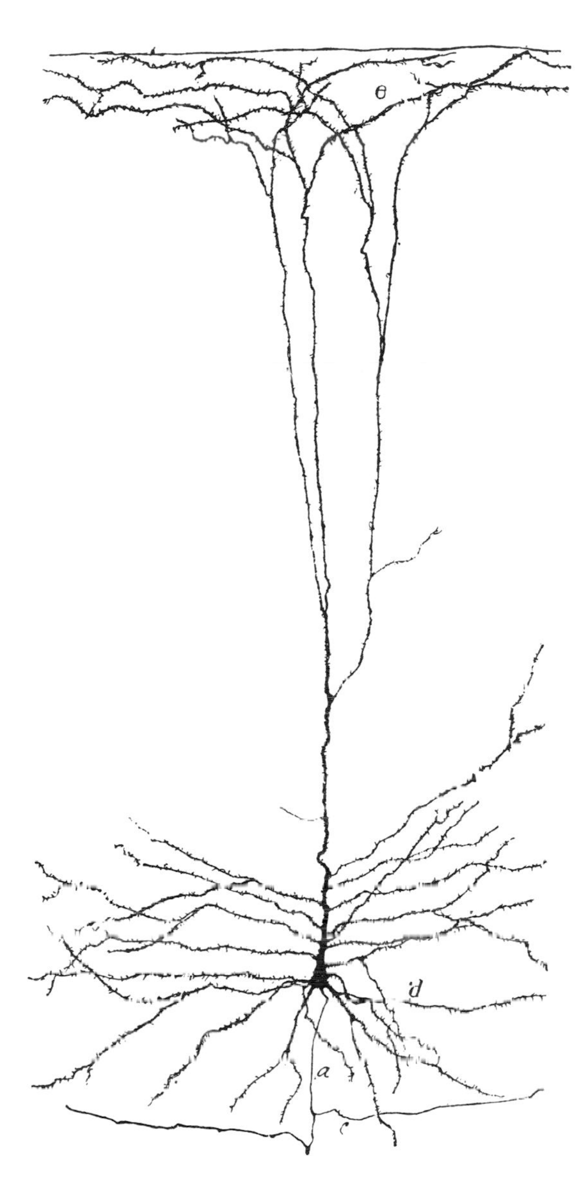

Fig. 12.18. Ramón y Cajal's depiction of an axon (a) with collateral processes (c) allowing spread of impulses at preterminal levels contributed importantly to the development of ideas about reverberating circuits. This drawing shows a Betz cell in the parietal convolution of 30-day-old infant. Golgi method. d—Long basal dendrites, e—terminal protoplasmic bouquet. (From Ramón y Cajal, 1909–1911, vol. 2, p. 566, Fig. 369, ×1.)

thus explicitly formulated, apparently for the first time" (Nauta and Kuypers, 1958, pp. 3–4).

The reticular formation was a minor interest of Ramón y Cajal, who devoted to it only 10 pages of his almost 1000-page *Histologie* (1909, pp. 949–959). He traced two groups of fibers coursing between brain stem and thalamus and noted the

Fig. 12.19. William F. Allen was 40 years old when he received his Ph.D. from the University of Minnesota in 1915. By then he had been assistant to E. P. Allis in France and Jacques Loeb in California and was an authority on the vascular and lymphatic systems of fishes.

abundance of collaterals (Fig. 12.18) from axons in the corticospinal tracts that projected to the reticular formation of the pons and medulla. No significant study of the reticular formation was carried out in the United States until that of William Finch Allen (1875–1951; Fig. 12.19) at the University of Oregon. Allen regarded the formatio reticularis as consisting of the "left over cells of the brain stem and spinal cord which are not concerned in the formation of motor root nuclei and purely sensory relay nuclei" (Allen, 1932, p. 498), and described its efferent and afferent fibers. He "presumed" that all sensory fibers communicated "in one way or another" with the reticular formation, concluding that it contained visceral centers as important to the organism as the hypothalamic nuclei at lower levels. From his studies of cerebellar stimulations he made an interesting and prescient speculation: "It may be that there are separate areas [of cerebellar cortex] for inhibition as well as for augmentation" (ibid., p. 494). Unfortunately, because it was published in an obscure local journal, Allen's work had low visibility beyond the Pacific Coast.

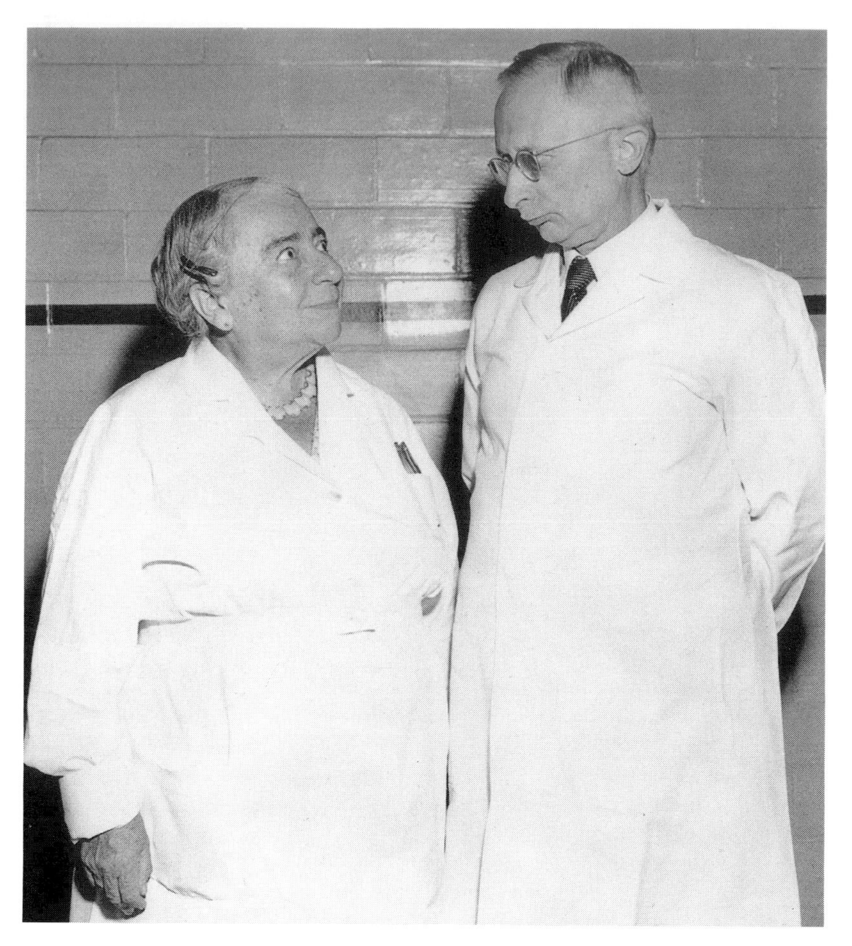

Fig. 12.20. Ernest A. Spiegel and his wife, Mona Spiegel-Adolf, photographed in 1982 at Temple University, Philadelphia, where they had long scientific careers. Spiegel, Wycis, Marks, and Lee (1947) collaborated on the first human stereotaxic apparatus.

The notion of an excitatory influence from higher levels on the motor outflow of the spinal cord had been put forward to explain decerebrate rigidity, which Sherrington (1898), in the Pavlovian tradition, attributed to a release phenomenon due to interruption of inhibitory impulses to the contracting muscles. That time-honored explanation did not satisfy everyone; among the contrarians, Ernest Adolf Spiegel (1895–1985, Fig. 12.20), a Viennese neurologist and neurosurgeon known chiefly for his later work in human stereotaxy, found that decerebrate rigidity was abolished completely only when the reticular formation was transected (Spiegel and Bernis, 1925), implicating its involvement in tonic posture control. That finding and Allen's speculative suggestion were apparently unknown to Magoun and Rhines (and to Ranson) when they continued the investigation of antigravity muscle control begun by Ranson and for which he had reintroduced the stereotaxic instrument, as described in Chapter 11. Magoun and Rhines studied postural and movement behaviors after stimulation of extrapyramidal spinoreticular pathways in cat and monkey encéphale isolé preparations and discovered that the regulatory (in contrast to the initiating) influence was inhibitory or facilitatory according to which part of the brainstem reticular core was stimulated (Fig. 12.21). The senior author later explained the relation of those findings to spasticity as follows:

> The inhibitory or facilitatory effects of reticular stimulation are exerted as markedly upon pos-

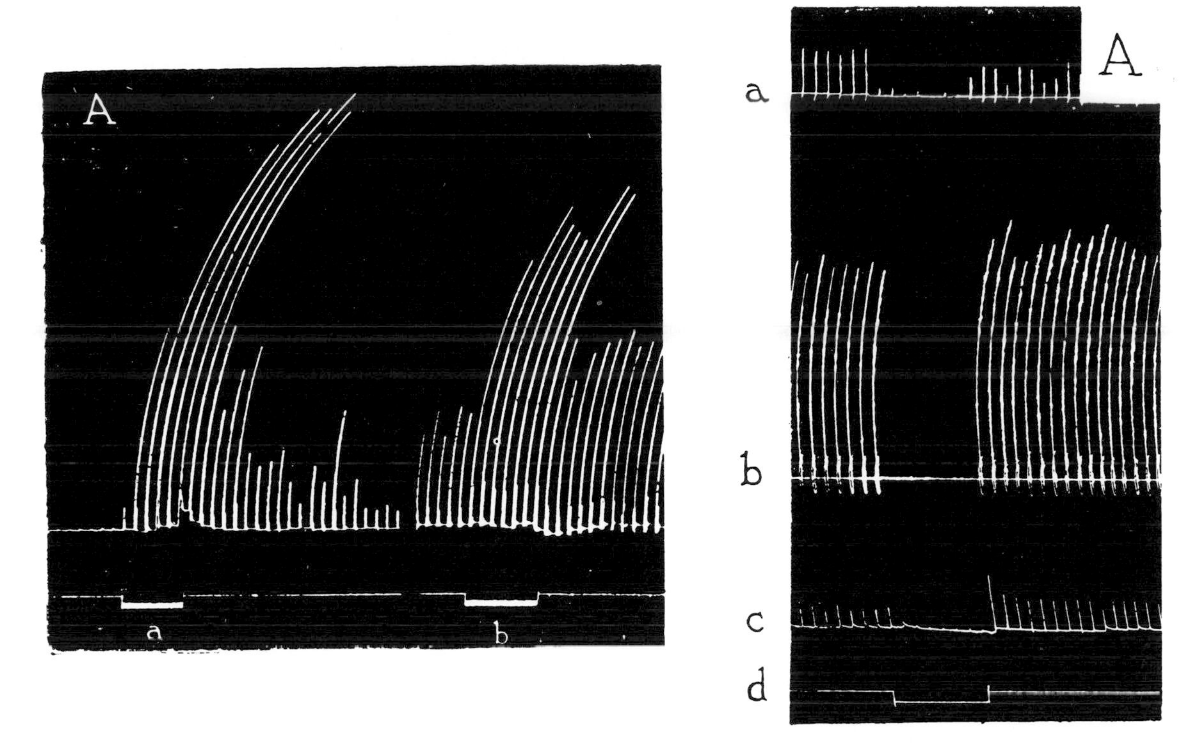

Fig. 12.21. Kymograph smoked-paper records of reflexes evoked by electrical shocks at two-second intervals. Right—inhibition in cat by bulbar stimulation. a—Flexor, b—patellar, c—blink, d—signal; ×1. (From Magoun and Rhines, 1946, p. 166, Fig. 1A.) Left: Facilitation by midbrain stimulation (signal). a—monkey hind leg response to cortical stimulation, b—patellar reflex. (From Rhines and Magoun, 1946, p. 220, Fig. 1A; ×1½.)

> tural or stretch reflexes as upon phasic motor activity and, following chronic ablation of cerebellar and cortical regions which project to the reticular region, a pronounced exaggeration of stretch reflexes ensues. . . . Excitability of the inhibitory component of the reticulo-spinal mechanism may be dependent upon bombardment by those cerebellar and cortical regions whose ablation is followed by spasticity and become deficient in their absence (Magoun, 1963, pp. 24,25).

Their series of papers (summarized in Magoun and Rhines, 1947; Magoun, 1950) constituted an in-depth analysis that validated the switch from an inhibitory to an excitatory view of postural control and provided a background for the later interpretation of the "chance observation" of an arousal pattern in the EEG on stimulation of the brain-stem reticular formation that initiated "one of the greatest booms in the history of neurology" (French, 1958, p. 97).

With the advantage of hindsight, it was possible to detect other early evidence of reticular-core function. In a pointed search for antecedents, Brazier (1980, p. 48) claimed as an unrecognized clue a report from Bremer's laboratory of cortical "arousal" in response to stimulation of the vestibular cortical projection area (Gerebtzoff, 1940b). More relevant to the concept of an ascending reticular system was the earlier identification on the cortex of a slow "secondary response" to sensory stimulation made by Forbes and his associates at Harvard. Although Forbes (1936) thought it was due to unit discharges, he later suggested that the secondary response was delayed by indirect passage over intervening synapses (Forbes, Renshaw, and Rempel, 1937), in contrast to the classical, monosynaptic pathway to cortex.

The observations of Caton in 1875 and of Beck in 1905 (*see* Chapter 5), that electrocorticogram oscillations cease when any afferent nerve is strongly stimulated, was confirmed by Neminsky in 1913 and found in the human EEG by Berger

Fig. 12.22. Horace Winchell Magoun shown ca. 1957 holding the second unit of the Horsley-Clarke stereotaxic instrument, built in London for Ernest Sachs in 1908. The instrument was in use at Washington University while Ranson and Hinsey were there and was presented by Dr. Sachs to Magoun as "the next torchbearer."

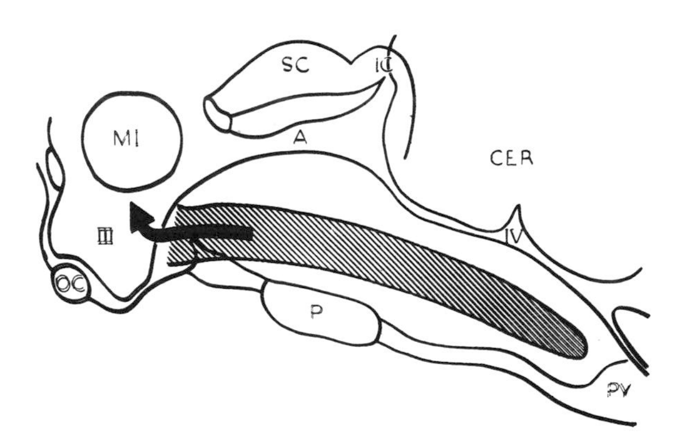

Fig. 12.23. Reconstruction of midsagittal plane of cat's brain showing (crosshatching) the ascending reticular activating system. A—Aqueduct; CER—cerebellum; IC—inferior colliculus; MI—massa intermedia; OC—optic chiasma; P—pons; PY—pyramidal crossing; SC—superior colliculus; III and IV—ventricles. (From Moruzzi and Magoun, 1949, p. 457, Fig. 3, ×3/4.)

(1929). Additional evidence of a diffuse effect of peripheral stimulation was provided by the work of Bremer, whose findings from encéphale isolé preparations (*see* Fig. 9.18, p. 196) prompted him to recognize some global influence of the brain-stem reticular core in mediating a change in state: "Modification of the cortical oscillogram during passage from a state of sleep to one of wakefulness does not represent a local sensory effect of the stimulus. . . . It can be observed identically, whatever the localization of the recording electrode, or the mode of stimulation. . . . One is dealing here with a general modification of cortical activity" (Bremer, 1936, pp. 466–467). Relatedly, in unanesthetized cats with electrodes on the dura, Rheinberger and Jasper (1937) found patterns of activity recorded from motor, sensory, auditory, and visual areas were related to behavioral state: "The electrogram from all regions was characterized by low amplitude higher frequency potentials when the animal showed behavioral indications of being generally aroused or activated. . . ." (ibid., p. 195).

The confluence of two new techniques—stereotaxis as a means of precise localization in the deep brain (and spinal cord) and electroencephalography in the service of an objective "end point" in probes of the systems's electrical activity—coupled with classic histological verification, facilitated a flood of new work on the interior regions of the brain, as noted. When Horace Winchell Magoun (1907–1991; Fig. 12.22) at Northwestern University Medical School was joined in late 1948 by Giuseppe Moruzzi (*see* Chapter 9), an Italian neurophysiologist intent on improving his technical skills, they combined their respective interests in brain stem and cerebellum and looked for corticopetal effects of stimulation of those two regions in their anesthetized cats; they found that with low-frequency stimulation the EEG was flattened to resemble that of an alerted animal (Moruzzi and Magoun, 1949). Initially they thought they were seeing either an artifact or inhibition, but careful elimination of technical errors and with sufficient amplification, they identified the low-voltage, fast waves characteristic of arousal.[2] The lead paragraph of the paper's discussion section stated:

> The evidence given above points to the presence in the brain stem of a system of ascending reticular relays, whose direct stimulation activates or desynchronizes the EEG, replacing high-voltage slow waves with low-voltage fast activity. This effect is exerted generally upon the cortex and is mediated, in part at least, by the diffuse thalamic projection system. Portions of this activating system, . . . have previously been identified (Moruzzi and Magoun, 1949, p. 468).

In addition to numerous electroencephalograms, the authors illustrated their results with a schematic reconstruction of the stimulated points (Fig. 12.23). The initial findings were corroborated by the effects of lesions: the arousal response was absent in animals in which the region of stimulation was electrocoagulated (Lindsley, Bowden, and Magoun, 1949) and the sleep pattern appeared in the EEG after damage to the midbrain reticular formation (Lindsley, Schreiner, Knowles, and Magoun, 1950).

Intuitively, the presence of two afferent pathways conveying sensory information to the neocortex would seem to be redundant, but each has its role:

> As [afferent] signals ascend [the classical] paths, they contribute polysensory excitation to parallel ascending nonspecific connections, distributed through the central core of the brain. The functions served by these specific and nonspecific cortical input channels are supplementary. The specific one conveys the informational content of the afferent message, for its signals are both modality- and locality-related. The core system, lacking these features, provides instead for behavioral and EEG arousal underlying an orientation and attention to the message (Magoun, 1969, p. 179).

The concept of diffuse projections from thalamus and brain-stem reticular formation, although backed by clear physiologic evidence, did not enjoy easy acceptance. Challenged by that evidence, W. J. H. Nauta, the consummate neuroanatomist, and his associates at MIT set out to definitively locate the possible pathways and inaugurated a rigorous series of "true" Wallerian degeneration studies (not retrograde or transneuronal). First investigating the specific thalamic fibers, they classified them into three neat groups: intrathalamic, subcortical, and cortical (Nauta and Whitlock, 1954). As for the diffuse projections, Nauta and Kuypers (1958) confirmed many of the earlier histologic findings using the Nauta-Gygax stain and added details of reticular interconnections at all levels of the brain stem. As they wrote: "knowledge of the pathways was in place waiting for a function. . . . [The] notion of an ascending reticular activating system was novel only in its striking function" (ibid., p. 3). Acceptance of the notion of a functioning system with an anatomic base, however, was dishearteningly slow: none of the ascending sensory collaterals which disperse widely in the reticular formation

> can offer an obvious explanation for the phenomenon of diffuse cortical arousal so clearly demonstrated by physiologic experimentation. Indeed, it may be logical to ask; Do any of the pathways here traced actually ascend beyond the confines of the "reticular formation"? . . . It must . . . be emphasized that at the present time no structures outside the "specific" thalamic nuclei of the thalamus have been definitively demonstrated to project significantly to the neocortex (ibid., pp. 26,27).

Papez, on the other hand, was not hesitant to identify pathways from brain stem to thalamus taken by

[2]For Moruzzi's recollection of the discovery, *see* L. H. Marshall (1987; an account by Magoun was published in 1985.

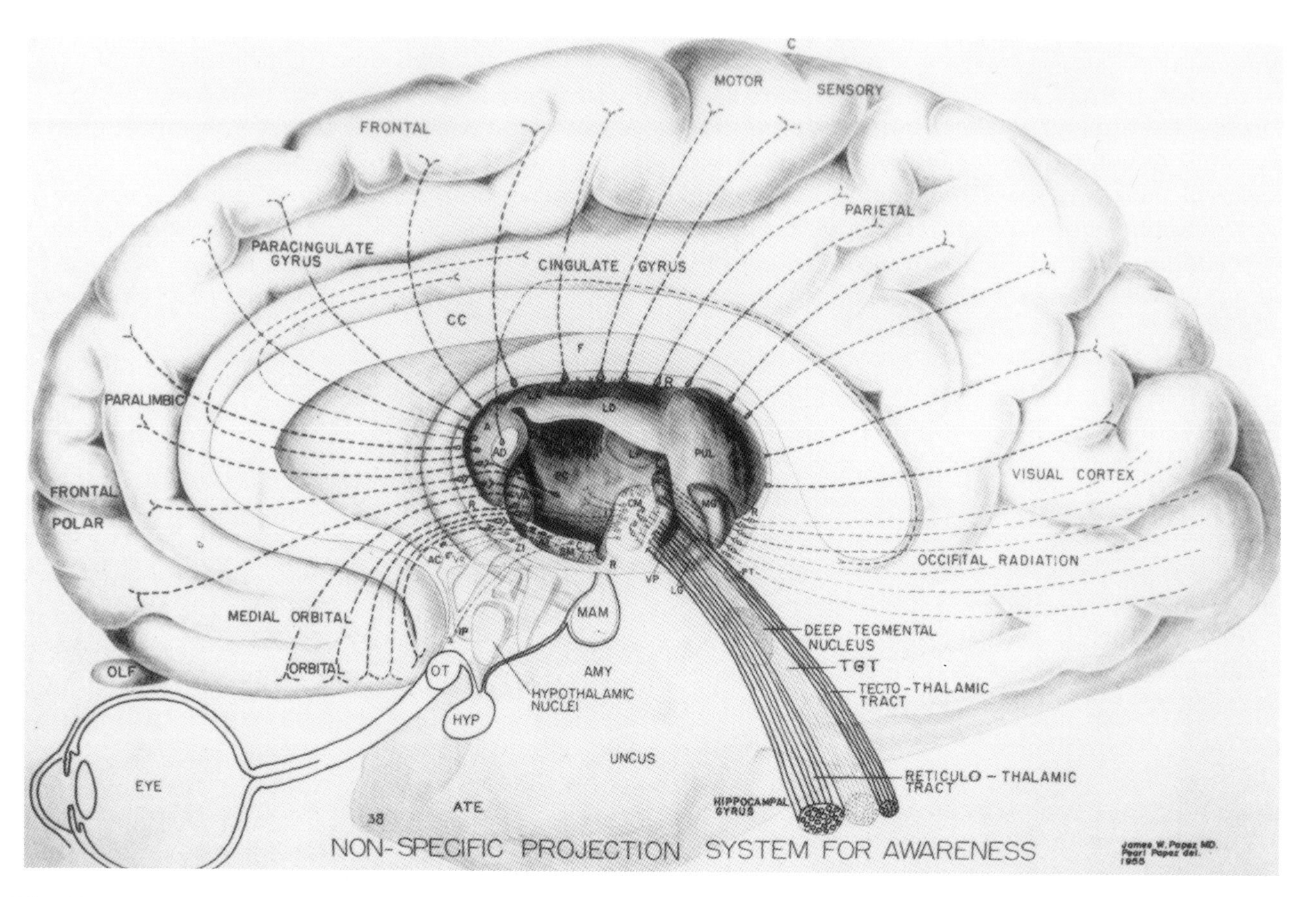

Fig. 12.24. Semidiagram of projections to the right thalamus and cerebrum of nonspecific impulses related to the EEG and consciousness. RTT,TGT—reticulo- and tegmentothalamic tracts; CM—centrum medianum; PF,L—parafascicular, limitans fibers; R—reticular nuc.; I—intralaminar bundle; SM—submedial nuc.; PC—paracentral nuc.; CL—central lateral nuc.; IP—inf. thalamic peduncle; NB—ant. perforated substance; A—anterior nuc.; VM—ventral medial nuc.; VA—ventral ant. nuc. (From Papez, 1956, p. 118, Fig. 1, ×3/4.)

nonspecific impulses and to diagram their abundant projections to the cortex. Compulsive for neuronal detail, he supervised his wife's drawing of the "Non-specific Projection System for Awareness" (Fig. 12.24).

Efforts to extend knowledge of the diffuse system were of course pursued most vigorously by the principals themselves, at the temporary Long Beach laboratories of the group assembled by Magoun when he moved to the University of California, Los Angeles in 1950, and at the Instituto Fisiologia of the University of Pisa, in what was "a magic period of research and study" (Zanchetti, 1981, p. xiv) initiated by Moruzzi's energy and vision on his return to Italy after two years in Chicago. In a summary of the unitary studies carried out with colleagues, Moruzzi (1954) described, among other findings, two responses of the bulbar reticular system to stimulation: the phasic, mediating EEG arousal, and the tonic, for maintenance of wakefulness. Although supportive data continued from many laboratories, an American neurophysiologist visiting the laboratory in Pisa reported informally:

> As Moruzzi puts it, the reticular story is going through a period of "crisis." The original concept . . . is being seriously challenged by many different kinds of evidence, particularly by the work of Adametz, Sprague, Huttenlocher, Bot[e]s, and by the work in low-voltage fast sleep. I think it is greatly to Moruzzi's credit that he recognizes this and is trying to come to terms with these new data (Spencer, 1961).

In addition to the skeptics, another major obstacle to advance of the arguments was the recurring annoyance of nomenclature, due largely to superimposing terms from structural and functional domains. As Brodal (1969, p. 306) later emphasized with italics, *"It should be made perfectly clear that the 'activating system' is a functional concept, the 'reticular system' a morphological one, and it has been obvious for many years these do not correspond."* This thorough Swedish investigator, whose Golgi preparations revealed that the short-axon and long-axon reticular cells are aggregated in clusters, argued that the organization of the reticular formation was not diffuse nor did it lack order.

The implications of an ascending reticular system, formulated from laboratory animal experimentation, for the human domains of such phenomena as sleep and wakefulness, epilepsy, and consciousness, were of great interest to investigators in many branches of biomedicine. Bremer's fundamental observations after transection of the upper brain stem (*see* Chapter 9, this volume) had signaled the beginning of attention to the problem of sleep and arousal (Magoun, 1954a, p. 6) even though the new data undermined his widely held theory of "deafferentation" of the specific paths as the cause of sleep. The ascending reticular system filled a gap in neurosurgical theorizing, as Hugh Cairns (1952, p. 142) suggested in a lecture on consciousness, because with its far-reaching collateral connections it resolved the seeming paradox of unconsciousness resulting from too little or too much afferent stimulation from the periphery. In psychiatry, "The functions of the brain stem . . . are related to the integration of the organism in three-dimensional space in one or another pattern of interaction. . . . These highly complex differential functions can be carried out only if the brain stem integrative functions are stable. . . ." (Rioch, 1954, p. 477).

Colleagues and visitors at Magoun's laboratory knew they were on to something big and pursued it accordingly. With the equipment in use around the clock, the ever-changing groups working on various aspects of the brain stem diffuse system added to the mass of data from that and other laboratories in such volume that attention to the specific thalamic tracts was refocused as collaterals from them to the diffuse system threatened to become more important than the "direct corticipetal paths in EEG arousal induced by afferent stimulation" (Magoun, 1954a, p. 6). Deep pathways were found anatomically and physiologically between the limbic and activating systems (Adey, 1958). Figure 12.25 recapitulates the known effects of reticular core activity, effects that were elegantly summarized in *The Waking Brain* (Magoun, 1963).

The pervasive interest in the concept of an ascending reticular system and the accumulation of knowledge of its role in the problem of how the brain works can be traced between 1954 and 1980 through a series of major conferences centered on that topic. The proceedings of the first conference, dubbed "the Laurentian" for its Canadian locale, were published as *Brain Mechanisms and Consciousness* (Delafresnaye et al., eds., 1954), a title linking the aroused cortex with neuronal activity. Magoun's paper was on wakefulness, and Jasper set the adversary tone of the discussion: "We are now afforded a rare opportunity to put Dr. Magoun on the carpet. . . ." (Magoun, 1954a, p. 15). And the audience did just that—the published discussion is lengthier than the text of the talk; there were comments about definitions, e.g., behavioral arousal versus EEG flattening; about experimental methods; and anatomic pathways—queries pertinent to a just-emerging concept. The second paper read at the Laurentian conference was by Moruzzi, who discoursed on the physiologic properties of the puzzling reticular system as revealed by work from his laboratory in Pisa. His concluding words attested to the still uncertain state of knowledge at the time: "The microphysiology of the central nervous system is just beginning and it is about one century younger than microscopic anatomy. There is no reason to be surprised, therefore, that many basic problems have not been approached experimentally and that the meaning of much of our data remains unclear (Moruzzi, 1954, p. 48).

During a discussion period (notorious for generating some of the most significant ideas at any meeting) the question of where do specific and nonspecific systems interact was posed by Magoun, who sensed a "merging attitude" on the part of investigators to recognize that interaction between the systems takes place at the diencephalic level in addition to that shown at a cortical level (Jasper and Ajmone-Marsan, 1952). From studies already noted, Nauta contributed the opinion that "we have anatomical evidence of such connections between the diffuse and specific parts of the thalamus as well as of connections of both parts with the reticular nucleus" (Nauta and Whitlock, 1954, p. 113).

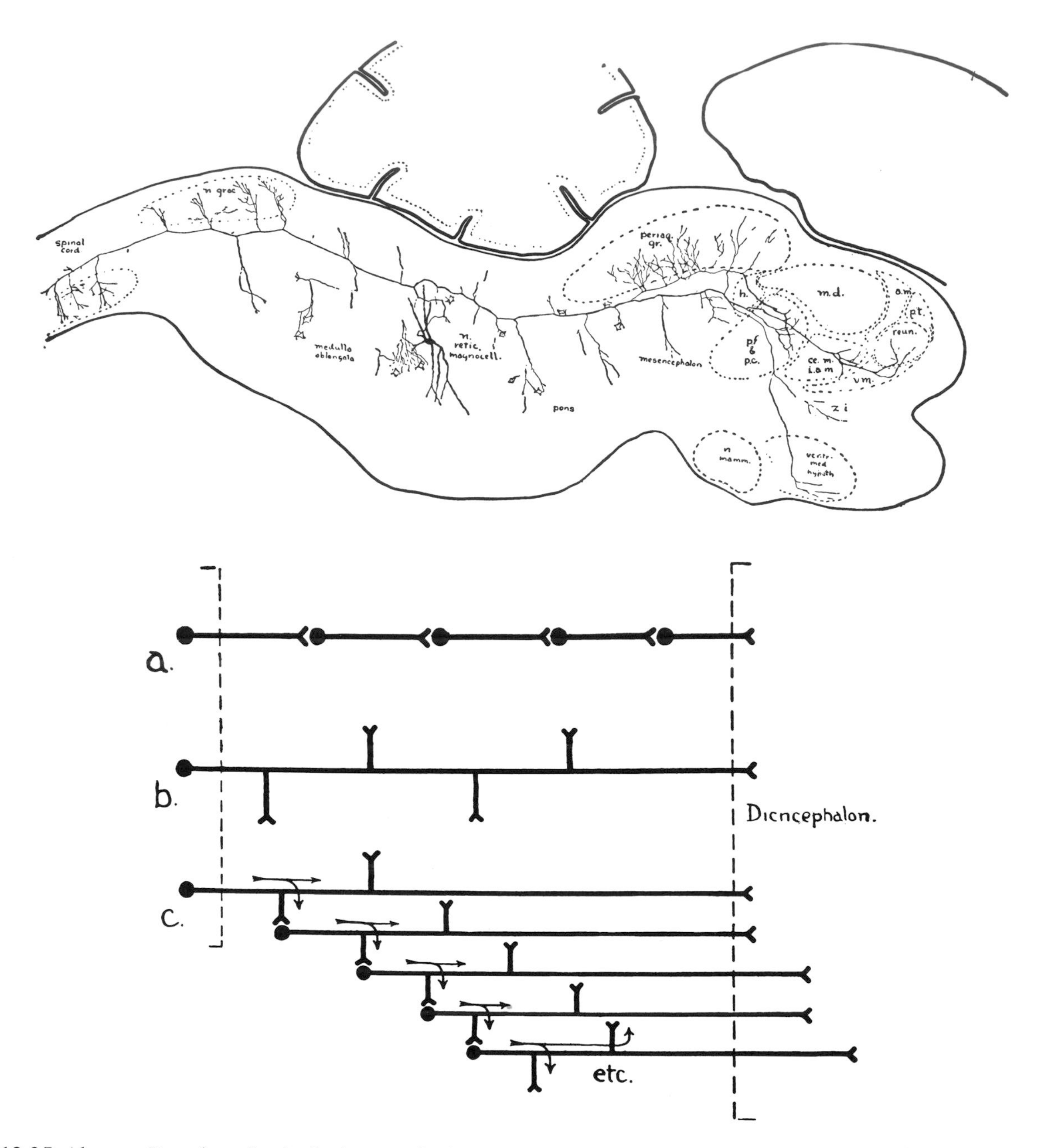

Fig. 12.25. Above—Drawing of a single, large reticular cell of the magnocellular nucleus in a two-day-old rat. The axon bifurcates, the caudal segment giving off collaterals to the reticular formation, nucleus gracilis, and spinal cord; the rostral segment gives off collaterals to the reticular formation and periaqueductal gray and appears to supply nuclei of the thalamus and hypothalamus. (From Scheibel and Scheibel, 1958, p. 46, Fig. 12, ×1.) Below—Diagram of possible conduction circuits through the reticular core of the brain stem. a—Short-axon cells hypothetized by Moruzzi and Magoun, b—single, long-axon cell from bulb (left) to diencephalon found predominantly, c—branching of collaterals in b forming circuitous paths through the reticular core producing longer latencies and conduction times. (From ibid., p. 44, Fig. 10, ×4/5.)

Four years after the Laurentian conference, a second major meeting was aimed directly at the morphology and function of the reticular system; it was held at the Henry Ford Hospital in Detroit and was published as *Reticular Formation of the Brain* (Jasper, Proctor, Knighton, Noshay, and Costello, eds., 1958). The papers reveal not only the focal interests of the active investigators and their acceptance of the concept, but also demonstrate the enormity and potential scope of reticular formation influence. As Brodal (1957, p. 23) had noted elsewhere, the early descriptions of long ascending connections from the reticular formation had been "largely forgotten" until their stimulation was shown to influence cortical electrical activity, and thus a reinvestigation of the microstructure of the region was not surprising. In addition to the anatomical studies of Nauta and Kuypers (1958) already described, the Scheibels offered both histologic and physiologic data that showed convergence of "heterogenous afferents on single elements of the brain stem reticular core" (Scheibel and Scheibel, 1958, p. 32; Fig. 12.25, above). Those authors also postulated various circuits to account for latency, conduction times, and lateral dispersion of impulses traversing the reticular core (Fig. 12.25, below). Amassian and Waller described (1958) clear evidence from individual brainstem reticular neurons of the relation of different firing patterns ("coding") to their receptive fields. Magoun (1958, p. 109) related those findings to Sharpless and Jasper's (1956) report of brain stem components at distinct levels and activities. From microelectrode studies of the cortical arousal system in behaving monkeys, Jasper suggested "that the rapid switching of local activation or inhibition may occur not primarily through unspecific thalamocortical circuitry, but by effects upon specific thalamocortical projection systems at a subcortical level" (Jasper, 1958, p. 328).

Other reports of research-in-progress at the Detroit symposium included, but were not limited to, drug actions, stress, gonadotrophin release, motor activity and muscle spindles, cortical circulation, vision and perception, conditioned and visceral reactions, and learning. That wide array, a measure of the clarifying and interpretative potential of the notion of the ascending arousal system, was summarized eloquently from a broad perspective:

> A quotation from Claude Bernard [says]: "The stability of the milieu intérieur is the condition of a free life". . . . This freedom to act, to play, to carry out intellectual work, one of the main achievements of evolution, is precarious. . . . That is to say, to be active or to sleep, to preserve and protect the integrity of our internal organization. Even more, the way in which we apprehend the outside world depends on the actual balance of our internal milieu and its repercussion on the brain stem reticular activity (Dell, 1958, pp. 377-378).

Although not centered on the brain stem reticular formation, the timing of the Moscow Colloquium on the EEG of higher nervous activity—in 1958 and only a year after the conference in Detroit and a meeting of the World Federation of Neurology in London—ensured that the topic was a major item of earnest discussion. Not surprisingly, the full weight of Pavlovian conditioning was brought to bear on "higher" nervous activity in the first sentence of the presentation by the eminent Russian physiologist, Petr Kuzmich Anokhin (1898–1984): "An analysis of the present-day situation in the physiology of the nervous system shows that the conditioned reflex is the nodal point at which the different trends in the physiology of the nervous system meet" (Anokhin, 1960, p. 257). Establishment of the International Brain Research Organization (IBRO)[3] was a permanent outcome of the successful Moscow meeting of 1958. A large part of the enthusiasm of the participants lay in the frank and open exchanges and Magoun praised Moruzzi's paper (Moruzzi, 1960) on a bulbar mechanism for synchronization of cortical activity as "the second major contribution to reticular physiology presented at this colloquium" (Magoun, 1960, p. 253). In Magoun's opinion, the first major paper was by Bremer, followed by Dell who gave a detailed discussion of his own and Bremer's simultaneous and independent demonstrations of facilitation of sensory evoked potentials by reticular activation (Dumont and Dell, 1958; Bremer and Stoupel, 1959). And yet another "landmark" singled out by Magoun was the "pronounced modification" of unit discharges in the reticular formation and elsewhere during spreading depression, demonstrated in the presentation by Bureš and Burésová (1960). With so much of the discussion centering on the reticular formation, this confluence of electroencephalography and behavioral conditioning was a waysign pointing the direction into the new terrain of neuroscience.

Another gathering of neuroscientists, this time organized specifically to consider the brain-stem

[3]*See* Jasper (1991) and L. H. Marshall (1996).

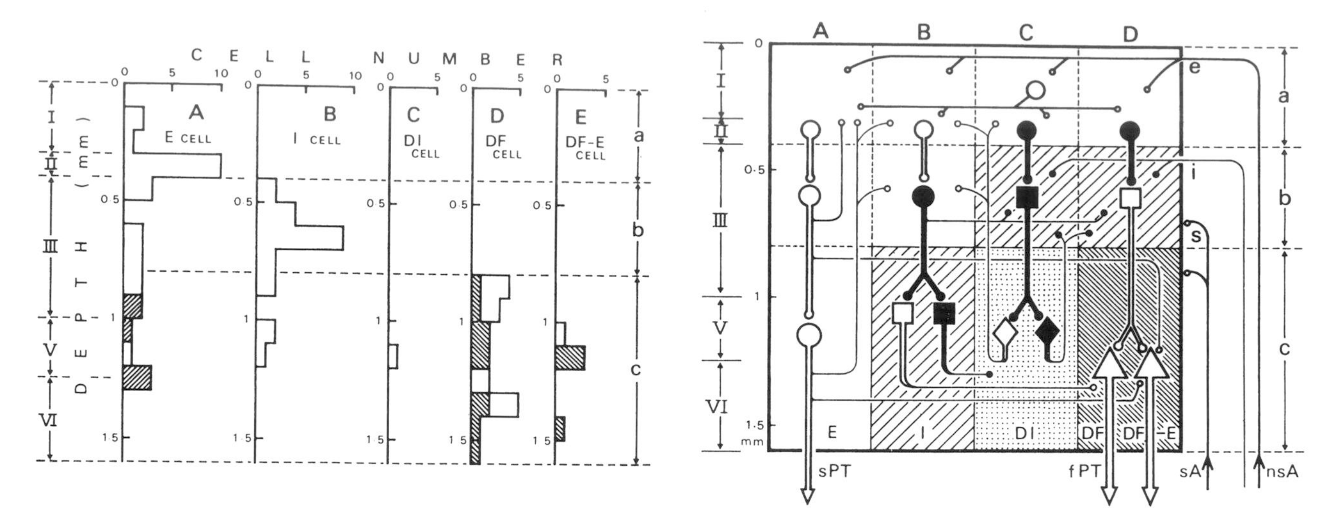

Fig. 12.26. Left—Distribution chart of four different types of action plus a mixed type found in cortical cells of laminae I through VI. The data were obtained on cats in an attempt to elucidate the electrical events of EEG arousal. Hatched columns—pyramidal cells; blank columns—nonpyramidal cells; E—excitatory; I—inhibitory; DI—disinhibition; DF—disfacilitation. (From Inubushi, Kobayashi, Oshima, and Torii, 1978, p. 701, Fig. 8.) Right—Summary diagram of an "arousal" circuit model based on data shown in the previous figure. The parallel three-neuron relays represent four different mechanisms for membrane stabilization in EEG arousal. Large, small open circles—excitatory neurons, synapses; large, small filled circles—inhibitory neurons, synapses; nSA—nonspecific afferents; sPT—specific pyramidal tract. (From ibid., p. 703, Fig. 9.)

reticular core, took place three decades after the 1949 papers. Sponsored by the Society for Neuroscience and IBRO, in *The Reticular Formation Revisited* (Hobson and Brazier, eds., 1980) the amorphous issue of who first formulated the concept was submerged by the rich array of new supportive evidence. As the Canadian neurophysiologist Mircea Steriade declared elsewhere, "It is a pleasure to reread [Moruzzi and Magoun's] 1949 communication and see how later developments not only confirmed the data but fully justified some of the major theoretical issues" (Steriade, 1981, p. 327). After reviewing the main conclusions of 1949, Steriade recalled subsequent work showing the psychophysiologic correlates of experimental reticular activation: improved accuracy and reduced reaction time in tachistoscopic tasks carried out by monkeys (Fuster and Uyeda, 1962), and single-cell experiments which demonstrated that the original concept "is alive and well" (Steriade, 1981, p. 371).

The mechanisms of the brain-stem reticular formation remained a prominent neuroscience research area for many years and are not yet exhausted.[4] The use of intracellular recording and stimulation in interpreting the EEG arousal response was elevated to a rarified altitude by elegant studies from Japan which constitute the first systematic study of the responses at the cellular level, according to the authors, and attest to the tremendous progress since Jasper reported cellular studies in 1958. Inubushi, Kobayashi, Oshima, and Torii (1978) minutely analyzed the activity of cortical neurons during arousal by reticular stimulation. Laminar II neurons in motor cortex are excited initially, then activity spreads to excitation or inhibition of deeper neurons. They found desynchronization of the EEG to consist of a

[4]A recent report of dramatic results of brain imaging studies implies activation of the midbrain reticular formation by sensory stimulation in man (Kinomura, Larsson, Gulyás, and Roland, 1996).

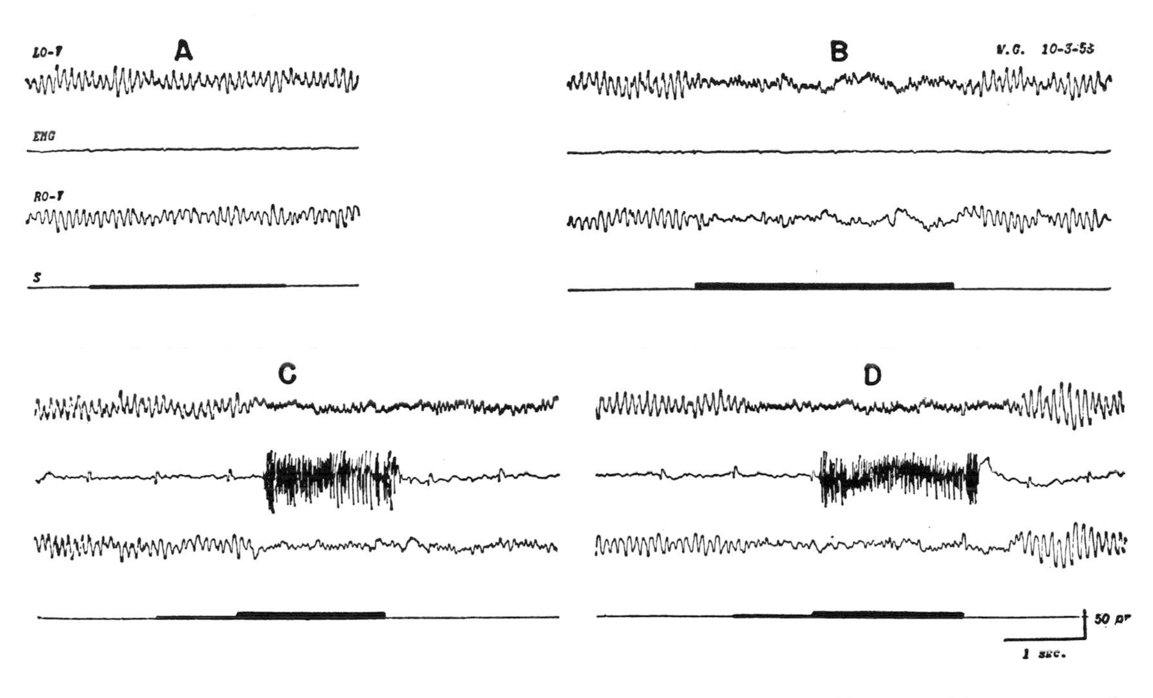

Fig. 12.27. Blocking of the alpha rhythm by conditioning (first, third traces) in normal human subject. A—No change with tone stimulus of 1024 c/s (thin signal line). B—Change with bright light (thick signal line). C—First paired trial showing no change until bright light is on. D—Ninth trial showing alpha desynchronization with tone before light stimulus. Second trace is right forearm EMG with an EKG artifact; fourth trace is signal. (From Morrell, 1958, p. 547, Fig. 1.)

sequential flow of information among five types of cortical cells characterized by the nature of their responses; Fig. 12.26, left, pictures a distribution chart of cell types among the cortical laminae. The authors offered a model "arousal" circuit (Fig. 12.26, right) and noted its relevance to the afferent nonspecific projection to lamina I and II demonstrated by Scheibel and Scheibel in 1958.

One of the most interesting derivatives of the ascending reticular system concept was its involvement in conditioning. It had been shown at the Collège de France, by the fundamental work of Fessard, one of the earliest neurophysiologists to make experimental use of the EEG, that in human subjects the click of the camera shutter became the conditioned stimulus for the "arrest reaction" to visual stimuli shown in the EEG (Durup and Fessard, 1935); that finding was immediately confirmed in the United States (Loomis, Harvey, and Hobart, 1936; Travis and Egan, 1938). Twenty years later the work was reinvestigated (Morrell and Ross, 1953) and extended, again in human subjects, to show that alpha-rhythm blocking could be conditioned by pairing a light with low-intensity sound, the latter being ineffective as an unconditioned stimulus. This implied that there are specific inhibitory and excitatory processes, seen by comparing rates of conduction between the striatal and precentral regions: there was a lengthening of conduction time whenever the inhibitory process was induced (Fig. 12.27). The authors concluded that "these results lend further support to the concept that the discrete, local activation pattern requires the participation of topographically organized diencephalic reticular formation" (Morrell, 1958, p. 558). Other conditioning experiments from the Bureš laboratory were mentioned above.

The findings on laboratory animals suggesting the usefulness of the diffuse reticular concept in the study of conditioning also opened the possibility that the formation of temporary connections fundamental to learning could be explained through the mediation of the brain-stem reticular formation.

> Involvement of the cephalic brain stem may make it easier to account for the important role which emotion and reward or punishment can play in the learning process. Exten-

> sion of these studies . . . which are directly exploring changes in the activity of the brain associated with learning are likely to form a major development interrelating neurophysiology and behavior (Magoun, 1958a, pp. 113-114).

It would be difficult to estimate the extent of the role played by the reticular system concept in furthering the emergence of neuroscience from a confluence of neural and behavioral disciplines, as though aroused to a new consciousness. Not so difficult to recognize, however, is a novel hypothesis stemming from the multiple circuits available in the reticular formation. "[O]ne of the very fundamental bases of species differences in animal behaviors" may lie in the number of relays and organization of pathways which all species nonetheless share (Nauta, 1958, p. 667). It remains for future developments to quantify that imaginative combination of form and function.

OVERVIEW OF THREE MAJOR INTEGRATIVE SYSTEMS

In the attempt to designate specific parts of the brain as constituting a "system," there is the risk of implying that boundaries and circuits are precise and neat. No implication could be more misleading. The three systems selected for historical description are characterized by many connections and functions that are in a state of continuing discovery and even obsolescence, already the fate of the limbic system (Swanson, 1987). Looking as though drawn up with a purse-string (Broca), the system was suggested as having something to do with emotion (Papez), an idea boosted by the Klüver-Bucy syndrome and expanded by its description as the paleomammalian brain (MacLean). The limbic involvement in learning and memory (Bekhterev) was demonstrated by careful testing in human patients (B. Milner). Ease of manipulation stimulated research on the hippocampal region, and, with single-unit recording, the seizure-active sites on dendrites were explored (Andersen et al.). The septal region was found to be the site of a reward center (Olds and P. Milner) and the amygdaloid body came to attention as being the most susceptible brain region tested to "kindling" by subthreshold stimulation (Goddard et al.). With varying boundaries and functions, the limbic system's strong connections to other parts of the brain, particularly to the hypothalamus, confer on it a basic role in behavior.

The presence of corticothalamic connections was projected in 1839 (Carpenter) and their topographic specificity noted soon after (Luys). By the end of the century, histologic and ablation studies had demonstrated the presence of two-way thalamocortical-corticothalamic circuits. The concept of feed-back and feed-forward mechanisms was a natural outcome (McCulloch and Pitts) and the application of biomathematics to create a novel subdiscipline, cybernetics (Wiener), was the result, opening the field of neural networks to research that did not hesitate to extend to artificial intelligence.

The best historical overview of the brain-stem reticular system and its myriad effects is framed in the words of one of its promulgators:

> Full attention did not return to the brain stem until recently, however, for it will be recalled that Edwardian contributions to the physiology of the central nervous system were focused largely upon reflex functions of the spinal cord below and upon activities of the sensorimotor cortex above, leaving the intervening stem of the brain unattended.
>
> Recent study has once more stressed the importance of this neural part, however, in identification of centrally placed, nonspecific mechanisms, which parallel the more lateral, specific systems of classical neurology and are richly interconnected with them. These nonspecific mechanisms are distributed widely through the central core of the brain stem and, as spokes radiate from the hub of a wheel to its peripheral working rim, so functional influences of these central systems can be exerted in a number of directions: caudally upon spinal levels which influence postural and other activity; rostrally and ventrally upon hypothalamic and pituitary mechanisms, concerned with visceral and endocrine functions; . . . and more cephalically and dorsally still, upon the cortex of the cerebral hemispheres. . . .
>
> . . . .Just as all spokes move together in the turning of a wheel, though they may bear weight sequentially, so the variously directed influences of these nonspecific reticular systems are closely interrelated in normal function (Magoun, 1963, pp. 18, 20).

POSTSCRIPT

As a part of culture and history...beyond...'social interests'...and insular intellectual history.
(Harrington, 1991, p. 372)

It is well apparent from the history presented here that neuroscience embraces many disciplines of knowledge and resists an attempt at definition. An eloquent modern champion of neuroscience has stated: "This novel multidisciplinary paradigm reflected the common conviction that the brain was not the domain of a single discipline" (Rosenblith, 1991, xi). A poet wrote: "The Brain–is wider than the Sky–" (Dickinson, ca. 1962). In the common wisdom of its largest organization, neuroscience consists of "nervous systems, including the part they play in determining behavior. . . ." (Bylaws, Society for Neuroscience, 1992). Implicit in that too concise description is the fundamental character of neuroscience: it is one of the basic biological sciences, and should not be distorted by identification with the clinical subdisciplines. Neuroscience refers to knowledge of how brain and behavior interact and is studied by scientists of many different disciplinary backgrounds. In naming the subdisciplines, anatomy is always first, not only because it leads the alphabet, but more importantly, because of its particular significance as the substrate of function (*see* Brodal, 1987, p. 769) in addition to the historical priority of the discoveries of structures. Neuroanatomists are joined by basic scientists from the three streams that have converged to form neuroscience: the neural, behavioral, and communicative sciences.

Some of the confusion about what constitutes neuroscience is reflected by the various approaches to writing its history. An historian sees it as a ". . . challenge of what it might mean to 'humanize' the sciences of the human brain" (Harrington, 1991, p. xiv). To a sociologist, the brain appears as a ". . . junction of two alien lands. . . ." (Star, 1991, p. 204). In the attempt to reconcile the psychology of the individual with its envelope of human biology, those two approaches–the historical and the sociological–to the development of knowledge of the brain have a tendency to create issues where none exist and to couch ideas in such convoluted language that the events and concepts become unfamiliar and difficult to fathom. Although there is movement toward "attention to language and meaning" (Harrington, 1991, p. 372) on the part of historians, we believe that neuroscientists who have experienced the "sense of wonder" in combination with the "street smarts" of their alley of science are best positioned to set down the history of their calling, thereby preserving it in terms of scientific literacy.

A shining example of historical literacy is provided by studies documenting the neuron doctrine. That fundamental concept, formally presented in Waldeyer's textbook in 1899, has been revisited many times in the succeeding century, and we cite four in this last half-century by as many outstanding neuroscientists. George Bishop (1956) could not write on the "Natural history of the nerve impulse" until he had considered "The Neurone as a Cell" and "The Cell as a Neurone," and included speculations about their evolution. Theodore Bullock (1959) considered the neuron doctrine in relation to electrophysiology, deeming the newly discovered networking of individual neurons a quiet revolution in conceptualization. In writing of the "generalized" vertebrate neuron, David Bodian (1962) followed his training and took the anatomists' stance, chiding his colleagues for not revising their terminology in concert with the new ideas about neuronal function and cytology. Capping those examples, Gordon Shepherd (1972) returned to the neuron doctrine itself and by application of then-current research findings, he perceived that there had been a need for a more flexible doctrine bolstered by more data before a clearer understanding of brain and behavior was forthcoming. The historical literacy of those four mature experimentalists in their areas of competence enabled them to reconstruct more than a categorization of events, and their readers are the richer for sharing their points of view.

In the endeavor to present neuroscience "as a part of culture and history," to use Harrington's wishful phrase (1991, p. 372), there is the lurking possibility of crowding so many trees between the finite covers of a book that the reader becomes lost in Ramón y Cajal's "thicket." We endeavored to avoid too dense a forest by following a theme, or postulate–function determines form–as a thread throughout phylogeny that bound together the separately discovered concepts. The phylogenetic viewpoint is useful in that it is one among many elements in the "ecology of discovery," if you will, that gradually increased in importance as comparative investigations became more precise and widely recognized. The very early philosopher-teachers, as they tried to make sense of what they saw (or did not see) in the human body, often drew on their imaginations, as Steno decried. More recent neurologists were tempted to extrapolate from lower to higher animals, as were Ferrier and Krause when they superimposed laboratory findings onto human cerebral topography. But by the time of their era, the phylogenetic evidence was at hand to excuse, if not sanction, such speculations. A historian of neurology expressed the power of function determines form in these words:

> We only reaffirm an age-old thought. . .by conceiving structures. . .as organs of life and function. In the history of human thought and individual conscious human existence the knowledge of our organs does not precede but succeeds our faculty to use them; in fact, nobody has to be a trained neuroanatomist to perform movements at will. . . . The facts of cerebral localization retain their full importance by conceiving the brain not as the creator, but only as the organ of mind (Riese, 1959, p. 151).

How has this evolutionary principle fared historically? We can seek the answer in the history itself rather than highlighting it in the separate events. As knowledge of the human brain unfolded, did functions become visible before their underlying form was recognized? Or was the behavior considered secondarily after the "organ" or substrate was described? Examples of both modes abound in a seemingly random fashion. Speech deficits were noted by Marc Dax and Broca before they began to collect cases of left frontal lobe damage; Moruzzi and Magoun identified the arousal effect of stimulation of a brain-stem reticular formation before Nauta sought and found nerve fibers to account for the phenomenon. Conversely, the corpus callosum was described by Galen long before its function was hinted at by Magendie. A common pattern was that clinicians often saw the functional deficits first and pointed the direction for the experimentalists to uncover the substrate. Thus, the coincidence of acromegaly with pituitary tumor noted by Marie and byMinkowski led ultimately to Philip Smith's experiments with growth hormone.

We have documented a few episodes in the history of neuroscience in which a solitary subject afflicted with one of "nature's accidents" has led through enlightened longitudinal study to significant insights of how the brain works. The longitudinal study of a single research subject is in contrast to today's dependence on large population studies analyzed by sophisticated statistical methods, and even then there is difficulty in arriving at a consensus. A few of the individual subjects who contributed so importantly to neuroscience progress include the epileptic patient of Joseph Bogen, W. J., whose intractable seizures led to Philip Vogel's carrying out a commissurotomy on him and the Gazzaniga–Sperry psychological measurements of minute changes in cerebral hemispheric functions. There were Brenda Milner's extensive studies of subject H. M., which revealed many obscure aspects of memory. Going back in time, there were Fröhlich's patient with pituitary disorder and Broca's examination of Tan with "aphémie." There are, of course, many more examples in biomedical fields outside neuroscience.

The story of the brain-stem reticular formation illustrates two previously unmentioned postulates. The first is that new knowledge depends on a constellation of improved experimental methods and techniques. Foremost among the new methodologies that replaced crude ablations and large electrodes were stereotaxic positioning of precise lesions and recording with microelectrodes that enabled data to be obtained from anatomically and physiologically identified neurons. Those mechanical advances were closely followed by histochemical methods that detected aminergic cells within the reticular system, and by tracing techniques using radioactive proteins and marker substances to reveal neuronalinterconnections. The second added postulate is that synthesis of isolated experimental data into a sweeping concept has rich rewards. The immediate example is the conceptualization of a mechanism for alerting the organism based on ancient neuropil and its integration into the central and autonomic nervous systems. Within the former resides control of degrees of consciousness, and within the latter is found those visceral functions associated with cardiovascular, respiratory, and other vital neural circuitry. The connectedness of neural structures and functions, far from being overwhelming, assures an ever-expanding search for how the brain works to provide appropriate behaviors for survival.

Reading in their own words what the investigators reported of their discoveries, it is possible to look beyond the quaintness of the language to the excitement that drove those early experimentalists to seek the utmost information with the methods at hand. We can appreciate the fact that because of the multidisciplinary nature of neuroscience, the methodology necessary for problem solving has become progressively formidable. This "autobiography" of neuroscience shows clearly the interdependence of form and function, though not always in that order, and should enrich the scientific experience and perhaps stabilize a love of science. When enunciated in the crisp précis of science, written history has a chance to attract those workers at the bench who are curious to learn how it all happened.

We borrow from William Feindel, Canadian neurosurgeon and historian at "The Neuro" in Montreal, who participated in the symposium, The Reticular Formation of the Brain (1958, 670) and creatively found the historically important "symmetrical reticular system of Herrick's tiger sala-mander" with Sherrington's "poetic description . . .of patterned flashing lights" in the lines of William Blake:

Tyger Tyger, burning bright,
In the forests of the night;
What immortal hand or eye,
Could frame thy fearful symmetry?

In what distant deeps or skies,
Burnt the fire of thine eyes?
On what wings dare he aspire?
What the hand, dare seize the fire?
. . . .

What the hammer? What the chain,
In what furnace was thy brain?
What the anvil? What dread grasp,
Dare its deadly terrors clasp?

CREDITS

We are deeply grateful to authors, publishers, repositories, and individuals for permission to reproduce figures or text not in the public domain. Their names are listed alphabetically in the case of illustrations followed by the figure number, the caption of which leads to the complete citation. For quotations, the author of the text cited and the page of the present volume on which the quotation appears are listed; the full citation is found in the list of references.

ACADEMIC PRESS, INC.: Figs. 3.12 and 7.30; Linderstrom-Lang p.175, McCleary p. 256

AMERICAN ASSOCIATION FOR THE ADVANCEMENT OF SCIENCE: Gazzaniga p. 121; Figs. 6.3 right, 6.5, 6.14, 7.9 below

AMERICAN MEDICAL ASSOCIATION: Bucy and Klüver p. 95, C. J. Herrick p. 182, Papez p. 251; Figs. 9.12 and 9.13

AMERICAN PHYSIOLOGICAL SOCIETY: Greep p. 232; Figs. 3.15, 5.25, 10.17, 10.18, 11.7, 11.21, and 12.21

AMERICAN SOCIETY FOR CLINICAL EVOKED POTENTIALS, INC.: Fig. 5.30

ANNUAL REVIEWS, INC.: Bard p. 245

ANNALES MÉDICALES DE NANCY ET DE L'EST: Fig. 4.12

APPLETON-CENTURY-CROFTS: Fig. 12.9

BLACKWELL SCIENCE, INC.: Fig. 3.14

BLACKWELL SCIENTIFIC PUBLICATIONS, LTD.: Fessard p. 120

BMJ PUBLISHING GROUP: G. Jefferson p. 266

CAMBRIDGE UNIVERSITY PRESS, North American Branch: Schiller p. 61, Head p. 70; Fig. 11.15

CAPRA PRESS: Todd p. 29

CENTER FOR ACADEMIC PUBLICATIONS JAPAN: Fig. 12.26

CENTRAL INSTITUTE FOR THE DEAF, WASHINGTON UNIVERSITY: Fig. 5.32

CHARLES C THOMAS, PUBLISHER: Fig. 9.20

CITY OF BAYEUX, FRANCE: Fig. 2.17 above

CLEVELAND MUSEUM OF NATURAL HISTORY: Fig. 2.18

COLLEGE OF PHYSICIANS OF PHILADELPHIA, LIBRARY OF: Fig. 12.20

CONSEJO SUPERIOR DE INVESTIGACIONES CIENTIFICAS, MADRID: Fig. 7.20 below

CORNELL UNIVERSITY PRESS: Galen p. 5

ÉDITIONS SKIRA, GENÈVE: Fig. 2.13

ELSEVIER SCIENCE-NL: Rose and Woolsey p. 217, Moruzzi and Magoun p. 271; Figs. 5.30, 7.23, 7.26, 10.19, 10.20, 12.23, 12.24

GONVILLE AND CAIUS COLLEGE, THE MASTER AND FELLOWS OF: Fig. 3.2

HACKETT PUBLISHING COMPANY, INC.: Descartes p. 31

HARCOURT, BRACE & CO.: Campion and Smith p. 204, Magoun p. 271

HARPER COLLINS PUBLISHERS, INC.: Jackson pp. 73, 189

HARVARD UNIVERSITY, PEABODY MUSEUM OF ARCHAEOLOGY AND ETHNOLOGY: Fig. 2.18

HARVARD UNIVERSITY PRESS: Fig. 6.8

INSTITUT FÜR GESCHICHTE DER MEDIZIN DER UNIVERSITAT–WIEN: Fig. 4.10

JOHN WILEY AND SONS, LTD.: Hubel and Wiesel p. 148; Figs. 7.8 left, 10.6, 11.8

JOHNS HOPKINS UNIVERSITY PRESS: Schiller p. 61

S. KARGER AG, BASEL: Figs. 6.4 above, 8.15

THE LANCET, LTD.: Fig. 7.29

LIPPINCOTT-RAVEN PUBLISHERS AND THE INTERNATIONAL SOCIETY FOR NEUROCHEMISTRY: Figs. 8.12, 8.16; AND THE UNIVERSITY OF MARBURG LIBRARY ARCHIVES: Figs. 8.1, 8.2, 12.15, 12.16

MACMILLAN MAGAZINES LTD.: Dart p. 17; Figs. 2.9 and 5.31

MASSACHUSETTS MEDICAL SOCIETY, PUBLISHING DIVISION OF THE: Fig. 6.4 below

MASSON EDITEUR, PARIS; Bremer, 1936 p. 270; Fig. 5.6

IL MINISTERO PER I BENI CULTURALI E AMBIENTALI, Firenzi: Fig. 9.2

NATIONAL ACADEMY OF SCIENCES, USA.: Fig. 6.7

NEUROSCIENCE HISTORY ARCHIVES, BRAIN RESEARCH INSTITUTE, UNIVERSITY OF CALIFORNIA AT LOS ANGELES: Figs. 3.1, 5.4, 5.7, 5.9 right, 5.10 above, 5.21, 5.22, 5.27–5.29, 5.34, 6.3 left, 6.6, 7.5, 7.20 above, 7.22, 7.27, 7.28, 8.10, 8.14, 9.7–9.10, 9.15–9.17, 10.3, 10.5, 10.6 left, 10.7, 10.21, 12.11, and 12.22

NEW YORK ACADEMY OF SCIENCES, ANNALS OF THE: Witelson p. 124; Fig. 1.4

OLDENDORF, WILLIAM H.: Fig. 5.24

ORIENTAL INSTITUTE OF THE UNIVERSITY OF CHICAGO: Fig. 2.16

OXFORD UNIVERSITY PRESS, INC.: DeFilipe and Jones pp. 17, 23, Head and Holmes p. 203, Cairns p. 205, Singer p. 199–200, Fulton p. 235; Figs. 5.26B, 9.15, 9.16

PATRIMONIO NACIONAL, MADRID: Fig. 5.2

THE PHYSIOLOGICAL SOCIETY, LONDON: Hubel and Wiesel p. 148; Figs. 11.6, 11.18

THE PIERPONT MORGAN LIBRARY, NEW YORK, M.240, f.8: Fig. 2.16 below

PLENUM PUBLISHING CORPORATION: Price p. 222, Guillemin p. 244; Fig. 11.10

PRENTISS, T.S., ESTATE OF: Fig. 2.8

PRINCETON UNIVERSITY PRESS AND THE LITERARY EXECUTORS OF THE PENFIELD PAPERS: Symonds p. 205

ROYAL COLLEGE OF PHYSICIANS, LONDON: Fig. 5.8 right

THE ROYAL SOCIETY, LONDON: Katz p. 165

SAWYER, CHARLES H.: Figs. 11.16, 11.17

SONNENSCHEIN, RALPH R.: Fig. 7.9 above

SPRINGER VERLAG, NEW YORK: Hyvärinen p. 93; Figs. 3.13, 7.24, 8.9, 8.11, 10.8, 11.13

SPRINGER-VERLAG GMBH & CO.: Fig. 11.12

STÄDTISCHE KUNSTHALLE, MANNHEIM: Fig. 10.14

SWETS & ZEITLINGER PUBLISHERS: Fig. 9.18

TOHOKU UNIVERSITY MEDICAL PRESS: Fig. 12.8

UNIVERSITY OF BRITISH COLUMBIA, VANCOUVER, WOODWARD BIOMEDICAL LIBRARY: Fig. 5.16

UNIVERSITY OF CALIFORNIA AT LOS ANGELES, LOUISE M. DARLING BIOMEDICAL LIBRARY, HISTORY AND SPECIAL COLLECTIONS DIVISION: 1.1, 1.3, 2.1, 2.15, 3.3, 3.6–3.10, 4.1–4.9, 4.13, 4.15, 4.17 through 4.22, 5.3, 5.13 left, 5.6 above, 5.33, 5.38, 6.9–6.13, 7.1–7.4, 7.6–7.8 right, 7.10, 7.14–7.17, 7.19, 7.20 above, 7.29, 8.5–8.7, 8.13, 9.1, 9.3–9.6, 9.11 left, 10.1, 10.2, 10.9–10.12, 11.1–11.3, 11.19, 11.20, 12.1, 12.2, 12.6, and 12.10

UNIVERSITY OF CALIFORNIA, RIVERSIDE, TOMAS RIVERA LIBRARY SPECIAL COLLECTIONS: Fig. 4.16

UNIVERSITY OF CALIFORNIA PRESS: Clarke and O'Malley pp. 30, 35, 44, 45, 46, 66, 116, 117, 128, 135, 137, 149, and 181; Schiller pp. 68, 251

UNIVERSITY OF ILLINOIS PRESS: Fig. 5.20

UNIVERSITY OF MINNESOTA ARCHIVES: Fig. 12.19

W. H. FREEMAN & CO.: Lewin p. 12, Ornstein p. 108

WASHINGTON UNIVERSITY SCHOOL OF MEDICINE, BERNARD BECKER MEDICAL LIBRARY: Figs. 5.39 and 5.40

WELLCOME INSTITUTE LIBRARY, LONDON: Fig. 5.3

WILEY-LISS, A SUBSIDIARY OF JOHN WILEY & SONS, LTD.: Bradbury p. 40

WILLIAMS & WILKINS: Figs. 12.4 right, 12.12; Papez p. 194, Sperry p. 208

YALE UNIVERSITY, HARVEY CUSHING/JOHN HAY WHITNEY MEDICAL LIBRARY: Sherrington p. 144.

References

Abel, J. J. 1903. On epinephrin and its compounds, with special reference to epinephrin hydrate. *Am. J. Pharm.*, 75: 301–325.

Adams, F. 1886. *The Genuine Works of Hippocrates,* Vol. II. New York: William Wood.

Adey, W. R. 1958. Organization of the rhinencephalon. Pages 621–644 in H. H. Jasper, L. D. Proctor, R. S. Knighton, W. C. Noshay, and R. T. Costello, eds., *Reticular Formation of the Brain.* Boston: Little, Brown.

Adrian, E. D. 1935. Discharge frequencies in the cerebral and cerebellar cortex. *J. Physiol.* (London), 83: 32P–33P.

Adrian, E. D. 1941. Afferent discharges to the cerebral cortex from peripheral sense organs. *J. Physiol.* (London), 100: 159–191.

Adrian, E. D. and B. H. C. Matthews. 1934a. The interpretation of potential waves in the cortex. *J. Physiol.* (London), 81: 440–471.

Adrian, E. D. and B. H. C. Matthews. 1934b. The Berger rhythm: potential changes from the occipital lobes in man. *Brain,* 57: 355–385.

Adrian, E. D. and G. Moruzzi. 1939. Impulses in the pyramidal tract. *J. Physiol.* (London), 97: 153–199.

Aldini, G. 1803. An account of the late improvement of galvanism. London: Cuthell and Martin.

Allen, W. F. 1932. Formatio reticularis and reticulospinal tracts, their visceral functions and possible relationships to tonicity. *J. Wash. Acad. Sci.*, 22: 490–495.

Alonzo de Florida, F. 1994. Dismissed observations: the kindling effect, a case study. *Perspect. Biol. Med.*, 37: 202–216.

Alonzo de Florida, F. and J. M. R. Delgado. 1958. Lasting behavioral and EEG changes in cats induced by prolonged stimulation of amygdala. *Am. J. Physiol.*, 193: 223–229.

Amacher, P. M. 1964. Thomas Laycock, I. M. Sechenov, and the reflex arc concept. *Bull. Hist. Med.*, 35: 38, 168–183.

Amin, A. H., T. B. B. Crawford, and J. H. Gaddum. 1954. The distribution of substance P and 5-hydroxytryptamine in the central nervous system of the dog. *J. Physiol.* (London), 126: 596–618.

Andersen, P. and S. A. Andersson. 1968. *Physiological Basis of the Alpha Rhythm.* New York: Appleton-Century-Crofts.

Andersen, P., T. W. Blackstad, and T. Lömo. 1966. Localization and identification of excitatory synapses in hippocampal pyramidal cells. *Exp. Brain Res.*, 1: 236–248.

Anderson, E. 1969. Earlier ideas of hypothalamic function, including irrelevant concepts. Pages 1–12 in W. Haymaker, E. Anderson, and W. J. H. Nauta, eds., *The Hypothalamus.* Springfield, Illinois: C. C Thomas.

Anderson, E. M. and J. B. Collip. 1933. Thyrotropic hormone of anterior pituitary. *Proc. Soc. Exp. Biol. Med.,* 30: 680–683.

Anker, M. V., Trans., *see* Wernicke, C. 1876.

Anokhin, P. 1960. On the specific action of the reticular formation on the cerebral cortex. *Electroenceph. Clin. Neurophysiol. Supp.*, 13: 257–270.

Anonymous. 1827. *A legacy of affection, advice, and instruction. Letter LXXV. Of the animal structure and the nervous system.* London: Printed for Sir Richard Phillips and Co.

Anonymous. 1897. La vie et l'oeuvre du Docteur Luys. *Chron. Med. Paris*, 4: 589–591.

Anonymous. 1952. Santiago Ramón y Cajal. *Cuadernos Biograficos II.* Madrid: Ministry of Culture.

Anrep, G. V. 1923. The irradiation of conditioned reflexes. *Proc. R. Soc. B* (London), 93-404-426.

Arantius, J. C. 1587. *De humano foetu...Ejusdem anatomicorum observationum liber....* Venetiis: Jacobum Brechtanum.

Aristotle. 1908. Historia animalium. Trans. by D'A. W. Thompson. Vol. 4 in J. A. Smith and W. D. Ross, eds., *The Works of Aristotle Translated into English.* Oxford: Clarendon Press.

Aschner, B. 1912. Über die Funktion der Hypophyse. *Pflügers Arch. Ges. Physiol.,* 146: 1–146.

Azoulay, L., Trans. *see* Ramon y Cajal, R. 1909–1911.

Bacq, Z. M. 1975a. *Chemical Transmission of Nerve Impulses. A Historical Sketch.* Oxford: Pergamon Press.

Bacq, Z. M. 1975b. Walter B. Cannon's contribution to the theory of chemical mediation of the nerve impulse. Pages 68–83 in C. McC. Brooks, K. Koizumi, and J. O. Pinkston, eds., *The Life and Contributions of Walter Bradford Cannon 1871–1945.* Brooklyn: State University of New York Downstate Medical Center.

Bailey, P. and F. Bremer. 1921. Experimental diabetes insipidus. *Arch. Int. Med.*, 28: 773–803.

Bailey, P. and G. von Bonin. 1951. *The Isocortex of Man.* Urbana: University of Ilinois Press.

Bailey, P. 1975. Moses Allen Starr 1854–1932. Pages 112–113 in D. Denny-Brown, ed., *Centennial Anniversary Volume of the American Neurological Association 1875–1975.* New York: Springer.

Baillarger, J. G. F. 1840. Recherches sur la structure de la couche corticale des convolutions du cerveau. *Mém. Acad. Roy. Méd. Paris,* 8: 149–183.

Bard, P. 1928. A diencephalic mechanism for the expression of rage with special reference to the sympathetic nervous system. *Am. J. Physiol.*, 84: 490–515.

Bard, P. 1968. Walter Bradford Cannon 1871–1945. Pages v–ix in W. B. Cannon, *The Way of an Investigator. A Scientist's Experiences in Medical Research.* New York: Hafner.

Bard, P. 1973. The ontogenesis of one physiologist. *Annu. Rev. Physiol.*, 35: 1–16.

Bargmann, W. 1949. Über die Neurosekretorische Verknüpfung von Hypothalamus und Neurohypophyse. *Z. Zellforsch.* 34: 610–634.

Barker, L. F. 1899. *The Nervous System and Its Constituent Neurones.* New York: Appleton.

Barlow, H. R., R. Fitzhugh, and S. N. Kuffler. 1957. Dark adaptation, absolute threshold and Purkinje shift in single units of the cat's retina. *J. Physiol.* (London), 137: 327–337.

Barrett, R., H. H. Merritt, and A. Wolf. 1967. Depression of consciousness as a result of cerebral lesions. *Res. Pub. Assoc. Res. Nerv. Men. Dis.*, 45: 241–272.

Bartholin, C. 1641. *Institutiones anatomicae ab auctoris filio Thomas Bartholino.* Leyden: F. Hack.

Bartholow, R. 1874. Experimental investigations into the functions of the human brain. *Am. J. Med. Sci.*, 67: 305–313.

Bartley, S. H. and G. H. Bishop. 1933. The cortical response to stimulation of the optic nerve in the rabbit. *Am. J. Physiol.*, 103: 159–172.

Bast, T. H. 1928. Karl Frederick Burdach. June 12, 1776–July 16, 1847. *Ann. Med. Hist.*, 10: 34–46.

Bataille, G. 1955. *Prehistoric Painting: Lascaux or the Birth of Art.* Translated by A. Weinhouse. Geneva: Skira.

Beevor, C. E. 1885. On Professor Hamilton's theory concerning thecorpus callosum. (Preliminary Note.) *Brain*, 8: 377–379.

Beevor, C. E. and V. Horsley. 1890. A record of the results obtained by electrical excitation of the so-called motor cortex and internal capsule in an orang-outang (Simia satyrus). *Phil. Trans. R. Soc. Lond. (B)*, 181: 129–158.

Békésy, G. v. 1967. *Sensory Inhibition.* Princeton: Princeton University Press.

Békésy, G. v. and W. A. Rosenblith. 1948. The early history of hearing—observations and theories. *J. Acoust. Soc. Am.*, 20: 727-748.

Bekhterev, W. V. 1900. Demonstration eines Gehirns mit Zerstörung der vorderen und inneren Theile der Hirninde beider Schläfenlappen. *Neurol. Zbl.*, 19: 990–991.

Bell, F. J., Trans., *see* Gegenbaur, C.

Belsky, S., Trans., *see* Sechenov, I. M.

Bence-Jones, H., Trans., *see* Du Bois-Reymond, É., 1849.

Boldman, C., Trans., *see* Ehrlich, P.

Bennett, H. S., J. H. Luft, and J. C. Hampton. 1959. Morphological classifications of vertebrate capillaries. *Am. J. Physiol.*, 196: 381–390.

Berengario da Carpi, J. 1535. *A Short Introduction to Anatomy (Isagogae brevis).* Translation and Introduction by L. R. Lind, 1959. Chicago: University of Chicago Press.

Berger, H. 1929. Über das Electrekephalogramm des Menschen. *Arch. Psychiatr. Nervenkr.* 87: 527–570.

Berkley, H. J. 1894. The nerve elements of the pituitary gland. *Johns Hopkins Hosp. Rep.*, 4: 285–295.

Berkley, H. J. 1898. Ricin-poisoning: Experimental lesions produced by the action of ricin on the cortical nerve cell of the guinea pig's and rabbit's brain, *Pathol. Soc. Philadelphia Trans.*, 18: 328–347.

Bernhard, C. G. 1964. Presentation speech. Pages 719–721 in *Nobel Lectures, Including Presentation Speeches and Laureates' Biographies. Physiology or Medicine, 1942–1962.* Amsterdam: Elsevier.

Bernstein, J. 1868. Über den zeitlichen Verlauf der negativen Schwankung des Nervenströms. *Pflügers Arch.*, 1: 173–207.

Bernstein, J. 1902. Untersuchungen zur Thermodynamik der bioelektrichen Ströme. *Pflügers Arch.*, 92: 521–562.

Betz, V. A. 1874. Anatomischer Nachweis zweier Gehirncentra. *Zbl. Med. Wiss.*, 12: 578–580, 595–599.

Betz, V. A. 1881. Über die feinere Struktur der Grosshirnrinde des Menschen. *Zbl. Med. Wiss.*, 19: 193–195, 209–213, 231–234. Also see French translation in *Rev. Anthropol.*, 4 (2nd ser.): 427–438.

Bianchi, L. 1895. The functions of the frontal lobes. Translated from the original manuscript by A. De Watteville. *Brain*, 18: 497–522.

Bianchi, L. 1922. *The Mechanisms of the Brain and Function of the Frontal Lobes.* Trans. by J. A. MacDonald. Edinburgh: Livingstone.

Bigelow, H. J. 1850. Dr. Harlow's case of recovery from the passage of an iron bar through the head. *Am. J. Med. Sci.*, 39, pt. 1; 12–22.

Bishop, G. H. 1956. Natural history of the nerve impulse. *Physiol. Rev.*, 36: 376–399.

Bishop, G. H. 1958a. The place of cortex in a reticular system. Pages 413–421 in H. H. Jasper, L. D. Proctor, R. S. Knighton, W. C. Noshay, and R. T. Costello, eds., *Reticular Formation of the Brain.* Henry Ford Hospital International Symposium. Boston: Little, Brown.

Bishop, G. H. 1958b. Physiology and phylogeny of the pain tracts. Unpublished abstract of talk delivered to the Washington University Medical Society, January 22, 1958. FP001, 7: 10, Washington University Medical Library.

Bishop, G. H. and M. H. Clare. 1952. Sites of origin of electrical potentials in striate cortex. *J. Neurophysiol.*, 15: 201–220.

Blinkov, S. M. and I. I. Glezer. 1968. *The Human Brain in Figures and Tables. A Quantitative Handbook.* Translated from the Russian by Basil Haigh. New York: Basic Books Inc., Plenum.

Bodian, D. 1962. The generalized vertebrate neuron. *Science*, 137: 323–326.

Bøler, J., F. Enzmann, K. Folkers, C. Y. Bowers, and A. V. Schally. 1969. The identity of chemical and hormonal properties of the thyrotropin releasing hormone and pyroglutamylhistidylproline amide. *Biochem. Biophys. Res. Commun.*, 37: 705–710.

Bonin, G. v., Trans., *see* Broca, P., 1861; Munk, H., 1881.

Bonin, G. v. 1934. On the size of man's brain as indicated by skull capacity. *J. Comp. Neurol.*, 59: 1–28.

Bonin, G. v. 1960. *Some Papers on the Cerebral Cortex.* Springfield, IL: C. C Thomas.

Bonin, G. v. 1970. Rudolf Albert von Kölliker (1817–1905). Pages 51–54 in W. Haymaker and F. Schiller, eds., *Founders of Neurology*, 2nd ed. Springfield, IL: C. C Thomas.

Booth, W. 1988. Research News: the social life of dolphins. *Science*, 240: 1273–1274.

Borelli, G. A. 1680–1681. *De motu animalum, Pars altera.* Rome: Bernabo.

Bouillaud, J.-B. 1825. Recherches cliniques propres à démontrer que la perte de la parole correspond à la lésion des lobules antérieurs du cerveau, et à confirmer l'opinion de M. Gall, sur le siège de l'organe du langue articule. *Arch. Gen. Med.*, 8: 25–45.

Bourgery, J.-M. 1845. Mémoire sur l'extrémité céphalique du grand sympathique dans l'homme et les animaux mammifères. *C. R. Acad. Sci.* (Paris) 20: 1014–1020.

Bowers, J. Z. 1972. *Western Medicine in a Chinese Palace: Peking Union Medical College, 1917–1951.* New York: Josiah Macy, Jr. Foundation.

Bradbury, M. W. B. 1972. Electrolyte disorders and the brain. Pages 1023–1042 in M. H. Maxwell and C. R. Kleeman, eds., *Clinical Disorders of Fluid and Electrolyte Metabolism.* New York: McGraw-Hill.

Bradbury, W. M. B. 1975. Ontogeny of mammalian brain–barrier systems. Pages 81–103 in H. F. Cserr, J. D. Fenstermachen, and V. Fencl, eds., *Fluid Environment of the Brain.* New York: Academic.

Bradbury, M. [W. B.] 1979. *The Concept of a Blood–Brain Barrier.* New York: Wiley.

Brazier, M. A. B. 1959. The historical development of neurophysiology. Pages 1–58 in J. Field, H. W. Magoun, V. E. Hall, eds., *Handbook of Physiology, Section 1: Neurophysiology,* Vol. I. Washington, D.C.: American Physiological Society.

Brazier, M. A. B. 1961. *A History of the Electrical Activity of the Brain.* New York: Macmillan.

Brazier, M. A. B. 1963. Role of the limbic system in maintenance of consciousness. *Anes. Anal.,* 42: 748–751.

Brazier, M. A. B. 1970. Felice Fontana (1730–1805). Pages 202–206 in W. Haymaker and F. Schiller, eds., *The Founders of Neurology,* 2nd ed. Springfield, Illinois: C. C Thomas.

Brazier, M. A. B. 1977. *Electrical Activity of the Nervous System..* 4th ed. London: Pitman Publishing.

Brazier, M. A. B. 1980. Trails leading to the concept of the ascending reticular system: the state of knowledge before 1949. Pages 31 52 in J. A. Hobson and M. A. B. Brazier, eds., *The Reticular System Revisited: Specifying Function for a Nonspecific System.* New York: Raven.

Brazier, M. A. B. 1982. The problem of neuromuscular action: Two 17th century Dutchmen. Pages 13–22 in F. C. Rose and W. F. Bynum, eds., *Historical Aspects of the Neurosciences.* New York: Raven.

Brazier, M. A. B. 1984. *A History of Neurophysiology in the 17th and 18th Centuries: from Concept to Experiment.* New York: Raven.

Breasted, J. H. 1930. *The Edwin Smith Surgical Papyrus.* Vol. II. Chicago: University of Chicago Press.

Bremer, F. 1922. Contribution à l'étude de la physiologie du cervelet. La fonction inhibitrice du paléo-cerebellum. *Arch. Int. Physiol.,* 19: 189–226.

Bremer, F. 1936. Nouvelles recherches sur le méchanisme du sommeil. *C.R. Soc. Biol.* (Paris), 122: 460–464.

Bremer, F., and N. Stoupel. 1959. Facilitation et inhibition des potentiels évoqués corticaux dans l'éveil cérébral. *Arch. Int. Physiol.,* 67: 240–275.

Bright, R. 1837. Cases and observations illustrative of diagnosis where tumors are situated at the basis of the brain; or where other parts of the brain and spinal cord suffer lesion from disease. *Guy's Hosp. Rep.* 2: 279–310.

Brightman, M. W. and S. L. Palay. 1963. The fine structure of ependyma in the brain of the rat. *J. Cell Biol.,* 19: 415–439.

Broca, P. 1861. Remarques sur le siège de la faculté du langage articulé, suivies d'une observation d'aphémie (perte de la parole). *Bull. Soc. Anat. Paris,* 36: 330–357. Trans. by G. von Bonin, pages 49–72 in G. von Bonin, *Some Papers on the Cerebral Cortex.* Springfield, Illinois: C. C Thomas, 1960.

Also pages 4–42 in P. Broca, 1888. *Memoires sur le Cerveau de l'Homme et des Primates.* Published with Introduction and Notes by S. Pozzi. Paris: Reinwald.

Broca, P. 1863. Localisation des fonctions cérébrales—Siège du langage articulé. *Bull. Soc. Anthrop. Paris,* 4: 200–202, 208.

Broca P. 1865a. Éloge funèbre de Pierre Gratiolet. *Mem. Soc. Anthropol. Paris,* Ser. 1, 2: cxii–cxviii.

Broca, P. 1865b. Sur le siège de la faculté du langage articulé. *Bull. Soc. Anthropol. Paris.* 6: 377–393.

Broca, P. 1878a. Anatomie comparée des circonvolutions cérébrales. Le grand lobe limbique et la scissure limbique dans le série des mammifères. *Rev. Anthropol. Paris,* Ser. 2, I: 385–498.

Broca, P. 1878b. Nomenclature cérébrale. Dénomination des divisions et subdivisions des hémisphères et des anfractuosités de leur surface. *Rev. Anthorpol. Paris,* Ser. 2, I: 193–256.

Broca, P. 1879. Recherches sur les centres olfactifs. *Rev. Anthropol. Paris,* Ser. 2, II: 385–455.

Brodal, A. 1947. The hippocampus and the sense of smell. A review. *Brain,* 70: 179–222.

Brodal, A. 1957. *The Reticular Formation of the Brain Stem.* Edinburgh: Oliver and Boyd.

Brodal, A. 1969. *Neurological Anatomy in Relation to Clinical Medicine,* 2nd ed. New York: Oxford University Press.

Brodal, A. 1987. Neuroanatomy, some reflections on its place in the neurosciences. Pages 769–773 in G. Adelman, ed., *Encyclopedia of Neuroscience,* Boston: Birkhäuser.

Brodmann, K. 1908. Beiträge zur histologischen Lokalisation der grosshirnrinde. VI. Mittenlung: Die Cortexgliederung des Menschen. *J. Psychol. Neurol.* (Leipsig), 10: 231–246.

Brodmann, K. 1909. *Vergleichende Lokalisationslehre der Grosshirnrinde in ihren Prinzipien dargestellt auf Grund des Zellenbaues.* Leipzig: J. A. Barth.

Broman, T. 1941. The possibilities of the passage of substances from the blood to the central nervous system. *Acta Psychiat. Neurol.,* 16: 1–25.

Brookhart, J. M., G. Moruzzi, and R. S. Snider. 1950. Spike discharges of single units in the cerebellar cortex. *J. Neurophysiol.,* 13: 465–486.

Brooks, C. McC., T. Ishikawa, K. Koizumi, and H. H. Lu. 1966. Activity of neurones in the paraventricular nucleus of the hypothalamus and its control. *J. Physiol.* (London), 182: 217–231.

Brouwer, B. 1933. Centrifugal influence on centripetal systems in the brain. *J. Nerv. Ment. Dis.,* 77: 621 627.

Brown, J. W. 1977. *Mind, Brain, and Consciousness: the Neuropsychology of Cognition.* New York: Academic.

Brown, S. and E. A. Schaefer. 1888. An investigation into the functions of the occipital and temporal lobes of the monkey's brain. *Phil. Trans. R. Soc. Lond. B.,* 179: 303–327.

Brown–Séquard, C. E. The dual character of the brain. *The Toner Lectures, Lecture II. Smithsonian Miscellaneous Collections.* WashingtonBAD AD *Physiol. Behav.,* 5: 207–210.

Bucy, P. C. and H. Klüver. 1940. Anatomic changes secondary to temporal lobectomy. *Arch. Neurol. Psychiatr.,* 44: 1142–1146.

Bullock, T. H. 1959. Neuron doctrine and electrophysiology. *Science,* 129: 997–1002.

Bunge, M. B., R. P. Bunge, and G. D. Pappas. 1962. Electron microscopic demonstrations of connections between glia and myelin sheaths in the developing mammalian central nervous system. *J. Cell Biol.,* 13: 448–453.

Burdach, K. F. 1819–1826. *Vom Baue und Leben des Gehirns.* 3 vols. Leipzig: Dyk'schen.

Burgus, R., T. Dunn, D. Disiderio, and R. Guillemin. 1969. Structure moléculaire du facteur hypothalamique hypophysiotrope TRF d'origine ovine; mise en evidence par spectrométrie de masse de la séquence PCA-His-Pro-NH_2. *C. R. Acad. Sci.* (Paris), 269: 1870–1873.

Burnham, J. C. 1977. The mind–body problem in the early twentieth century. *Perspect. Biol. Med.,* 20: 271–284.

Buchsbaum, M. and P. Fedio. 1970. Hemispheric differences in evoked potentials. *Physiol. Behav.,* 5: 207-210.

Busk, G., Trans., *see* Kölliker, R. A. v., 1854.

Bykov, K. M. 1924–1925. Versuche an Hunden mit Durchschneiden des Corpus Callosum. *Zentralbl. Neurol. Psychiatr.,* 39: 199–211. Translated in Vol. 1, *Fiziol. Lab. I.P. Pavlova, Akad. Nauk SSSR,* 1925.

Bykov, K. M. and A. D. Speranski. 1925. Dog with severed corpus callosum. *Fiziol. Lab. I.P. Pavlova,* 1: 47–59. Translated in *Collection of Translated Works, Vol. 2, Akad. Nauk SSSR.*

Cairns, H. 1952. Disturbances of consciousness with lesions in the brainstem and diencephalon. *Brain,* 75, part 2: 109–146.

Campbell, A. W. 1903. Histological studies on cerebral localisation. *Proc. R. Soc.* (London), 72: 488–492.

Campbell, A.W. 1905. *Histological Studies on the Localisation of Cerebral Function.* Cambridge: University Press.

Campbell, B. H. 1974. *Human Evolution,* 2nd ed. Chicago: Aldine Publishing Co.

Campion, G. G. and G. E. Smith. 1934. *The Neural Basis of Thought.* New York: Harcourt Brace.

Camus, J. and G. Roussy. 1920. Experimental researches on the pituitary body. Diabetes insipidus, glycosuria and those dystrophies considered as hypophyseal in origin. *Endocrinology,* 4: 507–522.

Cannon, W. B. 1929. *Bodily Changes in Pain, Hunger, Fear and Rage.* New York: Appleton.

Cannon, W. B. 1931. Again the James-Lange and the thalamic theories of emotion. *Psychol. Rev.,* 38: 281–295.

Cannon, W. B. 1945. *The Way of an Investigator. A Scientist's Experience in Medical Research.* New York: Hafner.

Cannon, W. B. and Z. M. Bacq. 1931. A hormone produced by sympathetic action on smooth muscle. *Am. J. Physiol.,* 96: 392–412.

Carmichael, S. W. 1989. The history of the adrenal medulla. *Rev. Neurosci.* 2: 83–99.

Carpenter, W. B. 1859. *Principles of Human Physiology.* 2nd. American ed. Philadelphia: Blanchard and Lea.

Casserio, J. 1627. *Tabulae anatomicae LXXIIX. . . .* Venice: E. Deuchinum.

Caton, R. 1875. The electric current of the brain. *Br. Med. J.,* 2: 278.

Caton, R. 1891. [Letter to the Editor.] *Central-blatt Physiol.,* 4: 785–786.

Chambers, G. H., E. V. Melville, R. S. Hare, and K. Hare. 1945. Regulation of the release of pituitrin by changes in the osmotic pressure of the plasma. *Am. J. Physiol.,* 144: 311–320.

Chan-Palay, V. and S. L. Palay. 1970. Interrelations of basket cell axons and climbing fibers in the cerebellar cortex of rats. *Z. Anat. Entwickl.-Gesch.,* 132: 191–227.

Chance, F., Trans., *see* Virchow, R.

Chang, T.-H. 1954. Interaction of evoked cortical potentials. *J. Neurophysiol.,* 16: 133–144.

Choulant, L. 1945. *History and Bibliography of Anatomic Illustration.* Trans. and annotated by Mortimer Frank. New York: Schuman's.

Clapham, C. 1881–1882. The noble forehead. *J. Ment. Sci.,* 27: 623–624.

Clark, W. E. LeGros. 1932. The structure and connections of the thalamus. *Brain,* 55: 406–470.

Clark, W. E. LeGros. 1948. The connexions of the frontal lobes of the brain. *Lancet,* 1: 353–356.

Clarke, E., Trans., *see* Neuburger, M., 1981.

Clarke, E. and K. Dewhurst. 1972. *An Illustrated History of Brain Function.* In Memoriam Charles Donald O'Malley (1907–1970). Berkeley/Los Angeles: University of California Press.

Clarke, E. and L. S. Jacyna. 1987. *Nineteenth-Century Origins of Neuroscience Concepts.* Berkeley/Los Angeles: University of California Press.

Clarke, E. and C. D. O'Malley. 1968. *The Human Brain and Spinal Cord. A Historical Study Illustrated by Writings from Antiquity to the Twentieth Century.* Berkeley/Los Angeles: University of California Press, 1968.

Clarke, R. H. and V. Horsley. 1905. On the intrinsic fibres of the cerebellum, its nuclei and its efferent tracts. *Brain,* 28: 13–29.

Cobb, S. 1952. On the nature and locus of mind. *Arch. Neurol. Psychiatr.,* 67: 172–177.

Cohn, R. 1971. Differential cerebral processing of noise and verbal stimuli. *Science,* 172: 599–601.

Cole, F. J. 1949. *A History of Comparative Anatomy from Aristotle to the Eighteenth Century.* London: Macmillan.

Cole, K. S. 1949. Dynamic electrical characteristics of the squid axon membrane. *Arch. Sci. Physiol.,* 3: 253–258.

Cole, K. S. and H. J. Curtis. 1938. Electric impedance of the squid giant axon during activity. *J. Gen. Physiol.,* 22: 649–670.

Collip, J. B., E. M. Anderson, and D. L. Thomson. 1933. The adrenotropic hormone of the anterior pituitary lobe. *Lancet,* 2: 347–348.

Combe, G., Trans., *see* Gall, Vimont, and Broussais.

Cotugno, D. F. A. 1764. *De ischiate nervosa commentarius.* Naples: Fratres Simonii.

Cournand, A. 1982. Air and blood. Pages 3–70 in A. P. Fishman and D. W. Richards, eds., *Circulation of the Blood. Men and Ideas.* Bethesda, MD: American Physiological Society.

Courville, C. B. 1970. Santiago Ramón y Cajal (1952–1934). Pages 144–151 in W. Haymaker and F. Schiller, eds., *The Founders of Neurology,* 2nd. ed. Springfield, IL: Thomas.

Craigie, E. H. 1938. The comparative anatomy and embryology of the capillary bed of the central nervous system. *Res. Pub. Assoc. Nerv. Ment. Dis.,* 18: 3–28.

Craigie, E. H. 1955. Vascular patterns of the developing nervous system. Pages 28–51 in H. Waelsch, ed., *Biochemistry of the Developing Nervous System.* New York: Academic.

Critchley, MacD. 1953. *The Parietal Lobes.* New York: Hafner.

Crombie, A. C. 1959. Descartes. *Sci. Am.,* 201 (4): 160–173.

Crone, C. 1960. The diffusion of some organic non-electrolytes from blood to brain tissue. *Acta Physiol. Scand.,* 50. Suppl. 175. Abstr.

Crone, C. 1965. The permeability of brain capillaries to nonelectrolytes. *Acta Physiol. Scand.,* 64: 407–417.

Crone, C. 1986. The blood–brain barrier: A modified tight epithelium. Pages 17–40 in A. J. Suckling, M. G. Rumsby, and M. W. B. Bradbury, eds., *The Blood–Brain Barrier in Health and Disease.* Chichester, UK: Ellis Horwood.

Crosby, E. C. 1917. The forebrain of Alligator mississippiensis. *J. Comp. Neurol.*, 27: 325–402.

Crosby, E. C. 1972. Discussion. Pages 88–90 in T. L. Frigyesi, E. Rinvik, and M. D. Yahr, Eds., *Corticothalamic Projections and Sensorimotor Activities.* New York: Raven.

Cross, B. A. and J. D. Green. 1959. Activity of single neurones in the hypothalamus: effect of osmotic and other stimuli, *J. Physiol.* (London), 148: 554–569.

Cserr, H. F. and M. Bundgaard. 1984. Blood–brain interfaces in vertebrates: a comparative approach. *Am. J. Physiol.*, 246(3): R277–R288.

Cunningham, D. J. 1890. The complete fissures of the human cerebrum, and their significance in connection with the growth of the hemisphere and the appearance of the occipital lobe. *J. Anat. Physiol.*, 24: 309–345.

Cunningham, D. J. 1892. Contributions to the surface anatomy of the cerebral hemispheres. *Mem. Irish Acad.*, (Dublin), VII: 131–136.

Curtis, H. J. and P. Bard. 1939. Intercortical connections of the corpus callosum as indicated by evoked potentials. *Am. J. Physiol.*, 126: 473.

Curtis, H. J. and K. S. Cole. 1940. Membrane action potentials from the squid giant axon. *J. Cell Comp. Physiol.*, 15: 147–157.

Cushing, H. 1906. Sexual infantilism with optic atrophy in cases of tumor affecting the hypophysis cerebri. *J. Nerv. Ment. Dis.* 33: 704–716.

Cushing, H. 1909. A note upon the faradic stimulation of the postcentral gyrus in conscious patients. *Brain*, 32: 44–53.

Cushing, H. 1912. *The Pituitary Body and Its Disorders; Clinical States Produced by Disorders of the Hypophysis Cerebri.* Philadelphia: Lippincott.

Cushing, H. 1932. *Papers Relating to the Pituitary Body, Hypothalamus.* Springfield, IL: Thomas.

Cuvier, G., Baron. [1800]–1805. *Leçons d'anatomie comparé de G. Cuvier.* 5 Vols. Paris: Baudouin. Vols. 1–2 (1800) edited by C. Duméril and Vols. 3–5 by G. L. Duvernoy.

Dale, H. H. 1906. On some physiological actions of ergot. *J. Physiol.* (London), 34: 165–206.

Dale, H. H. 1914. The action of certain esters of choline and their relation to muscarine. *J. Pharmacol. Exp. Ther.* 6: 147–190.

Dale, H. H. 1937. Transmission of nervous effects by acetylcholine. *The Harvey Lectures*, 32: 229–245.

Dale, H. H. 1937–1938. Du Bois-Reymond and chemical transmission. *J. Physiol.* (London), 91: 4P.

Damassio, H., T. Grabowski, R. Frank, A. M. Galaburda, and A. R. Damassio. 1994. The return of Phineas Gage: clues about the brain from the skull of a famous patient. *Science*, 264: 1102–1105.

Dandy, W. E. 1918. Ventriculography following the injection of air into the cerebral vesicles. *Ann. Surg.*, 68: 5–11.

Dandy, W. E. 1919. Experimental hydrocephalus. *Ann. Surg.*, 70: 129–142.

Dandy, W. E. 1946. The location of the conscious center in the brain – the corpus striatum. *Bull. Johns Hopkins Hosp.* 79: 34–58.

Dandy, W. E. and K. D. Blackfin. 1914. Internal hydrocephalus. An experimental, clinical and pathological study. *Am. J. Dis. Child.*, 8: 406–482.

Daremberg, C. *see* Galen; Rufus of Ephesus.

Dart, R. A. 1925. *Australopithecus africanus*: the man-ape of South Africa. *Nature*, 115: 195–199.

Darwin, C. 1859. *On the Origins of the Species by Means of Natural Selection, or the Preservation of Favored Races in the Struggle for Life.* London: John Murray.

Darwin, C. 1872. *The Descent of Man, and Selection in Relation to Sex.* 2 Vols. New York: D. Appleton & Co.

Dasheiff, R. M. 1994. The first American epileptologists: William P. Spratling, M. D., and Roswell Park, M. D. *Neurology*, 44: 171–174.

Davidson, M. B. 1962. *The Horizon Book of Lost Worlds.* New York: American Heritage Publishing Co.

Davis, H., P. A. Davis, A. L. Loomis, E. N. Harvey, and G. Hobart. 1939. Electrical reactions of the human brain to auditory stimulation during sleep. *J. Neurophysiol.*, 2: 500–415.

Davis, P. A. 1939. Effects of acoustic stimuli on the waking brain. *J. Neurophysiol*, 2: 494–499.

Davson, H. 1955. A comparative study of the aqueous humour and cerebrospinal fluid in the rabbit. *J. Physiol.* (London), 129: 111–133.

Davson, H. 1956. *Physiology of the Ocular and Cerebrospinal Fluids.* London: Churchill.

Davson, H. 1967. *Physiology of the Cerebrospinal Fluid.* London: Churchill.

Davson, H. and W. H. Oldendorf. 1967. Transport in the central nervous system. *Proc. R. Soc. Med.*, 60: 326–328.

Dawson, G. D. 1947. Cerebral responses to electrical stimulation of peripheral nerve in man. *J. Neurol. Neurosurg. Psychiatr.* Ser. 2, 10: 134–140.

Dawson, G. D. 1954. A summation technique for the detection of small evoked potentials. *Electroencephal. Clin. Neurophysiol.*, 6: 65–84.

De Beer, G. 1937. *The Development of the Vertebrate Skull.* Oxford: Clarendon Press.

De Lacoste-Utamsing, C. and R. L. Holloway. 1982. Sexual dimorphism in the human corpus callosum. *Science*, 216: 1431–1432.

De Robertis, E. and H. M. Gerschenfeld. 1961. Submicroscopic morphology and function of glial cells. *Int. Rev. Neurobiol.*, 3: 1–65.

De Watteville, A., Trans., *see* Bianchi, L., 1895; Luciani, L., 1884–1885.

DeFelipe, J. and E. G. Jones. 1988. *Cajal on the Cerebral Cortex. An Annotated Translation of the Complete Writings.* New York/Oxford: Oxford University Press.

Deiters, O. F. K. 1865. *Untersuchungen über Gehirn und Rückenmack das Menschen und der Säugetheire.* Braunschweig: Vieweg und Sohn.

Dejerine, J. 1901. *Antatomie des Centres Nerveux.* 2 vol. Paris: J. Reuff.

Dejerine, J. and G. Roussy. 1906. Le syndrome thalamique. *Rev. Neurol.* 14: 521–532.

Dejerine, J. and M. Vialet. 1893. Contribution á l'étude de la localization anatomique de la cécité verbale pure. *C. N. Soc. Biol. Paris.*, 45: 790–791.

Delafresnaye, J. F., E. D. Adrian, F. Bremer, and H. H. Jasper, eds. 1954. *Brain Mechanisms and Consciousness.* Oxford: Blackwell.

Dell, P. C. 1958. Humoral effects on the brain stem reticular formations. Pages 365–379 in H. H. Jasper, L. D. Proctor, R. S. Knighton, W. C. Noshay, and R. T. Costello, eds.,

Reticular Formation of the Brain. Boston: Little, Brown and Company.

Dempsey, E. W. and R. S. Morison. 1942a. The production of rhythmically recurrent cortical potentials after localized thalamic stimulation. *Am. J. Physiol.*, 295–300.

Dempsey, E. W. and R. S. Morison. 1942b. The interaction of certain spontaneous and induced cortical potentials. *Am. J. Physiol.*, 135: 301–308.

Dempsey, E. W. and R. S. Morison. 1943. The electrical activity of a thalamocortical relay system. *Am. J. Physiol.*, 183: 283–296.

Dempsey, E. W. and G. B. Wislocki. 1955. An electron microscopic study of the blood–brain barrier in the rat, employing silver nitrate as a vital stain. *J. Biophys. Biochem. Cytol.*, 1: 245–256.

Denenberg, V. H. 1981. Hemispheric laterality in animals and the effects of early experience. *Behav. Brain Sci.*, 4: 1–21, commentary 21–49.

Denny-Brown, D. 1970. Henry Head (1861–1940). Pages 449–452 in W. Haymaker and F. Schiller, eds., *Founders of Neurology*, 2nd ed., Springfield, Illinois: Thomas.

Descartes, R. 1649. *Les passions de l'âme.* Paris: Henri Le Gros.

Descartes, R. 1664. L'Homme de Rene Descartes et un traitté de la formation du foetus...Paris: Angot.

Descartes, R. 1989. *The Passions of the Soul.* Translated by S. Voss. Jr. from the 1649 edition. Indianapolis, IN: Hackett Publishing.

Dewhurst, K. 1968. Willis and Steno. Pages 43–48 in G. Scherz, ed., *Steno and Brain Research in the Seventeenth Century.* Proceedings of the International Symposium. Copenhagen 18–20 August 1965.

Dewhurst, K. 1980. *Willis's Oxford Lectures.* Oxford, Sandford Publications.

D'Hollander, F. 1922. Recherches anatomiques sur les conches optiques. *Arch. Biol.* (Paris), 32: 249–344.

Diamond, I. T., E. G. Jones, and T. P. S. Powell. 1969. The projection of the auditory cortex upon the diencephalon and brain stem in the cat. *Brain Res.*, 15: 305–340.

Diamond, M. C., A. B. Scheibel, and L. M. Elson. 1985. *The Human Brain Coloring Book.* New York: Harper & Row.

Diels, H. 1952. *Die Fragmente der Vorsokratiker, I.* Berlin: Weidmann.

Dieng, S. 1994. Extending the neuron doctrine: Carl Ludwig Schleich (1859–1922) and his reflections on neuroglia at the inception of the neural-network concept in 1894. *Trends Neurosci.*, 17: 449–452.

Dimond, S. J. 1977. Evolution and lateralizaiton of the brain: concluding remarks. Pages 477–501 in S. J. Dimond and D. A. Blizard, eds., *Evolution and Lateralization of the Brain.* New York: New York Academy of Sciences.

Djørup, F. 1968. Steno's ideas on brain research. Pages 111–114 in G. Scherz, ed., *Steno and Brain Research in the Seventeenth Century.* Oxford: Pergamon Press.

Donders, F. C. 1850. De Bewegingen der hersenen en de veranderingen der vaatrulling van de *pia mater*, ook bij gesloten' onuitzetbaren schedel regtstreeks onderzocht. *Ned. Lancet, The Hague,* 5 (2nd ser.): 521–553.

Douglas, G., Trans., *see* Steno, N., 1950.

Douglas, S. H. 1935. A note on the work of v. La Valette St. George, the discoverer of the Golgi apparatus and mitochondria of modern cytology. *R. Micros. Soc. J.*, 55: 28–31.

Douglas, W. W. 1968. Stimulus-secretion coupling: the concept and clues from chromaffin and other cells. *Br. J. Pharmacol.*, 34: 451–474.

Douglas, W. W. and A. M. Poisner. 1964. Stimulus–secretion coupling in a neurosecretory organ: the role of calcium in the release of vasopressin from the neurohypophysis. *J. Physiol. (London),* 172: 1–18.

Dow, R. S. 1939. Cerebellar action potentials in response to stimulation of various afferent connections. *J. Neurophysiol.*, 2: 543–555.

Dow, R. S. and G. Moruzzi. 1958. *The Physiology and Pathology of the Cerebellum.* Minneapolis: University of Minnesota Press.

Drabkin, D. L. 1958. *Thudichum, Chemist of the Brain.* Philadelphia: University of Pennsylvania Press.

Drake, W. E., Jr. 1968. Clinical and pathological findings in a child with a developmental learning disability. *J. Learn. Disabil.*, 1: 486–502.

Du Bois–Reymond, E. 1843. Vorläufiger Abriss einer Unter-suchung über den sogenannten Froschstrom und über die elektromotorischen Fische. *J. C. Poggendorff's biograph-isch-literarisches Handwörterbuch für Physik. Chemie. (Leipzig)*, 58: 1–3.

Du Bois-Reymond, E. 1849–1884. *Untersuchungen über Thierische Elektricität.* 2 vol. Berlin: G. Reimer. Translated by H. Bence-Jones. London: Churchill, 1852.

Du Vigneaud, V́., H. C. Lawler, and E. A. Popenoe. 1953a. Enzymatic cleavage of glycinamide from vasopressin and a proposed structure for this pressor–antidiuretic hormone of the posterior pituitary. *J. Am. Chem. Soc.*, 75: 4880–4881.

Du Vigneaud, V., C. Ressler, J. M. Swan, C. W. Roberts, P. G. Katsoyannis, and S. Gordon. 1953b. The synthesis of an octapeptide amide with the hormonal activity of oxytoxin. *J. Am. Chem. Soc.*, 75: 4879–4880.

Dufy, B., L. Dufy-Barbe, and D. Poulain. 1974. Gonadotropin release in relation to electrical activity in hypothalamic neurons. *J. Neurol. Transm.*, 34: 47–52.

Dumont, S., and P. Dell. 1958. Facilitation réticulaire des mécanismes visuels corticaux. *Electroenceph. Clin. Neurophysiol.*, 12: 769–796.

Durup, G. and A. Fessard. 1935. L'électrencephalogram de l'homme. Observations à l'action des stimuli visuels and auditifs. *L'Anné Psychol.*, 36: 1–32.

Dusser de Barenne, J. G. 1916. Experimental researches on sensory localizations in the cerebral cortex. *Q. J. Exp. Physiol.*, 9: 355–390.

Dusser de Barenne, J. G. 1924. Experimental researches on sensory localization in the cerebral cortex of the monkey (*Macacus*). *Proc. R. Soc. Lond. B,* 96: 272–291.

Dusser de Barenne, J. G. 1934. Some aspects of the problem of "corticalization" of function and of functional localization in the cerebral cortex. Pages 85–106 in S. T. Orton, J. F. Fulton, and T. K. Davis, eds., *Localization of Function in the Cerebral Cortex, an Investigation of the Most Recent Advances.* Baltimore, MD: Williams and Wilkins.

Dusser de Barenne, J. G. 1935. Central levels of sensory integration. Pages 274–288 in C. A. Patten, A. M. Frantz, and C. C. Hare, eds., *Sensation: Its Mechanisms and Disturbances.* Research Publications of the Association for Research in Nervous and Mental Diseases, Vol. 15.

Dusser de Barenne, J. G. and W. S. McCulloch. 1938. The direct functional interrelation of sensory cortex and optic thalamus. *J. Neurophysiol.*, 1: 176–186.

Eberstaller, [O.] 1884. Zur Oberflächen-Anatomie der Grosshirn-Hemisphären. *Wien. Med. Bl.*, 7: 479–482.

Eccles, J. C. 1936. Synaptic and neuromuscular transmission. *Ergeb. Physiol.*, 38: 339–444.

Eccles, J. C., ed., 1966. *Brain and Conscious Experience*. Study Week September 28 to October 4, 1964, of the Pontificia Academia Scientiarum. New York: Springer-Verlag.

Eccles, J. C. 1979. *The Human Mystery. The Gifford Lectures, University of Edinburgh, 1977–1978*. Berlin: Springer International.

Ecker, A. 1873. *On the Convolutions of the Human Brain*. Translated by J. C. Galton. London: Smith, Elder.

Economo, C. v. and L. Horn. 1930. Über Windungsrelief, Masse, und Rindenarchitektonik der Supratemporalflache. *Z. Ges. Neurol. Psychiat.*, 130: 678–757.

Economo, K. F. v. and J. v. Wagner-Jauregg. 1937. *Baron Constantin von Economo. His Life and Work*. Translated by R. Spillman. Burlington: Vermont Free Press.

Edinger, L. 1896. *Vorlesungen über der Bau der nervösen Centralorgane des Menschen und der Theire. Für Ärzte und Studirende*. Leipzig: Vogel.

Edinger, L. 1899. *Anatomy of the Central Nervous System in Man and of Vertebrates in General*. Philadelphia: F. A. Davis & Co.

Edinger, L. 1908. *Bau und den Verrichtungen des Nervensystems*. Leipzig: Vogel.

Eggert, G. H. 1977. *Wernicke's Works on Aphasia, a Sourcebook and Review*. The Hague, Mouton.

Ehrenberg, C. G. 1833. Nothwendigkeit einer feineren mechanischen Zerlegung des Gehirns und der Nerven vor der chemischen, dargestellt aus Beobachtungen von C. G. Ehrenberg. *J. C. Poggendorff's biographische-literarisches Handwörterbuch für Physik Chemie Leipzig*, 28: 449–473.

Ehrlich, P. 1906. Über die Beziehungen von chemischer Constitution, Vertheilung, und pharmakoligischer Wirkung. Reprinted and translated by C. Bolduan. Pages 404–442 in *Collected Studies in Immunity by P. Ehrlich and His Collaborators*, 1st ed. New York: Wiley.

Eldredge, N. and S. J. Gould. 1972. Punctuated equilibria: an alternative to phyletic gradualism. Pages 82–115 in T. J. M. Schopf, ed., *Models in Paleobiology*. San Francisco: Freeman, Cooper and Co.

Elliott, T. R. 1904. On the action of adrenalin. *J. Physiol.* (London) 31: xx–xxi.

Engel, J., Jr. 1993. Historical perspectives. Pages 695–705 in J. Engel, Jr., ed., *Surgical Treatment of the Epilepsies*, 2nd. ed., New York: Raven.

Erlanger, J. and H. S. Gasser, with the collaboration. . .of G. H. Bishop. 1924. The compound nature of the action current of nerve as disclosed by the cathode ray oscilloscope. *Am. J. Physiol.*, 70: 624–666.

Euler, C. v., J. D. Green, and G. Ricci. 1958. The role of hippocampal dentrites in evoked responses and after-discharges. *Acta. Physiol. Scand.* 42: 87–111.

Euler, U. S. v. 1933. On the presence of "novadrenine" in suprarenal extracts. *J. Physiol.* (London), 78: 462–466.

Euler, U. S. v. and J. H. Gaddum. 1931. An unidentified depressor substance in certain tissue extracts. *J. Physiol.* (London). 72: 74–87.

Eustachio, B. 1563. *Tabulae anatomicae clarissimi viri Bartholomaei Eustachii quas*. Republished in 1714 by J. M. Lancisius. Rome: F. Gonzagae.

Evans, H. M., M. E. Simpson, K. Meyer, and F. L. Reichert. 1933. Treatment of normal and achondroplastic dogs with growth hormone. *Mem. Univ. Calif.*, 11: 425–441.

Evans, M. J. and J. B. Finean. 1965. The lipid composition of myelin from brain and peripheral nerve. *J. Neurochem.*, 12: 729–734.

Exner, S. 1894. *Entwurf zur physiologischen Erklärung der Psychischen Erscheinungen*. Leipzig: Deuticke.

Faivre, E. 1857. Études sur le conarium et les plexus choroïdes chez l'homme et les animaux. *Ann. Sci. Nat., Heme Ser. Zool.*, 7: 52–90.

Farini, F. 1913. Diabete insipido ed opoterapia ipofisaria. *Gazz. Osped. Clin.*, 34: 1135–1139.

Farquhar, M. G. and G. E. Palade. 1963. Functional complexes in various epithelia. *J. Cell Biol.*, 17: 375–412.

Feindel, W. 1958. Discussion. Pages 669–670 in H. H. Jasper, L. D. Proctor, R. S. Knighton, W. C. Noshay, and R. T. Costello, eds., *Reticular Formation of the Brain*. Boston: Little, Brown and Co.

Feldberg, W. and J. H. Gaddum. 1934. The chemical transmitter at synapses in a sympathetic ganglion. *Am. J. Physiol.*, 81: 305–319.

Ferraro, A. 1970. Camillo Golgi (1843–1926). Pages 35–39 in W. Haymaker and F. Schiller, eds., *Founders of Neurology*. Springfield, IL: Charles C Thomas.

Ferrier, D. 1873. Experimental researches in cerebral physiology and pathology. *West Riding Lunatic Asylum Med. Rep.*, 3: 30–96.

Ferrier, D. 1874. On the localisation of the functions of the brain. *Br. Med. J.*, 2: 766–767.

Ferrier, D. 1876. *Functions of the Brain*. New York: Putnam.

Ferrier, D. 1888. Schäfer on the temporal and occipital lobes. *Brain*, 11: 7–30.

Ferris, H. B. 1922. The natural history of man. Chapter II in G. A. Baitsell, ed., *The Evolution of Man*. New Haven: Yale University Press.

Ferrus, A. 1989. Neuroscience in Spain: the Cajal Institute looks ahead. *IBRO News*, 17(1): 1.

Fessard, A. E. 1954. Mechanisms of nervous integration and conscious experience. Pages 200–236 in J. F. Delafresnaye, ed., *Brain Mechanisms and Consciousness*. Springfield, IL: C. C Thomas.

Fisher, C., W. R. Ingram, and S. W. Ranson. 1938. *Diabetes Insipidus and the Neuro-Humoral Control of Water Balance: a Contribution to the Structure and Function of the Hypothalamic-Hypophyseal System*. Ann Arbor, MI: Edwards Brothers.

Flanigan, S., E. R. Gabrieli, and P. D. MacLean. 1957. Cerebral changes revealed by radioautography with S^{35}-labeled l-methionine. *Arch. Neurol. Psychiat.*, 77: 588–594.

Flechsig, P. 1876. *Die Leitungsbahnen im Gehirn und Rückenmark des Menschen auf Grund Entwickelungsgeschichtlicker Untersuchungen*. Leipsig: Engelmann.

Flechsig, P. 1896. *Gehirn und Seele*, Leipzig: Veit.

Flechsig, P. 1901. Developmental (myelogenetic) localisation of the cerebral cortex in the human subject. *Lancet*, ii: 1027–1029.

Flechsig, P. 1904. Einige Bemerkungen über die Untersuchungs methoden der Grosshirnrinde, insbes-ondere des Menschen. *Ber. Verh. sächs. Ges. Wiss. Leipz., Math.-Phys. Klasse,* 56: 50–104, 177–248.

Flourens, P., Trans., *see* Rolando, L., 1823.
Flourens, P. 1823. Recherches physiques sur les propriétés et les fonctions du système nerveux dans les animaux vertébrés. *Arch. Gen. Med.,* 2: 321–370.
Flourens, P. 1824. *Recherches experimentales sur les propriétés et les fonctions du système nerveux dans les animaux vertébrés.* Paris: Crevot.
Flourens, P. 1842. *Examen de la phrénologie.* Paris: Paulin.
Flourens, P. 1846. *Phrenology examined.* Translated by D. L. Meigs. Philadelphia: Hogan and Thompson. Reprinted in D. N. Robinson, ed., *Significant Contributions to the History of Psychology, 1750–1920.* Washington, D.C.: University Publications of America, 1978.
Flourens, P. 1863. *De La Phrénologie et des Études Vraies sur le Cerveau.* Paris: Garnier Frères.
Foderá, M. 1823. Recherches expérimentales sur le système nerveux. *J. Physiol. Exp. Pathol.,* (Paris) 3: 191–217.
Forbes, A. 1936. Discussion of paper by R. W. Gerard. *Cold Spring Harb. Sym. Quant. Biol.*, 4: 298–299.
Forbes, A. and C. Thacher. 1920. Amplification of action currents with the electron tube in recording with the string galvanometer. *Am. J. Physiol.,* 52: 409–471.
Forbes, A., B. Renshaw, and B. Rempel. 1937. Units of electrical activity in the cerebral cortex. *Am. J. Physiol.*, 119: 309–310.
Forbes, H. S. and S. Cobb. 1938. Vasomotor control of cerebral vessels. *Proc. Assoc. Res. Nerv. Ment. Dis.,* 18: 201–217.
Forel, A. H. 1877. Untersuchungen über die Haubenregion. *Arch. Psychiat. Nervenkrank.* 7: 393–495.
Forel, A. H. 1887. Einige hirnanatomische Betrachtungen und Ergebnisse. *Arch. Psychiat. Nervenkrank.*, 18: 162–198.
Fontana, F. 1781. *Traité sur le vénin de la vipère* 2 vol. Florence: (s.n.) et se trouve à Paris chez Nzon l'Ainé, à Londrès chez Emsley.
Foster, M. and C. S. Sherrington. 1897. *Textbook of Physiology, III.* London: Macmillan.
Foville, A. L. 1844. *Traité complet de l'anatomie, de la physiologie et de la pathologie du système nerveux cérébro-spinal. Première Partie, Anatomie. Atlas.* Paris: Fortin, Masson.
Fowler, O. S., ed. 1850. Frontispiece. *Am. Phrenol. J.*, 2.
Frank, M., Trans., *see* Choulant, L.
Frank, R. G., Jr., L. H. Marshall, and H. W. Magoun. 1976. The neurosciences. Pages 552–613 in J. Z. Bowers and E. F. Purcell, eds., *Advances in American Medicine: Essays at the Bicentennial*, 2 vol. New York: Josiah Macy Foundation.
Franz, S. I. 1911. On the functions of the cerebrum: the occipital lobes. *Psychol. Mono.*, 13(4).
Franz, S. I. 1913. Observations on the preferential use of the right and left hands by monkeys. *J. Animal Behav.*, 3: 140–144.
Franz, S. I. 1915. Variations in distribution of the motor centers. Pages 80–161 in *On the Functions of the Cerebrum.* Psychological Monographs, 19 (1). Princeton, NJ: Psychological Review Co.
Frazier, C. H. and M. M. Peet. 1914. Factors of influence in the origin and circulation of the cerebrospinal fluid. *Am. J. Physiol.,* 35: 268–282.
Freeman, W. J. and J. W. Watts. 1942. *Psychosurgery in the Treatment of Mental Disorders and Intractable Pain.* Springfield, IL: Thomas.
French, J. D. 1952. Brain lesions associated with prolonged unconsciousness. *Arch. Neurol. Psychiat.*, 68: 727–740.
French, J. D. 1958. Special Article: The reticular formation. *J. Neurosurg.*, 15: 97–115.
Friedemann, U. and A. Elkeles. 1932. Untersuchungen über den Stoffaustausch zwischen Blut und Gehirn. *Klin. Wochensch.* pt. 2: 2026–2028.
Friedemann, U. and A. Elkeles. 1934. The blood–brain barrier in infectious diseases: its permeability to toxins in relation to their electrical charges. *Lancet*, 1: 719–724; 775–779.
Fritsch, G. and E. Hitzig. 1870. Über die elektrische Erregbarkeit des Grosshirns. *Arch. Anat. Physiol.* (Leipzig), 37: 300–332.
Fröhlich, A. 1940. Page 722 in *The Hypothalamus and Central Levels of Autonomic Function.* Proceedings of the Association for Research in Nervous and Mental Disease, vol. 20, Baltimore: Williams and Wilkins.
Fromman, C. 1864. Über die Färbung der Binde- und Nervensubstanz des Rückenmarkes durch Argentummtricum und über die Struktur der Nervenzellen. *Arch. Pathol. Anat. Physiol. Klin. Med.,* 31: 129–151.
Frost, R. 1940. The Master Speed. *Complete Works of Robert Frost.* New York: Holt Rinehart and Winston.
Fulton, J. F. 1932. New horizons in physiology and medicine: the hypothalamus and visceral mechanisms. *New Eng. J. Med.*, 207: 60–68.
Fulton, J. F. 1938. *Physiology of the Nervous System.* London: Oxford University Press.
Fulton, J. F. 1949. *Functional Localization in the Frontal Lobes and Cerebellum.* Oxford: Oxford University Press.
Fulton, J. F. 1951. *Frontal Lobotomy and Affective Behavior. A Neurophysiological Analysis.* New York: Norton.
Fulton, J. F. 1953. Discussion. *Epilepsia,* 2: 77.
Fulton, J. F., Comp. and completed by L. G. Wilson. 1966. *Selected Readings in the History of Physiology.* 2nd ed., Springfield, IL: C. C Thomas.
Fuster, J. M. 1980. *The Prefrontal Cortex: Anatomy, Physiology and Neuropsychology.* New York: Raven.
Fuster, J. M. and A. A. Uyeda. 1962. Facilitation of tachiscopic performance by stimulation of midbrain tegmental points in the monkey. *Exp. Neurol.* 6: 384–406.
Galaburda, A. M., M. LeMay, T. L. Kemper, and N. Geschwind. 1978. Right–left asymmetries in the brain. *Science*, 199: 852–856.
Galambos, R. 1961. A glia-neural theory of brain function. *Natl. Acad. Sci. Proc.* (Washington), 47: 129–136.
Galambos, R. 1976. Hallowell Davis: Father of the AER. Pages 381–391 in S. K. Hirsh, D. H. Eldridge, and I. J. Hirsh, eds., *Hearing and Davis, Essays Honoring Hallowell Davis.* St. Louis, Missouri: Washington University Press.
Galen. 1956. *On Anatomical Procedures. . . . De anatomicis administrationibus.* Translation. . .by C. Singer. Publication 7 of The Wellcome Historical Medical Museum, New Series. London: Oxford University Press.
Galen. 1968. *On the Usefulness of the Parts of the Body. De usu partium.* 2 vol. Translated from the Greek. . .by M. T. May. Ithaca, New York: Cornell University Press.
Gall, F. J. 1835. *On the functions of the Brain and of Each of its Parts....* 6 vols. Translated by W. Lewis, Jr. Boston: Marsh, Capen, and Lyon.
Gall, F. J. and J. C. Spurtzheim. 1809. *Recherches sur le système nerveux en général, et sur celui du cerveau en particulier.* Paris: Schoell and Nicolle.
Gall, F. J. and G. Spurzheim. 1810–1819. *Anatomie et physiologie du système nerveux en géné_al, et du cerveau en particulier.* Paris: F. Schoell.

Gall, [F. J.], Vimont, and Broussais. 1838. *On the functions of the cerebellum*, translated from the French by George Combe.... Edinburgh: MacLachlan & Stewart.

Galton, C., Trans., *see* Ecker, A.

Galvani, L. 1791. *De viribus electricitatis in motu musculari commentaris*. Bologna: Bononien Sci. Art. Instit. Acad. Comm., 7: 363–418. Translated by R. M. Green, Commentary on the Effect of Electricity on Muscular Motion. Cambridge: Licht, 1953.

Ganong, W. F. 1978. The brain and the endocrine system: a memoir. Pages 187–200 in J. Meites, B. T. Donovan, and S. M. McCann, eds., *Pioneers in Neuroendocrinology II*. New York: Plenum.

Gardner, D. 1992. A time integral of membrane current in nerve. *Physiol. Rev.*, 72 Suppl.: S1–S3.

Garey, L. J. 1976. Synaptic organization of afferent fibres and intrinsic circuits in the neocortex. Pages 2A–56–85 in A. Rémond, ed., *Handbook of Electroencephalography and Clinical Neurophysiology*, Vol. 2. Amsterdam: Elsevier Scientific Publishing Co.

Garner, R. L. 1896. *Gorillas & Chimpanzees*. London: Osgood, Mc Ilvaine & Co.

Gasser, H. S. 1952. Discussion of B. Franbenhaeuser, The hypothesis of saltatory conduction. *Cold Spring Harb. Sym. Quant. Biol.*, 17: 32–36.

Gasser, H. S. and H. S. Newcomer. 1921. Physiological action currents in the phrenic nerve. An application of the thermionic vacuum tube to nerve physiology. *Am. J. Physiol.*, 57: 1–26.

Gazzaniga, M. S. 1981. 1981 Nobel Prize for physiology or medicine. *Science*, 214: 517–518.

Gegenbaur, C. 1878. *Elements of Comparative Anatomy*. Translated by F. J. Bell. London: MacMillan.

Gennari, F. 1782. *De peculiari structura cerebri nonnullisque ejus morbis*. Parma: Ex Regio Typographeo.

Gerard, R. W. 1975. The minute experiment and the large picture. Pages 456–474 in F. G. Worden, J. P. Swazey, and G. Adelman, eds., *The Neurosciences: Paths of Discovery*. Cambridge, MA: MIT Press.

Gerard, R. W. and W. H. Marshall. 1933. Nerve conduction velocity and equilibration. *Am. J. Physiol.* 104: 575–585.

Gerard, R. W., W. H. Marshall, and L. J. Saul. 1936. Electrical activity of the cat's brain. *Arch. Neurol. Psychiat.* (Chicago), 36: 675–735.

Gerebtzoff, M.-A. 1940. Notes anatomiques. *J. Belge Neurol. Psychiat.* 8: 401–406.

Gerebtzoff, M.-A. 1940. Recherches sur la projection corticale du labyrinthe. I. Des effets de la stimulation labyrinthique sur l'activité électrique de l'écorce cérébrale. *Arch. Int. Physiol.*, 50: 59–99.

Geren, B. B. 1954. The formation from the Schwann cell surface of myelin in the peripheral nerve of chick embryos. *Exp. Cell Res.*, 7: 558–562.

Gerlach, J. V. 1858. *Mikroscopische Studien aus dem Gebiete der menschlichen Morphologie*. Erlangen: Enke.

Geschwind, N. 1965. Disconnexion syndromes in animals and man. *Brain*, 88 Part II: 237–294. Part III: 585–644.

Geschwind, N. and W. Levitsky. 1968. Human brain: left–right asymmetries in temporal speech region. *Science*, 161: 186–187.

Gibbs, F. A. and E. L. Gibbs. 1936. The convulsion threshold of various parts of the cat's brain. *Arch. Neurol. Psychiat.*, 35: 109–116.

Gibbs, F. A., E. L. Gibbs, and W. G. Lennox. 1938. Cerebral dysrhythmias of epilepsy. *Arch. Neurol. Psychiat.*, 39: 298–314.

Giblin, D. 1986. The Third "George Dawson Memorial Lecture." *ASCEP Bull.*, 4: 3–11.

Glaser, G. H. 1967. Limbic epilepsy in childhood. *J. Nerv. Ment. Dis.*, 144: 391–397.

Glick, S. D., D. A. Ross, and L. B. Hough. 1982. Lateral asymmetry of neurotransmitters in human brain. *Brain Res.*, 234: 53–63.

Glick, S. D., T. P. Jerussi, D. H. Waters, and J. P. Green. 1974. Amphetamine-induced changes in striatal dopamine and acetylcholine levels and relationship to rotation (circling behavior) in rats. *Biochem. Pharmacol.*, 23: 3223–3225.

Goddard, G. V. 1967. Development of epileptic seizures through brain stimulation of low intensity. *Nature*, 214: 1020–1021.

Goddard, G. V., D. C. McIntyre, and C. K. Leech. 1969. A permanent change in brain function resulting from daily electrical stimulation. *Exp. Neurol.*, 25: 295–330.

Goddard, G. V., B. L. McNaughton, R. M. Douglas, and C. A. Barnes. 1976. Synaptic change in the limbic system.... Pages 355–368 in K. E. Livingston and O. Hornykiewicz, eds., *Limbic Mechanisms: the Continuing Evolution of the Limbic System Concept*. New York: Plenum.

Goldmann, E. E. 1909. Die aussere ind innere Sekretion des gesunden und kranken Organismus im Lichte der "vitalen Farbung." *Beitr. Klin. Chirurg.*, 64: 192–265.

Goldmann, E. E. 1913. *Vitalfärbung am Zentralnervensystem; Beitrag zur Physico-pathologie des Plexus clorioideus und der Hirnhäute*. Berlin: G. Reimer.

Goldstein, G. W. and A. L. Betz, 1983. Recent advances in understanding brain capillary function. *Ann. Neurol.*, 14: 389–395.

Goldstein, K. 1970. Carl Wernicke (1848–1904). Pages 531–535 in W. Haymaker and F. Schiller, eds., *The Founders of Neurology*, 2nd ed. Springfield, IL: C. C Thomas.

Golgi, C. 1883. Recherches sur l'histologie des centres nerveux. *Arch. Ital. Biol.*, 3: 285–317, 4: 92–123.

Golgi, C. 1886. *Studi sulla fina anatomia degli organi centrali del sistema nervoso*. Milano: Hoepli.

Golgi, C. 1898. Sur la structure des cellules nerveuses des ganglions spinaux. *Arch. Ital. Biol.*, 30: 278–286.

Golgi, C. 1908. La doctrine du neurone. Théorie et faits. Pages 1–31 in *Les Prix Nobel en 1906*. Stockholm: Norstedt.

Goltz, F. L. 1869. *Beiträge zur Lehre von den Functionen der Nervencentren des Frosches*. Berlin: Hirschwald.

Goltz, F. L. 1888. Über die Verrrichtungen des Grosshirns. *Pflügers Arch.*, 42: 419–467. Also transl. by G. v. Bonin, 1960.

Goodman, L.S. and A. Gilman. 1970. *The Pharmacological Basis of Therapeutics*. 4th ed. New York: Macmillan.

Gorski, R., J. H. Gordon, J. E. Shryne, and A. M. Southam. 1978. Evidence for a morphological sex difference within the medial preoptic area of the rat brain. *Brain Res.*, 148: 333–346.

Gould, S. J. 1977. *Ontogeny and Phylogeny*. Cambridge, MA: Harvard University Press.

Gould, S. J. 1995. *Dinosaur in a Haystack. Reflections in Natural History*. New York: Crown Publishing Co.

Grafstein, B. 1971. Transneuronal transfer of radioactivity in the central nervous system. *Science*, 172: 177–179.

Graham, R. C., Jr. and M. J. Karnovsky. 1966. The early stages of absorption of injected horseradish peroxidase in the proximal tubules of the mouse kidney: Ultrastructural cytochemistry by a new technique. *J. Histochem. Cytochem.*, 14: 291–302.
Granit, R. 1966. *Charles Scott Sherrington: An Appraisal.* London: Nelson.
Granit, R. 1981. Tribute to Moruzzi. Pages 459–461 in O. Pompeiano and C. Ajmone Marsan, eds., *Brain Mechanisms of Perceptual Awareness and Purposeful Behavior*, International Brain Research Organization Monograph Series. Vol. 8. New York: Raven.
Grass, A. M. 1980. Interview 11 November 1980 CON GRA in the Neuroscience History Archives, Brain Research Institute, University of California, Los Angeles.
Grass, A. M. 1984. *The Electroencephalic Heritage.* Quincey, MA: Grass Instrument Co.
Gratiolet, P. 1854. *Memoire sur les pliés cérébreaux de l'homme et des primates.* Paris: Bertrand.
Gray, E. G. 1959. Axo-somatic and axodendritic synapses of the cerebral cortex: an electron microscope study. *J. Anat.*, 93: 420–433.
Gray, E. G. 1961. Ultra-structure of synapses of cerebral cortex and certain specialization neuroglial membranes. Pages 54–73 in J. D. Boyd, F. R. Johnson, and J. D. Lever, eds., *Electron Microscopy in Anatomy.* London: Edward Arnold.
Green, J. D. 1956. *Hypothalamic-Hypophysial Interrelations.* Springfield, IL: C. C Thomas.
Green, J. D. 1959. Some recent electrophysiological and electron microscope studies of Ammon's horn. Pages 266–271 in D. B. Tower and J. P. Schadé, eds., *Structure and Function of the Cerebral Cortex.* Proceedings of the Second International Meeting of Neurobiologists, Amsterdam. Amsterdam: Elsevier.
Green, J. R. 1985. The beginnings of cerebral localization and neurological surgery. *Barrow Neurol. Inst. Q.*, 1: 12–28.
Green, R. M., Trans., *see* Galvani, L.; Pupilli, G. C.
Greep, R. O. 1974. History of research on anterior hypophysial hormones. Pages 1–27 in E. Knobil and W. H. Sawyer, eds., *Handbook of Physiology, Section 7: Endocrinology, Vol. IV, the Pituitary Gland and its Neuroendocrinological Control, Part 2.* Washington, D.C.: American Physiological Society.
Gross, C. G. 1995. Aristotle on the brain. *Neuroscientist,* 1: 245–250.
Gross, C. G. and M. S. A. Graziano. 1995. Multiple representations of space in the brain. *Neuroscientist,* 1: 43–50.
Grünbaum, A. S. F. and C. S. Sherrington. 1903. Observations on the physiology of the cerebral cortex of the anthropoid apes. *Proc. R. Soc. Lond.*, 72: 152–155.
Grundfest, H. 1957. Excitation at synapses. *J. Neurophysiol.*, 20: 316–328.
Gudden, B. v. 1870. Experimentaluntertsuchungen über das peripherische und centrale Nervensystem. *Arch. Psychiat.*, (Berlin) 2: 693–723.
Guillemin, R. 1955. A re-evaluation of acetylcholine, adrenaline, nor-adrenaline and histamine as possible mediators of the pituitary adrenocorticotrophic activation by stress. *Endocrinology*, 56: 248–255.
Guillemin, R. 1978. Pioneering in neuroendocrinology, 1952–1969. Pages 220–239 in J. Meites, B. T. Donovan, and S. M. McCann, eds., *Pioneers in Neuroendocrinology II.* New York: Plenum.
Guillemin, R., E. Yamazaki, M. Jutisz, and E. Sakiz. 1962. Présence dans un extrait de tissus hypothalamiques d'une substance stimulant la sécrétion de l'hormone hypophysaire thyréotrope (TSH). Première purification par filtration sur gel Sephadex. *C. R. Acad. Sci.*, 255: 1018–1020.
Haeckel, E. H. P. A. 1866. *Generelle Morphologie der Organism. Allgemeine Grundzüge der organischen Formen-Wissenschaft, mechanisch begründet durch die von Charles Darwin reformirte Descendeztheorie.* Vol. 2. Berlin: G. Reimer.
Haeckel, E. H. P. A. 1879. *The Evolution of Man: a Popular Exposition of the Principal Points of Human Ontogeny and Phylogeny.* Vol. 1. Translated from the German of the 1866 edition. New York: Appleton.
Haeckel, E. H. P. A. 1891. *Anthropogenie, oder, Entwickungsgeschichte des Menchen: Keimes und Stammes-Geschichte.* Leipsig: W. Engelmann.
Haigh, B., Trans., *see* Blinkov, S. M. and I. I. Glezer.
Haines, D. E. 1991. The contributors to Volume 1 (1891) of *The Journal of Comparative Neurology:* C. L. Herrick, C. H. Turner, H. R. Pemberton, B. G. Wilder, F. W. Langdon, C. J. Herrick, C. von Kupffer, O. S. Strong, T. B. Stowell. *J. Comp. Neurol.*, 314: 9–33.
Halstead, W. C. 1947. *Brain and Intelligence. A Quantitative Study of the Frontal Lobes.* Chicago: University of Chicago Press.
Hamilton, C. R. and M. S. Gazzaniga. 1964. Lateralization of learning of colour and brightness discrimination following brain bisection. *Nature*, 201: 220.
Hamilton, C.R. and B. A. Vermeire. 1988. Complementary hemispheric specialization in monkeys. *Science*, 242: 1691-4.
Hamilton, D. [J.] 1884–1885. On the corpus callosum in the adult human brain. *J. Anat. Physiol.*, 19: 385–414.
Hamilton, D. J. 1885. On the corpus callosum of the embryo. *Brain,* 8: 145–163.
Hannover, A. 1840. Die Chromsäure, ein vorzügliches Mittel bei mikroskopischen Untersuchungen. *Arch. Anat. Physiol. Wissensch. Med.*, 549–558.
Harlow, J. M. 1869. Recovery from the passage of an iron bar through the head. *Publ. Mass. Med. Soc. Boston*, 2: 327–346.
Harrington, A. 1987. *Medicine, Mind and the Double Brain.* Princeton: Princeton University Press.
Harrington, A. 1991. *So Human a Brain: Knowledge and Values in the Neurosciences.* Boston: Birkhäuser.
Harris, G. W. 1944. *The Secretomotor Innervation and Actions of the Neurohypophysis; an Investigation Using the Method of Remote Control Stimulation.* Thesis, Cambridge University.
Harris, G. W. 1948. Electrical stimulation of the hypothalamus and the mechanism of neural control of the adenohypophysis. *J. Physiol.* (London), 107: 418–429.
Harris, G. W. 1958. The reticular formation, stress, and endocrine activity. Pages 207–221 in H. H. Jasper, R. S. Knighton, W. C. Noshay, and R. T. Costello, eds., *Reticular Formation of the Brain.* Boston: Little, Brown & Co.
Harris, G. W. 1972. Humours and hormones. The Sir Henry Dale Lecture for 1971. *J. Endocrinol.*, 53: ii–xxiii.

Harris, G. W. and D. Jacobsohn. 1952. Functional grafts of the anterior pituitary gland. *Proc. R. Soc. Lond. B*, 139: 263–276.

Harrison, R. G. 1907. Observations on the living developing nerve fiber. *Anat. Rec.*, 1: 116–118.

Harrison, R. G. 1908. Embryonic transplantation and development of the nervous system. *Anat. Rec.*, 2: 385–410.

Haymaker, W. 1970. Pio del Rio-Hortega (1882–1945). Pages 154–160 in W. Haymaker and F. Schiller, Comp. and Eds., *The Founders of Neurology*, 2nd ed., Springfield, IL: C. C Thomas.

Haymaker, W. and F. Schiller, Eds. *The Founders of Neurology*, 2nd ed., Springfield, IL: C. C. Thomas, 1970.

Head, H. 1915. Hughlings Jackson on aphasia and kindred affections of speech. *Brain* , 38, Parts I and II: 1–190.

Head, H. 1920. Aphasia: an historical review. *Brain* 43: 390–411.

Head, H. 1926. *Aphasia and Kindred Disorders of Speech*, 2 vol. Cambridge, UK: The University Press.

Head, H. and G. Holmes. 1911. Sensory disturbances from cerebral lesions. *Brain*. 34: 102–254.

Heimer, L. 1987. Neuroanatomical tracing methods. Pages 766–769 in G. Adelman, ed., *Encyclopedia of Neuroscience*, 2 vols. Boston: Birkhäuser.

Heller, H. 1940. The action of the antidiuretic principle of posterior pituitary extracts on the urine excretion of anaesthetized animals. *J. Physiol.* (London), 98: 405–418.

Heller, H. 1974. History of neurohypophysial research. Pages 103–117 in E. Knobil and W. H. Sawyer, eds., *Handbook of Physiology, Section 7: Endocrinology, Vol. IV, The Pituitary Gland and Its Neuroendocrine Control, Part I.* Washington, D.C.: American Physiological Society.

Hendrickson, A. 1969. Electron microscopic radioautography. identifiction of origin of synaptic terminals in normal nervous tissue. *Science*, 165: 194–196.

Henschen, S. E. 1893. On the visual path and centre. *Brain*, 16: 170–180.

Hernandez-Peon, R. 1960. Neurophysiological correlates of habituation and other manifestations of plastic inhibition (internal inhibition). Pages 101–114 in H. H. Jasper and G. D. Smirnov, eds., Moscow Colloquium on Electroencephalography of Higher Nervous Activity. *Electroenceph. Clin. Neurophysiol.*, Suppl. 13.

Herrick, C. J. 1914a. The cerebellum of Nicturus and other urodele Amphibia. *J. Comp. Neurol.*. 24: 1–29.

Herrick, C. J. 1914b. The medulla oblongata of larval Amblystoma. *J. Comp. Neurol.*. 24: 343–427.

Herrick, C.J. 1924. Origin and evolution of the cerebellum. *Arch. Neurol. Psychiat.* 11: 621–652.

Herrick, C.J. 1926. *Brains of Rats and Men. A Survey of the Origin of the Cerebral Cortex.* Chicago: University of Chicago Press.

Herrick, C.J. 1933. The functions of the olfactory parts of the cerebral cortex. *Proc. Nat. Acad. Sci.*, 19: 7–14.

Herrick, C.J. 1961. Nervous mechanisms in behavior. *Fed. Proc.*, 20: 628–631.

Herrick, C.J. and E.C. Crosby. 1918. *A Laboratory Outline of Neurology.* Philadelphia: W.E. Saunders.

Herrick, C.L., Trans., *see* Lotze, H.

Herrick, C.L. 1885. *Outlines of Psychology: Dictations from Lectures by Hermann Lotze.* Trans. with the addition of a chapter on the anatomy of the brain. Minneapolis: S.M. Williams.

Herrick, C.L. 1891a. The problems of comparative neurology. *J. Comp. Neurol.*, 1: 93–105.

Herrick, C.L. 1891b. The evolution of the cerebellum. *Science*, 18: 188–189.

Herrick, C.L. 1891c. Contributions to the comparative morphology of the central nervous system. I. Illustrations of the architectonic of the cerebellum. *J. Comp. Neurol.*, 1: 5–14.

Herrick, C.L. 1892a. Neurologists and neurological laboratories — No. 1. Professor Gustav Fritsch. With portrait. *J. Comp. Neurol.*, 2: 84–88.

Herrick, C.L. 1892b. Contribution to the morphology of the brain of bony fishes, II. *J. Comp. Neurol.*, 2: 21–72.

Herring, P.T. 1908. The histological appearances of the mammalian pituitary body. *Q. J. Exp. Physiol.*, 1: 121–159.

Herrlinger, R. 1971. *Die Nobelpreisträger der Medizin: ein Kapitel aus der geschichte der Medizin.* 2nd ed. Gräfelfing vor München: Heinz Moos Verlag.

Hess, H.H. and A. Pope. 1960. Intralaminar distribution of cytochrome oxydase activity in human frontal isocortex. *J. Neurochem.*, 5: 207–217.

Hill, A., Trans., *see* Obersteiner, A.

Hinsey, J.C. and J.E. Markee. 1933. Pregnancy following bilateral section of the cervical sympathetic trunk in the rabbit. *Soc. Exp. Biol. Med. Proc.*, 31: 270–271.

Hirsh, S.K., D.H. Eldredge, I.J. Hirsh, and S.R. Silverman, Eds. 1976. *Hearing and Davis, Essays Honoring Hallowell Davis.* St. Louis, Missouri: Washington University Press.

His, W. 1889. Die Neuroblasten und duren Entstehung im embryonalen Mark. *Arch. Anat. Physiol., Anat. Abt.*: 249–300.

His, W. 1893. Vorschläge zur Einteilung des Gehirns. *Arch. Anat. Physiol., Anat. Abt.*: 172–179.

His, W. 1895. Die anatomische nomenclatur. Nomina anatomica. *Arch. Anat. Physiol., Abt. Anat. Ent. Suppl.*: 1–180.

Hitzig, E. 1874. *Untersuchungen über das Gehirn. Abhandlungen physiologischen und pathologischen. Inhalts.* Berlin: Hirschwald.

Hobson, J. A. and M. A. B. Brazier, eds. 1980. *The Reticular System Revisited: Specifying Function for a Nonspecific System.* New York: Raven Press.

Hodgkin, A. L. and A. F. Huxley. 1939. Action potentials recorded from inside a nerve fibre. *J. Physiol.* (London), 144: 710–711.

Hodgkin, A. L. and A. F. Huxley. 1945. Resting and action potentials in single nerve fibres. *J. Physiol.* (London), 104: 176-195.

Hodgkin, A. L. and A. F. Huxley. 1952a. Currents carried by sodium and potassium ions through the membrane of the giant axon of *Logido. J. Physiol.* (London), 116: 449–472.

Hodgkin, A. L. and A. F. Huxley. 1952b. A quantitative description of membrane-current and its application to conductance and excitation in nerve. *J. Physiol.* (London), 117: 500–544.

Hodgkin, A. L., A. F. Huxley, and B. Katz. 1952. Ionic currents underlying activity in the giant axon of the squid. *Arch. Sci. Physiol.*, 3: 129–150.

Hodgkin, T., and J. J. Lister. 1827. Notice of some microscopic observations of the blood and animal tissues. *Phil. Mag.*, II: 130–138.

Hodgson, E. S. 1977. The evolutionary origin of the brain. Pages 23–25 in S. J. Dimond and D. A. Blizard, eds., *Evolu-*

tion and Lateralization of the Brain. Ann. New York Acad. Sci., vol. 299. New York: New York Academy of Sciences.

Hohlweg, W. 1975. The regulatory centers of endocrine glands in the hypothalamus. Pages 160–172 in J. Meites, B. T. Donovan, and S. M. McCann, eds., *Pioneers in Neuroendocrinology.* New York: Plenum Press.

Hohlweg, W. and K. Junkmann. 1932. Die hormonal-nervöse Regulierung der Funktion des Hypophysenvorderlappens. *Klin. Wochenschr.*, 11: 321–323.

Holland, H. 1839. *Medical Notes and Reflections.* Philadelphia: Haswell, Barrington, and Haswell.

Holloway, R. L. 1968. The evolution of the primate brain; some aspects of quantitative relations. *Brain Res.*, 7: 121–172.

Holloway, R. L. 1974. The casts of fossil hominid brains. *Sci. Am.*, 231(1): 106–115.

Holloway, R. L. 1976. Paleoneurological evidence for language origins. Pages 330–348 in S. R. Harnad, H. D. Steklis, and J. Lancaster, eds., *Origins and Evolution of Language and Speech,* Ann. New York Acad. Sci., vol. 280. New York: New York Academy of Sciences.

Holloway, R. L. 1981. Exploring the dorsal surface of the hominoid brain endocasts by stereoplotter and discriminant analysis. *Phil. Trans. R. Soc. Lond. B.*, 292: 155–166.

Holmes, G. 1918. Disturbances of visual orientation. *Br. J. Ophthalmol.*, 2: 449–468, 506–516.

Holmes, G. 1954. *The National Hospital Queen Square 1860–1948.* Edinburgh: E. & S. Livingston.

Holmes, G. and W. T. Lister. 1916. Disturbances of vision from cerebral lesions, with special reference to the cortical representation of the macula. *Brain,* 39: 34–73.

Holtz, P., R. Heise and K. Lüdtke. 1938. Fermentativer Abbau von l-Dioxyphenylalanine (DOPA) durch Nieve. *Arch. Exp. Pathol. Pharmak.*, 191: 87–118.

Horsley, V. A. 1886. Brain-surgery. *Br. Med. J.*, 2: 670–675.

Horsley, V. A. and R. H. Clarke. 1908. The structure and functions of the cerebellum examined by a new method. *Brain*, 31: 45–124.

Horsley, V. A. and E. A. Schafer. 1888. A record of experiments upon the functions of the cerebral cortex. *Phil. Trans. R. Soc. Lond. (B)*, 179: 1–45.

Hotton, N., III. 1978. *The Evidence of Evolution.* The Smithonian Library. New York: American Heritage Publishing Co.

Houssay, B. A. and E. A. Molinelli. 1928. Excitabilité des fibres adrénalinosécrétoires de nerf grand splanchnique; Fréquences seuil et optimum des stimulus. Role de l'ion calcium. *C. R. Soc. Biol.*, 99: 172–174.

Houssay, B. A., A. Biasotti, and R. Samimartino. 1935. Modifications fonctionnelles de l'hypophyse après les lesions infundibulo-tubériennes chez le crapaud. *C. R. Soc. Biol.*, 120: 725–727.

Howell, W. H. 1898. The physiological effects of extracts of the hypophysis cerebri and infundibular body. *J. Exp. Med.*, 3: 245–258.

Howell, W. H. 1906. Vagus inhibition of the heart in its relation to the inorganic salts of the blood. *Am. J. Physiol.*, 15: 280–294.

Hubel, D. H. 1982. Exploration of the primary visual cortex 1955–1978. *Nature,* 299: 515–524.

Hubel, D. H. and T. N. Wiesel. 1959. Receptive fields of single neurones in the cat's striatal cortex. *J. Physiol.* (London), 148: 574–591.

Hubel, D. H. and T. N. Wiesel. 1962. Receptive fields, binocular interaction and functional architecture in the cat's visual cortex. *J. Physiol.* (London), 160: 106–154.

Hubel, D. H. and T. N. Wiesel. 1972. Laminar and columnar distribution of geniculo–cortical fibers in the macaque monkey. *J. Comp. Neurol.*, 146: 421–450.

Hubel, D. H. and T. N. Wiesel. 1979. Brain mechanisms of vision. *Sci. Am.*, 241(3): 150–162.

Hubel, D. H., T. N. Wiesel, and S. LeVay. 1977. Plasticity of ocular dominance columns in monkey striate cortex. *Phil. Trans. R. Soc. Lond. B*, 278: 377–409.

Hughes, A. F. W. 1959. *A History of Cytology.* London: Abelard-Schuman.

Hunter, J. 1825. Case of fungus haematodes of the brain. *Med. Chir. Trans.*, 13: 88–96.

Hunter, J. and H. H. Jasper. 1949. Effects of thalamic stimulation in unanesthetized animals: the arrest reaction and petit mal-like seizures, activation patterns and generalized convulsions. *Electroenceph. Clin. Neurophysiol.*, 1: 305–324.

Hunter, R., Trans., *see* Lower, R.

Huschke, E. 1854. *Schaedel, Hirn und Seele des Menschen und der Thiere nach Alter, Geschlecht und Race.* Jena: Frederich Mauke.

Huxley, A. F. 1995. Electrical activity of nerve: the background up to 1952. Pages 3–10 in S. G. Waxman, J. D. Kocsis, and P. K. Stys, *The Axon: Structure, Function and Pathophysiology.* Oxford: Oxford University Press.

Huxley, T., Trans., *see* Kólliker, R.A. v., 1854.

Hydén, H. 1969. Biochemical aspects of learning and memory. Pages 95–125 in K. H. Pribram, ed., *On the Biology of Learning.* New York: Harcourt, Brace & World.

Hyndman, O. R. and W. Penfield. 1937. Agenesis of the corpus callosum. *Arch. Neurol. Psychiat.*, 37: 1251–1270.

Hyvärinen, J. 1982. *The Parietal Cortex of Monkey and Man.* Berlin: Springer-Verlag.

Ingram, W. R. 1975. A personal neuroscientific development with remarks on other events and people. Pages 175–191 in J. Meites, B. T. Donovan and S. M. McCann, eds., *Pioneers in Neuroendocrinology.* New York: Plenum Press.

Ingvar, D. H. and N. A. Lassen. 1961. Quantitative determination of regional cerebral blood flow in man. *Lancet,* ii: 806–807.

Iverson, E. 1947. Some remarks on the terms. *J. Egyptian Archaeol.*, 33: 47–51.

Inouye, T. 1909. *Die Sehstörungen bei Schussverletzung der korticalen Sehsphäre nach Beobachtungen an Verwund-eten der letzten japanischen Kriege.* Leipzig: Engelmann.

Inubushi, S., T. Kobayashi, T. Oshima, and S. Torii. 1978. An intracellular analysis of EEG arousal in cat motor cortex. *Jpn. J. Physiol.* 28: 689–708.

Ironside, R. and M. Guttmacher. 1929. The corpus callosum and its tumours. *Brain*, 52: 442–483.

Isaacson, R. L. and K. H. Pribram. 1975. *The Hippocampus.* 2 vol. New York: Plenum Press.

Ito, M. and M. Yoshida. 1966. The origin of cerebellar-induced inhibition of Deiters neurones. I. Monosynaptic imitation of the inhibitory postsynaptic potentials. *Exp. Brain Res.*, 2: 330–349.

Jackson, J. H. 1866. Notes on the physiology and pathology of language. *Med. Times Gaz.*, I: 659–662.

Jackson, J. H. 1868. Notes on the physiology and pathology of the nervous system, Hemispherical co-ordination. *Med. Times Gaz.*, II: 358–359.

Jackson, J. H. 1874. On the nature of the duality of the brain. *Med. Press Circ.*, i: 19 (Jan. 14), 41 (Jan. 21), 63 (Jan. 28). Reprinted in *Brain*, 38: 80–103, 1915.

Jackson, J. H. 1884a. The Croonian Lectures on evolution and dissolution of the nervous system. Lecture II. *Br. Med. J.*, 1: 660–663.

Jackson, J. H. 1884b. The Croonian Lectures on evolution and dissolution of the nervous system. Lecture III. *Br. Med. J.*, 1: 703–707.

Jackson, J. H. 1888. Discussion of paper read by A. H. Bennett. *Brain*, 10: [288–312], 312–318.

Jackson, J. H. 1889. On the comparative study of diseases of the nervous system. *Br. Med. J.* 2: 355–362.

Jackson, J. H. 1898. Relations of different divisions of the central nervous system to one another and to parts of the body. *Lancet*, 79–87.

Jacobsohn, D. 1975. My way from hypophysectomy to hypophyseal portal vessels (1934–1954). Pages 194–202 in J. Meites, B. T. Donovan, and S. M. McCann, eds., *Pioneers in Neuroendocrinology*. New York: Plenum Press.

Jacyna, L. S. 1982. Somatic theories of mind and the interests of medicine in Britain, 1850 1879. *Med. Hist.*, 26: 233–258.

James, M. R. 1907. *A Descriptive Catalogue of the Manuscripts in the Library of Gonville and Caius College*. Cambridge, England: University Press.

James, W. 1899. *The Principles of Psychology*, 2 vols. New York: Henry Holt.

James, W. 1904. Does consciousness exist? *J. Phil.*, 1: 477 491.

Jasper, H. H. 1949. Diffuse projection systems: the integrative action of the thalamic reticular system. *Electroenceph. Clin. Neurophysiol.* 1: 405–419.

Jasper, H. H. 1958. Recent advances in our understanding of ascending activities of the reticular system. Pages 319–331 in H. H. Jasper, L. D. Proctor, R. S. Knighton, W. C. Noshay, and R. T. Costello, eds., *Reticular Formation of the Brain*. Boston: Little, Brown.

Jasper, H. H. 1991. The International Brain Research Organization, a brief historical survey. Pages v–xiii in *IBRO Membership Directory 1991*. Published for IBRO by Pergamon Press.

Jasper, H. H. and C. Ajmone-Marsan. 1952. Thalamocortical integrating mechanisms. *Res. Publ. Assoc. Nerv. Ment. Dis.*, 30: 495–512.

Jasper, H. H. and G. D. Smirnov, eds. 1960. The Moscow colloquium on electroencephalography of higher nervous activity. *Electroenceph. Clin. Neurophysiol.*, supp. 13: 1–420.

Jasper, H. H., L. D. Proctor, R. S. Knighton, W. C. Noshay, and R. T. Costello, eds. 1958. *Reticular Formation of the Brain*. Henry Ford Hospital International Symposium. Boston: Little, Brown.

Jaynes, J. 1976. *On the Origin of Consciousness in the Breakdown of the Bicameral Mind*. Princeton: Princeton University.

Jefferson, G. 1935. John Hughlings Jackson. *Manchester Univ. Med. Sch. Gaz.*, 14(5), no pagination.

Jefferson, G. 1957. Sir Victor Horsley, 1857–1916. Centenary Lecture. *Br. Med. J.*, 1: 903–910.

Jefferson, G. 1958. Reticular formation and clinical neurology. Pages 729–738 in H. H. Jasper, L. D. Proctor, R. S. Knighton, W. C. Noshay, and R. T. Costello, eds., *Reticular Formation of the Brain*. Boston: Little, Brown.

Jefferson, G. and R. T. Johnson. 1950. The cause of loss of consciousness in posterior fossa compressions. *Folia Psychiat. Neurol. Neurochir. Neerl.*, 53: 306–319.

Jefferson, M. 1952. Altered consciousness associated with brain-stem lesions. *Brain*, 75: 55–67.

Johanson, D. C. and M. A. Edey. 1981. *Lucy, the Beginnings of Humankind*. New York: Simon & Schuster.

Jones, E. G. 1984a. Laminar distribution of cortical efferent cells. Pages 521–553 in A. Peters and E. G. Jones, eds., *Cerebral Cortex, vol. 1: Cellular Components of the Cerebral Cortex*. New York: Plenum Press.

Jones, E. G. 1984b. History of cortical cytology. Pages 1–32 in A. Peters and E. G. Jones, eds., *Cerebral Cortex, vol. 1: Cellular Components of the Cerebral Cortex*. New York: Plenum Press.

Jones, E. G. 1985. *The Thalamus*. New York: Plenum Press.

Jouvet, M. and F. Michel. 1958. Recherches sur l'activité électrique cérébrale au cours du sommeil. *C. R. Soc. Biol.*, 152: 1167–1170.

Joynt, R. J. 1974. The corpus callosum: history of thought regarding its function. Pages 117–125 in M. Kinsbourne and W.L. Smith, eds., *Hemispheric Disconnection and Cerebral Function*. Springfield, IL: C.C Thomas.

Joynt, R. J. and A. I. Benton. 1964. The memoir of Marc Dax on aphasia. *Neurology*, 14: 851–854.

Jung, R. 1975. Some European neuroscientists: a personal tribute. Pages 476–511 in F. G. Worden, J. P. Swazey, and G. Adelman, eds., *The Neurosciences: Paths of Discovery*. Cambridge, MA: MIT Press.

Kaada, B. R. 1951. Somato-motor, autonomic and electrocorticographic responses to electrical stimulation of 'rhinencephalic' and other structures in primates, cat and dog. *Acta Physiol. Scand.* 24, Suppl. 83: 1–258.

Kaau-Boerhaave, A. 1745. *Impetum faciens dictum Hippocrati*. Leyden: S. Luchtmans.

Kappers, C. U. A., G. C. Huber, and E. C. Crosby. 1936. *The Comparative Anatomy of the Nervous System of Vertebrates, Including Man*. 2 vol. New York: Macmillan.

Karlsson, U. 1966. Comparison of the myelin period of peripheral and central origin by electron microscope. *J. Ultrastruc. Res.*, 15: 451–468.

Kato, G. 1924. *The Theory of Decrementless Conduction in Narcotized Regions of Nerve*. Tokyo: Nankodo.

Katz, B. 1982. Stephen William Kuffler. 24 August 1913–11 October 1980. *Biog. Mem. Fell. R. Soc.* 28: 225–259. Reprinted in U. J. McMahan, Comp., *Steve: Remembrances of Stephen W. Kuffler*. Sunderland, MA: Sinauer Associates.

Kellie, G. 1824. An account of the appearances observed in the dissection of two or more individuals. . .with some reflections on the pathology of the brain. Parts I, II. *Trans. Med. Chir. Soc., Edinburgh,* 1: 84–124, 125–169.

Kety, S. S. 1982. The cerebral circulation. Pages 703–742 in A. P. Fishman and D.W. Richards, eds., *Circulation of the Blood. Men and Ideas*. Bethesda, MD: Americn Physiological Society.

Kety, S. S. and C. F. Schmidt. 1945. The determination of cerebral blood flow in man by the use of nitrous oxide in low concentrations. *Am. J. Physiol.*, 143: 53–66.

Kety, S. S. and C. F. Schmidt. 1948. The effects of altered arterial tensions of carbon dioxide and oxygen on cerebral blood flow and cerebral oxygen consumption in normal young men. *J. Clin. Invest.*, 27: 484–492.

Kety, S. S., W. M. Landau, W. H. Freygang, Jr., L. P. Rowland, and L. Sokoloff. 1955. Estimation of regional circulation in the brain by uptake of an inert gas. *Fed. Proc.*, 14: 85.

Key, G. and A. Retzius. 1875–1876. *Studien in der Anatomie des Nervensystems und des Bindegewebes.* Stockholm: P. A. Norstedt.

Kibjakow, A. W. 1933. Über humorale Übertragung der Erregung von einem Neuron auf das andere. *Pflügers Arch.* 232: 432–443.

Kimura, D. 1961. Cerebral dominance and the perception of verbal stimuli. *Can. J. Psychol.*, 15: 166–171.

Kinomura, S., J. Larsson, B. Gulyás, and P. E. Roland. 1996. Activation by attention of the human reticular formation and thalamic intraluminar nuclei. *Science,* 271: 512–515.

Kisch, B. 1954. Forgotten leaders in modern medicine. *Trans. Am. Philos. Soc.* series 2, 44, pt. 2: 139–317.

Kleijn, A. de, and R. Magnus. 1920. Über die unabhängigkeit der Labyrinthreflexe vom Kleinhirn und über die Lage der Zentren für die Labyrinthreflexe im Hirnstamm. *Pflügers Arch.,* 178: 124–178.

Klüver, H. and P. C. Bucy. 1938. An analysis of certain effects of bilateral temporal lobectomy in the rhesus monkey, with special reference to "psychic blindness." *J. Psychol.*, 5: 33–54.

Knoefel, P. K. 1984. *Felice Fontana Life and Works.* Trento: Societá di Studi Trentini di Scienze Storiche.

Kolle, K. 1959. *Grosse Nrosservenärzte; Lebensbilder.* Vol. 2. Stuttgart: Thieme.

Kölliker, R. A. v. 1849. Neurologische Bermerkungen. *Z. Wiss. Zool.*, 1: 135–163.

Kölliker, R. A. v. 1852. *Handbuch der Gewebelehre des Menchen für Aerzte und Studirende.* Leipzig: Engelmann.

Kölliker, R. A. v. 1854. *Manual of Human Histology.* Vol. 2. Translated and edited by G. Busk and T. Huxley. London: Sydenham Society.

Koshtoyants, K. 1960. I.M. Sechenov (1829–1905). Pages 7–27 in G.Gibbons, ed., *I. Sechenov, Selected Physiological and Psychological Works.* Moscow: Foreign Languages Publishing House.

Krashen, S. D. 1976. Cerebral asymmetry. Pages 157–191 in H. Whitaker and H.A. Whitaker, eds., *Studies in Neurolinguistics.* New York: Academic.

Krause, F. 1912. *Surgery of the Brain and Spinal Cord Based on Personal Experiences*, 3 vol. Translated by M. Thorek. New York: Rebman.

Krieg, W. J. S. 1970. Bernard Luys, 1828–1897. Pages 54–57 in W. Haymaker and F. Schiller, *The Founders of Neurology.* Springfield, Illinois: C.C Thomas.

Krieg, W. J. S. 1975. *Stereotaxy.* Evanston, IL: Brain Books.

Krivánek, J. and O. Buresova. 1972. Cortical acetylcholinesterase and handedness in rats. *Experientia*, 28: 291–292.

Kruger, L. and T. A. Woolsey. 1990. Rafael Lorente de Nó: 1902–1990. *J. Comp. Neurol.* 300: 1–4.

Kuffler, S. W. 1946. A second motor nerve system to frog skeletal muscle. *Proc. Soc. Exp. Biol. Med.*, 63: 21–23.

Kuffler, S. W. 1952. Neurons in the retina: organization, inhibition and excitation problems. *Cold Spring Harbor Sym. Quant. Biol.*, 17: 281–292.

Kuffler, S. W. 1953. Discharge patterns and functional organization of mammalian retina. *J. Neurophysiol.*, 16: 37–68.

Kuffler, S. W. and J. G. Nicholls. 1966. The physiology of neurological cells. *Ergeb. Physiol.*, 57: 1–90.

La Valette St. George, A. v. 1867. Über die Genese der Samenkörper. *Arch. Mir. Anat.*, 3: 263–273.

Lacey, J. I. 1985. The visceral systems in psychology. Pages 721–736 in S. Koch and D.E. Leary, eds., *A Century of Psychology as Science.* New York: McGraw-Hill.

Lain Entralgo, P. 1978. Ramón y Cajal 1852–1934. Madrid: Expediéntes Administrativos de Grandes Españoles.

Langley, J. N. 1905. On the reaction of cells and nerve-endings to certain poisons, chiefly as regards the reaction of striated muscle to nicotine and curari. *J. Physiol.* (London), 33: 374–413.

Lasek, R. J., L. Phillips, M. J. Katz, and L. Autilio-Gambette. 1985. Function and evolution of neurofilament proteins. Pages 462–478 in E. Wang, D. Fischman, R. K. H. Liem, and T.-T. Sun, eds., *Intermediate Filaments.* Ann. New York Acad. Sci., 455.

Lashley, K. S. 1917. Modifiability of the preferential use of the hands in the rhesus monkey. *J. Animal Behav.*, 7: 178–186.

Lashley, K.S. 1938. The thalamus and emotion. *Psychol. Rev.*, 45: 42–61.

Lashley, K.S. 1950. In search of the engram. *Symp. Soc. Exp. Biol.*, 4: 454–482.

Lassen, N.A. and D.H. Ingvar. 1972. Radioisotopic assessment of regional blood flow. Pages 376–409 in E.J. Potchen and V.R. McCready, eds., *Progress in Nuclear Medicine,* v. 1.

Latta, H. and J.F. Hartmann. 1950. Glass knives for EM tissue section. *Proc. Soc. Exp. Biol. Med.*, 74: 436–439.

LaVail, J.H. and M.M. LaVail. 1972. Retrograde axonal transport in the central nervous system. *Science*, 176: 1416–1417.

Laycock, T. 1845. On the reflex function of the brain. *Br. For. Med. Rev.*, 19: 298–311.

Laycock, T. 1860. *Mind and Brain*, 2 vol. Edinburgh: Sutherland and Knox.

Le Boë, F. de (Sylvius). 1679. Disputationum medicarum IV. Pages 18–21 in *Opera medica. . . .* Amsterdam: D. Elsevirium et A. Wolfgang.

Leake, C.D. 1970. Henry Dale (1875–1968). Pages 282–285 in Haymaker, W. and F. Schiller, Eds., *The Founders of Neurology*, 2nd ed. Springfield, Illinois: C.C Thomas.

Leakey, R.E. 1981. *The Making of Mankind.* New York: E.P. Dutton.

Leão, A.A.P. 1944. Speading depression of activity in the cerebral cortex. *J. Neurophysiol.*, 7: 359–390.

Leeuwenhoeck, A. van. 1675. Microscopical observations... concerning the optic Nerve, communicated to the Publisher in Dutch, and by him made English. *Phil. Trans. R. Soc.*, 10: 378–380.

Leeuwenhoek, A. van. 1677. Mr. Leewenhoek's [*sic*] letter the 14th of May, 1677. *Phil. Trans. R. Soc.*

Leeuwenhoek, A. van. 1719. *Epistolae physiologicae super compluribus naturae arcanis;...* Epistola xxxii, 309–317. Delft: Beman.

LeMay, M. and A. Culebras. 1972. Human brain—morphologic differences in the hemispheres demonstrable by carotid arteriography. *New Engl. J. Med.*, 287: 168–170.

LeMay, M. and N. Geschwind. 1975. Hemispheric differences in the brains of great apes. *Brain, Behav. Evol.*, 11: 48–52.

Lennox, W.G. 1951. Phenomena and correlates of the psychomotor triad. *Neurology*, 1: 357–371.

Lettvin, J.Y., H.R. Maturana, W.S. McCulloch, and W.H. Pitts. 1959.

Leuret, F. et L.P. Gratiolet. 1839. *Anatomie comparée du système nerveux considéré dans ses rapports avec l'intelligence*, vol. I. Paris: Baillière.

Leuret, F. et P. Gratiolet. 1859–1857 [sic]. *Anatomie comparée...Atlas de 32 planches dessignées d'après nature et gravées.* Paris: Baillière.

Levi-Montalcini, R. 1964. Growth control of nerve cells by a protein factor and its antiserum. *Science*, 143: 105–110.

Levi-Montalcini, R. 1975. NGF: An uncharted route. Pages 244–265 in F.G. Worden, J.P. Swazey, and G. Adelman, Eds., *The Neurosciences: Paths of Discovery.* Cambridge, Massachusetts: The MIT Press.

Levy, J. 1974. Psychological implications of bilateral asymmetry. Pages 121–183 in S.J. Dimond and J.G. Beaumont, Eds., *Hemispheric Function in the Human Brain*, London: Elek Science.

Lewin, R. 1984. *Human Evolution, an Illustrated Introduction.* New York: Freeman.

Lewis, E. and H.H. Hess. 1965. Intralaminar distribution of cerebrosides in human frontal cortex. *J. Neurochem.*, 12: 213–220.

Lewis, J. 1981. *Something Hidden: A Biography of Wilder Penfield, 1891–1976.* Toronto: Doubleday Canada.

Lewis, W., Jr., Trans., *see* Gall, F., 1835.

Leyton, A.S.F. and C.S. Sherrington. 1917. Observations on the excitable cortex of the chimpanzee, orang-utan and gorilla. *Q. J. Exp. Physiol.*, 11: 135–222.

Li, C.-L., H. McClennan, and H.H. Jasper. 1952. Brain waves and unit discharges in cerebral cortex. *Science*, 116: 656–657.

Libet, B., E.W. Wright, Jr., B. Feinstein, and D.K. Pearl. 1979. Subjective referral of the timing for a conscious experience. A functional role for the somatosensory specific projection system in man. *Brain*, 102: 193–224.

Lichtheim, L. 1885. On aphasia. *Brain*, 7: 433–484.

Lind, L.R., Trans., *see* Berengario da Carpi, 1535.

Linderstrøm-Lang, K. 1939. Distribution of enzymes in tissues and cells. *Harvey Lecture* 34: 214–245.

Lindsley, D.B., J. Bowden, and H.W. Magoun. 1949. Effect upon the EEG of acute injury to the brain stem activating system. *Electroenceph. Clin. Neurophysiol.* 1: 475–486.

Lindsley, D.B., L.H. Schreiner, and H.W. Magoun. 1949. An electromyographic study of spasticity. *J. Neurophysiol.*, 12: 197–205.

Ling, G.N. 1986. Interview 18 July 1986 CON LNG in the Neuroscience History Archives, Brain Research Institute, University of California, Los Angeles.

Ling, G. and R.W. Gerard. 1949. The normal resting potential of frog sartorius muscle. *J. Cell Comp. Physiol.*, 34: 383–396.

Liske, E., H.M. Hughes, and D.E. Stowe. 1967. Cross-correlation of human alpha activity: normative data. *Electroenceph. Clin. Neurophysiol.*, 22: 429–436.

Locke, F.S. 1895. The action of sodium oxalate on voluntary muscle. *J. Physiol.* (London) 15: 119–120.

Loeb, J. 1900. *Comparative Physiology of the Brain and Comparative Psychology.* New York: Putnam.

Loewi, O. 1921. Über humorale Übertragbarkeit der Herznervenwirkung. I. Mitteilung, *Pflügers Arch.*, 189: 239–242.

Lorente de Nó, R. 1933. Studies on the structure of the cerebral cortex. I. The area entorhinalis. *J. Psych. Neurol.*, 45: 381-438.

Löwenthal, M.S. and V.A.H. Horsley. 1897. On the relations between the cerebellar and other centres (namely cerebral and spinal) with especial reference to the action of antagonistic muscles (Preliminary account). *Proc. R. Soc.*, 61: 20–25.

Loewi, O. 1953. *From the Workshop of Discoveries.* Porter Lectures, Series 19. Lawrence: University of Kansas Press.

Loomis, A.L., E.N. Harvey, and G. Hobart. 1936. Electrical potentials of the human brain. *J. Exp. Psychol.* 19: 249–279.

Lorente de Nó, R. 1922. La corteza cerebral del ratón. Primera contribución.—La corteza acústica. *Trab. Lab. Invest. Biol.* (Madrid), 20: 41–78.

Lorente de Nó, R. 1935. Facilitation of motor neurones. *Am. J. Physiol.* 113: 505–523.

Lorente de Nó, R. 1938. Cerebral cortex: architecture, intracortical connections, motor projections. Pages 288–315 in J.F. Fulton, *Physiology of the Nervous System.* London: Oxford University Press.

Lorente de Nó, R. 1947. A study of nerve physiology, parts 1 and 2. *Stud. Rockefeller Inst. Med. Res.* 131: 1–496; 132: 1–548.

Lorente de Nó, R. 1949. Cerebral cortex: architecture, intracortical connections, motor projections. Pages 288–330 in J. F. Fulton, ed., *Physiology of the Nervous System*, 3rd ed. New York: Oxford University Press.

Lorente de Nó, R. 1976. Audiotaped interview. Oral History Project CON LOR, Neuroscience History Archives, Brain Research Institute, University of California, Los Angeles.

Lorry, A. C. 1760. Sur les mouvements du cerveau et de la dure-mère. Première mémoire, sur le mouvement des parties contenues dans le crâne, considérées dans leur état naturel. *Mem. Math. Physi.* (Paris), 3: 277–313.

Lotze, H. 1882. *Outlines of Psychology.* Translated by C. L. Herrick. Minneapolis: S. M. Williams.

Lower, R. 1672. *De catarrhis.* Reproduced in facsimile and for the first time translated from the original Latin together with bibliographical analysis by Richard Hunter and Ida MacAlpine. London: Dawsons, 1963.

Lubinska, L. 1964. Axoplasmic streaming in regenerating and in normal nerve fibers. *Prog. Brain Res.*, 13: 1–66.

Luciani, L. 1884–1885. On the sensorial localisations in the cortex cerebri. Translated by A. de Watteville. *Brain*, 7: 145–160.

Luciani, L. 1891. *Il cervelletto, Nuovi studi di fisiologia normale e patologica.* Florence: Successori le Monnier.

Luciani, L. 1911. *Fisiologia dell'uomo.* Milano: Societa Editrice Libraria. English trans. of 5th ed. by F. A. Welby. London: Macmillan.

Lugaro, E. 1907. Sulle funzioni della neuroglia. *Riv. Pat. Nerv. Ment.*, 12: 225–233.

Luys, J. 1865. *Recherches sur le système nerveux cérébro-spinal: sa structure, ses fonctions et ses maladies.* Paris: Baillière.

Lyell, C. 1830–1833. *Principles of Geology; Being an Attempt to Explain the Former Changes of the Earth's Sur-*

face by Reference to Causes Now in Operation. 3 vols. London: John Murray.

Lynch, G. and C. W. Cotman. 1975. The hippocampus as a model for studying anatomical plasticity in the adult brain. Pages 123–154 in R. L. Isaacson and K. H. Pribram, eds., *The Hippocampus*, vol. 1. New York: Plenum.

Lynch, J. C. 1980. The functional orgnization of posterior parietal association cortex. *Behav. Brain Sci.*, 3: 485–534.

MacDonald, J. A., Trans., *see* Bianchi, L., 1922.

Machne, X., I. Calma, and H. W. Magoun. 1955. Unit activity of central cephalic brain stem in EEG arousal. *J. Neurophysiol.*, 18: 547–558.

MacLean, P. D. 1949. Psychosomatic disease and the "visceral brain." Recent developments bearing on the Papez Theory of emotion. *Psychosom. Med.*, 11: 338–353.

MacLean, P. D. 1952. Some psychiatric implications of physiological studies on frontotemporal portion of limbic system (visceral brain). *Electroenceph. Clin Neurophysiol.*, 4: 407–418.

MacLean, P. D. 1954. Studies on limbic system ("visceral brain") and their bearing on psychosomatic problems. Pages 101–125 in E. D. Wittkower and R. A. Cleghorn, eds., *Recent Developments in Pschosomatic Medicine.* Philadelphia: J. B. Lippincott.

MacLean, P. D. 1967. The brain in relation to empathy and medical education. *J. Nerv. Ment. Dis.*, 144: 374–382.

MacLean, P. D. 1978. Challenges of the Papez heritage. Pages 1–15 in K. Livingston and O. Hornkiewicz, eds., *Limbic Mechanisms.* New York: Plenum.

MacLean, P. D. 1990. *The Triune Brain in Evolution. Role in Paleocerebral Functions.* New York: Plenum.

Magendie, F. J. 1824a. *An elementary compendium of physiology; for the use of students.* Translated from the French by E. Milligan. Phildelphia: James Webster.

Magendie, F. J. 1824b. Mémoire sur les fonctions de quelques parties du système nerveux. *J. Physiol. Exp. Pathol.* (Paris), 4: 399–407.

Magendie, F. J. 1825. Mémoire sur un liquide qui se trouve dans le crâne et le canal vertébral de l'homme et des animaux mammifères. *J. Physiol. Exp. Pathol.* (Paris), 5: 27–37.

Magendie, F. J. 1827. Troisième et dernière partie du second mémoire sur le liquide qui se trouve dans le crâne et l'épine de l'homme et des animaux vertébrés. *J. Physiol. Exp. Pathol.* (Paris), 7: 66–82.

Magendie, F. J. 1842. *Recherches physiologiques et cliniques sur le liquide cephalo-rachidien ou cérébro-spinal.* Paris: Méquignon-Marvis.

Magnus, R. and E. A. Schäfer. 1901. The action of pituitary extracts upon the kidney. *J. Physiol.* (London), 27: ix–x.

Magoun, H. W. 1949. Discussion. *Electroencephal. Clin. Neurophysiol.*, 1: 419–420.

Magoun, H. W. 1950. Caudal and cephalic influences of the brain stem reticular formation. *Physiol. Rev.*, 30: 459–474.

Magoun, H. W. 1952. An ascending reticular activating system in the brain stem. *Arch. Neurol. Psychiat.*, 67: 145.

Magoun, H. W. 1954a. The ascending reticular system and wakefulness. Pages 1–20 in J. F. Delafresnaye, E. D. Adrian, F. Bremer, and H. H. Jasper, eds., *Brain Mechanisms and Consciousness.* Oxford: Blackwell.

Magoun, H. W. 1954b. General discussion. Page 114 in J. E. Delafresnaye, E. D. Adrian, F. Bremer and H. H. Jasper, eds., *Brain Mechanisms and Consciousness.* Oxford: Blackwell.

Magoun, H. W. 1958a. *The Waking Brain.* First edition. Springfield, IL: C. C Thomas.

Magoun, H. W. 1958b. Early development of ideas relating the mind with the brain. Pages 4–27 in G. E. W. Wolstenholme and C. M. O'Connor, eds., *Ciba Foundation Symposium on the Neurological Basis of Behavior.* Boston; Little, Brown & Co.

Magoun, H. W. 1960. Evolutionary concepts of brain function following Darwin and Spencer. Pages 187–209 in S. Tax, ed., *Evolution After Darwin* Chicago: University of Chicago Press.

Magoun, H. W. 1963. *The Waking Brain.* Second edition. Springfield, IL: C. C Thomas.

Magoun, H. W. 1969. Advances in brain research with implications for learning. Pages 170–190 in J. Kagan, ed., *On the Biology of Learning.* New York: Harcourt, Brace & World.

Magoun, H. W. 1977. Highest nervous activity in the lower brian? *Cont. Psychol.*, 22: 880–882.

Magoun, H. W. 1985. The Northwestern connection with the reticular formation. *Surg. Neurol.*, 24: 250–252.

Magoun, H.W. and R. Rhines. 1946. An inhibitory mechanism in the bulbar reticular formation. *J. Neurophysiol.* 9: 165–171.

Magoun, H.W. and R. Rhines. 1947. *Spasticity: The Stretch Reflex and Extra-Pyramidal Systems.* Publication no. 9 in American Lecture Series, R.W. Pitts, ed., American Lectures in Physiology. Springfield, IL: C.C Thomas.

Magoun, H.W., L. Darling, and J. Prost. 1960. The evolution of man's brain. Pages 33–126 in M.A.B. Brazier, ed., *The Central Nervous System and Behavior: Transactions of the Third Conference.* New York: Josiah Macy, Jr. Foundation.

Malacarne, M.V.G. 1776. *Nuova esposizione della struttura del cervelletto umano.* Turin: Briolo.

Manni, E. 1973. Luigi Rolando 1773–1831. *Exp. Neurol.*, 35: 1–5.

Marie, P. 1886. Sur deux cas d'acromegalie. *Rev. Med.* (Paris) 6: 297–333.

Marie, P. and C. Chatelin. 1914–1915. Les troublés visuels dus aux lésions des voies optiques intracerebrales et de la sphère visuelle. *Rev. Neurol.* (Paris), 28: 882–925.

Marin-Padilla, M. 1984. Neurons of Layer I. A developmental analysis. Pages 447–478 in A. Peters and E.G. Jones, eds., *Cerebral Cortex*, Vol, 1. Cellular Components of the Cerebral Cortex. New York: Plenum.

Markee, J. E., C. H. Sawyer, and W. H. Hollingshead. 1946. Activation of the anterior hypophysis by electrical stimulation in the rabbit. *Endocrinology*, 38: 345–357.

Marshall, L. H. 1983a. More on Burt Green Wilder at Cornell. *Physiologist*, 26(6) Suppl: 361–363.

Marshall, L. H. 1983b. The fecundity of aggregates: the axonologists at Washington University, 1922–1942. *Persp. Biol. Med.*, 26: 613–636.

Marshall, L. H. 1987. An annotated interview with Giuseppe Moruzzi, 1910–1986. *Exp. Neurol.*, 97: 225–242.

Marshall, L. H. 1996. Early history of IBRO: the birth of organized neuroscience. I. The antecedent ground swell. *Neuroscience,* in press.

Marshall, W. H. and R. W. Gerard. 1933. Nerve impulse velocity and fiber diameter. *Am. J. Physiol.*, 104: 586–589.

Marshall, W. H., C. N. Woolsey, and P. Bard. 1937. Cortical representation of tactile sensibility as indicated by cortical potentials. *Science*, 85: 388–390.

Marshall, W. H., C. N. Woolsey, and P. Bard. 1941. Observations on cortical somatic sensory mechanisms of cat and monkey. *J. Neurophysiol.*, 4: 1–24.

Martensen, R. L. 1992. "Habit of reason:" anatomy and Anglicanism in Restoration England. *Bull. Hist. Med.*, 66: 511–535.

Matteucci, C. 1838. Sur le courant électrique ou propre de la grenouille. *Bibl. Univ. Genève*, 7: 156-168.

Maudsley, H. 1889. The double brain. *Mind*, Ser. 1, 14: 161–187.

Mauthner, L. 1890. Zur Pathologie und Physiologie des Schlafes nebst Bemerkungen über die "Nona." *Wien. Med. Woch.*, 40: 961–963.

Maxwell, D. S. and D. C. Pease. 1956. The electron microscopy of the choroid plexus. *J. Biophys. Biochem. Cytol.*, 2, pt.2: 467–474.

May, M. T., Trans., *see* Galen, 1968.

May, R. M., Trans., *see* Ramón y Cajal, S., 1928.

Maynard, E. A. and D. C. Pease. 1955. Electron microscopy of the cerebral cortex of the rat. *Anat. Rec.*, 121: 440 441.

Maynard, E. A., R. L. Schultz, and D. C. Pease. 1957. Electron microscopy ofthe vascular bed of rat cerebral cortex. *Am. J. Anat.*, 100: 409–433.

McAlpin, I., Trans., *see* Lower, R.

McCleary, R. A. 1967. Response-modulating functions of the limbic system: initiation and suppression. Pages 209–272 in E. Stellar and J. M. Spragne, eds., *Progress in Physiological Psychology*, Vol. 1. New York: Academic.

McCulloch, W. S. 1944. Cortico-cortical connections. Pages 211–242 in P. C. Bucy, ed., *The Precentral Motor Cortex*. Urbana, IL: University of Illinois Press.

McIlwain, H. 1958. Chemical contributions, especially from the nineteenth century, to knowledge of the brain and its functioning. Pages 167–186 in *The History and Philosophy of Knowledge of the Brain and its Functions*. Oxford: Blackwell.

McRae, D. L., C. L. Branch, and B. Milner. 1968. The occipital horns and cerebral dominance. *Neurology*, 18: 95–98.

Mehée de la Touche, J. 1773. *Traité des lesions de la tête, par contre-coup, avec des expériences propres à en éclairer la doctrine*. Meaux: L. A. Courtois.

Meigs, D. L., Trans., *see* Flourens, P., 1846.

Mettler, F. A. 1972. The corticothalamic projection: the structural substrate for the control of the thalamus by the cerebral cortex. Pages 1–19 in T. L. Frigyesi, E. Rinvik, and M. D. Yahr, eds., *Corticothalamic Projections and Sensorimotor Activities*. New York: Raven.

Meyer, A. 1910. The present status of aphasia and apraxia. *Harvey Lectures*, 5: 228–250.

Meyer, A. C. 1970. Karl Friedrich Burdach and his place in the history of neuroanatomy. *J. Neurol. Neurosurg. Psychiat.*, 33: 553–561.

Meyer, A. C. 1971. *Historical Aspects of Cerebral Anatomy*. London: Oxford University Press.

Meyer, M. F. 1933. That whale among the fishes are the theory of emotions. *Psychol. Rev.*, 40: 252–300.

Meynert, T. 1868. *Der Bau der Grosshirnrinde und seine örtlichen Verschiedenheiter, nebst einem pathologisch-anatomischen Cordlarium*. Leipzig: Engelmann.

Meynert, T. 1872a. Eine diagnose auf Schügeleskrankung. *Medizinische Jahrbücher*, 188–204.

Meynert, T. 1872b. Vom Gehirne der Säugethiere. Pages 694–808 in Stricker, ed., *Handbuck der Lehre von den Geweben des Menschen und der Thiere*, Vol. II. Leipzig: Engelmann.

Meynert, T. 1885. *Psychiatry: A Clinical Treatise on Diseases of the Fore-Brain Based Upon a Study of Its Structure, Functions, and Nutrition*. B. Sachs, Translator. New York: Putnam, 1968.

Milligan, E., Trans., *see* Magendie, F. J., 1824a.

Mills, C. K. and W. G. Stiller. 1907. Symptomatology of lesions of the lenticular zone with some discussion of the pathology of aphasia. *J. Nerv. Ment. Dis.*, 34: 558–588.

Milner, B. 1974. Hemispheric specialization: scope and limits. Pages 75–89 in F. O. Schmitt and F. G. Worden, eds., *The Neurosciences: Third Study Program*. Cambridge: MIT Press.

Minkowski, O. 1887. Über einen Fall von Akromegalie. *Berl. Klin. Wochensch.*, 24: 371–374.

Mohr, 1840. Mittheilungen für neuropathologische studien. *Wochenschr. Ges. Heilk.*, 565–571.

Molfese, D. L., R. B. Freeman, and D. S. Palermo. 1975. The ontogeny of brain lateralization for speech and nonspeech stimuli. *Brain Lang.*, 2: 356–368.

Monakow, C. v. 1882. Über einige durch Exstirpation circumscripter Hirnrinden-regionen bedingte Entwickelungs-hemmungen des Kaninschengehirns. *Arch. Psychiat. Nervkrankh.*, 12: 141–156.

Monakow, C. v. 1895. Experimentelle und pathologisch-anatomische Untersuchungen über die Haubenregion, der Schlügel und die Regio subthalamica nebst Beiträgen zur Kenntris Früh erworbener grossund Kleinhirndefecte. *Arch. Psychiatr. Nervenkr.* 27: 1–128, 386–478.

Monakow, C. v. 1914. *Die Lokalisation in Grosshirn und der Abbau der Funktion durch kortikale Herde*. Wiesbaden: Bergmann.

Mondino da Luzzi. fl. 1315. Anathomia overo dissectione del corpo humano, in Petrus de Montagnana's *Fasciculo di medicina*. Venice: English translation by Charles Singer, 1925. Florence: Lier.

Moniz, E. 1936. *Tentatives opératoires dans le traitment de certaines psychoses*. Paris, Masson.

Monroe, B. G. 1967. A comparative study of the ultrastructure of the median eminence, infundibular stem and neural lobe of the hypophysis of the rat. *Z. Zellforsch.*, 76: 405–432.

Monroe, R. R. and R. G. Heath. 1954. Psychiatric observations. Pages 345–382 in Tulane Department of Psychiatry and Neurology, *Studies in Schizophrenia: A Multidisciplinary Approach to Mind–Brain Relationships*. Cambridge, MA: Harvard University.

Moore, C. R. and D. Price. 1932. Gonad hormone functions and the reciprocal influence between gonads and hypophysis with its bearing on the problem of sex hormone antagonism, *Am. J. Anat.*, 50; 13–71.

Moore, R. E. 1954. *Man, Time and Fossils: the Story of Evolution*. London: J. Cape.

Moore, R. E. and the Editors of Time-Life Books. 1962, 1964. *Evolution*. New York: Time, Inc.

Morgagni, G. B. 1761. *De sedibus et causis morborum*, I. Venice: Remondiniana.

Morison, R. S. 1954. Discussion. Page 110 in J. F. Delafresnaye, E. D. Adrian, F. Bremer, and H. H. Jasper, eds., *Brain Mechanisms and Consciousness*. Oxford: Blackwell.

Morison, R. S. and E. W. Dempsey. 1942. A study of thalamocortical relations. *Am. J. Physiol.*, 135: 381–292.

Morell, V. 1995. *Ancestral Passions: the Leakey Family and the Quest for Humankind's Beginnings*. New York: Simon and Schuster.

Morrell, F. 1958. Some electrical events involved in the formation of temporary connections. Pages 545–560 in H. H. Jasper, L. D.Proctor, R. S. Knighton, W. C. Noshay, and R. T. Costello, eds., *Reticular Formation of the Brain*. Boston: Little, Brown.

Moruzzi, G. 1930. La rete nervosa diffusa (Golgi) dello strato di granuli dei cervelletto. *Arch. Ital. Anat. Embriol.*, 28: 238–252.

Moruzzi, G. 1938. Azione del paleocerebellum sui riflessi vasomotori. *Arch. Fisiol.*, 38: 36–78.

Moruzzi, G. 1950. *Problems in Cerebellar Physiology*. Springfield, IL: C. C. Thomas.

Moruzzi, G. 1954. The physiological properties of the brin stem reticular system. Pages 21–53 in J. F. Delafresnaye, E. D. Adrian, F. Bremer, and H. H. Jasper, eds., *Brain Mechanisms and Consciousness*. Oxford: Blackwell.

Moruzzi, G. 1980. In memoriam Lord Adrian (1889–1977). *Rev. Physiol. Pharmacol.*, 87: 1–24.

Moruzzi, G. and H. W. Magoun. 1949. Brain stem reticular formation and activation of the EEG. *Electroenceph. Clin. Neurophysiol.*, 1: 455–473.

Mott, F. W., Trans., *see* Munk, H., 1890.

Mountcastle, V. B. 1957. Modalities and topographic properties of single neurons in sensory cortex. *J. Neurophysiol.*, 20; 408–434.

Mountcastle, V. B. 1977. Philip Bard. 1898–1977. *John Hopkins Med. J.*, 141: 296–298.

Muir, A. R. and A. Peters. 1962. Quintuple-layered membrane junctions at terminal bars between endothelial cells. *J. Cell Biol.*, 12: 443–448.

Munk, H. 1881. Über die Funktionen der Grosshirnrinde. Berlin: A. Hirschwald. Translation of pages 28–52 by G. von Bonin, *Some Papers on the Cerebral Cortex*, Springfield, IL: C. C. Thomas, pp. 97–117.

Munk, H. 1890. Of the visual area of the cerebral cortex, and its relation to eye movements. *Brain*, 13: 45–70. Translated by F. W. Mott.

Myers, R. E. and R. W. Sperry. 1956. Contralateral mnemonic effects with ipsilateral sensory inflow. Abstract. *Fed. Proc.*, 15: 134.

Nakayama, T. 1955. Hypothalamic electrical activities produced by factors causing discharge of pituitary hormone. *Jap. J. Physiol.*, 5: 311–316.

Nauta, W. J. H. 1958. General discussion. Pages 666–667 in H. H. Jasper, L. D. Proctor, R. S. Knighton, W. C. Noshay, and R. T. Costello, eds., *Reticular formation of the Brain*. Boston: Little, Brown.

Nauta, W.J.H. and G.J.M. Kuypers. 1958. Some ascending pathways in the brain stem reticular formation. Pages 3–30 in H.H. Jasper, L.D. Proctor, R.S. Knighton, W.C. Noshay, and R.T. Costello, Eds., *Reticular Formation of the Brain*. Boston: Little, Brown.

Nauta, W. J. H., and D. G. Whitlock. 1954. An anatomical analysis of the non-specific thalamic projection system. Pages 81–116 in J. F. Delafresnaye, E. D. Adrian, F. Bremer and H. H. Jasper, eds., *Brain Mechanisms and Consciousness*. Oxford: Blackwell.

Neher, E. and B. Sakmann. 1976. Single-channel currents recorded from membrane of denervated frog muscle cells. *Nature*, 260: 799–802.

Nemminsky [sic.], W.W. 1913. Ein Versuch der Registrierung der elektrischen Gehirnerscheinungen. *Zentrabl. Physiol.*, 27: 951–960.

Neubuerger, K. T. 1970. Carl Weigert (1845–1904). Pages 388–391 in W. Haymaker and F. Schiller, Comps. and Eds., *The Founders of Neurology*, 2nd ed. Springfield, Illinois: C.C Thomas.

Neuburger, M. 1981. *The Historical Development of Brain and Spinal Cord Physiology Before Flourens*. Trans. and ed. with additional material by Edwin Clarke. Baltimore: Johns Hopkins University Press.

Nielsen, J. M. 1940. Dominance if the right occipital lobe. Report of two cases with autopsy. *Bull. Los Angeles Neurol. Soc.*, 5: 135–145.

Nikitovitch-Winer, M. and J.W. Everett. 1958. Functional restitution of pituitary grafts re-transplated from kidney to median eminence. *Endocrinology*, 63: 916–930.

Nissl, F. 1889. Die Kerne das Thalamus beim Kaninchen. *Tagbl. 62 Vers Deutsch. Naturforsch. Ärzte*, 62: 509.

Noël, J.-F. 1978. La vie et l'oeuvre du Docteur Francois Leuret (Nancy, 1797–1851). *Ann. Med. Nancy*, 17: 1397–1406.

Nottebohm, F. 1971. Neural lateralization of vocal control in a passerine bird. I. Song. *J. Exp. Zool.*, 177: 229–261.

Nottebohm, F. and A.P. Arnold. 1976. Sexual dimorphism in vocal control areas of the songbird brain. *Science*, 194: 211–213.

Obata, K., M. Ito, R. Ochi, and N. Sato. 1967. Pharmacological properties of the postsynaptic inhibition by Purkinje cell axons and the action of j-amino-butyric acid on Deiters neurones. *Exp. Brain Res.*, 4: 43–57.

Obersteiner, H. 1890. *The Anatomy of the Central Nervous Organs: A Guide to their Study in Health and Disease*. Translated, with annotations and additions by A. Hill. London: Griffin.

Ochs, S. 1979. The early history of material transport in nerve. *The Physiologist*, 22: 16–19.

Ojemann, G.A. 1982. Interrelationships in the localization of language, memory, and motor mechanisms in human cortex and thalamus. Pages 157–175 in R.A. Thompson and J.R. Green, eds., *New Perspectives in Cerebral Localization*. New York: Raven Press.

Ojemann, G.[A.] and C. Mateer. 1979. Human language cortex: localization of memory, syntax, and sequential motor-phoneme identification systems. *Science*, 205: 1401–1403.

Ojemann, G.A., P. Fedio, and J.M. Van Buren. 1968. Anomia from pulvinar and subcortical parietal stimulation. *Brain*, 91: 99–116.

O'Leary, J.L. 1956. Speculative trends in electrophysiology. *Arch. Neurol. Psychiat.*, 76: 137–197.

O'Leary, J.L. and S. Goldring. 1976. *Science and Epilepsy: Neuroscience Gains in Epilepsy Research*. New York: Raven Press.

Oldendorf, W.H. and W.J. Brown. 1975. Greater number of capillary endothelial cell mitochondria in brain than in muscle. *Proc. Soc. Exp. Biol. Med.*, 149: 736–738.

Oldendorf, W.H. and H. Davson. 1967. Brain intracellular space and the sink action of cerebrospinal fluid. *Arch. Neurol.*, 17: 196–205.

Olds, J. and P. Milner. 1954. Positive reinforcement produced by electrical stimulation of septal area and other regions of the rat brain. *J. Comp. Physiol. Psychol.* 47: 419–427.

Oliver, G. and E.A. Schäfer. 1895. On the physiological action of extracts of pituitary body and certain other glandular organs. *J. Physiol.* (London) 18: 277–279.

Olninck, I.N.W. 1970. Rudolf Magnus (1873–1927). Pages 240–243 in W. Haymaker and F. Schiller, eds., *The Founders of Neurology*. Springfield, Illinois: C.C Thomas.

Olszewski, J. 1954. The cytoarchitecture of the human reticular formation. Pages 54–80 in J.F. Delafresnaye, ed., *Brain Mechanisms and Consciousness*. Oxford: Blackwell.

O'Malley, C.D. 1964. *Andreas Vesalius of Brussels 1514–1564*. Berkeley and Los Angeles: University of California Press.

O'Neill, Y.V. 1980. *Speech and Speech Disorders in Western Thought before 1600*. Westport, Connecticut: Greenwood Press.

Oppenheimer, J.M. 1977. Studies of brain asymmetry: historical perspective. Pages 4–17 in S.J. Dimond and D.A. Blizard, Eds., *Evolution and Lateralization of the Brain*. Ann. New York Acad. Sci., Vol. 299.

Orbach, J., B. Milner, and T. Rasmussen. 1960. Learning and retention in monkeys after amygdala-hippocampus resection. *Arch. Neurol.*, 3: 230–251.

Ornstein, R.E. 1972. *The Psychology of Consciousness*. San Francisco: W.H. Freeman.

Ott, I. and J.C. Scott. 1911. The action of infundibili upon the mammary secretion. *Proc. Soc. Exp. Biol. Med.*, 8: 48–49.

Overton, E. 1902. Beitrage zur allgemeinen Muskel- und Nervenphysiologie. II. Mittheilung. Über die Unenthehrlichkeit von Natrium- (oder Lithium) Ionen für den Contractionsact des Muskels. *Pflügers Arch.*, 92: 346–386.

Pacchioni, A. 1705. *Dissertatio epistolaris de glandulis conglobatis durae meningis humanae,....* Rome: Pagliarini.

Paget, S. 1919. *Sir Victor Horsley; a Study of His Life and Work*. London: Constable.

Pagni, C.A. 1963. Étude électro-clinique des post-discharges amydalo-hippocampignes chez l'homme par moyen d'électrodes de profondeur placées avec méthode stéréotaxique. *Confin. Neurol.* 23: 477–499.

Palay, S.L. 1977. Introduction. Pages 3–6 in D.A. Rottenberg and F.H. Hopberg, Eds., *Neurological Classics in Modern Translation*. New York: Hafner.

Pandya, D.N., E.A. Karol, and D. Heilbronn. 1971. The topogrphical distribution of interhemispheric projections in the corpus callosum of the rhesus monkey. *Brain Res.*, 32: 31–43.

Panizza, B. 1855. Osservazioni sul nervo ottico. *Gior. I. Reale Inst. Lombardo*, 7: 237–252.

Papez, J.W. 1937. A proposed mechanism of emotion. *Arch. Neurol. Psychiat.*, 38: 725–743.

Papez, J.W. 1938. Thalamic connections in a hemi-decorticated dog. *J. Comp. Neurol.* 69: 103–120.

Papez, J.W. 1942. A summary of fiber connections of the basal ganglia with each other and with other portions of the brain. Pages 21–68 in *The Diseases of the Basal Ganglia*. Baltimore: Williams and Wilkins.

Papez, J.W. 1955. Central reticular path to intralaminar and reticular nuclei of thalamus for activating EEG related to consciousness. *Electroenceph. Clin. Neurophysiol.*, 8: 117–128.

Papez, J.W. 1970. Bernard von Gudden (1824–1886). Pages 43–47 in W. Haymaker and F. Schiller, Eds., *The Founders of Neurology*, 2nd ed., Springfield, Illinois: C.C Thomas.

Park, R. 1913. The conclusions drawn from a quarter century's work in brain surgery. *NY Med. J.*, 13: 303–309.

Parker, G.H. 1929. The neurofibril hypothesis. *Q. Rev. Biol.*, 4: 156-178.

Patterson, A. and O.L. Zangwill. 1944. Disorders of visual space perception associated with lesions of the right cerebral hemisphere. *Brain*, 67: 331–358.

Patterson, T.S. 1931. John Mayow in contemporary setting. A contribution to the history of respiration and combustion. *Isis*, 15: 47–96; 504–546.

Penfield, W., Ed. 1932. *Cytology and Cellular Pathology of the Nervous System*. 2 vol. New York: Hoeber.

Penfield, W. 1936–1937. The cerebral cortex and consciousness. *Harvey Lectures*, 32: 35–69.

Penfield, W. 1960. A surgeon's chance encounter with mechanisms related to consciousness. *J. R. Coll. Surg. Edinburgh*, 5: 173–190.

Penfield, W. 1975. *The Mystery of the Mind, a Critical Study of Consciousness and the Human Brain*. Princeton: Princeton University Press.

Penfield, W. and T. Rasmussen. 1949. Vocalization and arrest of speech. *Arch. Neurol. Psychiat.*, 61: 21–27.

Penfield, W. and A.T. Rasmussen. 1950. *The Cerebral Cortex of Man*. New York: Macmillan.

Penfield, W. and L. Roberts. 1959. *Speech and Brain Mechanisms*. Princeton: Princeton University Press.

Peters, A. 1988. Neuroanatomical research techniques. Pages 764–768 in G. Adelman, ed., *Encyclopedia of Neuroscience*. Boston: Birkhäuser.

Peterson, G. M. and J. V. Devine. 1963. Transfers in handedness in the rat resulting from small cortical lesions after limited forced practice. *J. Comp. Physiol. Psychol.*, 56: 752–756.

Pfaff, D. W. 1966. Morphological changes in the brains of adult male rats after neonatal castration. *J. Endocrinol.*, 36: 415–416.

Pfeifer, R. A. 1928. *Die Angioarchitektonik der Grosshirnrinde*. Berlin: Springer.

Pitts, W. and W. McCulloch. 1947. How er know universals: the perception of auditory and visual forms. *Bull. Math. Biophysics*, 9: 127–147.

Polyak, S. 1932. *The Main Afferent Fiber Systems of the Crebral Cortex in Primates*. Berkeley: University of California Press.

Popa, G. and U. Fielding. 1930. A portal circulation from the pituitary to the hypothalamic region. *J. Anat.*, 65: 88–91.

Pope, A. 1959. The intralaminar distribution of dipeptidase activity in the human frontal isocortex. *J. Neurochem.*, 4: 31–41.

Pope, A., W. F. Caveness, and K.E. Livingston. 1952. Architectonic distribution of acetylcholinesterase of psychotic and nonpsychotic patients. *Arch. Neurol. Psychiat.* 68: 425–443.

Pordage, S., Trans., *see* Willis, T., 1681.

Porter, R. 1988. Penfield's supplementary motor area reexamined: associations between an area of cerebral cortex and motor performance. Unpublished lecture to the Society for Neuroscience, Toronto, 17 November, 1988.

Porter, R. W., D. G. Conrad, and J. V. Brady. 1959. Some neural and behavioral correlates of electrical self-stimulation of the limbic system. *J. Exp. Anal. Behav.*, 2: 43–55.

Pourfour du Petit, F. 1710. *Lettres d'un médicin des hôspitaux du Roy, d'un autre médicin de ses amis.* Namur: C. G. Albert.

Pozzi, S. 1888. Introduction. Pages V–XXII in P. Broca, Memoires sur le cerveau de l'homme et des primates. Paris: Reinwald.

Prawdicz-Neminski, V. V., see Nemminsky [sic.], W. W.

Praxagoras, see Steckerl.

Preston, C. 1697. An account of a child born alive without a brain, and the observables in it on dissection. *R. Soc. Lond. Phil. Trans.*, 19: 457–467.

Price, D. 1975. Feedback control of gonadal and hypophyseal hormones: evolution of the concept. Pages 218–238 in J. Meites, B. T. Donovan, and S. M. McCann, eds., *Pioneers in Neuroendocrinology I.* New York: Plenum.

Procháska, G., see Unzer, J.A.

Prothero, J. W. and J. W. Sundsten. 1984. Folding of the cerebral cortex in mammals. A scaling model. *Brain, Behav. Evol.*, 24: 152–167.

Pupilli, G. C. 1953. [Quotation of E. du Bois-Reymond in Galvani], *Commentary on the effect of electricity on muscular motion,* trans. by R. M. Green. Cambridge, MA: Licht.

Purkyne, J. E. 1838. Nueste Untersuchungen aus der Nerven- und Hirn-anatomie. *Bericht über die Versammlung deutscher Naturforschen und Arzte.* Prag Sept. 1837: 177–180. Prague: Hasse.

Purpura, D. P. and R. J. Shofer. 1963. Intracellular recording from the thalamic neurons during reticulocortical activation. *J. Neurophysiol.*, 26: 494–505.

Quesnel, F. 1907. Präparate mit aktiven Zelidegenerationem nach Hirstamm-verbetzung bei Kaninchen. *Neurol. Zentralbl.*, 26: 1138–1139.

Rabi, R. 1958. Strukturstudien an der Massa intermedia des Thalamus opticus. *J. Hirnforsch.*, 4: 78–112.

Racine, R., F. Newberry, and W. M. Burnham. 1975. Post-activation potentiation and the kindling phenomenon. *Electroenceph. Clin. Neurophysiol.*, 39: 261–271.

Rafaelson, O. J. 1982. hudichum: the founder of neurochemistry. Pages 293–305 in F. C. Rose and W. F. Bynum, eds., *Historical Aspects of the Neurosciences* New York: Raven.

Raisman, G. and P. M. Field. 1971. Sexual dimorphism in the preoptic area of the rat. *Science,* 173: 731–733.

Raki, P. 1971. Neuron-glia relationship during granule-cell immigration in developing cerebellar cortex. A Golgi and electronmicroscopic study in Macacus rhesus. *J. Comp. Neurol.,* 141: 283–312.

Raki, P. 1988. Specification of cerebral cortical areas. *Science*, 241: 170–176.

Ramón y Cajal, S. 1892. El neuvo concepto de la histologia de los centros nerviosos. III. —Corteza gris del cerebro. *Rev. Cien. Med. Barcelona*, 18: 457–476. Translation by D. A. Rottenberg, pages 7–29 in D. A. Rottenberg and F. H. Hoffberg, eds., *Neurological Classics in Modern Translation.* New York: Hafner, 1977.

Ramón y Cajal, S. 1894. La fine structure des centres nerveux. Croonian Lecture, March 8, 1894. *R. Soc. Lond. Proc.*, 55: 444–468.

Ramón y Cajal, S. 1903. Plan de estructura del talamo optica. *Rev. Med. Cir. Pract.* 59: 329–348.

Ramón y Cajal, S. 1909–1911. *Histologie du Système Nerveux de l'Homme and des Vertébrés.* 2 Vol. Translated by L. Azoulay. Paris: Maloine.

Ramón y Cajal, S. 1928. *Studies on Degeneration and Regeneration of the Nervous System.* 2 Vol. Translated by R. M. May. [Reprinted 1968 New York: Hafner.] London: H. Milford.

Ranson, S. W. 1939. Sommolence caused by hypothalamic lesions in the monkey. *Arch. Neurol. Psychiat.*, 41: 1–23.

Ranson, S. W., S. W. Ranson, Jr., and M. Ranson. 1941. Fiber connections of corpus striatum as seen in Marchi preparations. *Arch. Neurol. Psychiat.*, 46: 230–249.

Rasmussen, A. T. 1947. *Some Trends in Neuroanatomy.* Dubuque, Iowa: William C. Brown.

Rasmussen, A.T. 1970. Wilhelm His (1831–1904). Pages 48–51 in W. Haymaker and F. Schiller, Eds., *The Founders of Neurology*, 2nd ed., Springfield, Illinois: C.C Thomas.

Ratclife, J., Trans., *see* Wernicke, C., 1876.

Ravenstein, E.G., Ed. 1901. *The Strange Adventures of Andrew Battell of Leigh, in Angola and the Adjoining Regions.* Reprinted from "Purchas his Pilgrimes." Edited with notes and a concise History of Kongo and Angola, by E.G. Ravenstein. London: The Haluyt Society, 1901. 2nd serial, No. 6.

Reese, T.S. and M.J. Karnovsky. 1967. Fine structural localization of a blood–brain barrier to exogenous peroxidase. *J. Cell Biol.*, 34: 207–217.

Regan, D. 1972. *Evoked Potentials in Psychology, Sensory Physiology and Clinical Medicine.* London: Chapman and Hall.

Reil, J.C. 1807–1808a. Untersuchungen über den Bau des kleinen Gehirns im Menschen. Zweyle Fortsetzung über die Organisation der Lappen und Läppchen. *Arch. Physiol. Halle*, 8: 385–426.

Reil, J.C. 1807–1808b. Fragmente über die Bildung des kleinen Gehirns im Menschen. *Arch. Physiol. Halle,* 8: 1–58.

Reil, J.C. 1809. Untersuchungen über den Bau des grossen Gehirns im Menschen. . .Vierte Fortsetzung VIII. *Arch. Physiol.* (Halle), 9: 136–146.

Reil, J.C. 1812. Mangel des mittleren und freien Theils des Balkens im Menschengehirn. *Arch. Physiol.* (Halle), 11: 341–344.

Reisch, G. 1503. *Margarita philosophica* Freiburg im Breisgau: J. Schott.

Remak, R. 1838. *Observationes anatomicae et microscopicae de systematis nervosi structura.* Berlin: Reimer.

Remak, R. 1844. Neurologische Erläuterungen. *Arch. Anat. Physiol. Wissen. Med.*, 463–472.

Renshaw, B., A. Forbes, and B.R. Morison. 1940. Activity in isocortex and hippocampus; electrical studies with microelectrodes. *J. Neurophysiol.,* 3: 74–105.

Retzius, G. 1896. *Das Menschenhirn Studien in der makroskopischen Morphologie.* Stockholm: P.A. Norstadt & Söner, 1896.

Rheinberger, M.B. and H.H. Jasper. 1937. Electrical activity of the cerebral cortex in the unanesthetized cat. *Am. J. Physiol.,* 119: 186–196.

Rhines, R. and H.W. Magoun. 1946. Brain stem facilitation of cortical motor response. *J. Neurophysiol.* 9: 219–229.

Riese, W. 1946. The 150th anniversity of S.T. Soemmerring's *Organ of the Soul*. The reaction of his contempararies and its significance today. *Bull. Hist. Med.*, 20: 310–321.

Riese, W. 1959. *A History of Neurology*. Md Monographs on Medical History, No. 2. New York: MD Publications.

Rinvik, E. 1972. Organization of thalamic connections from motor and somatosensory cortical areas in the cat. Pages 57–90 in T.L. Friggesi, E. Rinvik, and M.D. Yahr, Eds., *Corticothalamic Projections and Sensorimotor Activities*. New York: Raven Press.

Rio-Hortega, P. del. 1919. El "tercer elemento" de los centros nerviosos. I. La microglia en estado normal. Intervención de la microglia en los procesos patologicos. Naturaleza probable de la microglia. *Bol. Soc. Espan. Biol.*, 9: 69–120.

Rio-Hortega, P. del. 1920. La microglia y su transformación en células en basoncito y cuerpos gránulo-adiposos. *Trab. Lab. Invest. Biol. Madrid,* 18: 37–82.

Rio-Hortega, P. del. 1921. Estudios sobre la neuroglia. La glia de escasas radiciones (oligodendroglia). *Bol. Real Soc. Espan. Hist. Nat.*, 21: 63–93.

Rio-Hortega, P. del. 1932. Section X. Microglia. Pages 480–534 in W. Penfield, Ed., *Cytology and Cellular Pathology of the Nervous System*. New York: Hoeber.

Ritti, A. 1897. Nécrologie. Dr. J. Luys. *Ann. Med.-Psychol.* (Paris). 8th Serie, 6: 321–323.

Roberts, E. and S. Frankel. 1950. g-Aminobutyric acid in brain: its formation from glutamic acid. *J. Biol. Chem.*, 187: 55–63.

Robertson, W.F. 1900. A microscopic demonstration of the normal and pathological histology of mesoglia cells. *J. Ment. Sci.*, 46: 724.

Rokitansky, C. 1841. *Handbuch der Pathologischen Anatomie*. Bd. 3. Vienna. Publication of the Syndenham Society, London, 1849. Translated by E. Sieveking.

Rolando, L. 1809. *Saggio sopra la vera struttura del cervello dell'uomo e degl'animali e sopra le funzioni del sistema nervoso*. Sassari, 1809.

Rolando, L. 1823. Expériences sur les fonctions du système nerveux. Translated and annotated by P. Flourens. *J. Physiol. Exp. Pathol.*, 3: 95–113.

Rolando, L. 1830. *Della struttura degli emisferi cerebrali*. Torino: Stamperia Reale.

Roofe, P.S. 1963. Neurology comes of age. *J. Kansas Med. Soc.*, 64: 124–129.

Rose, J.E. and C.N. Woolsey. 1943. A study of thalamo-cortical relations in the rabbit. *Johns Hopkins Hosp. Bull.* 73: 65–128.

Rose, J.E. and C.N. Woolsey. 1949. Organization of the mammalian thalamus and its relationships to the cerebral cortex. *Electroencephal. Clin. Neurophysiol.*, 1: 391–404.

Rose, M. 1912. Histologische Lokalisation der Grosshirnrinde der kleinen Säugetiere. *J. Psychol. Neurol.* 19: 391–479.

Rosenblith, W.A. 1991. Foreword. Pages vii–ix in A. Harrington, Ed., *So Human a Brain: Knowledge and Values in the Neurosciences.* Boston: Birkhäuser.

Ross, D.A., S.D. Glick, and R.C. Meibach. 1981. Sexually dimorphic brain and behavioral asymmetries in the neonatal rat. *Proc. Natl. Acad. Sci. U.S.A.*, 78: 1958–1961.

Rottenberg, D. A., Trans., *see* Ramón y Cajal, S., 1892.

Rottenberg, D.A. and F.H. Hoffberg, Eds. 1977. *Neurological Classics in Translation.* New York: Haffner.

Ruelle, C. E., Trans., *see* Rufus of Ephesus.

Rufus of Ephesus. 1879. *Oeuvres de Rufus d'Èphèse.* 2 Vols. Trans. by C. Daremburg and C.E. Ruelle. Paris: J.B. Baillière.

Rushton, W.A.H. 1977. Some memories of visual research in the past 50 years. Pages 85–104 in A.L. Hodgkin, *The Pursuit of Nature: Informal Essays on the History of Physiology*, London: Cambridge University Press.

Russell, E.S. 1916. *Form and Function: A Contribution to the History of Animal Morphology*. London: A. Murray.

Russell, G.A. 1979. Historical introduction: The beginnings of commissure research. Pages xii–xvii in I. Steele-Russell, M. W. van Hof, and G. Berlucci, eds., *The Structure and Function of the Cerebral Commissures*. London: Macmillan.

Sachs, B., Trans., *see* Meynert, T., 1885.

Sachs, E. 1909. On the structure and functional relations of the optic thalamus. *Brain*, 32: 95–186.

Sachs, E. and E. F. Fincher. 1926. Anatomical and physiological observations on lesions in the cerebellar nuclein in *Macacus rhesus* (preliminary report). *Brain*, 50; 350–356.

Saffran, M., A. V. Schally, and B. G. Benfey. 1955. Stimulation of the release of corticotropin from the adenohypophysis by a neurohypophysial factor. *Endocrinology*, 57: 439–444.

Sarton, G. 1954. *Galen of Pergamon.* Logan Clendening Lectures on the History and Philosophy of Medicine, 3rd Series. Lawrence, Kansas: University of Kansas Press.

Saul, R. and R. W. Sperry. 1968. Absence of commissurotomy symptoms with agenesis of the corpus callosum. *Neurology*, 18: 307.

Sawyer, C. H. 1988. Anterior pituitary neural control concepts. Pages 23–39 in S. M. McCain, ed., *Endocrinology: People and Ideas*. Washington: American Physiological Society.

Sawyer, C. H., J. W. Everett, and W. H. Hollinshead. 1972. Joseph Eldridge Markee, 1903–1970. *Anat. Rec.*, 171: 131–133.

Sawyer, C. H., J. W. Everett, and J. E. Markee. 1949. A neural factor in the mechanism by which estrogen induces the release of luteinizing hormone in the rat. *Endocrinology*, 41: 218–233.

Sawyer, C. H., J. E. Markee, and J. W. Everett. 1950. Activation of the adenohypophysis by intravenous injection of epinephrine in the atropinized rabbit. *Endocrinology*, 46: 536–543.

Schäfer, E. A. 1888a. Experiments on special sense localisation in the cortex cerebri of the monkey. *Brain*, 10: 362–380.

Schäfer, E. A. 1888b. On the functions of the temporal and occipital lobes: a reply to Dr. Ferrier. *Brain*, 11: 145–165.

Schäfer, E. A. 1889. Experiments on the electrical excitation of the visual area of the cerebral cortex in the monkey. *Brain*, 11: 1–6.

Schäfer, E. A., ed. 1898. *Textbook of Physiology*. 2 vol. Edinburgh & London: Young J. Pentland.

Schally, A. V. 1978. In the pursuit of hypothalamic hormones. Pages 347–366 in J. Meites, B. T. Donovan, and S. M. McCann, eds. *Pioneers in Neuroendocrinology II*. New York: Plenum.

Schally, A. V., M. Saffran, and B. Zimmerman. 1958. A corticotrophin releasing factor: Partial purification and amino acid composition. *Biochem. J.*, 70: 97–103.

Scharrer, B. 1975. Neurosecretion and its role in neuroendocrine regulation. Pages 257–265 in J. Meites, B.T. Donovan, and S. M. McCann, eds., *Pioneers in Neuroendocrinology*. New York: Plenum.

Scharrer, E. 1928. Die Lichtempfindlichkeit blinder Elritzen (Untersuchungen über das Zwischenhirn der Fische. I.). *Z. Physiol.*, 7: 1–38.

Scharrer, E. 1945. Capillaries and mitochondria in neuropil. *J. Comp. Neurol.*, 83: 237–243.

Scharrer, E. 1952. The general significance of the neurosecretory cell. *Scientia*, 87: 176–182.

Scharrer, E. and R. Gaupp. 1933. Neuere Befunde am Nucleus supraopticus und Nucleus paraventricularis des Menschen. *Zeitsch. Ges. Neurol. Psychiat.*, 148: 766–772.

Scheibel, A. B. 1984. A dendritic correlate of human speech. Pages 43–52 in N. Geschwind and A. M. Galaburda, eds., *Cerebral Dominance: the Biological Foundations*. Cambridge, Massuchesetts: Harvard University Press.

Scheibel, A. B. 1987. Reticular formation, brain stem. Pages 1056–1059 in G. Adelman, ed., *Encyclopedia of Neuroscience*. Boston: Birkhäuser.

Scheibel, M. E. and A. B. Scheibel. 1958. Structural substrates for integrative patterns in the brain stem reticular core. Pages 31–55 in H. H. Jasper, L. D. Proctor, R. S. Knighton, W. C. Hoshay, and R. T. Costello, eds., *Reticular Formation of the Brain.* Boston: Little, Brown.

Scheibel, M. E. and A. B. Scheibel. 1966. Patterns of organization in specific and nonspecific thalamic fields. Pages 13–46 in D. P. Purpura and M. Yahr, eds., *The Thalamus.* New York: Columbia University Press.

Scheibel, M. E. and A. B. Scheibel. 1967. Structural organization of nonspecific thalamic nuclei and their projection toward cortex. *Brain Res.*, 6: 60–94.

Scheibel, M. E., A. B. Scheibel, and T. H. Davis. 1972. Some substrates for centrifugal control over thalamic cell ensembles. Pages 131–160 in T. L. Frigyesi, E. Rinvik, and M. D. Yahr, eds., *Corticothalamic Projections and Sensorimotor Activities*. New York: Raven.

Schiller, F. 1965. The rise of the "enteroid processes" in the 19th century. *Bull. Hist. Med.*, 39: 326–338.

Schiller, F. 1970a. Hermann Munk (1839–1912). Pages 247–250 in W. Haymaker and F. Schiller, eds., *The Founders of Neurology*, 2nd ed. Springfield, IL: C. C. Thomas.

Schiller, F. 1970b. Christian Reil (1759–1813). Pages 62–66 in W. Haymaker and F. Schiller, eds., *The Founders of Neurology*, 2nd ed. Springfield, IL: C.C. Thomas.

Schiller, F. 1970c. Franz Gall (1758–1828). Pages 31–35 in W. Haymaker and F. Schiller, eds., *The Founders of Neurology,* 2nd ed. Springfield, IL: C. C. Thomas.

Schiller, F. 1979. *Paul Broca: Founder of French Anthropology, Explorer of the Brain.* Berkeley/Los Angeles: University of California Press.

Schlag, J. and M. Waszak. 1970. Characteristics of unit responses in nucelus reticularis thalami. *Brain Res.*, 21: 256–288.

Schmidt, C.F. 1950. *The Cerebral Circulation in Health and Disease*. Springfield, Illinois: C.C Thomas.

Schneider, K.V. 1660. *Liber primus de catarrhis*. Wittebergae: Sumptibus haered. D.T. Mevii & E. Schumacheri, Excudebat Michael Wendt.

Schuetze, S. M. 1983. The discovery of the action potential. *Trends Neurosci.*, May, 164–168.

Schultz, R. L., E. A. Maynard, and D. C. Pease. 1957. Electron microscopy of neurons and neuroglia of cerebral cortex and corpus callosum. *Am. J. Anat.*, 100: 369–407.

Sechenov, I. M. 1863. [Reflexes of the brain]. *Meditsinsky Vestnik*, 47; 48. Pages 31–139 in G. Gibbons, ed., *I. Sechenov, Selected Physiological and Psychological Works*. S. Belsky, translator. Moscow: Foreign Languages Publishing House, 1960.

Sellier, J. and H. Verger. 1898. Recherches expérimentales sur la physiologie de la couche optique. *Arch. Physiol. Nor. Pathol.* (5th series), 10: 706–713.

Selye, H. 1979. *The Stress of My Life: a Scientist's Memoirs*. 2nd ed. New York: Van Nostrand Reinhold.

Serres, E.R.A. 1827. *Anatomie comparée du cerveau, dans les quatre classes des animaux vertébrés, appliquée à la physiologie et à la pathologie du système nerveux*. 2 Vol. Paris: Gabon.

Shannon, C. 1987. [Interview: Father of the electronic information age by A. Liversidge]. *Omni*, 9(11): 61–62, 64–66, 70, 72.

Shapiro, H.L. 1974. *Peking Man*. New York: Simon & Schuster.

Share, L. 1962. Vascular volume and blood level of antidiuretic hormone. *Am. J. Physiol.*, 202: 791–794.

Sharpless, S. and H.H. Jasper. 1956. Habituation of the arousal reaction. *Brain*, 79: 655–680.

Shatz, C.J., S. Lindstrom, and T.N. Wiesel. 1977. The distribution of afferents representing the right and left eyes in the cat's visual cortex. *Brain Res.*, 131: 103–116.

Shepherd, G.M. 1972. The neuron doctrine: a revision of functional concepts. *Yale J. Biol. Med.*, 45: 584–599.

Shepherd, G.M. 1991. *Foundations of the Neuron Doctrine.* New York: Oxford University Press.

Sherrington, C.S. 1897. Double (antidrome) conduction in the central nervous system. *Proc. R. Soc. Lond.*, 61: 243–246.

Sherrington, C.S. 1898. Decerebrate rigidity and reflex coordination of movement. *J. Physiol.* (London), 22: 319–332.

Sherrington, C.S. 1906. *The Integrative Action of the Nervous System.* New York: Scribner's Sons. Reprinted Cambridge University Press, 1947.

Sidtis, J.J., B.T. Volpe, J.D. Holtzman, D.H. Wilson, and M.S. Gazzaniga. 1981. Cognitive interaction after staged callosal section: Evidence for transfer of semantic activation. *Science*, 212: 344–346.

Sieveking, E., Trans., *see* Rokitansky, C.

Singer, C., Trans., *see* Galen, 1956; Mondino da Luzzi.

Singer, C. 1925. *The Fasciculo di medicina Venise 1495 with an introduction...*by Charles Singer. Part I. Florence: Lier.

Singer, C. 1952. *Vesalius on the Human Brain*. Introduction, Translation of Text, Translation of Descriptions of Figures, Notes to the Translations, Figures. London: Oxford University Press.

Skinner, A.H. 1949. *The Origins of Medical Terms*. Baltimore, Williams & Wilkins.

Slumberger, H.G. 1970. Rudolf Virchow (1821–1902). Pages 380–383 in W. Haymaker and F. Schiller, eds., *The Founders of Neurology* 2nd ed. Springfield, Illinois: C. C Thomas.

Smith, G.E. 1904. A preliminary note on an aberrant circumolivary bundle springing from the left pyramidal tract. *Rev. Neurol. Psychiat.*, 2: 377–383.

Smith, G.E. 1910. Some problems relating to the evoluton of the brain, the Arris and Gale lectures. *Lancet*, 1: 1–6, 147–153, 221–227.

Smith, G.E. 1919. The significance of the cerebral cortex. (Croonian Lectures abstracts.) *Br. Med. J.*, I: 758, 796–797; II: 11–12.

Smith, K.W. and A.J. Atelaitis. 1942. Studies on the corpus callosum. I. Laterality in behavior and bilateral motor organization in man before and after section of the corpus callosum. *Arch. Neurol. Psychol.*, 47: 519–543.

Smith, P.E. 1930. Hypophysectomy and a replacement therapy in the rat. *Am. J. Anat.*, 45: 205–275.

Smith, P.E. 1961. Postponed homotransplants of the hypophysis into the region of the median eminence in hypophysectomized male rats. *Endocrinology*. 68: 130–143.

Snider, R.S. 1950. Recent contributions to the anatomy and physiology of the cerebellum. *Arch. Neurol. Psychiat.*, 64: 196–219.

Snider, R.S. and A. Stowell. 1944. Receiving areas of the tactile, auditory, and visual systems in the cerebellum. *J. Neurophysiol.*, 7: 331–357.

Snider, R.S., W.S. McCulloch, and H.W. Magoun. 1947. A suppressor cerebello-bulbo-reticular pathway from anterior lobe and paramedian lobules. *Fed. Proc.*, 6: 207.

Snider, R.S., W.S. McCulloch, and H.W. Magoun. 1949. A cerebello-bulbo-reticular pathway for suppression. *J. Neurophysiol.*, 12: 325 334.

Sokoloff, L., M. Reivich, C. Kennedy, M.H. Des Rosiers, C.S. Patlak, K.D. Pettigrew, O. Sakurada, and M. Shinohara. 1977. The [^{14}C]deoxyglucose method for the measurement of local cerebral glucose utilization; theory, procedure, and normal values in the conscious and anesthetized albino rat. *J. Neurochem.*, 28: 897–916.

Somjen, G.G. 1988. Nervenkitt: notes on the history of the concept of neuroglia. *Glia*, 1: 2–9.

Sömmerring, S.T. 1796. *Über das Organ der Steele* Königsberg: F. Nicolovius.

Soury, J. 1899. *Le système nerveux central, structure et fonctions: histoire critique des théories et des doctrines*. Paris: Carré et Naud.

Spatz, H. 1933. Die Bedentung der vitalen Farbung für die Lehre vom Stoffaustausch zwischen dem Zentralnervensystem und dem übrigen Korper. *Arch. Psychiat. Nervenkrank.*, 101: 267–358.

Specht, S. and B. Grafstein. 1973. Accumulation of radioactive protein in mouse cerebral cortex after injection of 3H-fructose into the eye. *Exper. Neurol.*, 41: 705-722.

Spemann, H. 1902. Entwicklungsphysiologische Studien am *Triton*-ei. II. *Ebenda*, 15, H. 3.

Spencer, A. 1961. Letter to W.H. Marshall, August, 1961. W.H. Marshall Collection, Neuroscience History Archives.

Spencer, H. 1855. *The Principles of Psychology*. London: Longman, Brown, Green and Longmans.

Spencer, H. 1896. *The Principles of Psychology,* 3rd edition, 2 volumes in 3. New York: Appleton.

Sperry, R.W. 1961. Cortical organization and behavior. *Science,* 133: 1749–1757.

Sperry, R.W. 1968. Perception in the absence of the neocortical commissures. Pages 123–138 in *Perception and Its Disorders*. Proceedings of the Association for Research in Nervous and Mental Diseases, Pub. 48.

Spiegel, E.A. and W.J. Bernis. 1925. Die Zentren der statistchen Innervation und ihre Beeinflussung durch Klein- und Grosshirn. *Arb. Nerol. Inst. Weiner Univ.,* 27: 197–224.

Spiegel, E.A., H.T. Wycis, M. Marks, and A. J. Lee. 1947. Stereotaxic apparatus for operations on the human brain. *Science,* 106: 349–350.

Spillman, R., Trans., *see* Economo, K. F. and J. v. Wagner-Jauregg.

Spirov, M.S. 1962. [On the details of structure of the cerebral cortex of man.] *Arkh. Anat. Gistol. Embriol.*, 17: 106–111 (in Russian).

Starling, E. H. and E. B. Verney. 1925. The secretion of urine as studied on the isolated kidney. *Proc. R. Soc. Lond.*, 97: 321–363.

Starr, M. A. 1890. The pathology of sensory aphasia, with an analysis of fifty cases in which Broca's centre was not diseased. *Brain,* 12: 82–101.

Starzl, T. E. and H.W. Magoun. 1951. Organization of the diffuse thalamic projection systems. *J. Neurophysiol.* 14: 133–146.

Starzl, T. E. and D. G. Whitlock. 1952. Diffuse thalamic projection system in monkey. *J. Neurophysiol.*, 15: 449–468.

Steckerl, F. 1958. *The Fragments of Praxagoras of Cos and his School*. Leiden: Brill.

Stein, S. A. W. 1834. *De thalamo et orgine nervi optici in homine et animalibus vertebratis*. Copenhagen: S. Trier.

Steno, N. 1669. *Discours de Monsieur Stenon sur l'anatomie du cerveau. A Messieurs de l'Assemblée, qui se fait chez Monsieur Theuenot*. Paris: Ninville.

Steno, N. 1950. *A Dissertation on the Anatory of the Brain*. Trans. of the 1669 ed. by G. Douglas in 1733. Copenhagen: Nyt Nordisk Forlag, A. Busck.

Steriade, M. 1981. Mechanisms underlying cortical activation: neuronal organization and properties of the midbrain reticular core and intralaminar thalamic nuclei. Pages 327–377 in O. Pompeiano and C. Ajmone-Marsan, eds., *Brain Mechanisms and Perceptual Awareness*. New York: Raven.

Stoffels, J. 1939. La projection des noyaux antériurs du thalamus sur l'écorce interhémisphérique. Ëtude anatomo-expérimentale. *J. Belge Neurol. Psychiat.*, 39: 743–776; 783–833.

Stutinsky, F. 1953. Mise en évidence d'une substance antidiurétique dans le cerveau et le complexe rétrocérébral d'une blatte (*Blabera fusci* Brunn). *Bull. Soc. Zool. France*, 78: 202–204.

Swanson, L.W. 1987. Linbic system. Pages 589–591 in G. Adelman, ed., *Encyclopedia of Neuroscience*. Boston: Burkhäuser.

Swazey, J. F. 1974. *Chlorpromazine in Psychiatry. A Study of Therapeutic Innovation*. Cambridge, MA: The MIT Press.

Swedenborg, E. 1882. *The Brain, Considered Anatomically, Physiologically and Philosophically*. Edited, translated and annotated by R. L. Tafel. Vol. I. The cerebrum and its Parts. London: Speirs.

Sylvius, *see* Le Boë, F. de.

Symonds, C. 1975. Reflections. Pages 91–101 in W. Penfield, *The Mystery of the Mind*. Princeton: Princeton University Press.

Szentágothai, J. 1978. The neuron network of the cerebral cortex. A functional interpretation. *Proc. R. Soc.* (London) B., 201: 219–248.

Tadd, J. L. 1899. *New Methods in Education: Art, Real Manual Training, Nature Study.* New York: Orange Judd.
Tafel, R. L., Trans., *see* Swedenborg, E.
Taira, N. 1961. Origin of hippocampal seizure. *Tohoku J. Exp. Med.*, 191–199.
Takamine, J. 1901. Adrenalin, the active principle of the suprarenal glands and its mode of preparation. *Am. J. Pharm.*, 73: 523–535.
Talbot, S. A. and W. H. Marshall. 1841. Physiological studies on neural mechanisms of visual localization and discrimination. *Am. J. Ophthalmol.*, 24: 1255–1263.
Taylor, J. A., ed. 1958. *Selected Writings of John Huglings Jackson*, 2 vol. New York: Basic Books.
Tepperman, J. 1970. Horsley and Clarke: a biographical medalion. *Perspect. Biol. Med.*, 13: 295–308.
Terzian, H. and G. D. Ore. 1955. Syndrome of Klüver and Bucy. Reproduced in man by bilateral removal of the temporal lobes. *Neurology*, 5: 373–380.
Teuber, H.-L., W. S. Battersby, and M. B. Bender. 1960. *Visual Field Defects after Penetrating Missle Wounds of the Brain.* Cambridge: Harvard University Press.
Thomas, A. 1951. L'oeuvre de J. Dejerine. Pages 450–469 in *IV Congrès Neurologique International, Paris, 5–10 Septembre, 1949.* Vol. III. Paris: Masson.
Thompson, D'A. W., Trans., *see* Aristotle.
Thomson, E. H. 1981. *Harvey Cushing, Surgeon, Author, Artist.* New York: Neale Watson Academic Publications.
Thorek, M., Trans., *see* Krause, F.
Thudichum, J. L. W. 1884. *A Treatise on the Chemical Constitution of the Brain Based Throughout upon Original Researches.* London: Baillière, Tindall & Cox. Reprinted 1962 by Archon Books, Hamden, Connecticut.
Tiedemann, F. 1826. *The Anatomy of the Foetal Brain with a comparative exposition of its structure in animals.* Translated from the French editiion of W. Bennett. 1823 Edinburgh: Carfrae.
Tilney, F. 1927. The brain of prehistoric man. A study of the psychologic foundations of human progress. *Arch. Neurol. Psychiat.*, 17: 723–769.
Tilney, F. 1938. The hippocampus and its relation to the corpus callosum. *Bull. New York. Inst. Neurol.*, 7: 1–77.
Tobias, P. V. 1971. *The Brain in Hominid Evolution.* New York: Columbia University Press.
Todd, E. M. 1983. *The Neuroanatomy of Leonardo de Vinci.* Santa Barbara: Capra Press.
Todd, R. B., ed. 1839–1847. *Cyclopedia of Anatomy and Physiology.* Vol. 3. 722M–7220.
Tower, D. B. 1970. Johann Ludwig Wilhelm Thudicum (1829–1901). Pages 297–302 in W. Haymaker and F. Schiller, eds., *Founders of Neurology*, 2nd ed. Springfield, IL: C.C. Thomas.
Tower, D. B. 1983. *Hensing, 1719.* New York: Raven.
Travis, L. E. and J. P. Egan. 1938. Conditioning of the electrical response of the cortex. *J. Exp. Psychol.*, 22: 524–531.
Tschisch, W. v. 1886. Untersuchungen zur Anatomie der Grosshirnganglion des Menschen. *Ber. Verh. Kön. Säch. Gesellach. Wissensch. Leipzig,* 38: 95–101.
Twarog, B. M. and I. H. Page. 1954. Serotonin content of some mammalian tissues and urine and a method for its determination. *Am. J. Physiol.*, 175: 157–161.
Tyson, E. 1699. *Orang-Outang, sive Homo Sylvestris: or, the Anatomy of a Pigmie.* London: Printed for Thomas Bennet & Daniel Brown....
Unzer, J. A. 1851. *The Principles of Physiology by John Augustus Unzer; and a Dissertation on the Function of the Nervous System by George Procháska.* London: Syndenham Society.
Vale, W., J. Spiess, C. Rivier, and J. Rivier. 1981. Characterization of a 41-residue ovine hypothalamic peptide that stimulates secretion of corticotropin and b-endorphin. *Science*, 213: 1394–1397.
Valentin, G. G. 1836. Über den Verlauf und die letzten Ende der Nerven. *Nova Acta Phys-med. Acad. caes. Leopoldino Carolina Germanicae Naturae Curiosorum, Breslau*, 18 [1]: 51–240.
Van Breeman, V. L. and C. D. Clemente. 1955. Silver deposition in the central nervous system and the hemato-encephalic barrier studied with the electron microscope. *J. Biophys. Biochem. Cytol.*, 1: 161–166.
Van Valkenberg, C. T. 1913. Experimental and pathologico-anatomical researches on the corpus callosum. *Brain,* 36: 119–165.
Van Wagenen, W. P. and R. Y. Herren. 1940. Surgical division of commissural pathways in the corpus callosum: Relation to spread of an epileptic attach. *Arch. Neurol. Psychiat.*, 44: 740–759.
Vates, T. S., S. L. Bonting, and W. W. Oppelt. 1964. Na–K activated adenosine triphosphatase formation of cerebrospinal fluid in the cat. *Am. J. Physiol.*, 206: 1165–1172.
Velden, R. v. d. 1913. Die Nierenwirkung von Hypophysenextrakten bein Menschen. *Berl. Klin. Wochenschr.*, 50: 2083–2086.
Verney, E. B. 1947. The antidiuretic hormone and the factors which determine its release. *Proc. R. Soc. Lond. B*, 135: 25–106
Verzeano, M., D. B. Lindsley, and H. W. Magoun. 1953. Nature of recruiting response. *J. Neurophysiol.* 16: 183–195.
Vesalius, A. 1543. *De humani corporis fabrica.* Basle: [Ex Officina Joannis Operini].
Vicq d'Azyr, F. 1781. Recherches sur la structure du cerveau, du cervelet, de la moelle alongée, de la moelle épinière; & sur l'origine des nerfs de l'homme & des animaux. Paris: [detached from *Mem. Acad. R. Sci.*]
Vicq d'Azyr, F. 1786. *Traité d'anatomie et de physiologie.* Paris: Didot l'Aine.
Viets, H. R. 1938. West Riding, 1871–1876. *Bull. Inst. Hist. Med.*, 6: 477–487.
Vieussens, R. de. 1685. *Nevrographia universalis*, editio nova, Lyons: Certe.
Virchow, R. 1846. Über das granulierte Ansehen der Wandungen der Gehernventrikel. *Allg. Z. Psychiat.*, 3: 242–250. Translated by F. Chance (1863), reprinted 1971. New York: Dover.
Voetmann, E. 1949. On the structure and surface area of the human choroid plexuses. A quantitative anatomical study. *Acta Anat.*, 8, Suppl. 10: 1–116.
Vogt, M. 1943. The output of cortical hormone by the mammalian suprarenal. *J. Physiol.* (London), 102: 341–356.
Vogt, M. 1954. The concentration of sympathin in different parts of the central nervous system under normal conditions and after administration of drugs. *J. Physiol.* (London), 123: 451–481.
Voss, S., Jr., *see* Descartes, R, 1989.
Wada, J. A. 1973. Brain mechanisms underlying meditation. *X International Congress of Neurology, Barcelona, Sept.*

8–15, 1973. Excerpta Medica International Congress Series No. 296, page 251, Abstr. 803.

Wada, J. A. and T. B. Rasmussen. 1960. Intracarotid injection of sodium amytal for the lateralization of cerebral speech dominance. *J. Neurosurg.*, 17: 266–282.

Wada, J., R. Clarke, and A. Homm. 1975. Cerebral hemispheric asymmetry in humans. *Arch. Neurol.*, 32: 239–246.

Waldeyer-Hartz, H. W. G. v. 1891. Über einige neuere Forschungen im Gebeite der Anatomie des Central nervou systems. *Dt. Med. Wschr.*, 17: 1213–1218, 1244–1246, 1267–1269, 1287–1289, 1331–1332, 1352–1356.

Waller, A. V. 1850. Experiments on the section of the glossopharyngeal and hypoglossal nerves of the frog, and observations on the alterations produced thereby in the structure of their primitive fibres. *Phil. Trans. R. Soc.*, 140: 423-429.

Walker, A. E. 1936. An experimental study of the thalamocortical projection of the macaque monkey. *J. Comp. Neurol.*, 64: 1–39.

Walker, A. E. 1938. *The Primate Thalamus*. Chicago: University of Chicago Press.

Walker, A. E. 1949. Concluding remarks. *Electroenceph. Clin. Neurophysiol.*, 1: 451–454.

Walshe, F. M. R. 1957. The brain-stem conceived as the "highest level" of function in the nervous system, with particular reference to the "automatic apparatus" of Carpenter (1850) and to the "centrencephalic integrating system" of Penfield. *Brain*, 80: 510–539.

Walter, W. G. 1949. Coming to terms with brain waves. *Electroenceph. Clin. Neurophysiol.*, 1: 474.

Walter, W. G., R. Cooper, V. J. Aldridge, W. C. McCallum, and A. L. Winter. 1964. Contingent negative variation: an electric sign of sensorimotor association and expectancy in the human brain. *Nature*, 203: 380–384.

Wang, G. H. 1965. Johann Paul Karplus (1866–1936) and Alois Kreidl (1864–1928). Two pioneers in the study of central mechanisms of vegetative function. *Bull. Hist. Med.*, 39: 529–539.

Washburn, M. F. 1908. *The Animal Mind: A Textbook of Comparative Psychology*. New York: Macmillan.

Washburn, M. F. 1928. Emotion and thought: a motor theory of their relations. Pages 104–115 in M. L. Reymert, ed., *Feelings and Emotions: the Wittenberg Symposium*. Worchester, MA: Clark University Press.

Watanabe, E. 1936. Experimentelle Beitrage zur Histopathologie und Pathogenese des epileptischen Krampfanfalls (kurze Auszüge aus den Originalmitteilungen). Translated from the Japanese. *Folia Psychiat. Neurol. Jpn.*, 40: 1–36.

Weigert, C. 1882. Über eine neue Untersuchungsmethode des Centralnervensystems. *Z. Med. Wiss.*, 20: 753–757.

Weinhouse, A., Trans., *see* Bataille, G.

Weiss, P. A. and H. B. Hiscoe. 1948. Experiments on the mechanism of nerve growth. *J. Exp. Zool.*, 107: 315–395.

Welker, W. 1990. Why does cerebral cortex fissure and fold? A review of determinants of gyri and sulci. Pages 3–136 in E. G. Jones and A. Peters, eds., *Cerebral Cortex*, Vol. 8B. New York: Plenum.

Wernicke, C. 1874. *Der aphasische Symptomencomplex: Eine psychologiische Studie auf anatomisher Basis*. Breslau: Cohn and Weigert. Reprinted 1974, Berlin: Springer-Verlag.

Wernicke, C. 1876. Das Urwindungssystem des menschlichen Gehirns. *Arch. Psychiat. Nervenk.*, 6: 298–326. Translated by J. Ratclife and M. V. Anker.

Wernicke, C. 1903. Ein Fall von isolierter Agraphie. *Monatschr. Psychol. Neurol.*, 13: 241–265.

Wever, E. G. and C. W. Bray. 1930a. Auditory nerve impulses. *Science*, 71: 215.

Wever, E. G. and C. W. Bray. 1930b. Present possibilities for auditory theory. *Psychol. Rev.*, 37: 365–380.

White, A., H. R. Catchpole, and C. N. H. Long. 1937. A crystalline protein with high lactogenic activity. *Science*, 86: 82–83.

White, E. L. 1979. Thalamocortical synaptic relations: a review with emphasis on the projection of specific thalamic nuclei to the primary sensory areas of the neocortex. *Brain Res. Rev.*, 1: 275–311.

Wiener, N. 1948. *Cybernetics, or Control and Communication in the Animal and the Machine*. New York: John Wiley & Sons.

Wiesel, T. N. 1982. Postnatal development of the visual cortex and the influence of environment. *Nature*, 299: 583–591.

Wiesel, T. N., D. H. Hubel, and D. M. K. Lam. 1974. Autoradiographic demonstration of ocular-dominance columns in the monkey striate cortex by means of transneuronal transport. *Brain Res.*, 79: 273–279.

Wigan, A. L. 1844. *A New View of Insanity, the Duality of the Mind*. London: Longman, Browne, Green, and Longman.

Willis, T. 1664. *Cerebri Anatome*. London: J. Martyn and J. Allestry.

Willis, T. 1681. *The remaining medical works of that famous and renowned physician Dr. Thomas Willis*. Translated by S. Pordage. London: Dring, Harper, Leigh, and Martyn.

Windle, W. F. 1979. *The Pioneering Role of Clarence Luther Herrick in American Neuroscience*. Hicksville, New York: Exposition Press.

Winson, J. and C. Abzug. 1977. Gating of neuronal transmission in the hippocampus: efficacy of transmission varies with behavioral state. *Science*, 196: 1223–1225.

Wislocki, G. B. and L. S. King. 1936. The permeability of the hypophysis and hypothalamus to vitai dyes, with a study of the hypophyseal vascular supply. *Am. J. Anat.*, 58: 421–472.

Wislocki, G. B. and E. H. Leduc. 1952. Vital staining of the hematoencephalic barrier by silver nitrate and trypan blue, and cytological comparisons of the neurohypophysis, pineal body, area postrema, intercolumnar tubercle, and supraoptic crest. *J. Comp. Neurol.*, 96: 371–413.

Witelson, S. F. 1977. Anatomic asymmetry in the temporal lobes. Pages 328–354 in S. J. Dimond and D. A. Blizard, eds., *Evolution and Lateralization in the Human Brain*. Volume 299 of Annals of the New York Academy of Science.

Woollans, D. H. M. and J. W. Millen. 1954. Perivascular spaces of the mammalian central nervous system. *Biol. Rev.*, 29: 251–283.

Woolsey, C. N. 1943. "Second" somatic receiving areas in the cerebral cortex of cat, dog, and monkey. *Fed. Proc. Soc. Exp. Biol. Med.*, 2: 55.

Woolsey, T. A. and H. Van Der Loos. 1970. The structural organization of layer IV in the somatosensory region (SI) of mouse cerebral cortex. *Brain Res.*, 17: 205–242.

Yakovlev, P. I. 1970. Constantin von Monakow (1853–1930). Pages 484–488 in W. Haymaker and F. Schiller, *The Founders of Neurology*, 2nd ed. Springfield, IL: C. C. Thomas.

Yakovlev, P. J. and P. Rakic. 1966. Patterns of decussation of bulbar pyramids and distribution of pyramidal tracts on two sides of the spinal cord. *Trans. Am. Neurol. Assoc.*, 91: 366–367.

Yerkes, R. M. 1905. Animal psychology and criteria of the psychic. *J. Phil. Psychol. Sci. Meth.*, 2: 141–149.

Young, J. Z. 1936. The structure of nerve fibres in Cephalopods and Crustacea. *Proc. R. Soc., B,* 121: 319–337.

Young, R. M. 1970. *Mind, Brain, and Adaptation in the Nineteenth Century.* Oxford: Clarendon Press.

Zanchetti, A. 1981. Not farewell, but fare forward. Pages xiii–xvi in O. Pompeiano and C. Ajmone–Marsan, eds., *Brain Mechanisms of Perceptual Awareness and Purposeful Behavior.* New York: Raven.

Ziehen, T. 1906. Carl Werniche. *Monatss. Psychiat. Neurol.,* 18: i–v.

Zinn, J. G. 1749. Dissertation inauguralis. Gottingen. Cited in F. A. Elliott, the corpus callosum, cingulate gyrus, septum pellucidum, septal area and fornix. Pages 758–775 in P. J. Vinken and G.W. Bruyn, eds., *Handbook of Clinical Neurology*, vol. 2. Amsterdam: North Holland, 1969.

INDEX

C

D

E

F

G

H

I

J

K

L

M

N

O

Biographies

Although planning to pursue the romance languages when she entered Vassar College, **Louise Hanson Marshall** found the biological sciences more congenial, specifically a novel interdisciplinary major in euthenics that combined courses in physiology, psychology, and zoology. She subsequently earned a PhD in physiology at the University of Chicago. Dr. Marshall soon joined the aviation medicine unit at the National Institutes of Health in Bethesda and participated in determining human oxygen requirements at simulated altitudes and activities for the US Navy. After the Second World War, her research continued on the renal effects of decompression and artificial plasma expanders. In 1965 Dr. Marshall joined the National Academy of Sciences–National Research Council, where she was responsible for the newly constituted Committee on Brain Sciences. There she participated in the formation of the Society for Neuroscience, served as its acting secretary–treasurer, and initiated the historical program in its annual meetings. Retirement from the NRC brought Marshall to UCLA's Brain Research Institute as managing editor of *Experimental Neurology* and an opportunity to join H. W. Magoun in library research and writing about neuroscience history. At present she directs the BRI's Neuroscience History Archives, a center for the preservation and use of the evidentiary records documenting progress in knowledge of how the brain works.

Horace Winchell Magoun first became interested in history in high school. After graduation from Rhode Island State College, he continued in biology at Syracuse University, then received his doctorate in anatomy from Northwestern University Medical School. There he was associated with S. Walter Ranson for a decade. Dr. Magoun's research later focused on higher neural functions, culminating in his and Giuseppe Moruzzi's identification of an ascending arousal system. In 1950 Magoun founded the Department of Anatomy at the new UCLA Medical School, initiating a program that over a decade expanded into the Brain Research Institute. In time, he was dean of the graduate division at UCLA, and later became director of the fellowship office at the National Research Council. On his return to UCLA in 1974, he realized the fruition of his life-long love of history through research and writing on the confluence of the neural, behavioral, and communicative discoveries that formed today's neuroscience. Major honors included election to the National Academy of Sciences (1955) and the Academy of the Arts and Sciences (1960), the Lashley Prize of the American Philosophical Society (1970), and the Order of the Sacred Treasure of Japan (1971). This internationally acclaimed leader of American neuroscience died in 1991.